T0351006

MODULAR SYNTHESIS

Modular Synthesis: Patching Machines and People brings together scholars, artists, composers, and musical instrument designers in an exploration of modular synthesis, an unusually multifaceted musical instrument that opens up many avenues for exploration and insight, particularly with respect to technological use, practice, and resistance.

Through historical, technical, social, aesthetic, and other perspectives, this volume offers a collective reflection on the powerful connections between technology, creativity, culture, and personal agency. Ultimately, this collection is about creativity in a technoscientific world and speaks to issues fundamental to our everyday lives and experiences, by providing insights into the complex relationships between content creators, the technologies they use, and the individuals and communities who design and engage with them.

With chapters covering VCV Rack, modular synthesis, instrument design, and the histories of synthesizer technology, as well as interviews with Dave Rossum, Corry Banks, Meng Qi, and Dani Dobkin, among others, *Modular Synthesis* is recommended reading for advanced undergraduates, researchers, and practitioners of electronic music and music technology.

Ezra J. Teboul is a researcher and artist, and currently a student librarian at Concordia University. They obtained a PhD in Electronic Arts from Rensselaer Polytechnic Institute in 2020. In 2022, they were a scholar-in-residence at the Columbia University Computer Music Center. Their work focuses on the material histories of electricity, work, and music.

Andreas Kitzmann is an associate professor of Humanities at York University in Toronto, Canada. His research interests include modular synthesis, technology and culture, digital media and community, and memory studies. His self-authored books include *The Hypertext Handbook: The Straight Story* and *Saved from Oblivion: Documenting the Daily from Diaries to Web Cams*. He has co-edited two books, and his work has also been included in various edited collections and journals such as *A History of English Autobiography* (2016), *From Text to Txting: New Media in the Classroom* (2012), the *International Journal of Research into New Media Technologies* (2017), *First Monday* (2015), and *Organized Sound* (2023).

Einar Engström is a software engineer, modular synthesist, and computer musician. Creatively, he has been known to code in Lua, SuperCollider, Tidal Cycles, and the Teletype esolang, whilst professionally he primarily inhabits the BEAM ecosystem. Both practices are natural extensions to Einar's previous PhD research into the history and philosophy of computing music programming, which focused on the Acoustical and Behavioral Research Center at Bell Laboratories—the first behemoth of innovation in both telecommunications and computing. He also holds an MA in Visual Culture from Waseda University (Tokyo, Japan), is a former editor-in-chief of the bilingual international contemporary art magazine *LEAP* (Beijing, China) and technician and researcher at RE/Lab (Toronto Metropolitan University), and has hands in various electronic music record labels.

"Just when everyone thought digital had evolved into music's alpha species, modular analog synthesizers resurged after decades of dormancy. *Modular Synthesis* provides a valuable overview of how and why the technology that redefined the musical instrument in the 1960s became relevant and transformative again."

Nicolas Collins, *composer and author of* Handmade Electronic Music — The Art of Hardware Hacking

"It's no accident that the first academic collection on modular synthesizers touches on so many disciplines. Across its pages you will hear from musicians, artists, makers, and scholars from across the arts and human sciences. How many books on music discuss looms, switchboards, and electric fish? How many books on philosophy examine the material metaphors on which philosophers' ideas are based? How many books on electronics include studies of education or the gendered politics of naming? *Modular Synthesis* is as varied, multidisciplinary, and enchanting as its object of study."

Jonathan Sterne, *author of* Diminished Faculties, MP3, *and* The Audible Past

MODULAR SYNTHESIS

Patching Machines and People

Edited by Ezra J. Teboul, Andreas Kitzmann,
and Einar Engström

LONDON AND NEW YORK

Designed cover image: Courtesy of Seze Devres Photography
(www.sdphotography.net)

First published 2024
by Routledge
4 Park Square, Milton Park, Abingdon, Oxon OX14 4RN

and by Routledge
605 Third Avenue, New York, NY 10158

Routledge is an imprint of the Taylor & Francis Group, an informa business

British Library Cataloguing-in-Publication Data
A catalogue record for this book is available from the British Library

Library of Congress Cataloguing-in-Publication Data
Names: Teboul, Ezra J., editor. | Kitzmann, Andreas, editor. | Engström, Einar, editor.
Title: Modular synthesis : patching machines and people / edited by Ezra J.
Teboul, Andreas Kitzmann, and Einar Engström.
Description: Abingdon, Oxon ; New York : Routledge, 2024. | Includes
bibliographical references and index.
Identifiers: LCCN 2023051497 (print) | LCCN 2023051498 (ebook) | ISBN
9781032113470 (hardback) | ISBN 9781032113463 (paperback) | ISBN
9781003219484 (ebook)
Subjects: LCSH: Modular synthesizer (Musical instrument)--History. | Modular
synthesizer (Musical instrument)--Construction. | Modular synthesizer (Musical
instrument)--Performance. | Modular synthesizer music--History and criticism.
Classification: LCC ML1092 .M63 2024 (print) | LCC ML1092 (ebook) | DDC
786.7/419--dc23/eng/20240208
LC record available at https://lccn.loc.gov/2023051497
LC ebook record available at https://lccn.loc.gov/2023051498

ISBN: 978-1-032-11347-0 (hbk)
ISBN: 978-1-032-11346-3 (pbk)
ISBN: 978-1-003-21948-4 (ebk)

DOI: 10.4324/9781003219484

Typeset in Optima
by MPS Limited, Dehradun

CONTENTS

CONTRIBUTORS

Corry Banks is a prominent member of the Modbap community and the founder of Modbap Modular. Modbap is a term created by Banks to describe his exploration of modular synthesis combined with traditional hip-hop beat making methods. As a result, a thriving worldwide community has developed around the concept of modular synthesis and hip-hop beat creation. Corry is dedicated to merging hip-hop and technology, ensuring that hip-hop is represented in the technology industry and bridging the gap between hip-hop and music technology. With 25 years of experience as a technologist and a background in hip-hop and beat making, Banks holds an Associate's Degree in Electronics, a Bachelor's Degree in Technology Management, and a Master's Degree in Project Management.

Eliot Bates (they/them) is an ethnographer who researches the interface between people and sound/music technologies—including their design, materiality, instrumentality, and cultural milieus. An ethnomusicologist by training (PhD, UC Berkeley), from 2004 to 2016 they researched these within Istanbul's recording studios, instrument-building factories and music industries. Since 2013, their work has geographically broadened to consider European and North American gear cultures. Eliot has written two books: *Digital Tradition: Arrangement and Labor in Istanbul's Recording Studio Culture* (Oxford University Press, 2016), and *Music in Turkey: Experiencing Music, Expressing Culture* (Oxford University Press, 2011) and, with Samantha Bennett, co-edited *Critical Approaches to the Production of Music and Sound* (Bloomsbury Academic, 2018). The book *Gear: Cultures of Music and Audio Technologies*, co-authored with Bennett, will be published in 2024 by MIT

Press. When time permits, Eliot might be found performing or recording music on modular synthesizers (as the solo artist Makamqore, and in the duo Manifestoon Platoon), on an 11-stringed oud, or sounding nearly every object in the kitchen.

Anna Bockrath is an artist and educator based in Philadelphia. She is a current MFA candidate at the Tyler School of Art & Architecture at Temple University and received a BS in Art Education from Temple in 2016. Throughout her expanded practice, Anna is interested in ways that weaving can operate across a wide range of disciplines and materials, exploring topics such as memory, language, and technology. As an art teacher and museum educator, she has taught in various settings including Philadelphia schools, the Philadelphia Museum of Art, and the Delaware Art Museum.

Hannah Bosma is an interdisciplinary musicologist specialized in electroacoustic music, voice, gender and preservation, who loves to play with sound. In 2019–23, she worked at the University of Amsterdam (UvA) on her postdoc research project Preservation as performance: Liveness, loss and viability in electroacoustic music, funded by the Dutch Research Council NWO with a Veni grant, with STEIM and Willem Twee Studios among its case studies; as project leader of the research project *From archive to network: Syrian music in the Jaap Kunst audio collection and beyond*, funded with a NWO Hestia grant (2021–23); as co-organizer of the Lorentz Workshop Music beyond fixity and fluidity: Preservation and performance as instauration (Leiden, 12–16 September 2022); and as coordinator/lecturer of the MA-course Archiving Art, combining theoretical and project-based teaching in collaboration with various cultural organizations. Her research on preservation was inspired by her previous work at NEAR / Donemus / Music Center the Netherlands (1998–2012) and a research project at LiMA/DEN (2014). As a pioneering expert on gender, voice and electroacoustic music, she is invited for international publications, lectures and panels, such as in The Routledge Companion to Sounding Art (2017). She organized the conference The Art of Voice Synthesis (UvA 2016, www.artificialvoice.nl) and is involved in a film documentary project on the acoustic voice synthesizers, music machines and other installations of artist Martin Riches. Her doctoral dissertation The Electronic Cry: Voice and gender in electroacoustic music (UvA 2013) is available at https://hdl.handle.net/11245/1.400268.

Heidi Chan is a musician and researcher based in Toronto. She completed her PhD in Ethnomusicology at York University, researching the impact of digital sampling technologies on the development of world music over the past four decades, and is graduate assistant on a number of research projects at York University and the Responsive Ecologies Lab at Toronto Metropolitan University. An active member of Toronto's diverse music scene, she is principal bamboo flutist for Japanese folk ensemble *Ten Ten*, and produces and performs experimental electronic music under the name Bachelard. She is also a sound designer and composer for theatre, dance, and film, and has participated in productions across Asia, Europe, and North America. Chan is active as an instructor in digital music production and modular synthesizers, and has recently developed curriculum for the SOCAN Foundation, the Toronto Public Library, and Toronto Metropolitan University.

Brian Crabtree (US) creates objects, music, and objects that make music. In 2005 with Kelli Cain he founded monome, pioneering the grid-based performance interface. This open-source tool encourages people to envision and build their own musical systems, fostering an international community where people share code, sounds, and ideas. Brian and Kelli's work has shown at the Museum of Modern Art in New York and the Los Angeles County Museum of Art in addition to numerous international performances. They live in upstate New York, where time is shared with apple orchards, shiitake stacks, and birds of all size and color and song.

Jonathan De Souza is an associate professor in the Don Wright Faculty of Music, a core member of the Centre for Theory and Criticism, and an associate member of the Brain and Mind Institute at the University of Western Ontario. His research combines music theory, philosophy, and cognitive science. He is a co-editor for two forthcoming volumes, *The Oxford Handbook of the Phenomenology of Music* (with Benjamin Steege and Jessica Wiskus) and *The Neurosciences of Music: Interdisciplinary Insights* (with Jessica Grahn). Jonathan is the author of *Music at Hand: Instruments, Bodies, and Cognition* (Oxford University Press 2017), which received the Society for Music Theory's Emerging Scholar Book Award in 2020. The book asks how instruments mediate music's sounding organization and players' experience. His chapter in *The Oxford Handbook of Timbre* (edited by Emily I. Dolan and Alexander Rehding) examines attempts to synthesize the sounds of acoustic instruments, such as violins and trumpets, from Hermann von Helmholtz to Stevie Wonder. Jonathan's scholarly work draws on his lifelong experience as a multi-instrumentalist. He started Suzuki violin lessons at the age of three and had access to many other instruments through his father, a high-school music teacher with an interest in technology. Digital synthesizers were always part of Jonathan's musical world: as a child in the 1980s, he spent hours playing with Casio VL-1 and Roland Juno synthesizers. He has performed classical music, folk music, and jazz in Canada, the USA, and the UK and continues to play various instruments.

Dani Dobkin is a sound artist and composer from Philadelphia, PA. Dobkin received a BA in Experimental Music and the Electronic Arts from Bard College where they studied under Bob Bielecki, Marina Rosenfeld, and Richard Teitelbaum. Always in search of new ways to engage and interact with sound, recent artistic works play with installation, modular synthesis, handmade circuits, live electronic improvisation, and interdisciplinary projects with dancers, musicians, poets, and visual artists. Dobkin recently received an MFA in Sound Arts from Columbia University and is pursuing a Doctorate in Music Composition at Columbia.

David Dunn is a composer, performer, and audio engineer. Born in 1953, in San Diego, California, he was an assistant to the American composer Harry Partch and remained active as a performer in the Harry Partch ensemble for over a decade. While he has worked in a wide variety of audio media, inclusive of traditional and experimental music, installations for public exhibitions, video/film soundtracks, and bioacoustics research, he primarily engages in site-specific interactions or research-oriented activities. Much of his work is focused upon the development of listening strategies and technologies for environmental sound monitoring in both aesthetic and scientific contexts. Dunn is internationally known for his articulation of frameworks that combine the arts and sciences towards practical environmental activism and problem solving. As a pioneer in the fields of acoustic ecology, bioacoustics, interspecies communication, and scientific sonification, he has composed a body of innovative and experimental musical work while contributing to projects as diverse as sensory enhancement of healthcare environments, intervention strategies for forest and agricultural pests, reduction of sensory deprivation problems in captive animals, and research into cetacean communication. He has developed audio transducer systems to record such phenomena as the sounds of bark beetles within trees, underwater invertebrates in freshwater ponds, and the ultrasonic communication of bats. He has also designed self-organizing autonomous sound systems for interaction between artificial and natural non-human systems. He has been the recipient of over 30 awards and grants. His work and performances have appeared in over 500 international forums, concerts, broadcasts, and exhibitions, with recordings distributed by the Neuma, New World Records, Pogus, Non sequitur, EM Records, Earth Ear, and O.O. Discs labels. Since 2019 he has been a Professor Emeritus at the University of California, Santa Cruz. He resides in Santa Fe, New Mexico.

Einar Engström is a software engineer, modular synthesist, and computer musician. Creatively, he has been known to code in Lua, SuperCollider, Tidal Cycles, and the Teletype esolang. Professionally, he primarily inhabits the

BEAM ecosystem: the family of languages and thought deriving from Erlang, a programming language for telecommunications developed at Ericsson beginning in the mid-1980s. Both practices are natural extensions to Einar's previous PhD research in the history and philosophy of computing music programming, which focused on the Acoustical and Behavioral Research Center at Bell Laboratories–the first behemoth of innovation in both telecommunications and computing. He also holds an MA in Visual Culture from Waseda University (Tokyo, Japan); is a former editor-in-chief of the bilingual international contemporary art magazine LEAP (Beijing, China) and technician and researcher at RE/Lab (Toronto Metropolitan University); and has hands in various electronic music record labels.

Andrew Fitch completed a B.Eng (Hons) in 2010 and his PhD in 2015, at The University of Western Australia. His main research focused on memristor-based chaotic circuits and has published a book, journal articles and book chapters describing the results of this work. He has been building analog synthesizers since 1998 and currently runs Nonlinearcircuits, producing a wide variety of unique designs intended for use in experimental music. When not building synthesizers, Andrew enjoys bushwalking with his dogs and reading books by A.E. van Vogt.

Theodore (Ted) Gordon is a musicologist and musician whose work connects experimental music, critical organology, and science & technology studies. His current book project, *The Composer's Black Box: Cybernetics and Instrumentality in American Experimental Music*, shows how technoscientific concepts borrowed from cybernetics, information theory, and systems-thinking catalyzed differing musical organizations of performance practices, political processes, and embodied subjectivities in the late 1960s. His writing has been published by *Contemporary Music Review*, *Current Musicology*, *Portable Gray*, the Library of Congress, the American Musicological Association, and *Cultural Anthropology*, and his research has been supported by the New York Public Library and the Professional Staff Congress of the City University of New York. He has written program and liner notes for Unseen Worlds, and contributed exhibition texts for the 2019 exhibition Sounding Circuits at the New York Public Library for the Performing Arts. He received his PhD in the History and Theory of Music from the University of Chicago in 2018. From 2018 to 2020 he was a Mellon Postdoctoral Fellow at Columbia University, where he was also a Visiting Researcher at the Computer Music Center. Since 2020, he has served as assistant professor of Music at Baruch College, City University of New York. As an improviser, he performs with the viola and the Buchla Music Easel. With Matthew Mehlan, he has released music as the Chicagoland Electric Music Box Association, and since 2019, he has collaborated with New York-based artist and musician Marcia Bassett. He has performed at venues such as Elastic Arts, Experimental Sound Studio, Roulette, Rhizome DC, Pageant: Soloveev, the Emily Harvey Foundation, and FourOneOne in Brooklyn. He currently lives in Queens, New York.

Bana Haffar is an electronic music composer presently interested in sequencing and the materiality of sound. Through research-creation, she works with sample-based hardware and software musical sequencers, attempting to relate sound worlds and patterns to various non-musical systems. In *Shed,* a piece she composed for Third Coast Percussion Ensemble in 2020 for the Black Mountain College Museum, she began investigating the materiality of sound through the research and study of weaving. She developed a way to translate weaving drafts into rhythmic notation by transposing and programming them into a voltage controlled sequencer and created an accompanying graphic score. This pattern-based cross over continues to inform her work. A lifelong expatriate, she was born in Saudi Arabia in 1987 and spent much of her childhood in the Persian Gulf. She began her musical career as a session and touring bass player, switching to modular synthesizers shortly after discovering the infinite sustain switch on her Minimoog Voyager. Immersed in the electronic music underground of Los Angeles, she together with Eric Cheslak founded Modular on the Spot in 2014, a monthly generator series hosted on the banks of the LA River. A true believer in the potential of artist-led spaces of all forms, in 2021 she co-founded the Beirut Synthesizer center in Lebanon—an informal co-op that provides access to electronic instruments, resources, and community, inspired by the legacies of the San Francisco Tape Music Center and Black Mountain College. Her most recent solo works "Genera" and "Intimaa'" are released on Touch (UK).

Michael Johnsen is a circuit designer, performer, and researcher from Pittsburgh, USA. His recent research concerns the circuit-level understanding of David Tudor's "folkloric" homemade instruments and related luthierie. This work has resulted in restoration, cloning, and performance with vintage circuits, as well as publications/lectures. His own performance work is characterized by a relative lack of ideas per se, and an intense focus on observation, the way a shepherd watches sheep. As a performer/builder of live-electronics he cultivates an integrated menagerie of custom devices whose idiosyncratic behaviors are revealed through their complex interactions, producing teeming chirps, sudden transients and charming failure modes; embracing the dirt in pure electronics. He has shown work at singuhr (Berlin), INA GRM (Paris), MdM Salzburg, Kagurane (Tokyo), MoMA, SF Cinematheque, Radio France, Idiopreneurial Entrephonics, Kitchen (NYC), High Zero (Baltimore), and Musique Action. He co-edits ubu.com/emr, designs synthesizers for Pittsburgh Modular, and may be reached at johnsen.rahbek@gmail.com.

David Kant is a composer, educator, and researcher who works at the intersection of music and machine learning. David is the bandleader of the Happy Valley Band, an ensemble that plays AI interpretations of classic pop songs, and former co-founder of Indexical, a Santa Cruz non-profit dedicated to experimental music. David taught electronic music at the University of California, Santa Cruz, where he studied with Larry Polansky and David Dunn, and now leads an AI research team at Meta.

Andreas Kitzmann is an associate professor of Humanities at York University in Toronto, Canada. His research interests include modular synthesis, technology and culture, digital media and community and memory studies. He has co-edited two books: *Memory and Migration: Multidisciplinary Approaches to Memory Studies*. U of Toronto Press, 2014; *Memory Work: The Theory and Practice of Memory*. Peter Lang, 2005. His self authored books include: *The Hypertext Handbook: the Straight Story*, Peter Lang, 2005 and *Saved From Oblivion: Documenting the Daily from Diaries to Web Cams*, Peter Lang, 2004. His work has also been included in various edited collections and journals such as *A History of English Autobiography* (2016); *From Text to Txting: New Media in the Classroom* (2012); the *International Journal of Research into New Media Technologies* (2017); *First Monday* (2015) and *Organized Sound* (2023).

Jeff Lee spent 20 years working as a biomedical researcher while also nurturing a serious interest in electronic music and vintage analog synthesizers and drum machines. He founded System80 Inc., a boutique manufacturer of synthesizer modules, in 2017. Through System80, Jeff attempts to bring the historically and culturally significant analog sounds and synthesis methods from the 1970s and 1980s to the popular Eurorack modular synthesizer format. System80 has a particular focus on the circuits and interfaces of vintage Japanese instruments that were essential to the development of acid house, techno, and electro.

Melanie McBride is an adjunct professor and post-doctoral researcher at Toronto Metropolitan University investigating the material and physical contingencies of learning and skilled practice in domains of physically embodied and multimodal expertise. She is the founder of the Aroma Inquiry Lab at Toronto Metropolitan University's Responsive Ecologies Lab.

Dakota Melín is a synthesizer electronics designer and builder by avocation in Southern Ontario. He bought his first synthesizer, a Roland Juno 106, in

2014 and got into vintage synthesizer repair when it broke a few months later. He learned about modular synthesizers through online hobbyist forums and from the people he met there, becoming involved in various research projects focused on vintage modular synthesizer systems. He is inspired by the designs and instruments of Serge Tcherepnin, Don Buchla, Rob Hordijk, Arpad Benares and Bernie Hutchins. In 2020 he began to work on the hale modular system which melds musical ideas from earlier modular synthesizer systems with new concepts and approaches. The hale modular system allows for a high degree of flexibility and density of function by breaking complex functions down into simpler parts that can be rearranged and used in different ways.

Nyles Miszczyk is a Canadian audio engineer based out of Kitchener, Ontario. He currently teaches sound design and audio post-production at George Brown College. Nyles has been a beta tester for several synthesizer brands such as Modcan and Mystic Circuits and is currently working on his first Eurorack module design. He released his debut solo album *Thyrsis of Etna* in 2022 on the We Are Time record label.

Naomi Mitchell is a designer, builder, user, and educator whose work extensively incorporates modular synthesizers. Of particular interest is the interaction between artists and electronics and ways to generate and control randomness and unpredictability. She has been making Eurorack modules since 2016 as omiindustriies. Naomi has been involved with educational programs for FeM Synth Lab, Ableton LOOP, Patch Up!, Columbia and CalArts. She lives and works in Los Angeles.

Lori Napoleon is a musician, electronics artist and instrument builder, performing as Antenes. She resides in New York with her collection of self-made sequencers and synthesizers using repurposed vintage telecommunications equipment. She draws musical influence from the curious and ephemeral sound world of outdated telephone and telegraph networking systems, finding atmospheric and mechanical aberrations inherent in the materials and various forms of synthesis. Her productions integrate sounds reminiscent of pulsing analog relay switching systems, errant radio transmissions, cross-continental echo, signature drones and message interferences between wires. Her live performances weave between organically unfolding rhythms, ambiguously shifting textural elements and droning resonances. A devoted enthusiast of crossing between disciplines, she has held residencies for electronic arts at Willem Twee Studios (Den Bosch NL), ISSUE Project Room (NYC) NOKIA Bell Labs (Murray Hill, NJ), Harvestworks (NYC) and Signal Culture (Owego NY) and has appeared within numerous interdisciplinary platforms including Sound of Stockholm Festival, New York

Electronic Arts Festival, Fermi National Accelerator Laboratory's Art Gallery, Open House London's Sonic Visitations, and Trinity College's Science Gallery in Dublin.

Yoni Newman is a musician, artist, and sound designer based out of Toronto. As part of the electronic music and arts duo Karla, he has gigged regularly since 2019. Additionally, he has helped facilitate numerous events, including the monthly live electronic performance Series Hitplay, and most recently, Rogue Waves festival. Releasing the album "Karla is here" in 2019, Karla expanded beyond live music into installation art with the Music Gallery Residency project "Temple Block" in 2022. For many years Yoni has also consulted for independent Synth manufacturers Rabid Elephant and OK200. He has given talks on the subject of electronic instrument design and philosophy at Frequency Freaks, and Toronto Sound Festival, attempting to help bridge the cultural gaps between engineers, performing artists, and designers. Yoni was schooled in the Fine Arts through an MFA studio program at the Pennsylvania Academy of the Fine Arts where he built an AI-driven drawbot. He still maintains a painting and drawing practice, recently as Resident Artist at the MERZ program in Scotland. You can find him in random alleyways, in the winter, drawing trees, or hunched over some weird looking gear, making alien noises.

Jason Nolan is autistic. He is also the director of the Responsive Ecologies Lab (RE/Lab) and an associate professor in Early Childhood Studies at Toronto Metropolitan University. His present research focuses on reconceptualizing music education and exploration of acoustic sensory information in early learning environments from a social justice lens of equity, diversity and inclusion. Nolan's background in designing adaptations for disabled children stems from a perspective of design initiated by children in order to support their sensory exploration of the world around them and the communication of their goals, interests and needs to their careers. Informed by design work with disabled Indigenous children and their families in Bolivia, Nolan's work focuses on the missing modality of auditory sensory play and exploration with DIY electronic and found objects, and innovations in pedagogical approaches for marginalized communities.

Ryan Page is an electronic music composer. He holds a PhD in digital media (UCSC) and an MFA in Electronic Music (Mills College). His work focuses on human interaction with technology, particularly the limitations and biases of communications media and their relation to human perception. In 2017 they created Repairer of Reputations, a project aimed at exploring the relationship between deprecated communications technologies and electronic music.

Repairer of Reputations has produced film scores, video-game soundtracks, short films, and a modular synthesizer. In 2018, they founded the electronic musical instrument company Magus Instrumentalis with David Kant, Madison Heying and Mustafa Walker. Ryan is currently an Assistant Professor of Sound Design at Berklee College of Music. Recordings of his music have been released by The Path Less Traveled Records, Give Praise Records, The Electronic Music Foundation, Indexical, Full Spectrum, Self-Help Tapes and Vestige Recordings. Their PhD dissertation was supported by the Phi Beta Kappa Northern California Scholarship. In addition, their MFA thesis "Ideology As Material Force" received the Frog Peak Collective Experimental Music Award. He has performed and collaborated with David Dunn, Laetitia Sonami, Anna Friz, Alvin Lucier, David Behrman, Ikue Mori, Rhys Chatham, James Fei, and Trimpin. Page's music has been positively reviewed by The Wire, Pitchfork Magazine, Decibel Magazine, National Public Radio, Maximum Rock N' Roll, and Aquarius Records. They currently live in Providence, Rhode Island with their partner and daughter.

Michael Palumbo (MA, BFA) is a musician, community builder, teacher, and programmer. A PhD candidate in Digital Media at York, he is researching electroacoustic music improvisation in online, multiplayer virtual reality with the XR app "Mischmasch". As a composer, selected works include *Data Issues: Please See Attachment* (2017); performance video games *Recursive Writing* (2015) and *Stethoscope Hero* (2014); electroacoustic compositions *CrossTalk* (2013), *Soup Phase* (2013) and *Iron Harvest* (2012), and was an executive producer of the distributed performance concert *Telematic Embrace: You Had Me at Hello World* (2014) which was presented at the 137th Audio Engineering Society Convention in New York. Michael performs often with a modular synthesizer as a soloist and in ensembles. He is the artistic director of the Exit Points electroacoustic free-improvisation concert series and runs the record label also named Exit Points. He teaches sound and video art at York University, creative coding with javascript at the Ontario College of Art and Design, and private lessons in coding and music. Michael is increasingly interested in creating intersections between community building, mutual aid, and the performance of free-improvisation music. Please follow his work at https://linktr.ee/michaelpalumbo or on Instagram at @michaelpalumbo_

Meng Qi is a pioneering synthesizer designer and musician famous for his music and distinctive instruments, which are used by electronic artists all over the world. With dedication, research, and deep experience in electronic musical interfaces, he designs synthesizers with unique thoughts and aesthetics and performs synthesizers with a gestural, emotional edge. As a synthesis, creative coding, and instrument-building teacher, he has led courses at institutions like Beijing Contemporary Music Academy, Tianjin Academy of Fine Arts, Central Academy of Fine Arts, and other significant events throughout China and abroad. Meng Qi's two latest designs are Wing Pinger and Wingie. Wing Pinger is an analog musical instrument that merges chaos and melody on a highly optimized interface for sonic versatility and profound expressions; Wingie is a handheld stereo resonator with onboard microphones. Meng Qi maintains effective communication and cooperation with the whole community of synthesizer designers worldwide. Now he's working on his third standalone instrument with Trent Gill.

Justin Randell is Course Director of Music and Sound Design at London South Bank University (UK), leading the delivery of its production modules in electronic music making and film sound design. He has extensive experience in sound production and recording studio design, supervising postgraduate research in these areas. His interest in modular music is rooted in experimental approaches to electronic music that draw from algorithmic programming and hardware electronics. Previously, he has worked with sound across multiple disciplines, including art installations by artists such as Thomson & Craighead, and live audio-visual performances in collaboration with Weirdcore for artists such as Trevor Jackson and Aphex Twin. He has mixed music recordings for Mercury-nominated artist Ty and has recently been active in film sound design projects. He is currently developing new research in the field of immersive sound.

Hillegonda C. Rietveld, PhD, is Professor of Sonic Culture at London South Bank University (UK), where she introduced sound studies and electronic music production. She was Chief Editor of *IASPM Journal* in 2011–17. She has published extensively on electronic dance music and DJ cultures, including the monograph *This is Our House: House Music, Cultural Spaces and Technologies* (1998, 2020), the co-edited collection *DJ Culture in the Mix: Power, Technology, and Social Change in Electronic Dance Music* (2013), and a co-edited special issue on the dub diaspora for *Dancecult: Journal of Electronic Dance Music Culture* (Vol 7/2, 2015). The co-edited *Cambridge Companion to Electronic Dance Music* is due for publication in 2024. She additionally publishes research in concomitant aspects of electronic popular music, and co-edited special issues on game music for *GAME: The Italian Journal of Game Studies* (Vol 6, 2017) and *Journal of Sound and Music in Games* (Vol 4/3, 2023). In support of her doctoral research on house music cultures, she recorded in 1992 a twelve-inch single with Chicago house music pioneer Vince Lawrence for Rob's Records. During the 1980s, a previous era in modular hardware synthesis, she programmed, performed, and recorded electronic music, including an album, with Quando Quango for Factory Records.

Dave Rossum was born in San Jose, California in 1949. He earned a BS in molecular biology at Caltech. During his graduate studies at UC Santa Cruz, he discovered his passion for music synthesizers. In 1972, along with Scott Wedge, he founded E-mu Systems. At E-mu, Dave invented the modern polyphonic synthesizer keyboard, licensed to Tom Oberheim for use in his Four Voice synthesizer. Dave joined with Ron Dow at Solid State Music (SSM) to design the first custom analog integrated circuits for synthesizers. E-mu's 4060 polyphonic keyboard and sequencer was the first use of a microprocessor for electronic music. Dave Smith licensed these technologies for the Sequential Circuits Prophet 5. In 1980, Dave invented the technologies that made sampling synthesis practical in the form the E-mu Systems Emulator series, which Dave continued to enhance over the next two decades, and the iconic SP-1200 drum machine, beloved by many Hip-hop artists. In 1994, E-mu was acquired by Creative Technology, Inc. Dave became their Chief Scientist, leading to many advances in Creative's Sound Blaster products. In 2011, Dave left Creative to join Audience, Inc, as Senior Directory of Architecture, where he led the design of specialized digital signal processing chips for cell phone audio. Dave left Audience in 2015 to form Rossum Electro-Music LLC, returning to his modular roots, designing new analog and digital Eurorack modules. He also joined Universal Audio, Inc. as a Technical Fellow, where he describes his role as their "tame analog circuit designer." Today, at Rossum Electro-Music and Universal Audio, Dave is still inventing new synthesizer circuits and algorithms, and he also continues to design analog ICs for Sound Semiconductor, Inc. For fun, he runs marathons, SCUBA dives, and backpacks in the High Sierra with his standard poodle, Lily.

Paulo Sergio dos Santos is the founder of EMW. Born on June 20, 1964, in Amparo, São Paulo, Brazil, he was raised in a family deeply immersed in technology. The only child of an electronics technician and a housewife, Paulo's family relocated to Araras when he was just two years old, following his father's job with the now-defunct Tupi Television Network (TV TUPI). In 1975, they returned to Amparo to revive a longstanding family business in the construction ceramics industry. Paulo's childhood was rich with exposure to electronics, often accompanying his father in various technical endeavors. By the age of nine, he had already started building his first electronic circuits, and by eleven, he was experimenting with audio technology, creating devices like FM transmitters and sound effects generators. After training as an Electronics Technician, Paulo pursued a degree in Electronic Engineering at the National Institute of Telecommunications (INATEL). His university years marked his foray into the world of synthesizers, beginning with a Roland JX-3P and later a Prophet-5 from Sequential Circuits. This experience deepened his fascination with synthesis and audio technology. In 1991, Paulo returned to Amparo to start his own business. He founded PS DIGITAL Equipamentos Eletrônicos Ltda. in 1992, initially focusing on electronic equipment repair and developing automation software for radio stations. His 1994 launch of DIGIMÍDIA, an audio control and broadcasting software, achieved significant success, reaching over 300 radio stations worldwide. In 1999, facing stiff competition in the radio automation sector, Paulo established LASERLine, shifting his focus to laser technology, another area of great interest. This venture was instrumental in the 2002 launch of EMW. EMW, specializing in synthesizers, has become a leading name in both analog and digital modules, boasting over 150 models to date. Paulo is dedicated to emulating the sound of early analog synthesizers, favoring the use of discrete components over miniaturization. Preferring to let his work speak for itself, he shies away from personal promotion. He is the sole developer of his modules and the microcontroller software used in some EMW models. Paulo's ambition is to continue his journey of innovation, focusing on high-quality and authentic products. His dedication and passion for synthesizers and electronics are the legacy he aims to leave behind.

Sparkles Stanford (they/them) is a nonbinary PhD candidate in Philosophy at Duquesne University. Their academic research sits at the intersection of political philosophy, ontology, and sound studies. Stanford studies reactionary political theory, specifically the connection between fascism and racial biopolitics. Their research draws on late 20th-century French philosophy and various anti-fascist traditions. They are also interested in the history of sound art and electronic music. They have a soft spot in their heart for the E4XT sampler, magnetic tape, and pit bulls.

Asha Tamirisa [she/her/hers] is an artist and researcher primarily working with sound and video in performance and installation. Her work explores materiality, metaphor, history and archives, and gender and technology. Asha holds a PhD in Computer Music and Multimedia and an M.A. in Modern Culture and Media from Brown University, and has taught at Street Level Youth Media, Brown University, RISD, and Bates College.

Ezra J. Teboul (they/them) is a researcher and artist, currently student librarian at Concordia University. Previously they were scholar in residence at the Columbia University Computer Music Center. Their work focuses on the material histories of electricity, work, and music. Their first instrument was a plastic violin, and their first soldering project was a solid state emulation of the Orange MkII graphic preamp circuit. In 2016 they published their first article "Sonic Decay" with Sparkles Stanford, in the *International Journal of*

Žižek Studies; and released their first record, *Passive Tones*, with Karl Hohn on Afternoons Modelling. This is their first book.

Kurt Thumlert is an associate professor in the Faculty of Education at York University in Toronto, Canada, and is an executive member at York's Institute for Research on Digital Literacies (IRDL) as well as research associate at the Toronto Metropolitan University Responsive Ecologies Lab (RE/Lab). His current research focuses on informal learning, new media and technology studies, production pedagogy, and learning through making with sound-based technologies and other tools.

Arseni Troitski is a Tel Aviv-based musician with a background in linguistics. He releases music as himself (in collaborations) or as *black dingus* (solo). His instruments are modular synthesizers, field recorders, tape machines and software. Arseni is a part of the transoceanic networked duo Manifestoon Platoon (with Brooklyn-based ethnomusicologist and musician Eliot Bates). He is planning to continue his education and apply to a PhD program in ethnomusicology.

William J. Turkel is a professor of History at the University of Western Ontario and internationally recognized for his innovative work in digital history. He uses machine learning, text mining, and computational techniques in his study of the histories of science, technology and environment, drawing on many decades of programming experience. Author of *Spark from the Deep* (Johns Hopkins, 2013), *The Archive of Place* (UBC, 2007) and the open access textbook *Digital Research Methods with Mathematica*, (2nd ed 2019), he was Project Director of Digital Infrastructure for the SSHRC-funded Network in Canadian History & Environment (NiCHE) from 2004 to 2014. Dr. Turkel is a member of the College of New Scholars, Artists and Scientists of the Royal Society of Canada (2018–25). His first hardware synthesizer was a Casio VL-Tone he bought in 1979.

Graham Wakefield's research has evolved from computer music composition to the generation of open-ended environments for exploratory experience, emphasizing continuation over closure. This work is expressed through software design for creative coding, and immersive artworks of artificial ecosystems (both leveraging live system evolution through dynamic compilation). He is Associate Professor in the School of Arts, Media, Performance and Design and Canada Research Chair of Computational Worldmaking at York University, Toronto, where he runs the Alice Lab. He holds a BA in Philosophy from the University of Warwick UK, a Master in Composition from Goldsmiths College University of London, UK and a PhD in

Media Arts and Technology from the University of California Santa Barbara, USA. Graham played a central role in the development of software systems and authoring content for the AlloSphere: a three storey spherical multi-user immersive instrument in the California Nano-Systems Institute. Graham is also a software developer for Cycling '74, co-authoring the Gen extension for the widely-used media arts environment Max. His works and publications have been performed, exhibited and presented at international events including SIGGRAPH, ICMC, NIME, EvoWorkshops and ISEA. He also co-authored a successful textbook "Generating Sound & Organizing Time", an introduction to sample-level signal processing, which has a 4.9/5.0 score on Amazon, with volume 2 expected in 2024.

Jacob Weinberg is an interdisciplinary artist living and working in Philadelphia. He is currently an MFA candidate at the University of Pennsylvania, where he works in video, sound, installation, and new media. He also produces projections and sound design for performance. His work has been shown and produced throughout Philadelphia and Baltimore, including Slought, Asian Arts Initiative, Current Space, and others.

ACKNOWLEDGEMENTS

Ezra Teboul dedicates this book to Don Buchla, Pauline Oliveros, Trevor Pinch and Jon Appleton. For simply existing, they are also most appreciative of Rebecca Hanssens-Reed, Laela Teboul, Ruben Teboul, Max Ardito, Aphid Stern, Sparkles Stanford, Zack Batiste, Audrey Beard, Ada Bierling, Camille-Mary Sharp, Curtis McCord, Emily Doucet, Sadie Couture, Meesh Fradkin, Mike Groening, Margaret Banka, Mina Beckman, Andrew Feinberg, Molly Haynes, RJ Sakai, Megan Meo, Lise Bourbonniere, Garret Harkawik, Karl Hohn, Matthew D. Gantt, Adam Nikkel, Selin Altuntur, Stuart Jackson, Libi Striegl, Oskar Peacock, Marilla Cubberley, Natalie and Julia Castrogiovani, Romina Ignacia, Mia Adri Kang, Patricia Ekpo, Claire Sigal, and Yves Saint-Laurent and Anfisa and Spooky and Scully and Jules and Lubi and Carl and Apple and Guidoune and Tiff. They hope Miles is doing well and getting plenty of porch time.

For their camaraderie, they also thank the Write Thing (especially Dorothy Santos for being a mighty ray of sunshine), the winter luddite Zoom group, Jonathan Sterne and CATDAWG, and the CHSTM sound and technology group, especially co-conveners Eamonn Bell and Brian Miller, and regulars Magnus Schaeffer and Noah Kahrs. At Routledge, Emily Tagg, Hannah Rowe, Jennifer Hicks and Neha Shrivastava for shepherding us through the publication process. Also, the Queens University archivists Nicole Kapphahn and an unnamed colleague for their insight on the Le Caine archive, and Nick Patterson at the Columbia University Music Library for shepherding them through the CPEMC archives, which would not have happened without Seth Cluett's support—continued gratitude to him as well.

Andreas Kitzmann gives special thanks to the crew at the Responsive Ecologies Lab at Toronto Metropolitan University, especially Jason Nolan and

Kurt Thumlert, who continue to provide an intellectual and synth friendly refuge in the middle of the city and a sounding board for the ideas and ramblings that informed this and many other projects. Claes Thorén at the University of Uppsala also deserves hearty thanks for providing the spark that initiated this collection, as does Einar Engström who was also there at the beginning of this and many other projects. The editors also acknowledge and appreciate the financial support from the Faculty of Liberal Arts and Professional Studies and the Faculty of Graduate Studies at York University, Toronto.

INTRODUCTION

Andreas Kitzmann and Einar Engström

This edited volume brings together scholars, composers, and musical instrument designers in an exploration of modular synthesis, which is understood here as a material, cultural and theoretical approach to electronic sound synthesis for music and multimedia arts. Through historical, technical, social, aesthetic, and other perspectives, the contributions in this volume open a collective reflection on the powerful connections among technology, creativity, culture, and personal agency. Musical instruments present concrete windows onto the ways societies sensually organize through sound, technology, time, and beyond. In the 21st century, one of the most compelling of such windows is the modular synthesizer, an electronic musical instrument defined by customization, change, and open-endedness: users build their instruments to their own liking, with as many or as few components as their resources and preferences allow.

Modular synthesizers are electronic instruments which, as the term implies, are comprised of separate components such as oscillators, filters, effects and voltage controlled amplifiers that are housed tougher in some kind of box or rack and connected to each other with external patch cables, making the instrument flexible and reconfigurable to the user and visually similar to the telephone switchboards of the early 1900s.

The typical narrative for modular synthesizers often begins in the 1960s with the two foundational figures of Robert Moog and Don Buchla (Pinch and Trocco 2002) and the "styles" of synthesis they represent, with Moog being associated with "East Coast Synthesis" and Buchla with "West Coast Synthesis." On the West Coast, as the mythology asserts, additive synthesis is frequently cited as a defining approach with complexity, experimentation, non-conventional interfaces and elaborate cross patching hailed as the

DOI: 10.4324/9781003219484-1

primary aesthetic and technical operatives. Meanwhile, on the East Coast, synthesis is said to be subtractive, and associated with more conventional approaches to musical composition, standardized interfaces (i.e., keyboards) and generally an appeal to normative musical conventions and tastes.

That such a tidy narrative is reductive and historically inaccurate should come as no surprise to even the uninitiated. History and culture are complex issues, with many layers, paradoxes and blurred boundaries which make things messy when it comes to constructing lineages and mythologies or origin. Yet the impact of this origin story persists, not only in popularized accounts of modular synthesis but also in the design decisions made by developers. Searching the web for the "history of modular synthesis" quickly leads to sites such as a 2014 article in *Music Radar*, which proclaims that "legend has it that the first commercial synths were realized almost simultaneously by Robert Moog in New York, and Donald Buchla in San Francisco." The Wikipedia entry on the modular synthesizer tells a similar tale as do the sites of influential retailers such as Perfect Circuit and Reverb which both have pages devoted to these foundational approaches to musical synthesis.[1]

The manufacturers of modular synthesizers do their part in keeping the Moog/Buchla dichotomy alive, as in the case of Crea8 Audio's (in collaboration with Pittsburgh Modular) dynamic duo of the "East Beast," and "West Pest." Both are stand alone semi-modular synthesizers which tap into the well worn associations of East and West coast synthesis. The "East Beast" "evokes visions of big filter cutoff knobs, envelope generators, and importantly, big fat-sounding voltage-controlled oscillators," whereas the "West Pest" goes "all yoga mat, kale smoothie, and alfalfa sprout on the synth world… characterized by additive, derived, rich and complex sounds." While Crea8's descriptions are on the extreme side, other developers employ the East/West Coast divide in various ways, whether as a point of departure as in the case of the "0-Coast" synthesizer by Make Noise or as an homage, as in the case of AI Synthesis' "AI106 West Coast Mixer" which is described as taking its "inspiration from vintage mixer circuits from 60s and 70s West Coast synthesizers" (AI Synthesis).

The endurance of the East/West/subtractive/additive narrative speaks to more than just the lure of historical shorthand to reduce complexity, but also as a context to probe the role that such discursive constructions continue to play in the development of modular synthesis and, arguably in musical synthesis in general, whether hardware or software based. As a foundational myth or template, the divide elicits numerous reactions, ranging from a complete rejection of its relevance, an attempt to balance between the two or an affirmation of one or the other as a paradigm for development or use. The continued presence of "the foundational myth" has a kind of productive value in the sense of functioning as a persistent touchstone for shaping the industry and its related scenes in one way or another. Despite the many developments that have occurred since the 1960s and 70s the "ur-fathers" of Moog and Buchla continue to hover, sometimes in the foreground and at times as mere whispers.

Modular Synthesis: Patching Machines and People can, in part, be understood as a partial dialogue with the "fathers" and their respective stories and adherents—not as a reaction, affirmation or alternative, but rather as a means to weave connections between established, new and emerging patterns. Accordingly, one could think of this book as an attempt to extend the patch, to use a modular metaphor, and as a result introduce a set of divergent pathways and interconnections. One such extension is the route that musical synthesis took in the 1980s which is often described as the age where the analog synthesizer was usurped by its digital counterpart, and where patch cords and user defined signal paths were replaced by presets and programs and eventually by entirely software based instruments contained within personal computers and tablets. It is a common sentiment that the re-emergence of modular synthesis in the 1990s, with Dieter Doepfer leading the charge, functioned as a type of counter reaction to digital interfaces, particularly with respect to the lack of tactility and the paradoxical limitations of always having infinite possibilities at your disposal, which is the default for most software instruments. "The control of sound with just one parameter, brings about one of modulars' greatest contributions; that of the limited control of expressiveness and gesture" (O Connor, 2). Such deliberate imposition of limits is arguably the polar opposite of normative design trends when it comes to digital tools. It also draws attention to the importance and role of the interface, and that technological progress is not just a linear march forward, but also a process of returning to earlier incarnations and techniques. As such, we might want to rephrase the terms often used to describe modular synthesis in the 20th century and early 21st century. It is not a revival or a return to either one of the two coasts. It is not nostalgia or some hipster induced desire for aestheticized quirkiness. Nor is it a form of postmodern pastiche or revivalism. It *may* be an indication of the post digital, if one understands that term as "the messy state of media, arts and design after their digitization (or at least the digitization of crucial aspects of the channels through which they are communicated)" (Cramer, 17). This messy state is one where the teleology of progress ceases to be linear, with the old always being replaced by the new and where the distinction between old and new media does not really matter anymore. Instead, one uses "the technology most suitable to the job, rather than automatically defaulting to the latest new media device" (Cramer, 21). Modular synthesis then, or in fact, hardware based synthesis in general, stands not in opposition to screen based digital instruments, but is rather a hybrid form that provides users with the ability to navigate between the affordances of different interfaces and opportunities for sonic exploration.

The modular synthesizer can also be understood as representative of trends and phenomena outside its immediate context, and as a potent example of a technology that tangibly exhibits and represents major aspects of technologically-embedded existence in the 21st century. Accordingly, while this collection most assuredly exchanges knowledge about modular synthesizers, it simultaneously explores much larger ideas associated with the current state of technology, technology use, practice, and resistance. The modular synthesizer is representative of the blurred roles of production, consumption, and critique, namely in the dynamic relations between content creators, the technologies they use, and the individuals and communities who design and build them. Modular synthesis cultures thus emerge within social collaboration, equating a peculiar disruption of conventional patterns in the production, distribution, use and disposal of artistic tools. The lines between audiences, amateurs, and professionals (as well as technicians and artists) are not only blurred, but combined in a collective exploration that creates the tools, associated techniques, and aesthetic conventions used for sonic, musical, and even social practice.

Histories of modular synthesis—produced in everyday discourse as much as by professional historians—are rife with origin stories, idiosyncratic personae, myths of genius, and figures of innovation. Beyond the cursory outlines presented in companion volumes such as Chadabe (1997) and Manning (2004), the authoritative text in the field remains Pinch and Trocco's (2002) *Analog Days: The Invention and Impact of the Moog Synthesizer*. A classic implementation of social construction of technology (SCOT) methods within historical scholarship, the authors pit the development of the Moog modular synthesizer against the Buchla modular synthesizer from the early 1960s, assessing the respective "success" of each brand/name amongst musicians and composers through the late 1970s. Comparing the musical and extra-musical characteristics of so-called "East Coast" (Moog) and "West Coast" (Buchla) synthesis techniques, Pinch and Trocco find that Moog ultimately reached a larger audience due to founder Robert Moog's finesse in navigating business and music worlds alike; Don Buchla was doomed to obscurity as the consequence of his beliefs about and inimical behavior towards mass-market instruments—he was, simply put, too "avant-garde."

Nearly twenty years since the publication of *Analog Days*, the Moog brand is as celebrated as ever, and as commercially successful. However, a deep dive into modular cultures today—as opposed to *popular* electronic music cultures—reveals an intense, quasi-spiritual passion for Buchla among collectors, musicians, designers, documentarians, and DIYers. The bumper stickers "Don Buchla for President" and "My Other Car is a Buchla" are not an infrequent sighting in studios around the internet. The legacy of Buchla may begin in heady, countercultural California, but it now manifests as various singularities of value and meaning across the world, and therefore merits continued scholarly attention and analysis. The contemporary understanding of past phenomena in the world of modular synthesis is in need of multiplication, complication, and general re-questioning.

This volume contributes to a growing awareness of concepts of *modularity* as a constellation of engineering cultures, musical cultures, synthesis methods and psychoacoustic experimentation, economic and political contexts, and social collaboration. In shifting the focus away from technical feats and their timelines—as well as from the dichotomies of East Coast vs. West Coast or success vs. failure—the contributors to this section present a nuanced view of modular synthesis as a historically contingent framework for moulding and experiencing embodied perception; that is to say, of a full-fledged cultural history imbued with all the richness and complexity of an experimental past, right here in the present.

Opening the book is Ezra Teboul's pointed critique of the cultural, historical and ideological processes that have informed the development of synthesis technology within the general paradigms of the military/industrial/capitalist apparatus. By way of situating the technologies of modularity and control voltages within the larger sociotechnical systems in which they are embedded, Teboul not only reveals their debt to Western hegemonies but also challenges the assumption that musical instruments are somehow ideologically "neutral" given their association with artistic creativity and counter cultural leanings. The latter is especially the case with the "radical" uses of musical synthesis, as embodied by the Buchla and more generally the consciousness expanding exploits of electronica and experimental music. While all musical technologies are part of larger political and social systems, the modular synthesizer is unique in terms of its reliance on the Western technoscientific project and the global industrial infrastructure of communication and component manufacturing (and the extraction of raw materials upon which they depend). Teboul explores such assertions through a detailed and subtle reading of the technical and historical factors that are integral to the modular synthesizers development, exposing, as such, their "origin stories in war-time computing, imperially-motivated signal process, and class/race/gender discriminating industrial cultures" (internal page reference). Yet Teboul is not calling for the abandonment of the modular synthesizer or the rejection of using advanced technologies to create musical instruments. Instead he is advocating for a form of vigilance - an attentive eye "on the environmental, labor and cultural consequences" (internal page reference) of our reliance on nations-states, global supply infrastructures and military/industrial alliances that have allowed for us to order that shiny new modular from our favorite retailer. Such attention can provide a more nuanced frame for us to appreciate the "creative uses of electronics not especially designed for joyful experiences and meaningful connections" (internal page reference).

Teboul's preface is followed by a challenge to the very concept of modularity, which contributor Ted Gordon identifies as an "ideological fantasy" attached to the rise of neoliberal forms of life. Drawing on archival research, oral history, and personal performance experience, Gordon confronts the past origins and contemporary configurations of the legendary Buchla Easel. That Buchla abandoned production of this semi-modular instrument soon after creating it in the early 1970s is but the beginning of a decades-long legacy marked today by technological fetishism and myth-making, particularly in DIY communities. When the Easel was officially recreated by the rebranded Buchla and Associates in the late 2000s, high expectations were met with widespread disappointment in the build quality of the instrument, as well as with its inability to integrate with modern studios. In these overlapping stories of failure, Gordon identifies what he calls "productive friction," which points to fundamental flaws in the so-called open-endedness of the modular mindset.

Other historical case studies reveal the impact of modular synthesis on theoretical work in the fields of organology and media theory. In his contribution to this volume, Jonathan de Souza unearths the forgotten work of two important scholars in these fields: Herbert Heyde and Friedrich Kittler. In the 1970s, Heyde was inspired to imagine all musical instruments as so many kinds of modular synthesizers. A guitar's strings could be classified as oscillators, the mahogany as an amplifier, the player's fingers as alternating gates and filters, and so forth. For the media theorist Kittler, such reconceptualization could be taken even further, beyond the concerns of musicology and straight into the existential realm of technics. As de Souza shows, Kittler's private tinkering with synthesis first urged him to consider Wagnerian music drama as itself a kind of circuit, and the synthesizer as a sensible window onto the interplay of music and sound, the arts and media, and the symbolic and the real. In both cases, modular synthesis emerges as more than mere technology, as an epistemic tool: an active agent in the production of knowledge.

Much of the knowledge produced about modular synthesis, or popular historiographical work on the subject, is done by practitioners, from sharing information in online forums to collecting and preserving instruments for public access. For her chapter on "A Time-Warped Assemblage as a Musical Instrument," Hannah Bosma takes her primary research site as the synthesizer studio and archive Willem Twee Studios in Hertogenbosch, Netherlands, at which she identifies a case of "preservation as performance," or preservation motivated by contemporary creative interests. Unravelling these interests, she combines ethnographic methods of interview and observation with archival research to explore the historical effects of modular synthesis on electronic music wrought broadly—an analysis that she then takes to another Dutch institution, Amsterdam's STEIM (Studio for Electro-Instrumental

Music). Tracing this studio's seminal hardware innovations through to members' current work in software design allows Bosma to elaborate the implications of modular thought over time, from pre- to post-digital eras, and how these spill into and ripple through non-musical realms of human activity such as digital technology, management, and the arts.

The links between technical practice and creative work are further explored in the interview with the sound artist and composer Dani Dobkin, who works with modular synthesis, handmade circuits, and improvisational electronic music through an array of interdisciplinary and collaborative projects that span numerous genres and performance practices. In conversation with Ezra Teboul, Dobkin reflects on their teaching work, and the lessons arising from the communal nature of their exploration of these complex musical systems. This perspective is further informed by their participation in a restoration project for Columbia University's Buchla 100 system, yielding important insights into links between learning and teaching, repairing and making and improvising and performing. In the wide-ranging interview, Dobkin elaborates on the formative figures associated with electronic synthesis and how their personalities are reflected in arrangements of sounds and components. The result is a vision of modular synthesis as a historical narrative of contingencies, experiments, pleasant surprises and a generalized blurring of intention between the machinic and the human that sheds important light on the nature of creative modular practice.

Among the goals of this volume is to move beyond solely theoretical considerations of technology in an effort to engage directly and materially with technology itself. Michael Johnsen addresses this goal directly in his detailed history of Gordon Mumma's Sound Modifier Console created for the Osaka World Expo in 1970. The pavilion represents a curious and arguably contradictory confluence of experimental art, electronic engineering and corporate interests, given that the soft drink giant Pepsi, along with Bell Labs, provided the financial and technical means to realize the ambitious and at times bizarre vision of the project's lead, the visionary and idiosyncratic experimental musician and artist David Tudor. Johnsen's approach is to view Mumma's work through the twin lenses of musical and technological history during a time when the building of electronics was arguably as much of a "folk practice" as it was the domain for the military industrial complex and the emerging industries of Silicon Valley. A central component of the chapter is a detailed analysis of the circuitry itself, both in terms of specific design elements and its relationship with the wider contexts of the building of electronic instruments and related technologies. As Johnsen emphasises, "Historic circuits shouldn't (only) be studied while asleep—unpowered, and subjected to the dry postulates of circuit analysis." Rather they should be literally switched on and used and furthermore implemented into contemporary contexts, such as in the case of Johnsen's project to render selected elements of Mumma's designs as Eurorack compatible clones. In this manner, history is not just a static element intended for reflection, but an ongoing entity that reminds, informs and extends present and future possibilities.

For artists and developers engaged in modular synthesis, creative practice is not only bound up with an intimate relationship with their instrument in terms of understanding its various nuances, but is also driven by a wider engagement with the technical intricacies that lie behind the panels with their blinking lights, knobs and switches. To plug and play is one thing, but so is the ability to solder, to read technical papers, to scour the Web for obscure parts, to educate oneself on the peculiarities of various waveforms, and to immerse oneself in social networks where engineers, artists, musicians, developers and tinkerers share knowledge, advice, and stories. As such, modular synthesists blur the lines that normally divide the technical environments associated with other creative arts and genres where the technicians and engineers stand at a distance from the artists and any dialogue between the two is normally limited to calls for repair or an investment in a new instrument altogether. The wide terrain of creative practice that merges the technical with the creative is touched upon throughout the volume, but more clearly explored in the contributions by Lori Napoleon, Justin Randell and Hillegonda Rietveld, Dani Dobkin, Kurt Thumlert et al., and Arseni Troitski and Eliot Bates.

In "Switchboard Modulars," Napoleon explores the links between the telephone switchboards first developed towards the end of the 19th century and modular synthesis through a chance encounter with an exhibit at a historical museum in Escanaba, Michigan. Napoleon's interest in old switchboards goes beyond the strictly academic: it is integral to her recent artistic and creative practice which is centered around the transformation of old switchboards into working modular synthesizers and analog sequencers. Each one of the old devices is carefully deconstructed and then rewired and repainted, with the result being instruments that aesthetically blur the lines between history, technical and creative practice. Napoleon, who describes her artistic work as being situated within the tradition of *objet trouvé*, uses her work to "integrate science and technology" and to employ her "electronic studio as a laboratory for sculpting emergent patterns and textures" (http://www.meridian7.net/). As she notes in one of the videos uploaded to her website, the juxtaposition of old switchboards with contemporary modular synthesizers is one that characterizes the musician as something akin to an operator or technician, which for Napoleon yields important insights into the interrelated nature of creative and technological practice.

With an eye towards recognizing that modular synthesis relies as much on creative and performance practice as it does on technical expertise and labor, Justin Randell and Hillegonda Rietveld utilize an auto-ethnographic approach to explore specific patching techniques that enable what they describe as an embodied approach to creative musical synthesis. Among the central components of such an approach is the move away from the screen. As many modular enthusiasts are keen to point out, it is the tactile and physical nature of modular synthesis that is elemental to not only its distinctiveness from the digital, but also to the very nature of modular creative practice. Modular synthesizers have an affective and aesthetic dimension to them in so much that they invite touch and by design are created to be physically re-arranged and manipulated. As such, the performativity of modular synthesis extends beyond just the playing of them, but also in their assemblage and re-assemblage as users create and reconfigure their systems. Randell and Rietveld work through such ideas by taking the reader through the building of a modular patch to create a unique sound palette which they further explore in terms of how this process is informed by the "pre-packaged" elements of a commercial module and by the ability for users, to "hack" a given module either directly or by virtue of its intended use or integration within a system. The chapter concludes with insights into how modular synthesis has inspired new digital sound and music software and also how hybrid analog and digital modules intertwine in ways that elicit new forms of compositional performance in the post-digital age.

The modular synthesizer's role in challenging existing conventions aroud musical expression and practice is explored further by Thumlert, Nolan, Chan and McBride, who argue that music education remains rooted in exclusionary curricular forms and reified pedagogical coordinates. As a challenge to these forms and practices, they look to Eurorack modular synthesis as a dynamic site, practice and sociotechnical model for rethinking learning with sound and, by implication, for revisioning music learning within and beyond institutional spaces. The chapter begins with the observation that Eurorack synthesis has provided a unique and dynamic context for engaging sound, technology, and music making, thereby illuminating novel pathways for inquiry, learning and community participation. Drawing on the work of Allen Strange, an early electronic music composer and educator, Thumlert et al. employ a materials-centric approach to documenting and exploring the tools in play, and what and how people learn with, through and along side modular synthesizers. The authors make the compelling argument that modular synthesizers, as manifested in the Eurorack format, enable us to rethink learning with sound-based materials and tools, while also challenging dominant aesthetic systems and exclusionary pedagogies that persist in music education today.

Eliot Bates and Arseni Troitski's study of the llllll.com forum brings this question further by directly interacting with one of the active, mostly-anglophone communities of the online modular world. Working from previous publications on the co-construction of modular instruments and social interaction, the co-authors develop a nuance of interface design versus open-source sharing paradigms to investigate the interaction and co-dependence of both elements in the development, propagation and evolution of novel modules and approaches to modular synthesis. Prompted by Brian Crabtree (one of the originators of the monome instruments, whose community developed into the llllll.com forum) and his concept of "grid culture" Bates and Troitski offer the idea that modular culture crystallizes locally around *meta-interfaces*, that is, archetypical architecture which facilitate fluid interaction between open-source digital signal processing concepts and open interfaces, and, therefore, the development of artistically valuable tools in the eyes of their originating communities.

Modular synthesis provides an entry into music by way of the ability to work with sound directly and without the constraints and expectations of conventional music instruction and performance practice. For those coming from the traditional music world, modular synthesis prompts one to question the institutional conditioning of music education. Bana Haffar further challenged musical convention by taking electronic music outside of the studio through her modular on the spot series where music was made with the environment as opposed to being separated from it via the acoustically sealed confines of the electronic music studio. This desire to engage with the environment resonates with the Beirut Synthesizer Centre which in addition to providing space and equipment for those interested in musical synthesis also provides opportunities to connect and learn outside of normative educational institutions and paradigms. Haffar challenges us to think of synthesizers and synthesists as an emergent micro-species that is grounded in community access, experimentation, autonomy, and participation.

Creative individuals often have a deep-seated relationship with their tools that is as emotional as it is practical. Such a closeness speaks to broader discussions about our intimate relationships with technology and the often thin lines that separate the inner worlds of human consciousness and being from the relatively empirical conditions of material reality. The presence and often dominance of sound in such human-technology relations, for its part, gives sensible form to this intimacy, and opens up further discussion of how humans locate and express themselves in the contemporary technoscientific landscape.

William Turkel investigates technology's potential to elicit altered states of consciousness by probing the affordances provided by the modular synthesizer, notably in terms of auto-experimentation and its proclivity for self-generative and semi-autonomous musical/sonic composition. There are historical precedents here, given that sound and music have long been used, across human cultures, to create multiple forms of altered consciousness, ranging from the meditative to the ecstatic. Turkel's wide ranging explorations are informed by disability studies, and feminist, queer, trans*, utopian and (post)anarchist theories and relate auto-experimental practices to wider currents in occulture and esoterica.

That modular synthesis can be a conduit for eliciting alternative states of being is explored further by Asha Tamirisa, who seeks to remove the veneer of neutrality often associated with technology in general by rearticulating the cultural logics of sexual difference, human connectivity, and social organization of modular interfaces. As she argues, interfaces are instrumental in the production of an epistemology, a way of knowing the self via their ability to create and manipulate memory within a complex body of information. In this regard, interfaces are literally embedded with metaphorical and rhetorical weight that is reflective of larger cultural logics and biases. Given modular and electronic music synthesis' links to the history of science and engineering, Tamirisa argues that it affords a particularly resonant context from which to comment on the social relevance of technical design. Tamarisa weaves together a range of historical, technical and interpretive knowledge by an integration of archaeological methods and perspectives informed by feminist science and technology studies. Such methods are grounded in the visual, logical and poetic analyses of specific modules in order to reveal their potential to suggest or subvert normative notions of gender, power, and subjectivity.

The next three chapters utilize specific applications of technology, modular or otherwise, to investigate the dynamics between users and their instruments specifically and humans and technology more generally. Jacob Weinberg and Anna Bockrath bring together two technologies that at first consideration, come from opposite sides of the technological spectrum: the modular synthesizer and the floor loom. Understood as proto-digital technologies, these two technologies are recontextualized as tools that embody agency due to the manner in which they embody material practices that exist across multiple senses (sound, texture and form) which are further emphasized by conditions of material limitations, which is in marked contrast to the neverending options afforded by digital media. Specifically, their collaborative project connects the loom and the modular synthesizer through Max/MSP software which translates loom gestures into modular signals. Further parameter changes on the synthesizer feedback into the loom which then alters the textile patterns. The resulting artifacts, which are a musical composition, a textile weaving and a documented system of patches and connections, are used as descriptors of the remembered location of the "event," and as a means for Weinberg and Bockrath to work through their own experiences and material, technological practices.

David Dunn and David Kant are inspired by randomness in nature, but here it is a specific ecosystem which first motivated Dunn to experiment with electroacoustic chaos: the Atchafalaya basin of Louisiana. With its constantly evolving soundscape of environmental utterances, this temporary place of residence inspired a custom electronic system which, when interacted with, offered Dunn a place to contemplate and develop an intuition for the shifting and unpredictable dynamics of wildlife. David Kant, then, studied the system to develop a more explicit understanding of those electronics' complex behavior. In their contribution they offer an expansion of the original score and discussion for *Thresholds and Fragile States*, first detailing a digital model which could both be distributed and modified more easily, via software and dedicated hardware. In addition, they propose modes of analysis of the system in its analog and digital forms to a level of detail rarely seen in electronic instrument analysis. Their writing of this long process builds on previously unpublished material that was long-famous in the relevant research circle: it is exciting to see it come to wider light in such a context.

Ryan Page examines the tension between nostalgia and novelty in Eurorack-format modules that simulate analog media. Page explores in particular the modes of transmission common to modular synthesizers, where audio and control information is processed as analog signals. What is notable here is how such transmissions complicate our established means of classifying and distinguishing analog and digital systems, which arguably lies at the heart of the debates associated with post digital culture. Page focuses on the unit of the module itself and its status as an individual object which reveals what he describes as a precarious condition. Modules are designed to be dependent on external inputs in order to function and thus are difficult to understand or define outside their relationships to the other modules in an existing system. Such considerations are then used to analyse the nature of simulation and the manner in which the properties of analog media have been abstracted to inform the methods used to reproduce them in the simulated environments of digital environments. Page is careful to note that simulation is itself a fuzzy concept given that it is based on quantifiable abstractions of extant, continuous systems. Working from interviews with Eurorack designers/programmers and his own experiences as a scholar and practitioner, Page explores the manner in which notions of simulation both enable and constrain what is possible within the domains of modularity as a conceptual and material practice.

Technical practices in the arts offer a unique context within which to reconsider the traditional relationships people have with things. The definition of a good musical instrument is inherently contextual, as such it emphasizes the political nature of technology in general. Technology studies, although it is both acutely aware of the political nature of technology, and has had a long history of engaging with musical instruments, has nevertheless left the intersection of these discussions quite open. It is squarely these two modules that this book hopes to find interesting ways to patch together.

Accordingly, a number of chapters are dedicated to technical practices. These highlight the wide variety of theory available to the circuit-and-code-inclined experimentalist. DIY electronic music, and modular synthesis in particular, enables a wide range of engagement with every type of theory, from circuit theory to critical theory. By modularizing the functionalities of electronic music using everything from arguably-intuitive designs to intentionally indecipherable ones, modular designers offload part of the labor of music making onto their designs, the electronics they sell, and, indirectly, on the agency of the other modules by which their products might be surrounded. It is in this chaotic under-defined until it's over-defined context, between global supply chains of production and consumption and the promise of ever-better tools versus the reality of music practice and its unending challenges of access as much as inspiration, that modular has come to carve out its niche. In this niche scientific knowledge overlaps with folklore, personal experience with commercial expectations, offering a home for each person who can access it, a wild goose chase of potentially reinventing electronics for themselves, and, potentially, everyone who can afford their module.

Among modular synth designers, Meng Qi stands as an outlier both for his location—his homebase in Beijing, China being geographically and culturally distant from the rest of the predominantly Western modular world—and for his persistent foregrounding of the interface as the primary site of meaningful musical and technological discourse. In this wide-ranging interview with Einar Engström, Meng Qi elaborates his belief that haptic experience determines sonic experience, and back again. Along the way, he explains how such considerations specifically manifest in the knobs, potentiometers, spacing, size, and circuit layouts of his instruments. The implications of Meng's technical decisions have severe consequences, however: for them to appear as "sensible" to the user, they must first appear as a "burden." That is to say, the value of the instrument demands a mindset nearing asceticism, against the lure of immediate gratification, and a deep annexation of one's personal time. To design a musical instrument that carries real significance and leaves a legacy, the designer shouldn't seek to solve problems, but to create them.

Einar Engstrom's interviews are complemented by four additional short discussions between Heidi Chan and four Toronto-based designers. Together, this offers in the context of this section on technical work a snapshot of some of so-called North America's most active designers, illustrating both the imagined community at play over online networks, and the more physical spaces of a metropolitan region like Toronto and its creative sphere.

The design process is further explored through an in depth conversation with Dave Rossum, one of the foundational figures in the history of synthesizer design. Rossum describes himself as someone who is guided by a combination of scientific and intuitive approaches to instrument design. On the one hand, science guides his understanding of the nature of sound yet on the other the relationship that one has with an instrument is described as emotional and intertwined with the complex dynamics of human hearing and its bearing on cognition and perception. The combination of such approaches drives Rossum to continually seek out new approaches and technical possibilities, which in some cases come to him in the form of dreams. Not content with rehashing old ideas and past successes, Rossum speaks of his ongoing enthusiasm and unbridled joy in bringing new instruments to musicians. Yet despite the drive to constantly innovate, Rossum is careful to note that he strives to make instruments that are understandable, predictable and learnable. This is counter to design trends that favor more abstract and difficult interfaces and paradigms as a means to foster creative exploration. For Rossum, creativity comes from a deep understanding and relationship with an instrument and in that regard, the instrument must be purposefully designed. That said, no matter how intentional the design, musicians can still take an instrument in directions not imagined by their creator, such as in the case of Rossum's iconic SP-1200 drum sampler that effectively created much of the sonic palette of 1980s hip-hop.

The focus on designers continues with Paulo Santos of Electronic Music Works, one of the few modular synthesizer producers in South America. In this interview with Ezra Teboul, Santos charts the origins of the company, emphasizing in particular the role that classic analog technology and synthesis has had on the company's approach to modular synthesis in terms of maintaining and enriching the legacy of analog circuitry and sound design.

Further insights into the design process are provided by Corry Banks, the designer behind Modbap Modular who came to modular synthesizer design from his experiences in the Chicago hip-hop scene and his work in the fields of technical support and IT project management, which lead eventually to the creation of the blog and now Youtube channel "Bboytechreport" which gained early support from Moog and NAMM. Banks speaks to the sometimes conflicted relationship between the world of synthesizers and the world of beat making as practiced by hip-hop and R&B producers. Each has their own state of mind which tends to limit how technology is incorporated into musical practice. Banks seeks to bridge this divide through the creation of Modbap Modular, which blends the sound design parameters of modular synthesis into the form and aesthetics of beat making and hip-hop culture. His first modular, Osiris, is indeed described as a bridging device in so much that it affirms the value of bi-fidelity, which is to say the simultaneous presence of both high fi and low fi in a singular device. Modbap modules are known for their tactility and emphasis on performativity - they encourage the display of musical prowess front and center, as opposed to the more subdued displays of technical proficiency enacted by some modular synthesists. As such, Modbap verges on the political in so much that it enacts the hip-hop tenet of "taking stuff that wasn't made of us," and turning it into a powerful conduit for expression and the creation of community.

As hinted at above, generative and semi-autonomous machines widely construed seem to have been the inspiration for a number of metaphorical and literal implementations in the field of modular synthesis. Sparkles Stanford offers a philosophical connection via a critique of Gilles Deleuze with Felix Guattari's work on the concept of *modulations*, and their explicit discussion of electronic sound. Stanford elaborates on the value of their reflections in the context of the shift that Deleuze identified between societies of discipline to societies of control. Richard Pinhas, a synthesist better known in musical circles under the alias Heldon, was himself a student of Deleuze's, with numerous recorded interactions. On the basis of this practical perspective, Stanford comments on the nature of modern synthesis as a political act, one which operates between compositional and social media to enact and interact with existing and imagined power dynamics in a non-deterministic but occasionally locally predictable—in other words, chaotic—way.

It is therefore no wonder that modules are often designed with chaotic behavior in mind. Chaos can be implemented both at the level of control and signal, with modular being particularly good at blurring the boundary between the two. Naomi Mitchell's discussion of her module design and manufacturing as Omiindustriies reflects her understanding, ideas and ideals about the way randomness plays a role in playing an instrument. She details how, in her case, the synthesizer becomes a partially didactic machine, teaching us how it and those who made thought of sound prior to its use. She elaborates on the literal and metaphorical relationships of noise to the unknown, and how modular architectures might mediate this complex set of physical, cultural, and personal forces.

This negotiation between artistic ideas and ideals, as well as local and personal resources is also discussed in an interview with the circuit designer and active modular community member Andrew Fitch. Known to work under the Nonlinear Circuits or NLC brand, he discusses, in an interview with the editors, how scientific research papers influence his circuit design practice. Fitch details a number of wide ranging inspirations, between sound and electrical engineering, with a particular attention for practical and musical applications of chaos theory. An instance of the positive relationship-building via the shared mediums of online communities and electronic sound, Fitch's projects points to the distributed nature of the processes of invention and propagation underlying modular synthesis as a collective endeavor. The connection here between electronic chaos and the chaotic nature of human creativity may be mostly metaphorical, but nevertheless compelling: just as

David Dunn developed a modular synthesizer to better understand the natural context of acoustic ecosystems, NLC offers us another example of how chaos can be explicitly adopted into the everyday.

Michael Palumbo and Graham Wakefield's chapter is informed by the development of a VR modular synthesizer that allows multiple performers to interactively construct and modify modules and patches in a shared virtual space. The system is part of a larger project entitled "Mischmasch" which hails from the Alice Lab for Computational Worldmaking at York University in Toronto. Palumbo and Wakefield focus on the "Patch History Sequencer" subsystem within Mischmasch which can be understood as an experimental sequencer that affords such abilities as a more elegant and expansive 'undo' action; playback of patch edit histories; looping of specific history regions; and the ability to create alternative patch histories. Palumbo and Wakefield expand particularly on the use of the subsystem to recombine two patch histories and to what extent the concept of modulation can be applied in the sequencing of such histories. In doing so, they provide valuable insights not only into the technical natures of non-linear editing and version control systems, but also into the wider terrain of creative practice, specifically improvisation, distributed creativity, machine agency and the co-dependent nature of user/machine evolutionary histories.

The final chapter of this book is an interview with Brian Crabtree, one of the co-founders of Monome. As a purveyor of elegant electronic musical instruments, Monome's core identity is perhaps best aligned with the location of its "headquarters" in a small farmhouse in upstate New York where nature is defined as foundational to their way of being. In this interview, Monome's atypical approach to sound and music-making is explored in depth, often in ways that weave together an expansive set of concerns, approaches and paradigms—from the everyday observations of the meandering path of a stream and the intricacies of mushroom farming to the development of a bespoke coding language that is accessible to novice programmers. Crabtree notes that many of their machines aim to turn "work like computer stuff into creativity-facilitating playground-ecosystems." Among the tactics employed to realize such ambitions is avoiding the paths provided by mainstream capitalist infrastructures and to an extent some of the conventions associated with modular synthesis, notably the "anti-computer sentiment" that is often foregrounded by modular enthusiasts. Through instruments such as Teletype and Crow, Monome uses code as a means to strip away the layers of translation between music making and its notation and also to provide a much "weirder palette" of musical dimensions, especially in terms of time, pitch and timbre. At its core, Monome attempts to represent "others ways of being" via its instruments, creating tools that facilitate rather than provide, thereby harnessing what technology is good at and, most importantly, engendering a joyful way into its realm.

Note

1 Perfect Circuit (https://www.perfectcircuit.com/signal/what-is-west-coast-synthesis)
  Reverb (https://reverb.com/ca/news/the-basics-of-east-coast-and-west-coast-synthesis).

Works Cited

A Brief History of Modular Synthesis. https://www.musicradar.com/news/tech/a-brief-history-of-modular-synthesis-609975

AI Synthesis. "West Coast Mixer." https://aisynthesis.com/product/ai106-west-coast-eurorack-mixer/

Cramer, F. 2015. "What is 'Post-Digital'?" In Berry, D. M. and Dieter, M. (Eds.), *Postdigital Aesthetics: Art, Computation And Design*. London, UK: Palgrave Macmillan, 14–26.

O Connor, Neil. 2019. "EMAS - Electro Acoustic Music Association 40th Anniversary - University of Greenwich Reconnections: Electroacoustic Music and Modular Synthesis Revival." Conference Paper.

Perfect Circuit. https://www.perfectcircuit.com/signal/what-is-west-coast-synthesis

Pinch, Trevor and Frank Trocco. 2002. *Analog Days: The Invention and Impact of the Moog Synthesizer*. Cambridge: Harvard University Press.

Reverb. https://reverb.com/ca/news/the-basics-of-east-coast-and-west-coast-synthesis

PREFACE (ALL PATCHED UP: A MATERIAL AND DISCURSIVE HISTORY OF MODULARITY AND CONTROL VOLTAGES)

Ezra J. Teboul[1]

Modular synthesizers are unique in their nested geometry: electronic systems within systems within systems, matryoshka dolls of copper, silicon, resin and exotic minerals carefully layered. These are assembled primarily from industrially produced electronic components, communicate using control voltages, and powered by various standardized power supply voltages. When on, they produce signals which make them relevant in a variety of contexts, primarily in music composition, performance, and recording, and to a lesser extent, generative visual systems.[2] In practice this shapeshifting set of objects splinters into thousands of musical instruments, compositions, musical scenes, and artistic practices. This book provides a context for their study, as well as a sampler of meaningful instances from the present and recent past which highlight the variety, creativity, and politics of the practice. Indebted to this artistic endeavor's uneasy relationship to the profit motive, we consider it a form of cultural, material, social and technical exploration more than a form of knowledge production per se. As such there are no maps, only directions.

The chapters in this book discuss the motivations, creativity, curiosities, and politics of a variety of instances of modular synthesis. Ahead of those discussions it seems important to highlight the cultural and historical processes which connect the body of knowledge and the industrial apparatus introduced above to existing artifacts and artistic practices. Over the previous three decades, scholars such as Trevor Pinch, Kelli Smith-Biwer, Gascia Ouzounian, Hedley Jones, Georgina Born, Julian Henriques, and DeForrest Brown, Jr., have offered an intellectual trigger which has helped advance our understanding of the co-construction of circuitry and society, often using the modular synthesizer

DOI: 10.4324/9781003219484-2

and related artifacts as part of their case-studies. Few link, however, the co-construction of modularity and control voltages, which originate outside of musical electronics, back to the practices of making electronic music they are considered central to.

This is an opportunity to investigate how we got to our current environment, with hundreds of modular synthesizer manufacturers drawing on a complex history of precedents and resources, for thousands of people recording and performing with these systems. To address that, in this chapter I discuss modular synthesizers as the extremities of the large sociotechnical systems which make them possible as a starting point to highlight the way modularity, electronic sound, and control voltages were co-constructed from surplus components and creativity on the fringes of the Western technoscientific project of the 20th century and its imperial outposts. Doing so allows us to show that a signal in its electrical form can be extensively repurposed for creative and practical reasons. As I will detail, this relatively simple observation, that signals can be encoded and repurposed as needed, is both central to the history of electronics that make modularity and voltage control possible, and to modular synthesis itself.

Although I don't use their language, I follow Bowker and Star's suggestion of "infrastructural inversion," for "recognizing the depths of interdependence of technical networks and standards, on the one hand, and the real work of politics and [knowledge] production on the other." (1999, 34) How might we recognize these depths of interdependence and the real work of politics and knowledge production in electronic sound?

Modularity and Control Voltages in Electronic Sound

As a qualifier for objects, modularity blurs the boundaries and specificities of these systems. With mass production, technical people recognized both the need for and affordances of standardization and connectivity (Yates and Murphy 2019): if we are going to make a lot of relatively identical parts, there are benefits to helping users figure out which fit each other prior to assembly. Connectivity becomes both a marketing point and an industry expectation. Voltage controls both enable and are required by modular conceptions of electronic circuits. The two concepts encourage each other epistemologically, intellectually and materially.

Complementing various general, etymological histories of modularity (see for example Bosma or Tamirisa, this volume), historian of technology Andrew Russell highlights how "system architects [have] used modular concepts to order, coordinate and control." (2012, 260) He grounds the affordances of modularity into that industrial productive thrust of the machine age (Hulton 1968) and details its slow spread throughout electronics research and development. If architecture and furniture building benefitted from a labor and materials use perspective by modularizing designs, electronics offered a dynamic environment in which interconnectivity had wide-ranging interactivity and therefore strategic implications.

Russell paints modularity as inextricably embedded within the fabric of large scale private and public research and development efforts. These defined a 20th century of World Wars and Cold War scientific competition where control of manufacturing and interoperability offered tactical advantages. Strategic innovations would be re-branded as consumer commodities promising quality of life improvements.[3] Much of the same path is visible for electronic sound technology. Pre-synthesizer sound studios were built from interconnectable scientific test equipment such as precision oscillators, filter banks, and other signal processing devices rescued directly from the discard pile of university physics laboratory (Dunn 1992; Schaeffer 2023; Bernstein, 2008; or Bosma, this volume). Modular synthesis emerges, in no small sense, from the custom equipment developed to complement these configurations of test equipment.

The versatility of signals collapses pre-existing theoretical and professional boundaries by shifting across containers and meanings. Glinsky places the popularization of "synthesizers" as a term for electronic instruments to the early 1950s (2022, xiii). Paranoid for avoiding a lawsuit from the Musician's Union (which the union never even imagined), Louis and Bebe Barron had to be billed as providing "electronic tonalities" rather than music for the soundtrack of the 1956 science-fiction film *Forbidden Planet* (Rubin 1975); in parallel, engineers had, since the late 19th century, dabbled in the sonic arts when they realized that they could control various things like the frequency or amplitude of electronic signals with circuits borrowed from radio or telegraphy (Teboul 2017). Other previous work (Teboul 2018; 2020a) highlighted the fraught categories of composer, inventor, technician, and amateur in electronic music as a subset of gendered communications technical practices (Haring 2007). These frictions have proven to be a challenge for organology, the discipline formally dedicated to musical instruments. Indeed pre-electronic taxonomies of musical instruments were by 1950 primarily meant for museological applications, and used mechanical and morphological criteria for classifications.[4] As electronics and pickups enabled a decoupling of resonator and performance interface, the scheme lost its mapping potential, resulting in blanket "electrophone" categories (Teboul 2020b, 18–20). Although these taxonomic schemes remain in use, they have never gained relevance to what Nicolas Collins calls "silicon luthiers," the craftspeople of electronic music. And yet it is that exact decoupling of resonating bodies and control mechanisms which make modular synthesis so versatile. As De Souza (this volume) highlights, the modularity of systems thinking would be so inviting to Herbert Heyde that he reconceptualized an entire alternate organology based in cybernetics, the systems thinking of his particular time and place (Heyde 1975). Sadly, by that point, organology's relegation to primarily a museological practice meant that his theoretical contributions would remain untranslated from German to this day.

As Experiments

Perhaps synthesizer designers would not need an ordering theory from music when they were already drawing so heavily on those they borrowed—along with resistors, capacitors, or transistors—from modularized electronics. Hugh Le Caine, a nuclear physicist supported by the Canadian National Research Council, slowly shifted to electronic instrument design in the 1930s and 40s. With his article describing his Electronic Sackbut, he is likely the first to write about bringing voltage-controlled circuitry from systems engineering to electronic music contexts (1956, 469).[5] LeCaine would also retroactively use voltage control terminology to describe some of his own work dating back to 1937. Although neither his reed instruments from that year, nor the Sackbut he is more famous for (built 1945–48) are modular, they exhibit the earliest documented instances of both an exponential-scale, volt-per-octave subsystem and voltage-control mechanisms (Young 1984 and 1989, 35[6]).

Harald Bode's use of the word module in his 1961 Audio Engineering Society paper, "A New Tool for the Exploration of Unknown Sound Resources for Composers," is the nominal beginning of modular synthesis. From a technical perspective, Bode is likely to be preceded in spirit and perhaps in form by a variety of other experimentalists. His first electronic instrument, the 1937 Warbo-Formant Orgel, displayed polyphonic capacities (Palov 2011a), making it a contender for some early form of voltage control. For the explicit use of modularity or of its ordering capacity, Bode may well have been inspired by early electronic music studios. Building on the 1951 tape music experiments by future director Vladimir Ussachevsky, Columbia University began collecting interconnectable test equipment and building custom audio gear to expand on their use (Ussachevsky 1958; Mauzey 1958). In 1963 Ussachevsky hired Bode for a Klangumwandler frequency shifter unit, which resulted in an ongoing collaboration for additional equipment compatible with their existing collection (Bode 1976).

Through his time on the east coast of the United States, Bode might have heard of Max Mathews and Bell Labs' MUSIC-N series software, for which development started as early as 1957.[7] Retroactive qualifications of this foundational work as digital forms of modular synthesis by Risset, Howe, Park and others clearly identify MUSIC-N's undeniably modular software logic (Lindgren 2020). Considering how none of Mathews' publications or related discussions predating Bode's 1961 article formally discuss modularity, what might appear in the record as a controversy seems to be primarily a philosophical question about the nature of inventions from a later perspective. Establishing a connection between Mathews and any of the other usual figures aside from Moog[8] seems unlikely without uncovering new archival or technical materials displaying undeniable parallels. Such research might even seem counter-indicated, as Mathews himself stated that he had no contact with or influence on Moog or Buchla (Computer History Museum 2014, 1:44:27). By 2014, would Mathews have forgotten to mention contact with Bode, when discussions about Moog and Buchla kept coming back to him? Either way, it is not without irony that, simultaneously, the early published justification for BLODI and the Music Compiler (as MUSIC-N software was occasionally labelled) seems to have been the simulation of custom, analog scientific test equipment for perceptual testing because of its tedious manufacturing (David 1961), and that Moog's inaugural article on modularity specifically cites the time-consuming nature of Matthews' synthesis approach and its limitations as motivation for his own contributions (1964).

By 1977, modular conceptions of digital sound synthesis were also taking advantage of modular computing equipment (Alles 1977). From the point of view of the interface or the routing of signal, the virtual "unit generators" of sound synthesis environments (Matthews and Pierce 1987), inherited from the MUSIC-N series, are clearly morphologically relevant to modularity but have generally been discussed using other language. The use of "module" or "patch" therefore is, according to some, more of a linguistic holdover for user convenience (Boulez and Gerzso 1988), than a meaningful encapsulation of underlying programming philosophy. Quoting David Zicarelli, "modularity is a general phenomenon of computer science, but one that does not always work to one's advantage in artistic endeavors." (2002, 48)

Notwithstanding Matthews, Risset, Howe, Boulez, Gerszo or Zicarelli, the MUSIC series software and the blank starting windows of environments as varied as Max/MSP, CSound, Pure Data, VCV Rack, Supercollider or Tidal-Cycles share a deeper connection to modular components and hardware than simply running on modular personal computers or displaying modular characteristics in construction or interface design. In this context, signals are voltages and currents, physical quantities whose meanings are socially (re)constructed. Analog and digital signal processing, as niches within the wider electronics research thrust that defined the arts and sciences of the 20th century and continue to shape the 21st, are built on a shared curiosity for what is possible with clever recombination of past work as afforded not just by modularity, but by the modularity that the recontextualization of signals makes possible. Often these recontextualizations respond to scarcity, limitations, and impossibilities in the technical and material domain as much as they are motivated by artistic "what ifs?" (see Palumbo and Wakefield, this volume). Like the relay, the vacuum tube and the transistor discussed below allowed clever re-scripting of signals and their meaning through recontextualizing them in new circuits or new patches, sometimes out of curiosity but also often out of real need. Analog and digital sound synthesis very much responded to, as Mathews modestly identifies, a *zeitgeist* which included ideas about modularity, operability across signal systems, and creativity (Computer History Museum 2014). If anything, piping digital streams of data from one virtual outlet to a virtual inlet, as is done in modern digital sound synthesis environments, is the most content-agnostic and origin-unaware version of the analog signal manipulation and re-interpretation work made possible by Le Caine, Bode, Moog, Buchla, et al.'s analog circuit inventions.

Other precedents also deserve further investigation. Ricardo Dal Farra's work on South American histories of electronic music should encourage us to think of who is missing in the published, anglophone record on these topics. Raúl Pavón built in 1960 a "small electronic musical instrument that featured an oscillator with multiple waveform outputs, a variety of filters, an envelope generator, a white noise generator and a keyboard, among other modules." The Omnifon's frontplate clearly displays the patch cords of modular designs and voltage control, but Dal Farra does not elaborate on whether or not the transliterated discourse of modularity was used explicitly at the time of design and manufacturing (Dal Farra 2006), let alone providing any connection to Bode if it was not. Further archival research will be required on all points to establish any connections or precedents to Bode.

With his prototype and the discourse surrounding it, however, Bode currently stands as a pivotal actor who notably participated in collapsing the benefits of modularity as a manufacturing tactic into the hands of (some still very limited) users who would not *have* to learn to solder.[9] His influential work did not so much change the clientele for synthesizers, as much as it made the affordances of electronic sound relatively more accessible to composers and artists. The cumbersome process of sound by electronic means was no less dependent on privileged access to specialized tools, specific expertise and rare components, along with the extensive experimentation time required to achieve even modest goals, it just developed a more formal and arguably quite effective de-facto organizing logic through the contingent developments of commercial and scientific research. It established the potential of modular synthesis as a commercial, industrial practice in addition to the do-it-yourself research and creation practice it had been up to that point.

As Commodities

Bode's paper led to the reconceptualization of modular synthesizers as commodities with the 1964 projects of Robert Moog and Don Buchla (Pinch and Trocco 2002; Gordon 2018; Bernstein 2008). Within a few years both had commercialized their early prototypes and were explicitly using the concepts of modularity and control voltages to describe and market their products

(Moog 1964; 1965; Buchla Associates 1966; R.A. Moog Co. 1967). After Moog and Buchla, the only requirement for access to synthesizers was money, friendship with the right person, fame, institutional affiliation, or luck.[10]

Buchla, based in the San Francisco bay area and doing contract work for federal laboratories and agencies, encountered electronic analog computers prior to releasing his first batch of 100 series modules. These patchable[11] scientific calculation environments were an explicit influence on Buchla's design, if only through his collaborators Morton Subotnick and Ramon Sender (Pinch and Trocco 2002, 37). Pinch and Trocco continue:

> Buchla's familiarity with silicon transistors and his knowledge of analog computers (from working in physics) led him to voltage control: 'I had this idea you could take voltages and multiply and mix them and things like that. So it wasn't too far a cry for me to attach a voltage to the pitch of an oscillator or to the amplitude of a VCA… But as soon as I added voltage control to the elements of the synthesizer it became a different ball game because you could parametrize everything. You weren't limited by how fast you could turn a knob to get between two states of a parameter.' (39)

Buchla would go on to give additional insights for his own thinking on voltage control in two later publications. The first is a four page poem from 1971, "The Electric Music Box," which reads in part:

SETS OF CONTROL VOLTAGE
VALUES TO FUNCTION AS
CONTROL OF ANY PARAMETR
e.g., INTENSITY FREQUENCY
TIME FLOW RATE DEGREE OF
MODULATION BY FREQUENCY OR
AMPLITUDE FILTER BAND WIDTH
AND CENTER FREQUENCY PULSE
PERIOD PERCENTAGE AND AS
ANALOG VOLTAGE ADDRESS
OF SEQUENCER NUMBER OF VOICES
CHOICE OF LOGICAL DEVICE OR
SERIAL PROCESS
WHAT THE MACHINE CAN AFFORD
AND PRODUCE IS A NEW LANGUAGE
CULTIVATING AND REFINING
IN THE LISTENER A NEW SENSE
OF PERCEPTUAL MECHANICS

(Buchla and MacDermed 1971, 13)

This highlights Buchla's outward framing of his music boxes as instruments which could enable the recontextualizing of voltages as the bases for poetic experiences. In the 1982 issue of *Keyboard Magazine*, he offers additional insight as to his experiments with and perspective on control voltage mechanisms:

> It didn't take us long to arrive at the concept of voltage-controlling all of the parameters that could possibly be regarded as significant musical variables, including such things as voltage-controlled reverb and voltage-controlled degree of randomness. It did take us a few years to find truly general ways of dealing with the voltages that did the controlling. That turned out to be a bigger problem. (…) Our instruments incorporate what we call arbitrary function generators, with which you can generate any kind of shape you want. The instrument doesn't force you to make any assumptions about what are pleasing shapes and what are not. That has to be decided on a musical level. (Buchla and Aikin 1982, 12-13)

In that sense Buchla's arbitrary function generator mirrors the philosophy of a segment of electronics and systems engineering that wished to develop systems and user interfaces. These went beyond the canonical circuits of the disciplines and offered the user high-level decisions such as the designing of arbitrary waveforms.

Robert A. Moog, a New York native with a long interest in engineering physics (his PhD in engineering physics was completed at Cornell in eight years on a scholarship from the Radio Corporation of America (RCA), and his dual bachelor's in physics and electrical engineering included a summer internship at Sperry Rand), had been developing a small business of building theremins while living upstate. According to musician and collaborator Herbert Deutsch, Moog's design was influenced by Harald Bode's article, as well as his work on the Melochord system (Pinch and Trocco 2002, 335 fn.12). Realizing the possibilities offered by cheapening transistors, he transformed his small theremin business into a full-fledged modular synthesizer factory. Pinch and Trocco, in an interview some 40 years later, got him to highlight the impact of the transistor on his designs:

> Now I can't tell you how important it was that I could buy a silicon junction transistor for 25 cents. That's amazing … I can remember during that summer [working at Sperry] the technician over there undid this little box and took out thirty transistors … These are the first silicon junction transistors. In 1957 this is $1000 for the transistors. And there they were [in 1964] for 25 cents. (1998, 12)

In their history of the Moog, *Analog Days,* Pinch and Trocco elaborate on Moog and Deutsch's implementation of voltage control: "there was nothing particularly original in the notion of voltage control or in the design of the circuits Moog employed." (2002, 28) They go on to point out Moog's skill was in making some musical potential of the transistor accessible to musicians.

As Technical Concepts

From a technical perspective, control voltage is a difficult concept to define because, like modularity, it is context dependent. In addition, it is inherently concerned about the potential meaning of a voltage—something which can be redefined after the original voltage has been produced. However, in the history of electronics generally and in audio technologies specifically, there tend to be three identifiable elements beyond it defining physical dimensions, voltage ranges, connectors, and power supplies (which are more immediately recognizable as the material bases for modularity). These are:

- Impedance bridging, in which a circuit with a low output impedance is connected to another with high input impedance. Impedance is a complex concept outside of the scope of this introduction (for a detailed discussion see Robinson 2020, 220–224), but this facilitates consistent behavior from both circuits because it maximizes the amount of electrical energy transferred between the connected circuits and prevents operation outside expected conditions.
- Protection circuits, such as buffers, diodes, and regulators, which help minimize the chances that unexpected voltage, current, or frequency ranges will damage the modules involved or the power supply and sound system.
- Means of making voltages control things. The infinite number of ways to implement this have yielded the major building blocks of modular synthesis, from resistors used as voltage-to-current converters which change the gain on an operational trans-conductance amplifier (OTA), to using the p-n junction of a transistor or diode as a variable resistor doubling as a linear-to-exponential conversion circuit (typical of the Moog ladder filter and many others, see Werner and McClellan 2020), or sampling control voltages with a digital-to-analog converter (DAC) to change a parameter in a digital signal processing (DSP) routine which might produce new signals and / or control voltages.[12]

Modular sound offers a layer of mediation between the actual work of the circuit maker and the sound artist. That mediation might be fairly unnecessary, as demonstrated by the extensive number of artists working directly to compose with electronic components (e.g., Teboul 2020; Teboul and Werner 2021; Nakai 2021). Yet as shown by the present number of users, this mediation between the musical potential of circuitry and what is actually made available to musicians is very much desired. How is it satisfied?

Extending but nuancing Friedrich Kittler's provocative claim that rock music is a misuse of military equipment, modular electronic sound has relied on and continues to be shaped largely by the global industrial infrastructure of communications (Kittler 2013; Winthrop-Young 2002). At almost every turn of musical synthesis's early histories, the spectres of national defense, global warfare intrigue, or paychecks from industrial actors almost entirely tied to either World War 2 or Cold War era "warfare by other means" take it away from its countercultural experimental image and closer to the technoscientific project of the 20th century (for more on this in the specific case of Don Buchla, see Gordon, this volume).[13] Not all of these episodes are from Western perspectives, nor are they as dramatic as Leon Theremin's life, which brought him back and forth between the Soviet Union and the United States under literal espionage circumstances (Glinsky 2005; Spritzendhorfer and Tikhonova 2013), but these ties remain widespread on both sides. In the following section, I describe how the meaning-making capacity of the relay, the vacuum tube, the transistor, and various larger systems place modular synthesis as a fringe practice on the edges of 20th-century Western technoscience.

An Industrial History of Control/Power/Signal

The Relay as an Amplifier of Signals-Thinking

The electrical relay and its use acts as a convenient point of interruption to discuss the genesis and meaning of modularity and control voltages, because of its strategic role in telecommunications, power systems, and computing. Tracing these through the technological record helps us identify the main forces present in making possible modern synthesizer systems.

Signal-regenerating devices like the relay were necessary because early electrical systems suffered significant losses across stretches of bare wire, becoming indistinguishable from noise after some dozens of miles. These inefficiencies motivated a plethora of experiments as 19th-century inventors sought out clever implementations, profit, or both, in exploring this new medium (Burns 2004, chapter 3). As early as 1831 the Albany, N.Y. scientist Joseph Henry had publicly demonstrated his own implementation of an electromechanical telegraph system (Coulson 1950, 63). Later experiments included an electromechanical repeater to detect a noisy telegraphic impulse and retrigger a new click, effectively "relaying" the signal (Burns 2004, 84-85; Molella 1976, 1276).

This is where power, control and signal overlap in a single, dynamic, voltage difference. The series of dots and dashes which make up the telegraph's alphabet are created by interrupting the voltage continuously maintained across a telegraph system using a hand-actuated switch called a key. Pressing the key temporarily disconnects a line: a short interruption causes a dot, a long one makes a dash. In that sense, and because the entire system of the telegraph is a form of power network. It is not used for powering anything other than itself, but it is a power network nonetheless. The relay is effectively an automatic key that simply repeats the action of the human operator so the signal can be refreshed along a line. Its functionality leverages the affordances of electromagnetism, where a signal above a certain threshold voltage will actuate a mechanical motion (using what is now called solenoid) mimicking key actions which caused the original signal (Morse 1840). In other words, the human keyer generates a signal by interrupting a low-power direct current network, which controls a set of automated repeaters. Control over electrical power afforded by chemical energy was commercialized as the interactive signal processing system we call the telegraph. The relay is the prototypical electrical control system.

Codes like Morse's, the meaning-making bases for electrical telegraphy, are therefore notable for the way that they leverage the basic capacity of early electrical apparatus for communications. In their commercial implementations these took advantage of the fact that electrical power, control signals and messages could all be the same. This optimizing logic, of cleverly leveraging encoding, control and power from the same voltage, is significant, as it itself would be split and recombined—modularized—in various contexts to become a pervasive tactic not just in analog and digital sound synthesis but also computing and communications generally. The "computational metaphor" of media studies (Turner 2006; Chun 2011) owes much of its explanatory power to the signal metaphor that preceded it. Similarly, cybernetics owes much of its cultural capital to the powerful ways in which signals could be wielded and occupy physical or theoretical space across the 20th century. Electronic sound's epistemological edge comes from sharing with signals a relative mono-dimensionality. *Contra* Kittler, there technically isn't software, but more is revealed when one looks at the dynamic building block of signals, rather than the static information partially encoded in components and the inscriptions they temporarily hold (1993). Media theory, too, in that sense, has paid the price of its very composability. Rewritability is more of an organizing logic than writing. This is not simply grounded in the abstract precedent of the signal as a mode of thought, but also in the common material root of signals, cybernetics, and computations: telegraphy's infrastructure, shaped by codes like Morse's and built from anonymous components like the relay. If we must have a computational metaphor, it should be a quantized signal metaphor, one which situates the baggage of calculation in the incredibly robust denoising and therefore transmissibility afforded by a *schema franca* forged from the vaguely religious yet godless trinity of signal, power, and control.

The ubiquity of this recombinable signal strategy is visible in electrical engineering's histories. Eccles and Jordan's 1918 (patented 1921) vacuum tube flip-flop should remind us that the difference between an electrically actuated switch and an amplifier is primarily a question of the parts surrounding a vacuum tube (or a servo, or a transistor, or an op-amp, etc.). By shifting between scales of electronic circuitry, designers have creatively reused subassemblies for new applications, an affordance of standardized and systematic design methods developed in the 19th and 20th centuries. In 1934, MIT electrical engineering professor Harold Hazen realized that, using his background in analog computing, he could make "a critical conceptual leap: he stated that because they translated the low-power input from an instrument of perception into a high-power, articulated output, servomechanisms behaved fundamentally like amplifiers." (Mindell 2002, 141) Through *Between Human and Machine*, Mindell details how realizations such as this one would cascade into a number of reconceptualizations of voltage control within the burgeoning field of systems engineering. After World War 2, analog computers had offered these voltage control mechanisms in modular form to their buyers (Small 2013, 33), and it would not be long before these concepts would be transferred to digital computers. Indeed, Claude Shannon's 1938 *Symbolic Analysis of Relay and Switching Circuits* "brought the design of switching systems into the world of mathematical logic and network theory." (Mindell 2002, 174) The relay functioned as a common actor between the large-scale telephone switching centers that characterized the Bell system as it grew, and the early electromechanical implementations of digital computers that paralleled those centers (Hochheiser 2013; Aspray 1990, chapter 6).

The Relay as a Trigger for Modularity in the Electronics Industry

The use of relay as a control mechanism, as afforded by this temporary overlapping of signal, power, and action control at a distance, triggered the standardization and modularization of communications infrastructure. By 1899, telegraphy companies had implemented combinations of wires, relays, and batteries to such a large extent (Hochfelder 2016) that the term *relay rack* became used to describe the assembly of equipment along telecommunications lines. (Loewenthal 1899, 144). Modularity as an ordering concept from a user perspective predates the emergence of modular terminology as discussed by Russell. The DeForest Radio, Telephone and Telegraph Company's "Unit Sets" both implemented a notably early example of consumer-sided commercialization of modularity without actually managing to capitalize on it. Unit

Sets consisted of a grid of standardized square slots within an empty frame allowed customers to build radio circuits with components that could be swapped as the rapidly evolving consumer market made new experimental products available (*QST* December 1919, 68). From a manufacturing perspective, modularity became an ordering concept at an industrial scale with railroad expansions, and for the electronics industry, the American Telephone and Telegraph company. Owing to the scale of its monopoly over telecommunications it had the influence to force assembly standards in the field of electronics and signal processing. In a 1923 article for the *Bell Systems Technical Journal*, Bell employee Charles Demarest presented many tools for facilitating the building, maintaining and repairing of the Bell system's rapidly growing substation network and the implied labor (Demarest 1923). These included a 19.5″ rack format, which, thanks to the Bell system's stronghold as a consumer on the U.S. electronics supply chain throughout the 20th century, remains in extensive use in audio and computing equipment today in a simplified 19″ form (139).[14] By reusing hardware salvaged from telecom switchboards, instruments like Lori Napoleon's (this volume) offer poetic connections between modular synthesis and telecommunications.

The R.C.A. Mark I and II synthesizers, the latter of which was installed at Columbia University in 1957, were built by R.C.A., using the rackmount standards described by Demarest and housed in steel enclosures (Meredith 2012; Werner et al. 2022). Neither Demarest's article nor the technical, musical, or journalistic discourse surrounding the Mark II used the word modular, but the construction of the system was clearly indebted to the building and wiring techniques of telegraphy and telephony equipment, and effectively used almost entirely the same parts.

Here I can illustrate effectively the nuance between an internal form of modularity that helps appreciate the industrial grounding of experimental electronics for sound, and the user modularity of De Forests' Unit Sets or modern modular synthesizer systems. The Mark II's patch bays, although present in the system, were primarily for output routing and paper tape assignment for different parameters, with the synthesis routine itself hardwired. Its rack units, although they separate synthesis steps in discrete blocks, some of which are repeated as in a modern modular, do not have any user-accessible forms of control other than the paper punch tape used for note input and the dials on its front. From the point of view of user-modularity, it is closer to Le Caine's Electronic Sackbut than it is to a Buchla music box. It would not even qualify as "semi-modular" in modern parlance, as there are a meaningful number of connections which cannot be rerouted without beginning to hack at the custom cables chaining the units in the back of the system. Because it was built for musicians, not technicians, the modular logic visible inside the Mark II was not accessible to its users, only its makers and maintainers. This resulted in modifications of the circuit, as these users hit limitations to the bounds explorable within the factory settings for the Mark II (ibid). Modularity and control voltages in commercially available electronic music systems would require a shift in both technological capacity (with the cheapening of the transistor as a consumer commodity) and attitude towards musicians as more capable composers inside electronics than mid-1950s RCA engineers were willing to make of them.

The Vacuum Tube and the Transistor as Amplifiers of Modular Practices in the Electronics Industry

The transistor's introduction mirrors the relay's versatile applications and the multiple hats worn by the signals they would amplify / switch. Although the initial press release for the quite ragged first prototypes flaunts its amplifying potential (Bell Laboratories 1948), the transistor, like its predecessor the vacuum tube, would soon enough be used as an electrically-triggered switch, replacing the vacuum tube as the basis for computational experimentation in the 20th century.

Epistemologically, the meaning of the transistor, just like the vacuum tube or the relay, is not only made at the component level, but at every circuit level (Rekoff 1985). In many cases, the sub-assembly level is more important than the actual component (this is the entire premise for integrated circuits, where design compensates for variability in component qualities). For the general public, the meaning making happens at the level of the interface (see Tamirisa or Banks, this volume). Mirroring the rest of this chapter, the nesting of meaning-making should not be considered in a hard, discrete fashion, as component-meaning might overlap with sub-assembly and systems meaning, as is the case with the transistor.

The desire to modularize electronics assembly was in part motivated at an industrial level by the fabrication challenges of large military electronics projects afforded by these tubes and transistors. SAGE, an IBM / Boeing project produced in collaboration with the Department of Defense and the US Air Force, had emerged from a 1949 meeting of the Air Defense Systems Engineering Committee (Jacobs 1983, 323). Russell, discussing a 1956 promotional film on the project, summarizes the project's contributions to modularity: "the problems of space and weight, reliability, and maintenance, declared the film's narrator, 'were all finally solved by employing the basic principle of modular construction.'" (2012, 278) The numerous ramifications of the SAGE systems and the engineering achievements in its various subsystems are well discussed elsewhere (see the 1983 volume 5 issue 4 of the *Annals of the History of Computing*) but as Russell assesses SAGE is notable here for its cementing of modularity as an efficient ordering concept within the large sociotechnical system that is the midcentury electronics industrial and research project of the United States.

This is also visible in Project Tinkertoy, a US Navy research project undertaken at the National Bureau of Standards starting in 1950. Tinkertoy appears on the "basic research" end of modularity, as it was an open ended experiment in packaging rather than one dedicated to solving a specific high-level problem. In that sense, it represents some of the experimentation with the scales at which modularity could be implemented happening at the time. In its final output, it offered components with standard, mechanical connectors rather than the raw leads of those components (Henry 1956; Jurgen 1987). Although wrappers like these did not exactly become common beyond physics education toy sets (reminiscent of De Forests' own Unit Sets from the early vacuum tube era), the principal objectives of the project, the use of modularity as a solution for the "development of facilities or systems suitable for rapid mobilization in emergency periods" (National Bureau of Standards 1953, 169) contributed again to making modularity a pervasive concept across the technoscientific apparatus of the era and the technical culture that seeped into public life via the scale at which engineers were being enrolled into that apparatus.

Modularity in formal digital computing became an industrial actor most notably with IBM's System/360, a new line of hardware and software whose designs, begun in 1960, explicitly centered around the use of the concept to re-haul both the actual hardware produced and the supply chain that made it possible. Russell again:

> IBM's application of modular principles throughout all aspects of system design, production, management, and labor created a cohesive collection of components that could be adapted to meet the computational needs of many different users. From the standpoint of IBM executives and managers, the modular design of the System/360 streamlined design and manufacturing processes and generated new economies of scale and scope through the reduction of variety and incompatibility. (275)

This was not the first formal experiment with modularity for IBM, as some of its engineers had been involved in the SAGE project. Before the System/360, drawing on that experience, they had released in 1959 a "Standard Module System" for its 7030 Stretch line of mainframes (Boyer 2004). These standardized germanium transistor based cards (available in at least 1444 variations - see Shirriff n.d.) fulfilled a wide array of functions (IBM 1959) and could be stacked by the hundreds within 7030 or related systems (Shirriff 2021). The effect of IBM's embrace of standardized power schemes and connectors to stack circuits with various geometries in empty shells and racks on the computing supply chain as a whole cannot be understated. As Gerard O'Regan writes in *A Brief History of Computing*, the System/360 family (which ranged from small systems to mainframes used in U.S. missile defense control) "set IBM on the road to dominate the computing field for the next 20 years, up to the introduction of personal computers in the 1980s." (2021, 97). Just like the Bell system's influence effectively made the 19" rack size predominant in electronics manufacturing, and produced enough cheap surplus that both the Mark II and most recording and mastering studio effects were and are still built with that standard, IBM's investment in developing the System/360 effectively dominated its competition (Burroughs, Honeywell, Sperry-Rand) for two decades (98).

Computing Hardware as the Material Basis for the Eurorack Format

Eurorack, a currently popular modular synthesis format (Bates 2021; 2023),[15] is a modular standard derived in part from the Eurocard computing hardware standard by the German synthesizer manufacturer Doepfer. Eurocard was added to the formal specifications of the 19" rack standard in the 1984 revision of IEC 60297-3 (the general 60297 heading from 1969 includes the older 19" format definitions, originally made for nuclear electronics equipment).[16]

Beyond Eurorack, Eurocard and the 19" inch standard's industrial computing and telecommunications roots, its prefix makes the imperial connotations of contemporary electronic music almost caricatural. Yet Eurocard is a misleading name because the dimensions are a combination of both metric and imperial units. To grasp the unreasonable lengths manufacturers have made to accommodate historical practices which subtend the computing and communications industries, it is worth reproducing a written summary of the standard in full:

> The Eurocard packaging system is a complex mixture of English and metric dimensions. Although this may seem confusing, widespread conformance to the standard dimensions means that users are not troubled by these issues.
>
> Eurocard subracks have standardized sizes in all three dimensions. Height is specified by the unit 'U' (which stands for 'Unit'), with 1 U being 1.75 inches. Width is specified by the unit 'HP' (which stands for 'Horizontal Pitch'), with 1 HP being 0.20 inches. Subracks usually range in height from 1U to 9U.
>
> The height of a Eurocard is less than the height of rack by 33.35 mm to allow space for panels and card guides. The height of the card in a 3U rack is therefore 100 mm. As two stacked 3U cards are the same height as a 6U card, this scheme allows racks to be constructed which mix 3U and 6U cards. Front panels are also slightly smaller than the rack size, and the typical panel height for a 133.35 mm 3U rack is 130 mm. Eurocards come in modular depths that start at 100 mm and then increase in 60 mm increments. (Systems Integration Plus 2007)

Once again the component-level shared vocabulary of Eurorack and modern digital computing should be noted. Both tend to use surface mount devices (SMD), although a number of synthesis manufacturers will proudly advertize that they do not as part of their branding to capitalize on a fetishization of analog electronics within their customer base. Prior to the 2000s modular synthesizers had been assembled with mostly through-hole components, components whose metal leads went through the resin substrate of the printed circuit board (PCB) to be soldered to a trace on the other side. As the industrial production of personal electronics followed a miniaturization trend, through-hole components have slowly been replaced in part or wholly by these SMDs.[17] This has been accompanied by the emergence and increasing density of integrated circuits. These silicon wafers which combine up to billions of nano-meter scale transistors to perform complex analog and digital signal processing tasks (Prasad 2013; Maloberti and Davies 2016, 65-70). SMD has in part enabled an increasingly hybrid (analog and digital) modular synthesis ecosystem. As discussed by Ryan Page, Michael Palumbo and Graham Wakefield, as well as Corry Banks (this volume), digital signal processing (DSP) has allowed both the emulation of historically significant circuits and their sonic characteristics, as well as the development of new synthesis approaches. The latter have arguably become more accessible as additional tools for experimentation become commercially available, and resources for those forms of experimentation become more robust.

The explosion of Eurorack modules, as with the increased interest in digital musical instruments (DMIs) and new interfaces for musical expression (NIMEs) is in no small part correlated with the relatively robust supply of cheap SMDs, including increasingly capable integrated circuits for DSP applications (Paradiso 2017; Bates 2023). This robust supply is in no small part the contemporary development of a high-investment international semiconductor supply chain, whose foundations were laid out at the beginning of modularity's effect on the electronics industry. Notwithstanding the large scholarship on NIMEs and DMIs (Jensenius and Lyons 2017), the connection between the industrial basis for musical experimentation with these new technologies has tended to restrict itself to signposting or acknowledgements of the environmental consequences of runaway productivism as an unfortunate backdrop for the democratization of access to the latest musical systems (Masu et al. 2021; Morreale et al. 2020).

Reminiscent of the multiple ways the same component can be used across circuits (switches and amplifiers, flip flops and integrators, etc.), the popularity of modular synthesizers which use a single kind of connector throughout their ecosystems (as with Moog, Serge, and Eurorack systems) contributes to a long history of re-scripting the meaning of signals. In the process, however, we should not forget what it took and what it takes for these circuits to have appeared at our fingertips.

In "Hearing Aids and the History of Electronics Miniaturization" (2011) Mara Mills demonstrates that the miniaturization of electronics technology was in no small part motivated by the needs for personal assistive devices like hearing aids. Mills thereby complicates Kittler's wholesale qualification of creative uses of electronics as a re-scripting of military technology. Research into the production and miniaturization of transistors received massive investment from government sources with a vested interest in military applications, but electronic music can also be read from an infrastructural perspective as an extension of attempts at life-affirming uses of extracted materials. This more nuanced frame for tinkering offers more space for appreciating creative uses of electronics not especially designed for joyful experiences and meaningful connection, while keeping a clear eye on the environmental, labor, and cultural consequences of relying on nation-states, international geopolitical conflict-era investments crystallized as industrial infrastructure, and private capital for those electronics' supply chain.

From Patch Up to Punch Up: Synth-Ethics or the Stakes of Modular Electronic Sound

According to Richard Scott, "modular synthesizers are useless"—his provocation is not derogatory, but rather, points to the lack of expectations with which synthesizers tend to be purchased (Scott 2016b). Most do so because it seems fun, not because they expect to accumulate cultural or financial capital. Most synth purchases seem to overwhelmingly imply some sort of financial loss.[18] This has economic implication drawn out along the lines of class; not everyone can afford a synthesizer, let alone a modular one or the time to use them. But this is precisely what is at stake in studying electronic music instruments generally: they, with a few other technological commodities, occupy a strange space between the muted violence of the modern supply chain embodied in the dead labor of electronics, and the acceptance that they enable a form of exploratory play in which a temporary muting of profit imperatives can operate. Students and amateurs of modular electronic sound, in that sense, are in a privileged position to think of what our built environment might look like in worlds where capitalist extractivism would not hold such a strong grasp over our everyday precisely because we, along with a few other artists and scholars, look at what happens when technology doesn't have an immediately or consistently useful result. If modularity has taught us one thing, it is that looking at a complex system and cherry picking the parts we like and replacing the parts that are not working for us can often be a viable strategy. If control voltages have taught us anything else, it is that there is something generative about these parts talking to each other, even if it is in mismatched ways.

Synthesis offers an endlessly reflexive medium for experimentation between culture, technics, and sound. It is one of many ideal environments for various future projects documenting the land and labor in electronics and work / material cultures that bind us to them. These would contribute to addressing the fact that, as Sylvia Wynter quoting Gerald Barney discusses:

> thinking globally, what "we really have is a poverty-hunger-habitat-energy-trade-population-atmosphere-waste-resource problem," none of whose separate parts can be solved on their own. They all interact and are interconnected and thus, together, are constitutive of our species' now seemingly inescapable, hitherto unresolvable "global problematique." The main problem with respect to solving the cognitive contradiction with which we are now confronted is therefore how we can begin not only to draw attention to but also to mind about those outside our specific and particular referent-we perspectives and worldviews. (McKittrick 2015, 44)

To address this generalized problem, we will have to find consensus on which technologies we keep and which we keep in our histories. There we should be careful to not simply reiterate the discourse with which these technologies were sold to consumer markets by industrial actors, even though company copy is often one of the main sources that remains available. For example, a variety of museum shows documenting technologies and their impacts while being funded by the very companies that made those technologies have contributed to an ideology of positive production amongst the wider public (e.g., McCarty 1990; Barr and Johnson 1934; LACMA 1971; Wilson et al. 1986; Kardon et al. 1995; as discussed by Sharp in the context of oil companies and the Canadian museum economy 2020; 2022).

Identifying the fraught underlying politics of creative technologies does not make the joy they provide less real, but offers points of discussion for future work. Here I have, like Pinch and Trocco, not spent much time talking about the music we use these electronics for (2002, 11). However, music and electronics are both deeply social projects, regardless of how a lack of receipts might make this realization difficult (Clayton et al. 2013). To develop Tara Rodgers' four suggestions for ways to do sound activism, and especially the last one ("that detrimental environmental impacts resulting from creative uses of electronics and audio technologies [should be] minimized," Rodgers 2015, 82) I want to point to existing work which discuss "a proliferation of sonic-political acts that have local and far-reaching ripple effects." (Ibid) In the context of modularity this means asking: can modular synths punch up, and can we do so in a consensual way across the ecosystems affected by making and caring for them? What would a synthesizer designed with respect for every ecosystem involved look like? Can each component be assembled in a way that embodies the cultures and environments it is on loan from in a meaningful way? This is not a suggestion for more "fairphones" or "ethical" consumer electronics: these do not place the decision to produce technology in the hands of those most affected in a manner uncoerced by the pressures of living under capitalism. Conversely, the loudest absence in these records are the countless technicians, factory workers, and miners—primarily working class, of color, and eventually in the global South—who extracted the raw materials, processed them, sometimes implemented the prototypes designed by the above engineers, and assembled the commodities required for telegraphy and telephony's relatively unchecked refactoring of communication systems across the globe as permitted by the relay and the vacuum tube, the transistor and the microchip. Also evident in history of technology sources (e.g., Jones-Imhotep 2008; Dummer 2013; Maloberti and Davies 2016) is the extent to which invention, adoption of concepts and stabilization of technical forms, theoreticians, and originality have long been the focus of electronics history, when those inventors and theoreticians are not themselves the people writing that history (e.g., Black 1934; 1977; Darlington 1999). In addition to the systemic and documented exclusion of non-white cis males from technical disciplines in scientific research, this has been done at the expense of knowing exactly how much work, pounds of dirt, barrels of oil and value have been extracted in making, transporting, and selling all these circuits. This occulting of the material, mineral, and labor bases of communication and the creative uses has political implications, as the imperial connotations of the 19th-century rail networks in the United States made way for the larger-scale capitalism of intercontinental communications projects of the 20th century.

From more sociological and anthropological perspectives, a number of scholarly sources have highlighted the myriad forms the connection between mastery of electrical communication tools and occupied or exploited land and labor has taken (e.g., Pellow and Park 2002; Smith et al. 2006; Bronfman 2016; 2020; Cowie 1999; Garcia and Yuste 2010; Norberg 1976; Grossman 1979; Nakamura 2014; Vagnerova 2017; Adams 2017; Hossffeld 1988; Lecuyer 2017; Valdivia 2022; Sevilla 1992; Sturgeon 2003). A complementary effort can be seen in the recent scholarship on supply chain, which document both the paths and the labor/environmental consequences of commodity production and consumption (e.g., Lüthje 2013; Nieves 2018; Warren 2001; Gabrys 2011; Macquarie 2022; Boudreault-Fournier and Devine 2019; Burrington 2018; Hockenberry 2020; Hockenberry et al. 2021; Diaz et al. 2022; Radetzki 2009; Riofrancos 2020; Bady 2018; Labban 2019; 2022; Arboleda 2020a; 2020b). Leanne Betasamosake Simpson writes that "extracting is stealing—it is taking without consent, without thought, care or even knowledge of the impacts that extraction has on the other living things in that environment." (2016, 75) She also offers a sketch of an "alternative to extractivism:" it is "deep reciprocity. It's respect, it's relationship, it's responsibility, and it's local." (Ibid) What might this look like, how might it be informed by a historical and discursive understanding of niche art practices like modular synthesis?

Much remains to be done: acknowledging relationships and responsibility could begin with labor and material histories of each component, connector, of each assembly technique, of each maintenance/repair practice, of each factory, of each land, of each community from which the workers came and the industry's impact on their areas, of each waveform, of each symbol used in abstracting these products on the page and the ideologies built around them to make them feel like weightless pure progress straight in the palm of your hand. The present history of concepts and materials implied by modularity and control voltages should be read as nothing more than a very brief starting point from which the real work can be split and done. In that sense the present chapter is an encouragement: if we can find local, temporary utopias in communities of practice like modular synthesis, what can a deep and caring knowledge of this particular practice, and the infrastructure, history, people and things that make it possible as I have discussed here bring to the table as we must wrangle our way out of the double bind of capitalism and ecosystemic collapse?

I am not using this history to call for a wholesale abandonment of electrical music, and neither am I arguing for the vulgar luddism demonized by a variety of thinkers (Noble 1995, chapter 1 and appendix 2) or grossly misappropriated by others (Fleming 2021). It is not about refusing the machine, but imagining alternate worlds where we communally determine which machines are worth it. It is about refusing the machines that all those affected by their making and use do not deem worth the labor and land which will be mobilized in their making, caring, and decay. This is not having our cake and eating it too. Assistive and creative technologies are worth more than productive ones—the latter should be made only to the extent that they enable the former. Following Tara Rodgers' call for "cultivating activist lives in sound," I participated in editing and contributed to this collection in hopes of documenting the instances of electronic sound where, despite the deeply traumatic events and processes, the careless corporations and the bigoted actors which litter the history of electronics and in which modular synthesis is not an exception, we have time and again seen music rise as recognizable sets of technocultural practices which have brought critical reflections on the labor, tools and environments themselves as well as hope, joy, and glimpses of the world we could have instead. Despite repeated disappointment from music industries (Brown 2020), rampant cismasculine biases inherited from wartime technocultures (Tamirisa, this volume, Biwer-Smith 2023, Born and Devine 2016; Rodgers 2015; Haring 2007) and a variety of systemic process disfavoring people in resistance from participating in the ephemeral utopias they deserve emerging faster than academia and journalists can document them, testimonies of how experimenting with and listening to electronic sound has had radical potential to so many abound across communities (e.g., Haffar, this volume, and McKittrick and Weheliye 2017; Hé 2023; Cardenas et al. n.d.; Akomfrah 1996; Henriques 2011, amongst many others). Quoting Leah Tigers discussing noise music:

> To see noise as a trans woman and develop it into song, an album, a career, a scene, many scenes, and finally a history begins with the labor of articulating your own condition. Many parts of this condition will be and become shared. Through what is shared we become legible to culture. Broad affinities such as the one I have set out here to describe are, by definition, not universal; they are not felt by everyone and are not felt exclusively by anyone. Yet they can be powerful, powerfully spiritual, if only in letting us know we are not alone in our isolation. Perhaps we never were. (2019)

The noise made by modular synthesizers certainly does not mean or feel the same for everyone either. At the very least, as people curious about the multiplicity of meanings that signals—electrical or otherwise—can have, we can bring something more poetic than strict definitions to the work to come.

Notes

1 For references or support on this chapter, I owe warm thanks to Kelli Smith-Biwer, Ky Brooks, Luke DuBois, Kurt J. Werner, Matt Sargent, Will Mason, Andreas Kitzmann, Aphid Stern, Sparkles Stanford, Chris King, Bana Haffar, Alex Magoun, Tom Everett, Mara Mills, Ingrid Burrington, Leah Tigers, Eliot Bates, Camille-Mary Sharp, Ted Gordon, William Turkel, Sadie Couture, Jonathan Sterne, Magnus Schaeffer, Ken Shirriff, Rob Arcand, Anna Bonesteel, Sophie Ogilvie-Hanson, Allyson Rogers, and Andy Stuhl.

2 A reader not familiar with the terminology used here will find a wealth of resources on the topic, e.g., Austin (2016). The use of modular systems for visuals has been discussed extensively in High et al. (2014) and in "Archaeologies & Organologies: Towards a Pre-History of Early Synthetic Video & Image Processing Practices," an unpublished draft by Chris King (2024). King highlights the work of C.E. Burnett and Yoshiharu Mita, both of which developed modular video circuits which interacted via control voltages before Bode even though neither explicitly used "modular" terminology. Mita, however, did use "voltage control." See patents US2292045A (1942) and US2910681A (1959) respectively.

3 The extent to which additional commodities did not alleviate the drudgery of housework or increase leisure time for all is made evident in a number of landmark technology studies (e.g., Cowan, 1989).

4 Magnusson (2017) recently offered heterarchical classification schemes, with user-editable rhizomatic structures which could take advantage of digital collaborative platforms for dynamic evolution, but such as scheme remains to implemented publicly as of writing.

5 For the earliest instances of "control voltage" in the IEEEXplore database, see Hayes (1907) and Thomas (1909).

6 Citing Le Caine 1966 "Recherches au Temps Perdu: Some Personal Recollections of My Work in Electronic Music Written at the Request of a History Student, Queen's University." 27-31 - Queen's University Archive, Folder 5157, Box 10, File 10.

7 Evidence of computer music produced as early as 1951 is extensively discussed (Doornbusch 2004; 2005). It is unclear that the software underlying music production on the CSIRAC could be thought of as modular, but, as a general purpose computer from 1949, the hardware certainly was. The 704 mainframe on which Mathews ran the MUSIC compilers was also modular, but at the level of the tube racks. The dimensions of the 700/7000 series were, unrelated to relay racks, determined so that "every unit can be moved easily through conventional doorways and elevators and can be transported with normal trucking or aircraft facilities." (Frizzell 1953, 1285).

8 Their 1964 and 1965 exchanges are well documented: see Park 2009 and Dayal (2011). Moog's (1964) article "Voltage-Controlled Electronic Music Modules" cites James Tenney's 1963 article "Sound Generation by Means of a Digital Computer" which discusses Mathew's MUSIC software in some detail, without using the word modular. Bode and Moog's connection is documented as well (Chadabe 1997, 141). It's likely that Moog also was aware of the Mark I and II but just as he framed his own contributions as a response to the cumbersome nature of using the Music Compiler, Glinsky suggests that Moog's (1964) article is also in part a response to the tedious approach required to use the Mark II (Ibid, 70).

9 For more on Bode, see the 13.4 2011 issue of eContact! dedicated to him and his work (Palov, 2011).

10 Those going the route of handmade electronic music, in the wake of David Tudor or The Hub/The League of Automated Music Composers, bypassed the expertise embedded in the dead labor of Moog and Buchla and Bode's modular systems to learn to compose with soldering irons and components, and later, computer code. See Teboul (2020), Collins (2020), Nakai (2021), Gersham-Lancaster (1998), The Hub (2021).

11 In a 2017 SIGCIS panel at the SHOT conference in Philadelphia, I presented a short talk titled *Instruments as Computers: a Dialogue With Handmade Electronic Sound*, where I made an argument for considering modular synthesizers as a specialized and updated form of analog computer systems, where a patch on either form of system was roughly equivalent to a modern computer program. Thomas Haigh, in the audience, pointed out that setting up computers which used patch bays was only retroactively labeled programming, and that "it doesn't make sense to talk about programming an analog computer, as it does not carry out a series of distinct operations over time. Back in the 1930s and 1940s AFAIK nobody talked about programming analog computers – the term spread only after it was generally accepted in digital computing." (email with the author, 28 November 2017). As far as I can tell, Haigh is correct, and therefore a strong connection between analog signal processing and modern digital signal processing needs to be found elsewhere, as I outline in the rest of this chapter. Notwithstanding this level of scrutiny, the use of terms like "patch programming" in Serge and Serge-inspired modular synthesis systems will be connection enough to some (Serge 1976). See Qi, this volume. Readers may also appreciate this project which constructs a classical computer architecture in virtual analog using modules in a VCV rack system, by twitter user @thingkatedid: https://web.archive.org/web/20230307122046/https://github.com/katef/eurorack-cpu

12 This section includes lightly edited suggestions from Kurt Werner (email with the author, 25 June 2023). For more detailed descriptions, see David Kant's discussion of vactrols in David Dunn's chaotic synthesizer, or Johnsen' discussion of Gordon Mumma's cybersonic circuits (this volume).

13 In the case of Gordon Mumma, "military" pervaded the imaginary of the technology at hand in the late 60s so much that some components would be remembered as such even when they were likely consumer-market commodities (see Johnsen, this volume).

14 Western Electric, the manufacturing branch of AT&T / Bell Laboratories, also developed a 23 inch rack format still in use in telecommunications today. The corporate archival research required to clarify the mechanisms of the coexistence of these two formats is beyond the scope of this chapter.
15 Although Eurorack has makers and practitioners worldwide, the majority seem to be in North America and Europe (Wilson 2017). There are, as mentioned previously, plenty of other modular synthesis formats all of which deserve their own histories. Eurorack and Eurocard are here presented as an example and an acknowledgement of that format's current standardizing influence.
16 A review of modular synthesis' principal dimensions have been offered elsewhere and will not be repeated here (e.g., Bjorn and Meyer 2018, 30-33; or synthesizers.com, n.d.).
17 Chris King places the shift to SMDs in Eurorack to after 2013 (Twitter messages with the author, 1 July 2023).
18 "By definition, every hobby is a waste of time and money." (Hochfelder 2019, 46)

References

Adams, Stephen B. 2017. "Arc of Empire: The Federal Telegraph Company, the U.S. Navy, and the Beginnings of Silicon Valley." *Business History Review* 91 (2): 329–359.

Akomfrah, John, dir. 1996. *The Last Angel of History*. United Kingdom and Germany: Black Studio Film Collective, Zweites Deutsches Fernsehen, Channel 4 Television Corporation.

Alles, H.G. 1977. "A Modular Approach to Building Large Digital Synthesis Systems." *Computer Music Journal* 1 (4): 10–13.

Arboleda, Martín. 2020a. *Planetary Mine: Territories of Extraction Under Late Capitalism*. London: Verso.

Arboleda, Martín. 2020b. "From Spaces to Circuits of Extraction: Value in Process and the Mine/City Nexus." *Capitalism Nature Socialism* 31 (3): 114–133.

Aspray, William. 1990. *Computing Before Computers*. Ames: Iowa State University Press.

Austin, Kevin. 2016. "A Generalized Introduction to Modular Analogue Synthesis Concepts." *E-Contact* 17 (4): 2016.

Bady, Aaron. 2018. "Heavy Stuff: Lead is Useful, Lead is Poison." *Popula*. 2018. https://popula.com/2018/07/18/ingredients-lead/.

Barr, Alfred H. and Philip Johnson. 1934. *Machine Art: March 6 to April 30, 1934*. New York: Museum of Modern Art.

Bates, Eliot. 2021. "The Interface and Instrumentality of Eurorack Modular Synthesis." In Levaux, Cristophe and Hennion, Antoine (Eds.), *Rethinking Music through Science and Technology Studies*. London: Routledge.

Bates, Eliot. 2023. "Feeling Analog: Using Modular Synthesisers, Designing Synthesis Communities." In Cottrell, Stephen (Ed.). *Shaping Sound and Society: The Cultural Study of Musical Instruments*. London: Routledge.

Bell Laboratories. 1948. "Press Release: A.M. Papers of Thursday July 1st 1948 (Announcement of the Transistor)."

Bernstein, David, John Rockwell and Johannes Goebel. 2008. *The San Francisco Tape Music Center: 1960s Counterculture and the Avant-Garde*. Berkeley, California: University of California Press.

Bjørn, Kim and Chris Meyer. 2018. *Patch & Tweak: Exploring Modular Synthesis*. Nagle, Paul (Ed.). 3rd edition. Fredriksberg: Bjooks.

Black, H. S. 1934. "Stabilized Feedback Amplifiers." *The Bell System Technical Journal* 13 (1): 1–18.

Black, Harold S. 1977. "Inventing the Negative Feedback Amplifier: Six Years of Persistent Search Helped the Author Conceive the Idea 'in a Flash' Aboard the Old Lackawanna Ferry." *IEEE Spectrum* 14 (12): 55–60.

Bode, Harald. 1961. "A New Tool for the Exploration of Unknown Electronic Music Instrument Performances." *Journal of the Audio Engineering Society* 9 (4): 264–266.

Bode, Harald. 1976. "Frequency Shifters For Professionals." *dB*. Plainview, March 1976, 32–36, New York: Sagamore Publishing Company, Inc.

Born, Georgina and Kyle Devine. 2016. "Gender, Creativity and Education in Digital Musics and Sound Art." *Contemporary Music Review* 35 (1): 1–20.

Born, Georgina. 1995. *Rationalizing Culture: IRCAM, Boulez, and the Institutionalization of the Musical Avant-Garde*. Berkeley: University of California Press.

Boulez, Pierre and Andrew Gerzso. 1988. "Computers in Music." *Scientific American* 258 (4): 44–51.

Bowker, Geoffrey C. and Susan Leigh Star. 1999. *Sorting Things Out: Classification and Its Consequences*. Inside Technology. Cambridge, Mass: MIT Press.

Boyer, Chuck. 2004. *The 360 Revolution: System/360 and the New World of on Demand Business*. Armonk, NY: IBM.

Bronfman, Alejandra. 2016. *Isles of Noise: Sonic Media in the Caribbean*. Raleigh: University of North Carolina Press.

Bronfman, Alejandra. 2020. "Glittery: Unearthed Histories of Music, Mica and Work." In *Audible Infrastructures*, 75–92. New York: Oxford University Press.

Brown, DeForrest. 2022. *Assembling a Black Counter Culture*. Brooklyn, NY: Primary Information.

Buchla & Associates. 1966. "The Modular Electronic Music System." Berkeley, California.

Buchla, Donald and Charles MacDermed. 1971. "The Electric Music Box." *Synthesis* 1 (2): 13–16, January 1971.

Buchla, Donald and Jim Aikin. 1982. "An Interview With Don Buchla." *Keyboard Magazine*, 8–21, December 1982.

Burns, R. W. 2004. *Communications: An International History of the Formative Years*. Stevenage, U.K: Institution of Electrical Engineers.

Burrington, Ingrid. 2018. "Neodymium." Popula. 2018. https://popula.com/2018/07/30/neodymium/.

Caine, Hugh Le. 1956. "Electronic Music." *Proceedings of the IRE* 44 (4): 457–478.

Cardenas, Edgar, Randy Kemp, Raven Kemp, Rykelle Kemp, Cristobal Martinez and Meredith Martinez. n.d. "Radio Healer: About." Accessed June 26, 2023. https://cristobalmartinez.net/RadioHealer_Website/About.html.

Chun, Wendy Hui Kyong. 2011. *Programmed Visions: Software and Memory*. Cambridge: MIT Press.

Clayton, Martin, Trevor Herbert and Richard Middleton (Eds.). 2013. *The Cultural Study of Music: A Critical Introduction*. New York: Routledge.

Collins, Nicolas. 2020. *Handmade Electronic Music: The Art of Hardware Hacking*. Third Edition. New York: Routledge.

Computer History Museum. 2011. *Max Mathews & John Chowning - Music Meets the Computer*. https://www.youtube.com/watch?v=Hloic1oBfug.

Coulson, Thomas. 1950. *Joseph Henry His Life and Work*. Princeton: Princeton University Press.

Cowan, Ruth Schwartz. 1983. *More Work for Mother: The Ironies of Household Technology from the Open Hearth to the Microwave*. New York: Basic Books.

Darlington, Sidney. 1999. "A History of Network Synthesis and Filter Theory for Circuits Composed of Resistors, Inductors, and Capacitors." *IEEE Transactions on Circuits and Systems* 46 (1): 4–13.

David, Edward E. 1961. "Digital Simulation in Research on Human Communication." *Proceedings of the IRE* 49 (1): 319–329.

Demarest, Charles S. 1923. "Telephone Equipment for Long Cable Circuits." *Bell System Technical Journal* 2 (3): 112–140.

Devine, Kyle and Alexandrine Boudreau-Fournier (Eds.). 2020. *Audible Infrastructures: Music, Sound, Media*. Oxford: Oxford University Press.

Diaz, Francisco, Anastasia Kubrak and M. Otero Verzier (Eds.). 2022. *Lithium: States Of Exhaustion*. Idea Books.

Dummer, Geoffrey William Arnold. 2013. *Electronic Inventions and Discoveries: Electronics from Its Earliest Beginnings to the Present Day*. New York: Elsevier.

Dunn, David. 1992. "A History of Electronic Music Pioneers." *Eigenwelt Der Apparatewelt: Pioneers of Electronic Art*, 21–62.

Eccles, W. H. and F.W. Jordan. 1921. Improvements in Application of Thermionic Valves to Production of Alternating Currents and in Relaying. GB155854A, filed April 17, 1918, and issued January 6, 1921.

Farra, Ricardo Dal. 2006. "Un Voyage Du Son Par Les Fils Électroacoustiques: L'art Et Les Nouvelles Technologies En Amérique Latine." Ph.D. dissertation, Montreal: Universite du Quebec A Montreal.

Fleming, Sean. 2021. "The Unabomber and the Origins of Anti-Tech Radicalism." *Journal of Political Ideologies* 27 (2): 207-225.

Frizzell, Clarence E. 1953. "Engineering Description of the IBM Type 701 Computer." *Proceedings of the IRE* 41 (10): 1275–1287.

Gabrys, Jennifer. 2011. *Digital Rubbish: A Natural History of Electronics*. Ann Arbor: University of Michigan Press.

Garcia, Alejandro and Antonio Yuste. 2010. "The Role of the White House in the Establishment of a Governmental Radio Monopoly in the United States. the Case of the Radio Corporation of America." In *2010 Second Region 8 IEEE Conference on the History of Communications*, 1–6.

Glinsky, Albert. 2005. *Theremin: Ether Music and Espionage*. Urbana Chicago Springfield: University of Illinois Press.

Glinsky, Albert. 2022. *Switched On: Bob Moog and the Synthesizer Revolution*. New York: Oxford University Press.

Gordon, Theodore Barker. 2018. "Bay Area Experimentalism: Music and Technology in the Long 1960s." University of Chicago.

Gresham-Lancaster, Scot. 1998. "The Aesthetics and History of the Hub: The Effects of Changing Technology on Network Computer Music." *Leonardo Music Journal* 8 (1): 39–44.

Grossman, Rachael. 1979. "Women's Place in the Integrated Circuit." *South East Asia Chronicle* 1979 (4): 48-55.

Haring, Kristen. 2007. *Ham Radio's Technical Culture*. Cambridge: MIT Press.

Hayes, Stephen Q. 1907. "Switchboard Practice for Voltages of 60,000 and Upwards." *Transactions of the American Institute of Electrical Engineers* XXVI (2): 1333–1357.

Hé, Kirsten. 2023. "Living Sound Forever: The Genius of Wendy Carlos." *Xtra Magazine*, May 17, 2023. https://xtramagazine.com/culture/wendy-carlos-trans-profile-251085.

Henriques, Julian. 2011. *Sonic Bodies: Reggae Sound Systems, Performance Techniques, and Ways of Knowing*. New York: Continuum.

Henry, R. 1956. "Project Tinkertoy: A System of Mechanized Production of Electronics Based on Modular Design." *IRE Transactions on Production Techniques* 1 (1): 11.

Heyde, Herbert. 1975. *Grundlagen Des Natürlichen Systems Der Musikinstrumente*. Leipzig: Deutscher Verlag für Musik.

High, Kathy, Sherry Miller Hocking and Mona Jimenez (Eds.). 2014. *The Emergence of Video Processing Tools*. Two Volumes. Bristol: Intellect Books.

Hochfelder, David. 2016. *The Telegraph in America, 1832-1920*. Johns Hopkins Studies in the History of Technology. Baltimore: Johns Hopkins University Press.

Hochfelder, David. 2019. "A History of the Electrical Signal: From the Atlantic Telegraph Cable to the Quest for Artificial Intelligence." In *The Routledge Companion to Media Technology and Obsolescence*, 46–59. New York: Routledge.

Hochheiser, Sheldon. 2013. "STARS: Electromechanical Telephone Switching [Scanning Our Past]." *Proceedings of the IEEE* 101 (10): 2299–2305.

Hockenberry, Matthew Curtis, Nicole Starosielski and Susan Marjorie Zieger (Eds.). 2021. *Assembly Codes: The Logistics of Media*. Durham: Duke University Press.

Hockenberry, Matthew. 2020. "Techniques of Assembly: Logistical Media and the (Supply) Chaîne Opératoire." *Amodern* 9 (April). https://amodern.net/article/techniques-assembly/.

Hossfeld, Karen J. 1988. "Divisions of Labor, Divisions of Lives: Immigrant Women Workers in Silicon Valley." Ph.D., United States – California: University of California, Santa Cruz.

Hulton, Pontus. 1968. *The Machine, as Seen at the End of Mechanical Age*. New York: Museum of Modern Art.

IBM. 1959. *Standard Modular System Component Circuits*. Customer Engineering Manual of Instruction. Armonk, NY: IBM.

International Electrotechnical Commission. 1969. "IEC 60297:1969." 1969.

International Electrotechnical Commission. 1984. "IEC 60297-3:1984." 1984.

Jacobs, John F. 1983. "SAGE Overview." *Annals of the History of Computing* 5 (4): 323–329.

Jensenius, Alexander Refsum and Michael J Lyons. 2017. *A NIME Reader: Fifteen Years of New Interfaces for Musical Expression*. Cham: Springer.

Jones, Hedley. 2010. "The Jones High Fidelity Audio Power Amplifier of 1947." *Caribbean Quarterly* 56 (4): 97–107.

Jones-Imhotep, Edward. 2008. "Icons and Electronics." *Historical Studies in the Natural Sciences* 38 (3): 405–450.

Jurgen, Ronald K. 1987. "Whatever Happened to Project Tinkertoy?" *IEEE Spectrum* 24 (5): 20–21.

Kardon, Janet, Rosemarie Haag Bletter and American Craft Museum (New York, N.Y.). 1995. *Craft in the Machine Age, 1920-1945*. New York: H.N. Abrams in association with the American Craft Museum.

Kittler, Friedrich A. 2013. "Rock Music: A Misuse of Military Equipment." In *The Truth of the Technological World*, 152–164. Stanford University Press.

Kittler, Friedrich. 1995. "There is no Software." *Ctheory*. http://www.ctheory.net/articles.aspx?id=74/.

Labban, Mazen. 2019. "Rhythms of Wasting/Unbuilding the Built Environment." *New Geographies 10: Fallow*, January.

Labban, Mazen. 2022. "Mine/Machine." *Dialogues in Human Geography*, January.

Lécuyer, Christophe. 2017. "From Clean Rooms to Dirty Water: Labor, Semiconductor Firms, and the Struggle over Pollution and Workplace Hazards in Silicon Valley." *Information & Culture* 52 (3): 304–333.

Lindgren, Brian. 2020. "62 Years and Counting: MUSIC N and the Modular Revolution." Unpublished manuscript.

Loewenthal, Max. 1899. "The New Exchange of the Central New York Telephone and Telegraph Co. at Syracuse, N.Y." *Electrical Engineer* 27 (561): 142–147.

Los Angeles County Museum of Art. Art and Technology Program. 1971. *A Report on the Art and Technology Program of the LACMA, 1967-1971*. Los Angeles: Los Angeles County Museum of Art.

Lüthje, Boy. 2013. *From Silicon Valley to Shenzhen: Global Production and Work in the IT Industry*. Lanham: Rowman & Littlefield.

Macquarie, Charlie. 2022. "Open Up This Pit." Hypocrite Reader. 2022. http://hypocritereader.com/100/open-up-this-pit.

Maloberti, Franco and Anthony C Davies. 2016. *A Short History of Circuits and Systems*. Aalborg: River Publishers.

Masu, Raul, Adam Pultz Melbye, John Sullivan and Alexander Refsum Jensenius. 2021. "NIME and the Environment: Toward a More Sustainable NIME Practice." In *International Conference on New Interfaces for Musical Expression*.

Mauzey, Peter. 1958. "Specialized Equipment Used at the Columbia University Studio for the Production of Tape Music." In *Proceedings of the 10th AES Conference, Sept. 29 - Oct 3, 1958*. Audio Engineering Society. NY: AES.

McCarty, Cara. 1990. *Information Art: Diagramming Microchips*. New York: Harry N. Abrams with the Museum of Modern Art.

McKittrick, Katherine and Alexander G. Weheliye. 2017. "808s and Heartbreak." *Propter Nos* 2 (1): 13–42.

McKittrick, Katherine (Ed.). 2015. *Sylvia Wynter: On Being Human as Praxis*. Durham: Duke University Press.

Meredith, Kevin, dir. 2012. Featuring Alexander Magoun and Rebecca Mercuri. *The Story of the RCA Synthesizer*. https://www.youtube.com/watch?v=rgN_VzEIZ1I.

Mills, Mara. 2011. "Hearing Aids and the History of Electronics Miniaturization." *IEEE Annals of the History of Computing* 33 (2): 24–45.

Mindell, David A. 2002. *Between Human and Machine: Feedback, Control, and Computing Before Cybernetics*. Baltimore: John Hopkins University Press.

Molella, A.P. 1976. "The Electric Motor, the Telegraph, and Joseph Henry's Theory of Technological Progress." *Proceedings of the IEEE* 64 (9): 1273–1278.

Moog, Robert A. 1964. "Voltage-Controlled Electronic Music Modules." In *Audio Engineering Society Convention Proceedings, October 1st*.

Morreale, Fabio, A. Bin, A. McPherson, P. Stapleton and M. Wanderley. 2020. "A NIME of the Times: Developing an Outward-Looking Political Agenda For This Community." https://researchspace.auckland.ac.nz/handle/2292/51480.

Morse, Samuel F.B. 1840. Improvement in the Mode of Communicating Information by Signals by the Application of Electro-Magnetism. United States US1647A, issued June 20, 1840.

Nakai, You. 2021. *Reminded by the Instruments: David Tudor's Music*. Oxford: Oxford University Press.

Nakamura, Lisa. 2014. "Indigenous Circuits: Navajo Women and the Racialization of Early Electronic Manufacture." *American Quarterly* 66 (4): 919–941.

National Bureau of Standards. 1953. "Project Tinkertoy: Modular Design of Electronics and Mechanized Production of Electronics." Technical News Bulletin 37. Gaithersburg, Maryland: National Bureau of Standards.

Nieves, Evelyn. 2018. "The Superfund Sites of Silicon Valley." *The New York Times*, March 26, 2018, sec. Lens. https://www.nytimes.com/2018/03/26/lens/the-superfund-sites-of-silicon-valley.html.

Noble, David F. 1995. *Progress Without People: New Technology, Unemployment, and the Message of Resistance*. Toronto, Ont: Between the Lines.

Norberg, Arthur Lawrence. 1976. "The Origins of the Electronics Industry on the Pacific Coast." *Proceedings of the IEEE* 64 (9): 1314–1322.

O'Regan, Gerard. 2021. *A Brief History of Computing*. 3rd edition. Cham: Springer.

Ouzounian, Gascia. 2020. *Stereophonica: Sound and Space in Science, Technology, and the Arts*. Cambridge, Massachusetts: The MIT Press.

Palov, Rebekkah. 2011a. "CEC — eContact! 13.4 — Harald Bode." Online Journal. CEC | Canadian Electroacoustic Community. 2011. https://econtact.ca/13_4/index.html.

Palov, Rebekkah. 2011b. "Harald Bode — A Short Biography." CEC | Canadian Electroacoustic Community. July 2011. https://econtact.ca/13_4/palov_bode_biography.html.

Paradiso, Joseph A. 2017. "The Modular Explosion - Deja Vu or Something New?" In *Presented at the Voltage Connect Conference, March 10-11, 2017. Berklee College of Music, Boston MA.*

Pellow, David N. and Lisa Sun-Hee Park. 2002. *The Silicon Valley of Dreams: Environmental Injustice, Immigrant Workers, and the High-Tech Global Economy.* Critical America. New York: New York University Press.

Pinch, Trevor and Frank Trocco. 1998. "The Social Construction of the Early Electronic Music Synthesizer." *Icon*, 4: 9–31.

Pinch, Trevor and Frank Trocco. 2002. *Analog Days: The Invention and Impact of the Moog Synthesizer.* Cambridge: Harvard University Press.

Prasad, Ray. 2013. *Surface Mount Technology: Principles and Practice.* Springer Science & Business Media.

QST. 1919. De Forest Unit Receiving Set (Advertisement). Newington, Connecticut: American Radio Relay League.

R.A. Moog Co. 1967. "Electronic Music Composition-Performance Equipment - Short Form Catalog." R.A. Moog Co.

Radetzki, Marian. 2009. "Seven Thousand Years in the Service of Humanity—the History of Copper, the Red Metal." *Resources Policy* 34 (4): 176–184.

Rekoff, M. G. 1985. "On Reverse Engineering." *IEEE Transactions on Systems, Man, and Cybernetics* 15 (2): 244–252.

Riofrancos, Thea N. 2020. *Resource Radicals: From Petro-Nationalism to Post-Extractivism in Ecuador.* Radical Américas. Durham: Duke University Press.

Robinson, Kevin. 2020. *Practical Audio Electronics.* Abingdon, Oxon: Routledge, an imprint of the Taylor & Francis Group.

Rodgers, Tara. 2015. "Cultivating Activist Lives in Sound." *Leonardo Music Journal* 25: 79–83.

Rubin, Steve. 1975. "Retrospect: Forbidden Planet." *Cinefantastique* 4(1): 4–12.

Russell, Andrew L. 2012. "Modularity: An Interdisciplinary History of an Ordering Concept." *Information & Culture* 47 (3): 257–287.

Schaefer, Magnus. 2024. "New Protocols: Electronic Music Studios in Buenos Aires 1958–73." New York: Institute for Studies on Latin American Art.

Scott, Richard. 2016. "Back to the Future: On Misunderstanding Modular Synthesizers." *EContact! Journal of the Canadian Electroacoustic Community* 17 (4). https://econtact.ca/17_4/scott_misunderstanding.html.

Serge. 1976. "Introduction to the Use of the Serge Modular Music System." https://web.archive.org/web/20190423042216/http://www.serge.synth.net/documents/1976sergemanual.pdf.

Sevilla, Ramon C. 1992. "Employment Practices and Industrial Restructuring: A Case Study of the Semiconductor Industry in Silicon Valley, 1955-1991." Ph.D., United States – California: University of California, Los Angeles.

Sharp, Camille-Mary. 2020. "Decolonize and Divest: The Changing Landscape of Oil-Sponsored Museums in Canada." Doctoral Thesis. Toronto: University of Toronto.

Sharp, Camille-Mary. 2022. "Oil-Sponsored Exhibitions and Canada's Extractive Politics of Cultural Production." *Imaginations* 13 (1): 13–35.

Shirrif, Ken. 2021. "Germanium Transistors: Logic Circuits in the IBM 1401 Computer." March 2021. http://www.righto.com/2021/03/germanium-transistors-logic-circuits-in.html.

Shirriff, Ken. n.d. "IBM SMS (Standard Modular System) Card Database." Accessed June 27, 2023. https://static.righto.com/sms/index.html.

Simpson, Leanne Betasamosake. 2017. *As We Have Always Done: Indigenous Freedom Through Radical Resistance*. Indigenous Americas. Minneapolis London: University of Minnesota Press.

Small, James S. 2013. *The Analogue Alternative: The Electronic Analogue Computer in Britain and the USA, 1930-1975*. New York: Routledge.

Smith, Ted, David Allan Sonnenfeld and David N. Pellow (Eds.). 2006. *Challenging the Chip: Labor Rights and Environmental Justice in the Global Electronics Industry*. Philadelphia: Temple University Press.

Smith-Biwer, Kelli. 2023. "The Hi-Fi Man: Masculinity, Modularity, and Home Audio Technology in the U.S. Midcentury." Chapel Hill: University of North Carolina, Chapel Hill.

Spritzendhorfer, Dominik, and Elena Tikhonova, dirs. 2013. *Elektro Moksva*. Austria.

Sturgeon, Timothy J. 2003. "What Really Goes on in Silicon Valley? Spatial Clustering and Dispersal in Modular Production Networks." *Journal of Economic Geography* 3 (2): 199–225.

Synthesizers.com. n.d. "Modular Form Factors." Synthesizers.Com. Accessed June 25, 2023. https://shop.synthesizers.com/pages/form-factors.

Systems Integration Plus, Inc. 2007. "Sizing and Dimensions." 2007. https://web.archive.org/web/20160304050242/http://www.vme.com/index.php?L=2-3-0-1.

Teboul, Ezra J. 2017. "The Transgressive Practices of Silicon Luthiers." In Miranda, Eduardo (Ed.), *Guide to Unconventional Computing for Music*. New York: Springer, 85–120.

Teboul, Ezra J. 2018. "Electronic Music Hardware and Open Design Methodologies for Post-Optimal Objects." In Sayers, Jentery (Ed.), *Making Things and Drawing Boundaries: Experiments in the Digital Humanities*. Minneapolis: University of Minnesota Press, 177–184.

Teboul, Ezra J. 2020a. "Hacking Composition: Dialogues With Musical Machines." In Cobussen, Marcel and Bull, Michael (Eds.), *The Bloomsbury Handbook of Sonic Methodologies*. 807–820, New York: Bloomsbury Academic.

Teboul, Ezra J. 2020b. "A Method for the Analysis of Handmade Electronic Music as the Basis of New Works." Doctoral Dissertation, Troy, NY: Rensselaer Polytechnic Institute.

The Hub. 2021. *The Hub: Pioneers of Network Music*. Brümmer, Ludger (Ed.). Heidelberg: Kehrer.

Thomas, Percy H. 1909. "Output and Regulation in Long-Distance Lines." *Transactions of the American Institute of Electrical Engineers* XXVIII (1): 615–640.

Tigers, Leah. 2019. "A Sex Close to Noise: An Essay about Transgender Women and Music." NineBillionTigers. 2019. http://www.trickymothernature.com/asexclosetonoise.html.

Turner, Fred. 2006. *From Counterculture to Cyberculture: Stewart Brand, the Whole Earth Network, and the Rise of Digital Utopianism*. Chicago: University of Chicago Press.

Ussachevsky, Vladimir. 1958. "Musical Timbre Mutation by Means of the 'Klangumwandler,' a Frequency Transposition Device." Presented at the 10th Audio Engineering Society Convention.

Vagnerova, Lucie. 2017. "Nimble Fingers in Electronic Music: Rethinking Sound Through Neo-Colonial Labour." *Organised Sound* 22 (2): 250–258.

Valdivia, Ana. 2022. "Silicon Valley and the Environmental Costs of AI." *Political Economy Research Centre* (blog). 2022. https://www.perc.org.uk/project_posts/silicon-valley-and-the-environmental-costs-of-ai/.

Warren, Christian. 2001. *Brush with Death: A Social History of Lead Poisoning*. Hopkins, A Johns (Ed.) Baltimore: Johns Hopkins Univ. Press.

Werner, Kurt James and Ezra J. Teboul. 2021. "Analyzing a Unique Pingable Circuit: The Gamelan Resonator." In *Proceedings of the 151st Audio Engineering Society Convention*. New York: Audio Engineering Society. https://www.aes.org/e-lib/browse.cfm?elib=21506.

Werner, Kurt James and Russell McClellan. 2020. "Moog Ladder Filter Generalizations Based on State Variable Filters." In *Proceedings of the 23rd International Conference on Digital Audio Effects (DAFx2020), Vienna, Austria, September 2020-21*, 70–77.

Werner, Kurt James, Ezra J. Teboul, Seth Allen Cluett, and Emma Azelborn. 2022. "Modelling and Extending the RCA Mark II Synthesizer." In *Proceedings of the Digital Audio Effects (DAFX) Conference*. Vienna, Austria. https://dafx2020.mdw.ac.at/proceedings/papers/DAFx2020in22_paper_39.pdf.

Wilson, Richard Guy, Dianne H. Pilgrim and Dickran Tashjian. 2001. *The Machine Age in America, 1918-1941*. New York: Brooklyn Museum of Art in association with Abrams.

Wilson, Scott. 2017. "This Map Is a Directory of Eurorack Companies Around the Globe." 2017. https://www.factmag.com/2017/03/13/eurorack-companies-world-map/.

Winthrop-Young, Geoffrey. 2002. "Drill and Distraction in the Yellow Submarine: On the Dominance of War in Friedrich Kittler's Media Theory." *Critical Inquiry* 28 (4): 825–854. 10.1086/341236.

Yates, JoAnne and Craig Murphy. 2019. *Engineering Rules: Global Standard Setting Since 1880*. Hagley Library Studies in Business, Technology, and Politics. Baltimore: Johns Hopkins University Press.

Young, Gayle. 1984. "Hugh Le Caine: Pioneer of Electronic Music in Canada." *HSTC Bulletin: Journal of the History of Canadian Science, Technology and Medecine* 8 (1): 20.

Young, Gayle. 1989. *The Sackbut Blues: Hugh Le Caine - Pioneer in Electronic Music*. Ottawa: National Museum of Science and Technology.

Zicarelli, David. 2002. "How I Learned to Love a Program That Does Nothing." *Computer Music Journal* 26 (4): 44–51.

1

THE BUCHLA MUSIC EASEL

From Cyberculture to Market Culture

Theodore Gordon

Introduction

This is the Music Easel: an electronic music system designed by Donald Buchla in Berkeley, California in 1973 (Figure 1.1).[1] With its impressive array of various sockets, knobs, sliders, and lights, and what appears to be something like a musical keyboard, this system bears morphological similarities to other electrical and electronic instruments that by the early 1970s came to be known as synthesizers. The concretization of the synthesizer as a new type of musical instrument had only recently occurred (Rodgers 2015, 211); the national synthesizer industry was even younger, an emerging market for a new product which took advantage of miniaturized electronic components and the growing manufacturing capabilities of the globalizing American Cold War economy. Designed nearly concurrently with what would become two of the best-selling electronic musical instruments of the decade—the Minimoog and the ARP Odyssey—Buchla's Music Easel could be understood as an idiosyncratic example of a keyboard synthesizer designed by someone who was known to have eschewed both musical keyboards and the term "synthesizer."

However, looking carefully at the Easel's design suggests major differences not only in degree, but also in kind, from other keyboard synthesizers. What appears to be a musical keyboard is actually a flat panel of metallic touchpads, labeled a "Touch Activated Voltage Source"; what appears to be the control panel for this keyboard is labeled a "Stored Program Sound Source," routing the electronic signals from the Touch Activated Voltage Source to an increasingly complex matrix of sound-producing and sound-controlling functions, with its

DOI: 10.4324/9781003219484-3

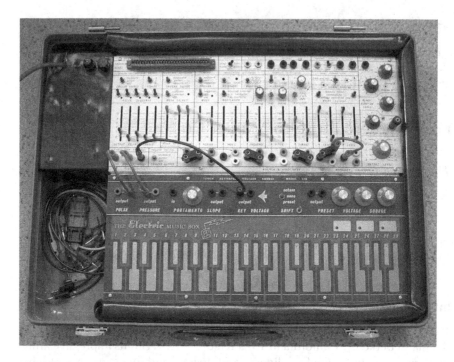

FIGURE 1.1 The Buchla Music Easel, 1973.

signal paths traced out on the panel and connectable by evenly spaced banana-plug sockets.

In this regard, the Music Easel also bears a morphological similarity to another very different kind of instrument: an analog electronic computer. These scientific instruments use continuously varying differences in electrical potential to execute programs, which are defined by routing the flow of electronic signal through discrete operational amplifiers laid out on the computer's front panel. Once interconnected in a manner that matches the user's desired function, and set with initial conditions determined using potentiometers, the instrument runs a program and outputs continuously varying voltage—which can be read via a voltage meter or an oscilloscope, or in theory transduced into sound.

The Easel's front panel presents the user with a similar array of semi-modular operational sections, studded with banana-plug sockets that allow them to be connected; the name of its top module also suggests that this instrument was designed to execute programs, rather than to play music. In the world of the Easel, the user does not "play" the keyboard and expect novel sounds to emerge; instead, the user both executes "stored programs" and changes the parameters and definitions of those programs in real time. Rather than synthesizing sound as audible air pressure waves with novel timbres, the user synthesizes a concept of "sound" as a signal that is at once electronic, physical, and perceptual. Buchla's Music Easel, among his other instrumental systems, was designed to explore new, unknown creative processes latent in the human mind.

The ethos of Buchla's instruments emerged within the expansive discourse of the Californian counterculture of the 1960s and 1970s, drawing vocabulary, metaphors, and technologies from the fields of cybernetics and information theory—often in contradictory and parascientific ways. As a full-time electronic instrument designer, a some-time psychonaut, and a part-time engineer for NASA and other governmental entities at the Lawrence Radiation Laboratory, Buchla embodied a major paradox of this counterculture: the ostensibly democratic new worlds of sensation, creativity, and consciousness he imagined through his instruments were only materially, financially, and technologically possible through the Cold War military-industrial complex. As Fred Turner and others have argued, this paradox was not only material, but perhaps more importantly conceptual: both top-down state apparatuses of control and bottom-up technologies of self-discovery relied on the same understanding of the world as "looping circuits of energy and information" (Turner 2006, 38).

Buchla's Music Easel can be understood as an attempt to operationalize one specific category of such looping circuits: the creative processes of the human mind. Buchla's model for these processes relied on what Fred Turner, after Kevin Kelley, has called the "computational metaphor"—a "new universal metaphor" that understood all phenomena through the "vocabulary and syntax" of computing (Turner 2006, 15). Specifically, in a poem written upon the introduction of the Easel, and in the Easel's user manual, Buchla and his associates embraced an idiosyncratic concept of the "human biocomputer" popularized by the notorious cetacean and psychedelics researcher John C. Lilly. Published in various forms during the period in which Buchla was designing the Easel, Lilly's manuscript *Programming and Metaprogramming in the Human Biocomputer* articulated an understanding of human consciousness that strongly appealed to Buchla and his associates—including the composer Allen Strange, who titled his manual for the Easel *Programming and Meta-Programming in the Electro Organism*.

Using Lilly's concepts of the human biocomputer, the Easel externalized what Buchla understood to be the processes of "programming" and "metaprogramming" human musical performance. Unlike a keyboard synthesizer, which would reliably operate with a single "program"—i.e., it would produce a sound at a set pitch with a novel timbre when a key was depressed—Buchla designed the Easel to execute programs containing instructions that would change the nature of the program itself, which Lilly called "metaprogramming." These "stored programs," specified in the name of the Easel's top module, were nonlinear, and could include unpredictable, random, and self-regulating elements. Because of the simultaneity of its own programming and metaprogramming, the Easel resists being understood as a keyboard synthesizer. Instead, understood through the paradigm of the human biocomputer, it presents the user with an externalized, operationalized interface for human creativity.

The Music Easel was only one of Buchla's many attempts to create such an interface; it was a specific configuration of what he called his "Series 200" systems, discussed later in this chapter. Indeed, as more advanced microcomputer components became increasingly available in the 1970s, Buchla quickly abandoned the Music Easel in order to keep up with the cutting edge of technology. Only a few dozen Easels were ever made during that decade, and the Easel was performed on only a handful of commercially-available recordings in the 20th century. In the broad scope of Buchla's long career, the Music Easel was a relatively minor instrument: a single articulation of a never-ending process of visionary design, technological innovation, and instrumental experimentation.

But in the decades after its discontinuation, the Music Easel became much more than an historical footnote. Indeed, in the midst of what Trevor Pinch and Frank Trocco have called the "analog revival" in the early 2000s (317), it came to retroactively connect Buchla to a category of musical instruments, and a musical industry, to which he never aspired to relate. In 2012, following the sale of his intellectual property to a group of industry investors who formed an enterprise called Buchla Electronic Musical Instruments (BEMI), Buchla's Music Easel again became available on the marketplace as an ostensible replica of the original 1973 design. And in the throes of a widespread renewed interest in analog synthesizers, the Music Easel became an object of desire that concentrated—and made salable—the surplus of interest in Buchla as an apparent patriarch in the emergent historical narrativization of the synthesizer.

The narrative into which the Music Easel was re-introduced in 2012 began to coalesce in the 1980s, when the synthesizer industry had exploded into the mainstream, and the timbres of Moog, ARP, Roland, and Yamaha instruments became commonplace in contemporary popular music production. Beginning in that decade, critics, journalists, and academics alike began to reevaluate the history of that category of instrument using a "great man" model, with many left-leaning commentators seeking out new patriarchs who destabilized the dominance of commercially-oriented instruments. Specifically, many sought to destabilize the apparent patriarch whose name appeared on some of the most popular keyboard synthesizers of the century: Robert Moog. Writing in *Keyboard* magazine in 1988, for example, Dominic Milano declared that the "American synth industry" was birthed by *two* "fathers, Don Buchla and Bob Moog;" however, whereas Buchla "[developed] the voltage-controlled synthesizer in 1963," Moog made it "commercially successful" in 1965 (Milano 1988, 42). Despite the historical inaccuracy of this timeline, Milano's narrative became the dominant understanding of Buchla's role in what had become a multi-billion dollar industry: an "unsung hero" (Collins 2006, 192) who was "anti-commercial" (Pinch and Trocco 2002, 52) and whose instruments evoked a "counter-cultural design philosophy" (Dunn 1992, 21) compared to the commercially oriented instruments designed by Moog.

With the reintroduction of the Music Easel into the synthesizer marketplace in 2012, anyone could now purchase an apparently anti-commercial, counter-cultural instrument for the relatively low price of $3,995.[2] At no point in history had this instrument ever been as affordable: when introduced, the Music Easel sold for nearly $20,000, adjusted for inflation.[3] As Buchla said in a 1991 interview with Woody and Steina Vasulka, his instruments were akin to "Maserati or Rolls Royce[s]," which reflected both the instruments' design and the high cost of specialized components that Buchla used in their manufacturing (Buchla 1991). By insisting on "replicating" the Music Easel to 1973 standards, BEMI was marketing more than the Easel's functionality: they were marketing the Easel's aura.

This chapter will first follow the emergence of the Music Easel in the early 1970s, showing how the instrument articulated Buchla's interest in the concept of the human biocomputer, and how the Easel was explicitly designed to be a biocomputer/electronic computer interface for creative discovery of sonic signal. After briefly discussing the short life of the Music Easel in the 1970s, it will follow Buchla's imbrication into a "pioneers narrative" of electronic music beginning in the 1980s, and how that narrative influenced the events of the 2000s "analog revival," during which Buchla sold his company to a group of investors who decided to revive the Easel. After documenting the many issues that arose during that revival, it will close with a meditation on the nature of the Easel as an object of musical desire, critiquing the ideological characterization of the Easel as countercultural.

"The Genesis of the Electric Music Box"

By 1973, the year in which Buchla began to manufacture the Music Easel, synthesizers were big business. As Tara Rodgers has documented, the term "synthesizer" rose to prominence in the 1950s through RCA's publicity campaign for their "Sound Synthesizer"; in the 1960s, Robert Moog further publicized the term through the marketing of his novel keyboard instruments (Rodgers 2015, 211). By the early 1970s, dozens of new manufacturers began to design synthesizers with the same basic set of features, mostly including a keyboard that controlled the frequency of one or more oscillators, whose signals could then be filtered, modified, amplified, and transduced, exploiting newly available miniaturized transistors, operational amplifiers, and integrated circuits. In 1973, the American Music Conference began to track this new industry, reporting that 7,000 synthesizers valued at $8 million were sold in North America that year (Thèberge 1997, 53); the next year, the *Wall Street Journal* estimated that synthesizers were used in 4% of American high schools ("Synthesizers Succeed" 1974), as well as dozens of institutions of higher education.

Although Buchla did benefit from the growing synthesizer industry, he never scaled his production to meet consumer demand. As he told Woody and Steina Vasulka in 1991, "I was interested in providing very exotic functionality that served my needs and those of the people around me that I associated with" (Buchla 1991). As a result, his instruments were expensive, uncommon, and difficult to obtain. Although this is easy to understand in terms of business logistics—Buchla wanted to maintain control over the production of his instruments and didn't want to outsource manufacturing or design—this decision also emerged from a more fundamental difference between Buchla and other designers of electronic musical instruments. As I will argue, Buchla's instrumental systems such as the Music Easel were different in kind, not degree, from keyboard synthesizers. Indeed, reading Buchla's instruments as "synthesizers" elides their connections to the parascientific discourses of biocomputing, psychedelics, and systems theory: they were designed not to be synthesizers, but rather analog, mind-manifesting music computers. The sounds generated through human-computer interactions with Buchla's systems were, in some sense, secondary to the experience they offered to their users in creative, consciousness-expanding self-discovery.

A new 1970 periodical entitled *Synthesis* illustrates the extent to which Buchla did not fit in with his supposed peers. Intended to succeed the recently defunct *Electronic Music Review,* and explicitly modeled after the *Whole Earth Catalog, Synthesis* was to be the first periodical that would collect information, opinions, and resources about the emergent, eponymous field of synthesis. For its first full issue, the editors asked major synthesizer manufacturers to each submit editorial content that would describe their products.[4] Buchla was invited to contribute; but unlike his ostensible "competitors," all of which submitted technical advertisements for their instruments, Buchla submitted a poem.

Titled "Genesis of the ELECTRIC MUSIC BOX," and co-authored with his friend Charles MacDermed, Buchla's poem begins:

February 9th 1971 / Scorpio rising / Sun and moon in perfect opposition / Earth interposed / the planetary house lights dimmed / at that moment with a blood red moon suspended in each witnesses' breath / the box switched on [...]

In ecstatic, breathless poetry, Buchla and MacDermed describe the genesis of Buchla's most recent instrumental system: the "Electric Music Box." This name playfully joins the concept of the music box—an early mechanical automatic music machine—with the field of electrical engineering. It also plays on the popular electrical engineering concept of the black box, in which a complex system is understood only through input and output, which had become popularized by Norbert Wiener, Ross Ashby, and other cyberneticians several decades prior.[5] Unlike a black box, however, Buchla's "Electric Music Box" was more akin to what Wiener might call a white box: a system in which the complexities of its inner workings were all visible by the user, and indeed controllable down to its intricate parameters.

Rather than a synthesizer designed to create new, synthetic sonic timbres, Buchla's new box was explicitly designed to create new pathways for the control of musical performance. Indeed, Buchla never used the term synthesizer to describe his instruments; although often dismissed as another of Buchla's unique idiosyncrasies, this was an intentional, and crucial, choice. Buchla modeled the flow of control through his systems on a very specific understanding of the human mind, shaped equally by cybernetics, psychedelics, and computing. Like Buchla's earlier designs from the 1960s, the "Electric Music Box" was envisioned as a psychedelic technology: a way of manifesting the mind through externalizing its creative processes and allowing for increased capacities for communication and control.

Buchla and MacDermed explicitly describe the mind-manifesting contours of this relationship between human and machine in their poem:

> What the machine can afford and produce is a new language cultivating and refining in the listener a new sense of perceptual mechanics / when the listener is also the operator-creator his visual and auto-muscular senses add to his interpretive model of the unfolding auditory process / the operator functions in a special relationship to his own perceptual feedback system / a system in which the machine functions as a critical interface within the circular circuit / critical thresholds exceeded enabling biocomputer to interlock with hardware computer / the program as a patch-up and set of settings of dials and switches represents a score but a score which is variable / gesture-expression-meaning-logic-structure-statement are arrived at through cycles of input and output / the scale of cyclic variations relies upon the principle of correlation by precession procession succession for there to be change […] (Buchla and MacDermed 1971)

This poem positions Buchla's Electric Music Box as a "hardware computer" and its human user as a "biocomputer," both of which function together as a "systematic circuit" that executes "programs"—the dial settings and switches on the hardware computer—enabling a cyclical process of discovery. Rather than a musical instrument that produces novel sounds, the "Box" was imagined as a human-computer interface that enabled both new sounds and new vectors of the control of sound to emerge through human-computer interaction. Buchla's poem suggests that unlike the "realization of ideation" that occurs with acoustic instruments, the realization made possible through his system was fundamentally different: the automatic processes with which humans previously created musical ideas could be internally modified with a level of complexity theretofore unknowable and unachievable with the limited parameters of a non-computer instrument.

The rarefied technoscientific discourses within which Buchla developed his "Electric Music Box" instruments, of which the "Music Easel" was only one configuration, had by 1971 become embraced by many involved with the Californian counterculture—especially those who subscribed to Stewart Brand's *Whole Earth Catalog*, a publication which embraced and popularized systems thinking. More specifically, they embraced a particular subset of systems thinking: the "computational metaphor," which, as Fred Turner (after Kevin Kelley) has argued, attracted a changing set of political and material valances from its emergence in the World War II military-industrial complex to its simultaneous development by both Big Science and the nascent 1960s counterculture (Turner 2006, 16). Although gracefully simple at its core—that any phenomenon could be described as a system of information exchange—this metaphor allowed people to follow those information exchanges in significantly diverging directions, with differing consequences for technology, sociality, and politics.

Buchla's invocation of the computational metaphor was idiosyncratic. It held not only that any perceptible phenomenon could be understood as information, but also that human observers of the flow of information were themselves computers. Indeed, Buchla's poem begins with a meditation on the concept of the biocomputer, which Buchla likely discovered through the work of John C. Lilly, whose writings had been listed in the *Whole Earth Catalog* beginning in 1969. That was the same year that Buchla had begun to design modules for his new Electric Music Box series.[6] At the core of Lilly's computational ontology of human life was the assertion that "All human beings [...] are programmed biocomputers. No one of us can escape our own nature as programmable entities. Literally, each of us may be our programs, nothing more, nothing less" (Lilly 1987, xii). Lilly described the "biocomputer" as having "*stored program* properties": it executes programs stored in its memory which are activated either by the computer itself or an external agent. But expanding beyond this "first order" understanding, Lilly also posited that biocomputers have "selfprogramming," "otherpersonprogramming [sic]," and "selfmetaprogramming [sic]" properties, which describe the changes in "programming" effected either by the mind itself, another person, or another technology—such as a sensory deprivation tank, LSD-25, or another computer.

Buchla conceptualized his "Electric Music Box" as a systematic assemblage of modules that produced such a mind-manifesting computer: "a system in which the machine functions as a critical interface within the circular circuit / critical thresholds exceeded enabling biocomputer to interlock with hardware computer" (Buchla and MacDermed 1971). Buchla's poem recalls Lilly's utilization of the concept of "interlock"—an engineering term that describes the phenomenon when two or more mechanisms become mutually dependent in order to function.[7] For Buchla, the "interlock" produced between the human biocomputer and his own "Electric Music Box"-as-computer facilitated not only the execution of programs, akin to scores produced by humans, but also the variation of those very programs in real time. In Lilly's terminology, this real-time, on-line variation is called "metaprogramming," and could serve to unconsciously disrupt the default programs so often executed by humans and acoustic instruments.

In Buchla and MacDermed's extended computational fantasy, the "Electric Music Box" could produce theretofore unknown, emergent kinds of musical performance:

> To pluck a string is to choose a program / to initiate a complex multiple and variable process which constitutes the oscillation of multiple frequencies in roughly the overtone series and their respective attack sustain and decay envelopes together with the vastly variegated intermodulations of amplitude and frequencies / to drop a ping pong ball on a marble floor is to set in motion an automatic rhythmic pattern / so to design extend and variably control the internal workings of an automatic process is inherent in the most simple of human acts / to be heard to be imagined to mature in experience to flower in perception to exist in symbiotic oceans of phenomena and noumena [...]

For Buchla, any physical action that produced sound was framed as a "program"; through his new Box, that program could be designed, extended, and variably controlled *by itself*—an act of cybernetic, self-governing "metaprogramming." Buchla's poem imagines a new kind of musical instrument that could metaprogram sonic events-as-"programs"; rather than plucking a string or dropping a ping-pong ball, the internal parameters of those programs could be instantaneously modified through the biocomputer-computer interlock. Buchla, like Lily, had experienced this kind of real-time, emergent self-modification through other technologies such as LSD-25; like that technology, Buchla's imagined Electric Music Box would be, as he described it in this poem, "a teacher [...] a friend that can provide a relationship on a high utility level."[8]

Manifesting the Electric Music Box

Buchla and MacDermed's poem was written in February 1971, amidst a time of transition in Buchla's life. Throughout that year, Buchla was working to back out of a contract he had signed with CBS Musical Instruments, Inc., a corporate entity that had purchased his previous designs from the mid-1960s and had employed him to design proprietary instruments for the growing educational market for electronic instruments. Throughout the period between roughly 1968 and 1971, Buchla worked on his "Electric Music Box" modules privately, awaiting the end of his contract in order to make them available to the public (Smith 2021). He began to produce data sheets as early as November 1971, and by February 1972, he had begun to advertise over 25 new modules in the "Series 200"—a numbering scheme which differentiated these new modules from his previous designs.[9] These included polyphonic computer-controlled keyboards and touch-controlled voltage sources, quadraphonic interfaces for spatialized sound, completely redesigned oscillators, gates, filters, and sequencers, as well as the infamous "Source of Uncertainty" Model 265—a unique module that produced white noise, random voltage output with voltage-controlled rate of change, and a series of unpredictable voltages with voltage-controlled correlation.

Not only did Buchla's new "Electric Music Box" series feature individual modules that could be assembled into large systems; it also featured "integrated" modules that condensed multiple functionalities into a single panel of knobs and switches, allowing for smaller, more portable configurations.[10] The first of these integrated modules was dubbed the "Dodecamodule" (Model 212), whose name playfully riffed on the concept of "dodecaphony" in serial composition. This module came close to the developing industry standard for a synthesizer: it contained three envelope generators, three filters, three low-pass gates (a combined voltage-controlled amplifier and low-pass filter), and three voltage-controlled amplifiers—which, when added up, equaled a total of twelve functions. This module was explicitly designed to work with Buchla's Model 237 keyboard, which featured three-voice polyphony, velocity sensitivity, and pressure sensitivity, all made possible through Buchla's use of cutting-edge computerized components; together, Buchla advertised them as the "System 101," "aptly suited to the needs of music educators, broadcast and recording studios, small electronic studios, filmmakers, and composers" ("Series 200 Data Sheet," 1971).

Priced at $2,850, Buchla's "System 101" balanced market appeal with his own more personal quest for developing psychedelic, musical human-computer interfaces. Indeed, the engineer and historian Richard Smith, who worked with Buchla beginning in the 1990s and who holds much of Buchla's paper archives, has theorized that Buchla explicitly designed the "System 101" (and the related "System 151," with additional oscillators) after visiting a National Association of Music Merchants (NAMM) trade show in 1971, in order to compete with the growing roster of synthesizer manufacturers such as Moog, EMS, and ARP.[11] But despite the market-based origins of the "System 101," Buchla likely only ever manufactured 6–8 "System 101" and "System 151" instruments between the period of late 1971 and early 1973; selling roughly one instrument every few months would keep him financially solvent, and he had no desire to scale up production.

The Model 208 Stored Program Sound Source

By the early 1970s, income from sales of the "System 101" and a steady roster of institutional customers allowed Buchla to return to first principles: he wanted to design a human-machine interface that would allow the human biocomputer to interlock with the hardware computer in a way that would manifest a new form of sonic consciousness. However, Buchla's vision of music computing differed in a fundamental way from many of his peers. Even as research into encoding sound into digital signal continued throughout the decade, Buchla instead desired to encode what he considered the "structural parameters of sound": signals of control flowing between human and instrument. In 1971, he wrote in *Source* magazine:

> Programming giant computers to pump out a million or so bits per second to be translated into sound is undoubtedly a worthwhile enterprise, but it doesn't seem as though most of us will ever get near one, much less actually play music on one. [...] Now, if computers dealt only with the structural parameters of sound and left the actual generation of sound to external hardware, we could drop the required information rate by a factor of around ten thousand and let a minicomputer assume the task of generating control voltages and timing pulses. This is still not a trivial task [...], but it is certainly made easier [...] by making *all* musical parameters voltage (and therefore computer) controllable. (Buchla 1971)

Buchla had, in fact, been designing such a computer since the late 1960s, which eventually came to be called the Series 500. Indeed, during his time under contract with CBS Musical Instruments, Inc., Buchla diverted much of the financial and technological resources at his disposal towards this project, even recruiting a computer programmer, Doug Crowe, to work on it surreptitiously. The Series 500 used a minicomputer—a new class of small computers that had emerged in the late 1960s—to create a hybrid digital-analog system, in which the digital computer controlled analog components. Complete with a cathode-ray tube monitor and computer keyboard input terminal, the Series 500 stored programs on cassette tapes; when loaded with a tape, in the words of programmer Doug Crowe, "the computer essentially does the patching for you" ("Introducing a preview to the Buchla 500"). The Series 500, however, was not completed until 1975; Buchla only ever made three systems, one of which sold for $60,000.[12]

Between 1971 and 1973, Buchla was stuck between market demand for keyboard instruments and his own fantasy of a music computer. The integrated instruments he had designed thus far, such as the System 101, seemed to satisfy the market, but did not satisfy his vision; the fully computerized Series 500 instrument was years away from completion, and was also extremely expensive to develop. So Buchla began to develop another kind of musical computer that did not require a separate general-purpose minicomputer: one which combined various basic functions of electronic music, just like the System 101, but one which could also store patches as "stored programs" and also allow the user to modify those programs in real time during musical performance. This module became the Model 208 Stored Program Sound Source. Paired with a Model 218 Touch-Controlled Voltage Source, and fitted into a specialized suitcase, these two modules constituted the "Music Easel" system, essentially a small-scale configuration of the Electric Music Box that could function as a "hybrid" analog-digital instrument controllable by both human and electronic computers (Figure 1.2).

The Music Easel was most likely designed and prototyped by the summer of 1973.[13] One of the first musicians to play an Easel was probably the composer and technologist Allen Strange; he and Buchla met in the San Bernardino National Forest during the "Expanded Ear" Festival in April 1973 (Strange 1997). A former student of and collaborator with Pauline Oliveros at the University of California, San Diego, Strange had published a textbook on

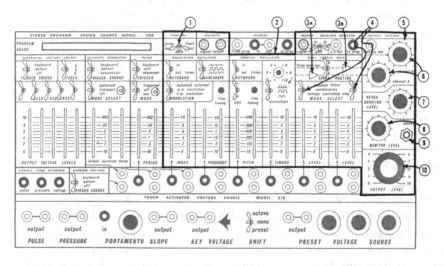

FIGURE 1.2 Diagram of the Model 208 Stored Program Sound Source and Model 218 Touch Activated Voltage Source (Strange 1974, 8).

electronic music in 1972; he had also formed a group of musicians who improvised using portable Synthi AKS synthesizers, named Biome, which he remembered as being inspired by the ecological thought of R. Buckminster Fuller (Strange 1997). Indeed, Biome was performing in the forest when Strange met Buchla; it is very likely that the two connected over their mutual interest not only in systems theory, but also their interest in small, portable electronic musical instruments.

After befriending Buchla, Strange proposed that he would write a user manual for the Music Easel if Buchla agreed to give him one, since he could not afford to purchase one outright. Buchla agreed, and Strange produced a manual entitled *Programming and Meta-Programming in the Electro Organism: An Operating Directive for the Music Easel* (Strange 1974). Directly referencing John C. Lilly's manuscript *Programming and Metaprogramming in the Human Biocomputer*, discussed earlier in this chapter, Strange's manual explicitly describes the easel in terms of its ability to store and recall "programs," each of which could be understood as what he called "instrument definitions." For Strange, these "programs" could themselves be "metaprogrammed" through a unique feature of the Model 208 Stored Program Sound Source: a 28-pin card-edge connector port on the upper-left corner of the instrument, which would allow any parameter to be externally controlled either by a "program board"—a printed circuit board with a matrix of inputs and outputs, connectable by resistors of varying values, or a computer.

The Model 208 Stored Program Sound Source was the core of the Music Easel. Much like with an analog computer, the numerical values, signal path connections, and various possible functions of this module were manually configured on its front panel; connections were made using shorting bars, values were set using slide potentiometers, and the amplitude of its electronic signals were indicated by LEDs. But Buchla and Strange extended this computational metaphor even further: performing with this instrument was conceptualized as executing a "program." And not only was this instrument capable of executing programs, it was also capable of "metaprogramming" those programs in real time. Although metaprogramming was in theory possible with any Series 200 system by simply re-patching the system, the Model 208 made patching more ergonomic: the shorting bars connecting the color-coded banana sockets could be quickly switched. And, more radically, the user could also use the 28-pin card slot on the upper left corner of the instrument to quickly switch programs by plugging and unplugging different program cards, instantly resetting the values and connections on the front panel to new minimums and maximums. (Figure 1.3)

The various functions involved with these programs included two oscillators, one of which could be internally routed to modulate the frequency or amplitude of the other; an envelope generator; a timing pulse generator; a five-stage voltage sequencer; two voltage-controlled gates that also could function

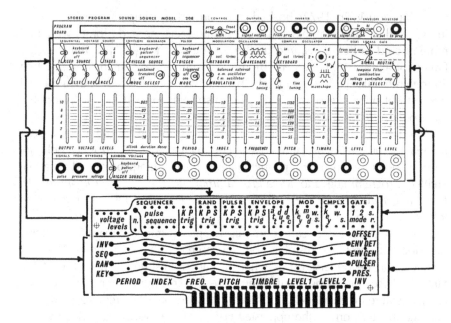

FIGURE 1.3 A diagram of the relationship between the functions of the Model 208 Stored Program Sound Source's front panel and its program cards (Strange 1974, 45).

as low pass filters; a voltage inverter; an internal spring reverb chamber; and outputs from both a headphone amplifier and line level connectors. Each major functional section in the Model 208 was color-coded and could be interconnected through a correspondingly color-coded patching matrix beneath a horizontal band of sliders, which controlled both the individual function's parameters and the amount by which those parameters would be affected by external control voltage signals. The horizontal band of sliders on the 208 echoed the front panel design of the Arp Odyssey, one of the best-selling keyboard synthesizer instruments introduced in 1972; but unlike that instrument, the 208's sliders, and their matrix, were computer-controllable via the program card slot; their positions on the front panel did not necessarily correspond to their actual states.

Buchla and Strange, joined by Strange's spouse Pat Strange quickly combined their interests in systems-thinking and electronic music and created a performance group they called the Electric Weasel Ensemble, sharing their name with the the cartoon "Easel Weasel" created by the Santa Cruz-based skateboard artist Jim Phillips for Strange's 1974 manual. This group functioned with the same fundamental philosophy of Strange's group Biome, which was a "totally closed system" in which the performers were "totally dependent on each other" (Means 1974). The "system" described by

Strange was practical: the ensemble would connect several Music Easels together via their program card slots, or their banana plug connectors, to send control signals back and forth between instruments, or to have each instrument controlled by a single timing pulse generator or sequencer. The ease of this systematic interconnection between instruments and humans operationalized a Lillian computational metaphor: the "interlock" between biological and electronic computers.

Performing with the Electric Weasel Ensemble in the mid 1970s quickly led Strange to describe a new, emergent musical agency that arose between each individual "computer" involved with the Ensemble. In an interview after an April 1974 concert, Strange responded to a question about the group's improvisation:

> The input is improvised. What we do as far as telling the system what kind of things we like is improvised, but the actual synthesizer decides for itself what's gonna get out to the speakers. Out input is improvised but the output, the structure of the piece that the performer listens to, is very carefully selected, not by the performer *but by the system itself.* (Means 1974, emphasis added)

Strange's response echoes Buchla and MacDermed's ecstatic poem: to perform with the easel, he suggests, is to allow the "system"—either a network of Music Easels, or even a single Music Easel—to "design, extend, and variably control the internal workings of an automatic process." Performing with the Music Easel would allow its user to "exist in symbiotic oceans of phenomena and noumena," to glimpse at the expanded consciousness made possible through the externalization and extension of the human biocomputer.

From Counterculture to Cyberculture (to Market Culture)

Although the Music Easel system began appearing in Buchla's catalog in 1973, actually obtaining one was difficult. While Buchla did not shun the consumer market entirely, he kept his customer base small, even as the sales of synthesizers by other manufacturers began to skyrocket that same year. Indeed, in 1974, asked to write an article on "electronic music" for the *Whole Earth Epilog* (an addendum to 1973's *The Last Whole Earth Catalog*), Buchla made absolutely no mention of his own instruments, and instead described keyboard synthesizers manufactured by his ostensible competitors: Moog, E-mu, ARP, EMS, and Korg (Buchla 1974). Even those close to him had trouble obtaining an instrument; Stephen Ruppenthal, a later member of the Electric Weasel Ensemble, remembered that initially asking Buchla for an instrument was like "approaching the great master," likening him to the 17th-century luthier Antonio Stradivari

(Ruppenthal 2021). Rather than a commercial manufacturer of products, able to scale production to make as much profit as possible, Buchla instead operated as what Ezra Teboul (after Nicolas Collins) has called a "silicon luthier" (2017), maintaining strict control over the manufacturing and sales of his instruments.[14]

Although Buchla's business model did not prioritize growth, it nevertheless was functional; Buchla priced his systems such that he would be able to both stay operational and afford to keep up with the rapidly changing technological and manufacturing landscape of the 1970s. Indeed, in a 1991 interview with Woody and Steina Vasulka, Buchla compared his instruments to "Maserati or Rolls Royce" automobiles (Buchla 1991). Although no sales or inventory records are publicly available, Richard Smith, an engineer and technician who worked with Buchla beginning in the 2000s and who has *de facto* custody of many of Buchla's records, has speculated that Music Easel sales were a way for Buchla to be able to finance his continual development of more complex instrumental systems such as the Series 500; he also notes that Buchla continued to revise the modules included in the Easel system throughout the decade, both replacing obsolescing components and redesigning the Touch Controlled Voltage Source to make it compatible with his ever-improving designs for computer control (Smith 2021). Between 1974 and 1977, Smith estimates that Buchla & Associates manufactured and sold an Easel every one to two months, and that 25–30 original Music Easel systems were likely ever built.[15]

By the end of the 1970s, computing technology had changed radically, and Buchla began devoting time to designing his new "Series 300," which extended the computer control of his "Series 200" systems—including the Easel—with completely custom-designed hardware and software, furthering his goal of an integrated biocomputer/music computer interface. Between 1977 and 1987, Buchla designed no less than four entirely new systems, each unique, and each utilizing the latest advances in computer technologies—employing multiple CPUs, customized programming languages, and multiple DACs and ADCs.[16] Throughout this period, a handful of educational institutions and individual musicians continued to use Buchla's Series 200 systems, including the Easel; but this community of users grew increasingly small and increasingly specialized.[17] The Music Easel system became a rarity used by a shrinkingly small number of performers, the most well-known of whom was likely Charles Cohen, a Philadelphia-based improviser who continued to use his a Easel system until his passing in 2017.[18] Buchla himself seems to have anticipated this obsolescence, even as he attempted to design the Easel system as a versatile, computer-programmable, expandable instrument: in his 1974 *Whole Earth Epilog* article, he laments the fact that most "synthesizers"—possibly including the Easel—are "limited-function musical instruments that are particularly vulnerable to early technological and aesthetic obsolescence" (Buchla 1974).

Even as Buchla's early computerized electronic music systems became increasingly obsolete and obscure both in terms of technology and market capacity in the 1980s, they also became increasingly venerated among those who wished to stake out a claim against the commercial and aesthetic explosion of "synthesis" in popular music discourse. Beginning in the 1980s and continuing into the 20th century, Buchla became interpolated into what Frances Morgan, echoing Tara Rodgers (2015), has called the "pioneers narrative" of electronic music, which draws upon colonial metaphors to celebrate rugged individualism, self-reliance, and territorial expansion (Morgan 2017, 1). In an attempted historical intervention into the market dominance of instruments designed by Yamaha, Roland, Korg, and others, Buchla became celebrated as an "unsung hero" (Collins 2006, 192), a foil to the commercial success of another American patriarch: Robert Moog.[19] A 1988 *Keyboard* magazine article by Dominic Milano about "American synthesizer builders," for example, constructs an ad hoc (and inaccurate) history of the "American synth industry" by beginning with "the advent of the voltage-controlled synthesizer developed by Don Buchla in 1963 and made commercially successful in 1965 by Bob Moog" (Milano 1988, 43).[20] In this narrativization, Buchla is framed as the technological innovator, and Moog as the savvy businessman:

> America happens to be the land of the synthesizer—its birthplace divided between East Coast and West Coasts. Its fathers, Don Buchla and Bob Moog, were dreamers whose workbenches full of circuit boards turned the world music community on its collective ear and spawned a two-billion dollar-a-year electronic music industry. (Milano 1988, 43)

Milano positions Buchla as a brother to Moog, two patriarchal figures who "spawned" a successful industry. At the same time, Milano's article is a lament: for him, Buchla, Moog, and their "offspring" had fallen victim to "bad business sense [...] bad luck [...] and bad timing." From Milano's vantage of the late 1980s—even as "synthesizers" reigned supreme in popular music—one of the ostensible patriarchs of synthesis was not getting the recognition he deserved.

Buchla's place in the developing pioneers narrative of American electronic music was also cemented by academics, although for different reasons. Rather than lament Buchla's lack of commercial success or recognition, composers such as David Dunn criticized the industry's apparent ideological steamrolling of Buchla's countercultural politics. In 1992, Dunn wrote:

> "What began in this century as a utopian and vaguely Romantic passion, namely that technology offered an opportunity to expand human perception and provide new avenues for the discovery of reality, subsequently

evolved through the 1960s into an intoxication with this humanistic agenda as a social critique and counter-cultural movement. The irony is that *many of the artists who were most concerned with technology as a counter-cultural social critique built tools that ultimately became the resources for an industrial movement that in large part eradicated their ideological concerns.* Most of these artists and their work have fallen into the anonymous cracks of a consumer culture that now regards their experimentation merely as inherited technical R & D" (Dunn 1992, 45, emphasis added)

Dunn cites an inherent conflict between the "ideological concerns" of instrument designers such as Buchla and the "consumer culture" that profited from Buchla's technological innovations without giving him proper credit.

But what, exactly, were Buchla's ideological concerns? Dunn implies that instruments designed by people like Buchla must have contained some element of social critique. But examining texts written by Buchla in the early 1970s, such as his 1971 poem and his 1974 *Whole Earth Epilog* article, suggests that Buchla's political ideology was not exactly social; it was instead essentially limited to the completely self-contained emergence of an individuated subject within a self-governing system. The scope of such liberation was limited to the individual musician and their instrument, or perhaps a small autonomous group such as the Electric Weasel Ensemble—but it could not be scaled up to a societal level. In the 1974 *Whole Earth Epilog*, Buchla described the ultimate goal of his instruments as allowing "the *musician* (not some design engineer) to establish the relationships between various input stimuli, functional constituents, and potential responses of the instrument" (Buchla 1974). And by focusing on the role of the individual musician as a subject capable of establishing every element within an instrumental system, Buchla set the stage for a new kind of consumerism in the 2000s, in which the acquisition of his modules would implicitly allow for a self-manifestation that was loaded with the overdetermined liberalism of the cybernetic and information-theoretical ideologies of the 1960s.

Buchla in the Analog Revival

In 2002, Trevor Pinch and Frank Trocco published *Analog Days*, their sweeping, celebratory study of Robert Moog and his instruments. In this book, written during what the authors called the "analog revival" (317)—a marked resurgence of interest in analog synthesizers in the 2000s, differentiated from the digital synthesis that had dominated the marketplace since the 1990s—Pinch and Trocco took what previous authors such as Milano saw as a "divide" between Moog and Buchla and transformed it into a dialectic. In this understanding, Buchla and Moog presented two opposing ideological and technical positions that ultimately served to produce the dialectical synthesis of the synthesizer itself. If Buchla had remained an "unsung hero," as Nicholas Collins wrote as late as 2006, his story was beginning to be sung once more.

Buoyed by the renewed interest in his modular electronic music systems, in 2002, Buchla returned to designing modular systems with his "200e" series of modules, interpreted by the industry to be "a feature-packed synthesizer with a staggeringly huge price tag" (Reid 2005). After over a decade of designing novel haptic and kinesthetic controllers for live musical performance—notably Thunder (1989), Lightning (1991), the Marimba Lumina (1999), and the PianoBar (co-designed with Bob Moog in 2002)—Buchla returned to his dream of the 1960s, this time utilizing a combination of analog and digital signal paths, controllable by voltage, MIDI, and Buchla's own proprietary standard for inter-module communication. Priced at tens of thousands of dollars for complete systems, and often over a thousand dollars for individual modules, the 200e system again saw Buchla embracing the business model of a "silicon luthier," designing specialized instruments for a customer base of private institutions and individual connoisseurs. By 2012, Buchla & Associates earned estimated annual gross sales of roughly $900,000.[21]

Even as Buchla was successfully swept up in the "analog revival," the 2000s were a challenging decade for him: after the death of his friend Robert Moog in 2005, Buchla was diagnosed with cancer, and he began to fear for his financial security (Buchla v. Buchla Electronic Musical Instrument LLC 2015). Although he had recovered from his bout with cancer in 2009, in 2011 Buchla made a calculated decision to sell his company—including all of his previous intellectual property—to a group of investors for what he hoped would be a large financial windfall, providing him with the support he needed both to continue to develop new instruments and to pay for anticipated expenses in his later years. The deal was brokered through an Oregon-based businessman named Michael Marans, who had worked with Buchla before; Marans connected Buchla to a group of investors based in Australia who provided the financial backing for this new venture. According to Eric Fox, one of Buchla's main retailers in North America, the Australian investors had "big boy jobs": they were involved with major live audio companies such as JBL and Soundcraft—and Buchla's ostensible competitor, Moog Music, Inc. (Fox 2021).

Rather than take over Buchla's existing business outright, these investors formed a new limited liability corporation, Buchla Electronic Musical Instruments (BEMI), which shielded their substantial assets and separated the enterprise from their other more mainstream endeavors (Buchla 2015). Buchla was hired on a two-year contract as the company's Chief Technology Officer, and in the original contract drawn up for Buchla's employment, BEMI emphasized the importance of Buchla's ongoing projects to his successful continuation as an employee: namely an "8-voice polyphonic synthesizer" and a "location sensing keyboard," two prototype projects which Buchla had been slowly developing over severalyears (Buchla 2015). However, after the first two years of this new arrangement, BEMI changed their priorities, and made the executive decision to focus on another product entirely: the Music Easel.

Michael Marans, the first President of BEMI, was explicit about the reasons he wanted to sell the Easel: it would introduce Buchla's brand into the consumer marketplace and put the brand on equal footing with competitors. As he said in a 2013 interview with Kent, the moderator of the popular ModWiggler internet forum:

> We want Buchla to be in everyone's rigs like Moog would be, or Nord would be, or Kurzweil and any other respected brand. [...] For us, the Easel is the first statement of 'We can actually get something out there. We can mass produce ('Mass Produce'!, ha!) enough of them at a reasonable cost so that the average user can make the choice to get into a Buchla. Then we just want to keep expanding that side of it and not at the expense of the high-end products. (BEMI Interview. April 24th, 2013)

Buchla seems to have gone along with this idea, working with his long-time engineer Joel Davel to revise the Easel's schematics to make it possible to manufacture the instrument 40 years after it was originally designed. According to Davel, Buchla even redesigned the "Electric Music Box" logo on the Model 218 Touch-Activated Voltage Source, perhaps to indicate that this instrument was a new revision—not an obsolete project from 40 years prior (Davel 2021).[22] But the element of the Easel that most excited Buchla to revisit was the same thing that most excited him in 1973: the ability for the instrument to be computer controlled. In 1973, that meant storing individual programs on program cards which could be inserted into the card slot on the Model 208 Stored Program Sound Source. But in 2012, with the rise of ubiquitous computing devices, it could also mean something much more mundane: controlling the Easel from an iPad. And, indeed, Buchla's new design fit easily into the Apple ecosystem: the "iProgram Card," a program card which fit into the 208's card slot and which could be wirelessly controlled via an iPad app. Even as BEMI sought to profit from the renewed interest in Buchla as an alternative to popular brands such as Moog and Kurzweil, Buchla instead focused on leveraging what he perceived to be cutting-edge technology to achieve his idiosyncratic vision of computerized musical performance.

Despite Buchla's slow progress on the iProgram Card, BEMI began to promote the Easel during the National Association of Music Merchants (NAMM) Winter trade show in January 2013, sending Jeffrey Vallier, BEMI's

Director of Engineering, to promote their new product. In a video interview conducted with a customer during the trade show, Vallier discussed the impetus for "re-issuing" the Easel: "we just wanted something accessible for customers [...] at a nicer price point" ("WNAMM13: Buchla Music Easel ReMake"). To attract potential customers eager to experience the aura of Buchla's rarity, Vallier claimed that these instruments would be an "exact replica from Don's original schematics" ("WNAMM13: Buchla Music Easel ReMake").[23]

What resulted from BEMI's focus on selling "exact replica" Easels was unsatisfactory to most parties directly involved. The decision to use 40-year-old schematics and specifications led to immense difficulties: instead of using a single printed circuit board with surface-mount components, for example, the decision to use the original configuration of "mother" and "daughter" cards with hand-soldered through-hole components created cascading problems for both manufacturing and repair. BEMI moved production from Buchla's long-standing facility on Curtis Street in Berkeley to Grants Pass, Oregon; they assigned a new engineer to finish the project, which disconnected the final design from Buchla himself. With BEMI's leadership mostly in Australia, the production, sales, and support for Easels was in disarray. Customers waited for months, or years, to receive their instruments, and when they did, they were rife with technical and manufacturing issues that were difficult to resolve.[24] Despite these issues, between 2012 and 2015, Buchla estimated that BEMI sold approximately 200 Easels, priced at $3,995 each—for a possible gross sales figure of roughly $800,000 (Buchla 2015).

Despite the apparently robust sales of the replica Easels, BEMI apparently did not provide Buchla with sales figures, and paid him much less than he had anticipated. With Buchla dependent on BEMI for income, yet beholden to them to work on projects such as the Easel replica instead of his new designs, interpersonal strife began to develop, taking a toll on Buchla's health. Their relationship deteriorated during the fall of 2013; as BEMI continued to promote the sales of replica Easels (and to discount Buchla's 200e modules), they terminated Buchla's employment.[25] In 2015, Buchla sued BEMI for breach of contract and damages, leading to private arbitration. Throughout this period, the Easel remained the flagship product marketed and sold by BEMI, greatly increasing Buchla's brand recognition in the second decade of the analog revival—even as Buchla himself was ailing.

Indeed, not long after Buchla's lawsuit was arbitrated in mid-2016, Buchla tragically passed away. Although the strife between Buchla and BEMI was well documented on community message boards such as ModWiggler, it apparently did not have a negative impact on BEMI's sales. Eric Fox, then one of Buchla's main North American distributors, estimated that between 2013 and 2019, BEMI sold approximately 400–700, up to possibly 1,000, Music Easels (Fox 2021). In 2020, the company restructured as Buchla USA, with

Fox as its new President, and unveiled a new product: the "Model 208c Easel Command," which had been redesigned by Buchla's long-time engineer Joel Davel. With schematics, specifications, and manufacturing processes no longer beholden to the 1973 standard, Davel modernized the Model 208 module, adding CV control connectors for more functions, and also, for the first time, making the numerical values of the sliders on the front panel both precise and accurate (Davel 2021). According to Fox, Buchla USA was able to sell over 1,000 Easel Commands in a year, setting the stage for the Easel to remain the flagship instrument of Buchla's legacy. The irony of the Easel Command's success was not lost on Fox: as he lamented in 2021, "we want to be cutting edge, but the market is for the past. How do we keep one foot in the past and one in the future?" (Fox 2021).

Conclusion: The Easel as Object of Desire

Fox's lament reflects a paradox at the heart of the Easel's 21st-century appeal: it represents a past defined by its vision of futurity. This is, of course, a common trope surrounding "vintage" keyboard synthesizers—one that often appeals to the *sound* of these instruments, either to timbre in a more abstract sense, or even to what Kodwo Eshun calls the "psychokinaesthetic" effect of those sounds on the human body (1998, 148). For Eshun, the sounds of keyboard synthesizers from 1972–73, even when they are heard decades later, continue to herald futures of varying politics, aesthetics, and socialities; their particular sound "snatches you into the skin you're in, abducts you into your own body, activates the bio-logic of thoughts, encourages your organs to revolt from hierarchy" (1998, 149). Even as these same instruments were also used to make music that did *not* herald such a future—or heralded a future with very different politics—for Eshun, it was the sound of these instruments that produced such revolutionary energy, and interest in revisiting that energy meant interest in revisiting those instruments.

But the Easel is different in kind, rather than simply in degree, from those instruments. Not only was it designed to be an idiosyncratic biocomputer/ electronic music computer interface, rather than a keyboard synthesizer; it was also largely absent from public musical life in North America (and, for that matter, anywhere else on the globe). The sound of the Easel is quite difficult to find on record, beyond archival recordings of live improvisations by Charles Cohen and a handful of recorded performances of the short-lived Electric Weasel Ensemble. The desire for the replica Easel was not primarily a desire for its sound. Instead, I argue that the desire for the replica Easel was a desire to revisit a promise of a future that never really panned out. That future, specifically, promised a new manifestation of creative consciousness through an exponentially expanded realm of computational possibility for electro-sonic signal, flowing recursively through humans and their instruments: the "online realization of ideation" that Buchla and MacDermed envisioned in their 1972 poem.

The degree of computational complexity that Buchla imagined in 1972 is, arguably, now possible: the speed of the "corrective feedback in one biochemical response cycle" has increased by orders of magnitude. Indeed, if the heart of the Music Easel—the Stored Program Sound Source—produces a set of computational possibilities, then those possibilities are realizable on even the most basic computer. This has not been lost on the music software industry: the Music Easel has been replicated in software as a VST plugin by the software company Arturia, offering users complete computerized control over all its parameters.[26] But if the functions of the Easel were easily made possible in software, why was there such an intense market desire for a replica of the Easel as a hardware instrument?

One response might be that this desire was not for functionality, but for the specificities of the Easel's interface. The Easel's functionality, after all, was effected through the knobs, sliders, switches, banana-plug connectors, and capacitive touch-plates of the front panel. Musical instruments are not mere containers for functionality; they are unique physical objects that interact with the human body in specific ways. As Eliot Bates has argued, the musical interfaces of electronic musical instruments are exceedingly variable and complex; he identifies at least six modalities of "interfaciality" that appear in the world of hardware electronic musical instruments (Bates 2021, 178). As Bates writes: "Multifaceted and multisensory experiences of interfaciality [...] are specifically what allow users, makers, and audiences to perceive modular synthesizer technical ensembles [...] as instruments" (Bates 2021, 182). The Easel's specific interface is arguably what makes it a musical instrument—one which cannot be simply recreated as software.

And yet, if the desire for the replica Music Easel was for its interface, why then did BEMI make the logistically and financially difficult decision to attempt to make "exact replicas" of the Easel as it existed in 1973? This decision suggests that the desire for the Music Easel was not one of functionality, nor of interface, but of the unique physical characteristics of the Easel's original components—for example, the idiosyncratic use of resistive opto-isolators (known as "vactrols") in the Easel's oscillators. These components carry with them a unique set of nonlinear behaviors, especially when pushed to extremes; many users believe that this gives analog electronic equipment a special kind of aura, more physical and unpredictable than digitally modeled functionality.[27]

The aura associated with these "replica" Easels also ties into another motivating factor for consumer demand: what guitarist Walter Becker (1996) cheekily named "G.A.S (Gear Acquisition Syndrome)." Indeed, because of the rarity of Buchla's instruments, a niche culture has developed among enthusiasts who build and sell "clones" of Buchla's modules and instruments, such as the Music Easel; there is an active community of builders and sellers who operate on online forums such as ModWiggler.[28] Sales of modules and complete systems built in the 1970s are often conducted in private among anonymous buyers and sellers, and sometimes involve modules which have been questionably sourced from aging electronic music studios and the recently deceased.[29] And because of the rarity of these instruments, public sales often verge on the spectacular.[30] But rarity alone is not enough to create value; nor are the physical characteristics of the 1973 Easel, which make the instrument unstable and fragile; nor is the Easel's interface, easily replicable in software; nor is the Easel's sound, which is also arguably replicable, and prior to 2012, exceedingly scarce on record.

Although a prospective Music Easel user might be motivated by any one of these elements, I argue that the Easel carries with it a surplus of desire: it represents a portal to a world in which human sensation, cognition, desire, and pleasure are laid bare, operationalized, and maximized. Yet even as it promises these things, it is also always-already obsolete, unable to completely fulfill the fantasy of maximization, unboundedness, freedom. Buchla seems to have recognized this as a constitutive feature of not only the Easel, but of his larger systems as well. Indeed, Buchla's constant drive to design, redesign, and refine his instrumental systems throughout his life suggests that his desire was also never fully satisfied. In the poem which announced his larger Series 200 Electric Music Box system, the use of the system is characterized as an iterative cycle, an ongoing encounter that needs to run continuously:

Experimental trial and error / Work and rework mold and / remold conceive and refine / concretize and audition / listen to the results of the idea in real time corrective / feedback in one biochemical / response cycle on line / realization of ideation (Buchla and MacDermed 1971)

Although the Easel can be understood as one stage of this perpetual cycle of technological innovation, its reemergence in 2012 located it as a part of a different kind of cycle: an economic cycle of neoliberalization, of a renewed interest in the idea of a musical instrument as a personal partner in self-discovery. Indeed, as Pinch and Trocco have argued, understanding the resurgence of interest in "analog synthesizers" designed by Buchla and Moog simply as "nostalgia" would be missing an important element: a desire for, as they put it, a musical instrument that is a "living-breathing entity that you can interact with and even fall in love with" (Pinch and Trocco 2002, 319). Pinch and Trocco named the qualities of interactivity, liveliness, and even libidinal desirability: a musical instrument as a projection and externalization of another self.[31]

For Pinch and Trocco's analog revivalist, the concept of "self" was understood as an entity that thought, sensed, and acted in a world defined by signal—electronic, sonic, cognitive, or otherwise. But it was also an entity that was constituted by a *lack* of functionality, which could be addressed through the construction of an "analog" to the self, a complex instrumental system with which one could play, dance, and fall in love. The world in which an "analog synthesizer" becomes an object of desire—indeed, libidinal, perhaps even erotic desire—is a world defined by the cybernetic, or self-governing, flow of signal, in which, perhaps, humans are biocomputers who desire "interlock" with other computers, biological or otherwise. But it is also a world in which knowing and loving one's self is set up as an unattainable goal: one needs to consistently build new and better analogs, more complex models. The Music Easel was one such model, quickly abandoned by Buchla in the 70s, but revived when it became legible as a "synthesizer" that offered an alternative path from the one taken by Moog.

As I hope to have shown in this chapter, the mainstream historical understanding of this path has been marked by two major errors: the first was that Buchla was designing "synthesizers," and the second is that the "countercultural" ethos embedded in Buchla's instruments expressed an anti-consumerist or anti-commercial politics. Buchla's Music Easel emerged from his desire to design computerized instruments that would serve as human biocomputer/analog electronic computer interfaces for the psychedelic discovery of possible new pathways for sonic and electronic signals. The actual sound of the instrument was secondary to the experience of playing *with* it; indeed, playing with it would never yield a consistent sonic result. If the ethos of this kind of play was countercultural, it was also always-already cybercultural: it allowed for emergent performances of self-governance. The scale of this emergent self-governance, however, was by definition limited to the individual human user of a single easel, or at most a small group of 4–5 performers connected together. Unlike Eshun's observation of the psychokinaesthetic power of the *sound* of 70s keyboard synthesizers, the political power of the Easel emerged not from its sound, but from the "special relationship to [the operator's] own perceptual feedback system" when "the listener is also the operator-creator" (Buchla and MacDermed 1971, 13). In other words, you had to actually play with an Easel in order to realize its liberatory potential; but playing with one was only possible for those with money, institutional/social access, or luck.

In this respect, making the Music Easel available to more people—not as an exact replica of Buchla's "Rolls Royce" instrument, but rather as a generic, living, functioning instrument—is a potential first step towards democratizing access to such a liberatory experience. In addition to the Easel, Buchla left a wealth of instrumental designs that is only beginning to be explored.[32] Although Buchla's intellectual property is now owned by Buchla USA, it is still possible for this new company, especially if it continues to employ engineers who worked closely with Buchla, to pursue such an exploration, and perhaps even such a democratization. In late 2021, Buchla USA, in partnership with Eurorack manufacturer Tiptop Audio, announced another "re-issue" of six modules from the same Electric Music Box Series 200 which included the Easel. However, rather than claiming to offer "exact replicas" of modules from the 1970s, Buchla USA chose to redesign them to modern manufacturing specifications, and to make them compatible with the increasingly popular Eurorack standards for size, power, and tuning; initial MSRP prices are significantly lower than Buchla's 200e series modules ("Buchla & Tiptop Audio"), suggesting that Buchla's instruments could become more affordable to more people in the coming years.

By attempting to recreate the sound, functionality, and interface of Buchla's instruments but not the internal components, these new instruments—unlike BEMI's replica easels—come closer to what Buchla in 1982 described as "designing from the outside in": focusing on "what the musician is going to encounter," rather than how the sound is produced (Buchla 1982). Buchla's focus on presence emphasizes that the psychedelic, liberatory potential of his instruments only happens during live performance. The challenge now facing the manufacturers of Buchla's instruments such as the Music Easel is to actually realize that potential: not by reducing them to the category of "synthesizers," and not by selling a fantasy of a countercultural past that was undergirded by the military-industrial complex and a liberalism that excluded those who could not afford it; but rather to bring instruments such as the Easel to as many people as possible, so that its revelations are truly accessible to anyone who wants to play.

Notes

1 Schematics of the Model 208 and Model 218 modules, which comprise the Easel system, are dated March through July 1973 (Drake 2017). See the discussion later in this chapter about dating the design of this system.
2 Although Music Easels were in theory available for purchase to anyone who wanted to buy one, as discussed later in this chapter, issues with manufacturing caused long wait times for instruments, thus potentially limiting the customer base. However, Easels were still available for "pre-order" through multiple retailers, many of whom took deposits while customers waited for delivery.
3 In Buchla's 1973 catalog, the Music Easel system listed for $2,850, which is equivalent to $19,558.57 at the time of this book's publication.
4 *Synthesis*'s first issue, published sometime in late 1970, was largely an advertisement for the content of the second issue, which was promised for January, 1971. However, Buchla and MacDermed's poem (see below) is dated 9 February 1971; it is unclear when Issue No. 2 was actually printed. (*Synthesis*, 1971).
5 The history of the "Black Box" metaphor has been explored by von Hilgers 2011.
6 Lilly's often gruesome experiments on dolphin communication and consciousness, combined with his own self-experimentation with sensory deprivation and

psychedelic drugs, led him by the late 1960s to focus nearly exclusively on developing the concept of the "human biocomputer." Although not his own original concept, Lilly began to understand all human and animal life through the language of computing, and to consider various human-made technologies (such as sensory deprivation tanks, LSD-25, and electronic general-purpose computers) as ways to expand the meta-levels of "programming" with which animal life operated. At a retreat at Esalen in 1969, Lilly met the human potential movement thinker W.W. Harman, and through Harman had his unpublished manuscripts listed in the *Whole Earth Catalog* (Lilly 1987, x). *Programming and Metaprogramming in the Human Biocomputer* was first published by the Julian Press in 1968, available as a mimeographed document from the *Whole Earth Catalog* in 1969, printed in a second edition in 1972, and released as a Bantam paperback in 1974. See Lilly 1987, x.

7 For Lilly, these mechanisms were computers—analog or electronic digital computers, animal and/or human biocomputers, or even extraterrestrial intelligences. As he wrote in a 1966 IEEE article about extraterrestrial communication, "The phenomenon of 'computer-interlock' facilitates mutual model construction and operation, each of the other. One biocomputer interlocks with one or more other biocomputers above and below the level of awareness any time the communicational distance is sufficiently small to bring the interlock functions above threshold levels" (Lilly 1987, 93).

8 Buchla and MacDermed's use of the term "teacher" to describe the mind-manifesting qualities of the Electric Music Box recall another extremely popular book, which Buchla had most likely read: Carlos Castañeda's 1968 *The Teachings of Don Juan,* which describes psychedelics themselves as "teachers."

9 After the development of the Series 200 modules, Buchla's previous designs began to be retroactively called the "Series 100," even though those modules did not bear that name when they were designed. Buchla's 1972 catalog for the Electric Music Box Series 200 is held, among other places, in the Robert Moog papers, #8629. Division of Rare and Manuscript Collections, Cornell University Library. Box 47, Folder 5.

10 Buchla had previously designed an "integrated" module for CBS Musical Instruments during the period in which he was still contractually obligated to work for them. This module, named "Module Cluster 410," was part of a system created by CBS called "Music System 3," which also included a touch-controlled voltage source, an oscillator, and a programmable "pulser" for initiating events. Chris Whitten, an English percussionist, is in the possession of one of the few known remaining "Music System 3" instruments and has uploaded several videos of himself playing this instrument on YouTube.

11 Conversation with Richard Smith, 24 May 2021. NAMM held their "Western Market" show in Los Angeles in March 1971; their larger Convention was held in Chicago in June. Based on the date of Buchla's catalog (November 1971), it is unclear if the instruments contained therein were designed after the June 1971 Convention, or if they were made in anticipation of the following "Western Market" show in March 1972.

12 The general purpose computer used in the system sold to the California Institute of the Arts was likely a Computer Automation, Inc. PDC-216 or PDC-808 ("Clarification on CalArts Buchla 500"), which itself retailed for $6,600 (*Computers and Automation* 1968, 52). As Doug Crowe opined to Charles Amirkhanian in 1975, "I'm not sure how the price of the instrument [$60,000] relates to the cost of producing it" ("Introducing a preview to the Buchla 500").

13 Buchla's paper archives are privately held by Richard Smith; there are no publicly available records that document his early instruments. However, in a 1971/2 catalog of

Series 200 Electric Music Box modules held in the Robert Moog papers at Cornell, the Music Easel is not advertised; the user's manual, copyright 1974 by Allen Strange, was likely written in 1973.

14 Even as Buchla's manufacturing operations grew modestly, he maintained strict control: several people with whom I have talked about their experience working for Buchla have described the intense level of scrutiny with which he approached his employees, often leading to a tense workplace environment.

15 Mark Vail's *Vintage Synthesizers* claims that only 13 Music Easel systems were ever built, although this claim lacks substantiation (Vail 2000).

16 These included the Series 300 (1976), the Touché (1978), the Buchla 400 (1982), and the Buchla 700 (1987). Each of these systems required completely custom-designed hardware and software; unlike music software written for general-use computers with a single CPU, Buchla's systems utilized multiple CPUs, often with idiosyncratic programming languages such as "FOIL" [Far Out Computer Language], co-written with David Rosenboom for the 1978 Touché system.

17 Although some educational institutions such as the California Institute of the Arts continued to purchase Buchla's new systems well into the 1980s, many continued to use Buchla's Series 200 systems as teaching instruments—including the Evergreen State College, Oberlin College, Stony Brook University, and others. Some, also, continued to use their Series 100 systems, as well, including Brandeis University, Harvard University, and Columbia University, which repaired their original 1965 system in 2019.

18 Despite Cohen's long career using the Easel, very few commercially available recordings exist of his performances; beginning in the 2010s, Cohen's recordings from the previous 40 years began to be reissued by experimental music label Morphine Records.

19 A "pioneers narrative" including Buchla is present in several academic histories of electronic musical instruments, written between the 1980s and 2000s: Thomas Holmes (1985), David Dunn (1992), Joel Chadabe (1997, 140), and Nicholas Collins (2006, 192) are notable examples.

20 As I have documented elsewhere (Gordon 2018), Buchla did not begin designing the modules of the Modular Electronic Music System until 1964—the same year that Moog was independently designing his earliest prototypes for what he jokingly called the "abominatron."

21 Although no business records exist in any public archive, this figure was included in the complaint in Buchla's 2015 lawsuit against Buchla Electronic Music Instruments.

22 This became a point of tension: rather than use Buchla's redesigned logo, BEMI recreated the original logo from 1973 in order to keep the Easel "authentic" (Davel 2021). Buchla's new logo can be seen on the prototype Music Easel displayed during the 2013 Winter NAMM conference, but did not appear on production models.

23 This claim was difficult, if not impossible to honor, given the obsolescence and discontinuation of many of the components used by Buchla after 40 years; the schematics of the Easel also changed significantly in the mid-1970s as Buchla kept abreast of the latest developments in electrical engineering components.

24 Joel Davel, the engineer who oversaw the Easel project, suggested that many of these issues resulted from a rushed production schedule and the direct hiring of a new engineer who had had no previous experience with Buchla to finish the project; as a result, early BEMI Easels had issues with faulty power supplies, cross-talk between audio channels, faulty components, and build quality issues that resulted in rattling, loose, and broken controls (Davel 2021).

25 Although no publicly available sales records exist for this period, in his 2015 complaint against BEMI, Buchla estimated that BEMI sold at least 200 Easels between 2013–15; on a ModWiggler forum post made by BEMI in early 2015, BEMI claimed that they had "shipped at least 60 times the amount of Easels that were originally made all those years ago" ("BEMI / Schneidersladen"). If that figure is accurate, then BEMI may have shipped between 1200–1800 Easels, although this is unlikely.

26 "Arturia - Buchla Easel V - Buchla Easel V," accessed 26 January 2022, https://www.arturia.com/products/analog-classics/buchla-easel-v/overview.

27 See Bates, forthcoming, for a discussion of "analog feel."

28 Roman Filippov is the most widely-known maker of Buchla clones; Filippov has launched a business, Sputnik Modular, which uses clones of Buchla's designs in the contemporary Eurorack format.

29 Rumors abound in this gray market; the provenance of Buchla instruments has become a specialty of a few dealers and repair technicians, such as Buchla's long time engineer Rick Smith, who keeps his client list private.

30 In January 2022, for example, the fascist-leaning actor and director Vincent Gallo began an eBay auction for a rare "System 101," discussed earlier in this chapter, with an initial asking price of $197,556.00.

31 Interestingly, these same qualities of interactivity, liveliness, and desirability are shared by a number of objects/products that have also seen a recent resurgence of market activity: AI assistants such as Siri, for example.

32 Although Buchla USA owns most of Buchla's intellectual property, many people have been documenting and studying Buchla's designs—with differing ideological and financial goals. On one end of the spectrum, "clone" manufacturers have sought to reverse-engineer and sell Buchla's designs commercially; some, like Roman Filippov, have even marketed Eurorack-compatible modules that largely replicate Buchla's designs without credit (under the name Sputnik Modular). Others, such as Bob Drake, maintain extensive websites with documentation of various Buchla modules, often linking to documentation such as schematics, PCB design, and build notes made by others within the Buchla enthusiast community. In addition to these resources, the academic-adjacent researchers Chip Flynn and Mark Milanovich have undertaken the "M.E.M.S. Project," which aims to document, reverse-engineer, and replicate unique copies of Buchla's instruments for academic and institutional study (The M.E.M.S. Project).

References

"Arturia - Buchla Easel V - Buchla Easel V." Accessed January 26, 2022. https://www.arturia.com/products/analog-classics/buchla-easel-v/overview.

Bates, Eliot. "'Feeling Analog: Using Modular Synthesisers, Designing Synthesis Communities.'" In *Shaping Sound and Society: The Cultural Study of Musical Instruments*, edited by Steven Cottrell. New York: Routledge, forthcoming.

Bates, Eliot. "'The Interface and Instrumentality of Eurorack Modular Synthesis.'" In *Rethinking Music through Science and Technology Studies*, edited by Antoine Hennion and Christophe Levaux, 170–188. London New York: Routledge, 2021.

Becker, Walter. "G.A.S.," 1996. http://sdarchive.com/gas.html.

BEMI. "BEMI / Schneidersladen." *ModWiggler*, January 13, 2015. https://modwiggler.com/forum/viewtopic.php?p=1766445#p1766445.

Buchla & Tiptop Audio. "Buchla & Tiptop Audio." Accessed January 31, 2022. https://tiptopaudio.com/buchla/.

Buchla, Don. "Series 200 Data Sheet." 1971. Box 47, Folder 5. Robert Moog papers, #8629. Division of Rare and Manuscript Collections, Cornell University Library.

Buchla, Donald. Buchla v. Buchla Electronic Musical Instrument, LLC et al, No. 3:2015cv00921 (US District Court for the Northern District of California 2015).

Buchla, Donald. "Electronic Music Software and Hardware." *The Whole Earth Epilog*, 1974.

Buchla, Donald. "The Horizons of Instrument Design: A Conversation with Don Buchla." *Keyboard*, December 1982. .

Buchla, Donald. "Transcription of Buchla Tape. Interview by Steina Vasulka and Woody Vasulka." December 1991. Vasulka Archive. http://www.vasulka.org/archive/RightsIntrvwInstitMediaPolicies/IntrvwInstitKaldron/61/BuchlaTranscription.pdf.

Buchla, Donald, and Charles MacDermed. "The Electric Music Box." *Synthesis*, January 1971. Box 32. Robert Moog papers, #8629. Division of Rare and Manuscript Collections, Cornell University Library.

Chadabe, Joel. *Electric Sound: The Past and Promise of Electronic Music*. Upper Saddle River, NJ: Prentice Hall, 1997.

Collins, Nicolas. *Handmade Electronic Music: The Art of Hardware Hacking*. New York: Routledge. 2006.

Davel, Joel. "Interview by Theodore Gordon." June 11, 2021.

Drake, Bob. "Historic Buchla Modules." Accessed February 1, 2022. https://fluxmonkey.com/historicBuchla/Buchla200Home.htm.

Dunn, David. "A History of Electronic Music Pioneers." In *Eigenwelt der Apparatewelt: Pioneers of Electronic Art, catalog of exhibition in Ars Electronica 1992 festival, Linz, Austria*, curated by W. and S. Vasulka, 1992. http://www.davidddunn.com/~david/Index2.htm.

Fox, Eric. "Interview by Theodore Gordon." June 10, 2021.

Gardner, James. "The Don Banks Music Box to the Putney: The Genesis and Development of the VCS3 Synthesiser." *Organised Sound* 22, no. 2 (August 2017): 217–227. 10.1017/S1355771817000127.

Gordon, Theodore . "Bay Area Experimentalism: Music and Technology in the Long 1960s." Ph.D. Dissertation, University of Chicago, 2018. https://search.proquest.com/docview/2111350819?accountid=10226.

Grove Music Online. "Moog, Robert A(Rthur)." Accessed September 15, 2021. https://www.oxfordmusiconline.com/grovemusic/view/10.1093/gmo/9781561592630.001.0001/omo-9781561592630-e-1002250264.

Hennion, Antoine, and Christophe Levaux, eds. *Rethinking Music through Science and Technology Studies*. London New York: Routledge, 2021.

Holmes, Thomas B. *Electronic and Experimental Music: History, Instruments, Technique, Performers, Recordings*. New York: Charles Scribner's Sons, 1985.

Introducing a Preview to the Buchla 500, 1975. "Other Minds Audio Archive." http://archive.org/details/AM_1975_07_27.

It's Psychedelic Baby Magazine. "Larry Wendt," February 20, 2018. https://www.psychedelicbabymag.com/2018/02/larry-wend.html.

Jasanoff, Sheila, and Sang-Hyun Kim. *Dreamscapes of Modernity: Sociotechnical Imaginaries and the Fabrication of Power*. Chicago, Il: University of Chicago Press, 2015.

Kent. "BEMI Interview. April 24th, 2013." *ModWiggler*, July 13, 2013. https://modwiggler.com/forum/viewtopic.php?t=88876.

Lilly, John Cunningham. *Programming and Metaprogramming in the Human Biocomputer: Theory and Experiments*. Miami, FL: Communication Research Institute, 1968. http://archive.org/details/programmingmetap00lill_0.

Lilly, John Cunningham. *Programming and Metaprogramming in the Human Biocomputer: Theory and Experiments*. 2nd edition. New York: Three Rivers Press/ Julian Press, 1987.

Means, Loren. "And Now, the Music Easel." *The Berkeley Barb*. May 10, 1974.

Matrixsynth. "Clarification on the CalArts Buchla 500," March 8, 2010. https://www. matrixsynth.com/2010/03/clarifiction-on-calarts-buchla-500.html.

Milano, Dominic. "American Synthesizer Builders: Triumph and Crises for an Industry in Transition." In *Keyboard Presents: Vintage Synthesizers: Groundbreaking Instruments and Pioneering Designers of Electronic Music Synthesizers*, edited by Mark Vail, 17–28. San Francisco: Miller Freeman Books, 1993.

Milano, Dominic. "American Synthesizer Builders: Triumphs And Crises For An Industry In Transition." *Keyboard*, May 1988.

"Modulations- 'Cinema For The Ear.'" Accessed September 29, 2021. http://media. hyperreal.org/modulations/.

Moog, Robert A. "Rôle of Electronics in Rock-and-Roll, or Popular, Music Will Never be the Same Again." *The Journal of the Acoustical Society of America* 44, no. 1 (July 1, 1968): 373–373. 10.1121/1.1970540.

Morgan, Frances. "Pioneer Spirits: New Media Representations of Women in Electronic Music History." *Organised Sound* 22, no. 2 (August 2017): 238–249. 10.1017/S1355771817000140.

NAMM.org. "NAMM Show Location & Date History 1901–2019." Accessed September 1, 2021. https://www.namm.org/library/blog/namm-show-location-date-history-1901-2018.

Pareles, Jon. "Don Buchla, Inventor, Composer and Electronic Music Maverick, Dies at 79." *The New York Times*, September 17, 2016, sec. Arts. https://www.nytimes. com/2016/09/18/arts/music/don-buchla-dead.html.

Pinch, T. J., and Frank Trocco. *Analog Days: The Invention and Impact of the Moog Synthesizer*. Cambridge, MA: Harvard University Press, 2002. http://site.ebrary. com/id/10312781.

Pinch, Trevor. "Why You Go to a Piano Store to Buy a Synthesizer." In *Path Dependence and the Social Construction of Technology*, edited by R. Garud and P. Karnoe, 381–400. New Jersey: LEA Press, 2001. http://pubman.mpiwg-berlin.mpg.de/pubman/ faces/viewItemFullPage.jsp;jsessionid=421972705AF0C038C18C8EF91D8977A1? itemId=escidoc%3A48704%3A5&view=ACTIONS.

Reid, Gordon. "Buchla 200e: Part 1." Accessed September 11, 2021. https://www. soundonsound.com/reviews/buchla-200e-part-1.

Rodgers, Tara. "Synthesis." In *Keywords in Sound*, edited by David Novak and Matt Sakakeeny, 208–221. Durham, NC: Duke University Press, 2015.

Ruppenthal, Steven. "Interview by Theodore Gordon." May 19, 2021.

Sack, E.A., R.C. Lyman, and G.Y. Chang. "Evolution of the Concept of a Computer on a Slice." *Proceedings of the IEEE* 52, no. 12 (December 1964): 1713–1720. 10.11 09/PROC.1964.3472.

Smith, Richard. "Interview by Theodore Gordon." May 24, 2021.

sonicstate. *WNAMM13:Buchla Music Easel ReMake*, 2013. https://www.youtube. com/watch?v=VB7RKpmxLnc.

Strange, Allen. "Interview by Eric Chasalow and Barbara Cassidy." VHS, June 25, 1997. The Video Archive of Electronic Music. https://ensemble.brandeis.edu/hapi/v1/contents/permalinks/i9EMf7z6/view.

Strange, Allen. *Programming and Meta-Programming in the Electro Organism: An Operating Directive for the Music Easel*. Buchla & Associates, 1974.

The Wall Street Journal. "Synthesizers Succeed in Gaining Greater Acceptance in the Music World," December 5, 1974. Box 32. Robert Moog papers, #8629. Division of Rare and Manuscript Collections, Cornell University Library.

Teboul, Ezra. "The Transgressive Practices of Silicon Luthiers." In *Guide to Unconventional Computing for Music*, edited by Eduardo Reck Miranda, 1st ed. 2017, 85–120. Cham: Springer International Publishing: Imprint: Springer, 2017. 10.1007/978-3-319-49881-2.

"The Birth of the Synthesizer | WIRED." Accessed September 28, 2021. https://www.wired.com/2005/06/the-birth-of-the-synthesizer/.

The Expanded Ear: Six Acre Jam, 1973. http://archive.org/details/AM_1973_04_28.

The M.E.M.S. Project. "The M.E.M.S. Project." Accessed January 31, 2022. https://www.memsproject.info/the-facts.

Théberge, Paul. *Any Sound You Can Imagine: Making Music/Consuming Technology*. Music/Culture. Hanover, NH: Wesleyan University Press: University Press of New England, 1997.

Turner, Fred. *From Counterculture to Cyberculture: Stewart Brand, the Whole Earth Network, and the Rise of Digital Utopianism*. Chicago: University of Chicago Press, 2006.

Vail, Mark. *Vintage Synthesizers: Pioneering Designers, Groundbreaking Instruments, Collecting Tips, Mutants of Technology*. 2nd edition. Miller Freeman Books, 2000.

Von Hilgers, Philipp. "The History of the Black Box: The Clash of a Thing and Its Concept." *Cultural Politics* 7, no. 1 (March 1, 2011): 41–58. 10.2752/175174311 X12861940861707.

2

MODULAR SYNTHESIZERS AS CONCEPTUAL MODELS

Jonathan De Souza

Introduction

Knobs, sliders, switches, and buttons, LED lights, inputs or outputs, and wires everywhere... Modular synthesizers are complex technological systems. That much is obvious even to an uninitiated viewer, who might not recognize many synthesizers as musical instruments. Their look often recalls 20th-century science fiction, a spaceship's cockpit or a mad scientist's lab. Many synthesized timbres sound futuristic too. After all, synthesizers involve a distinctive mode of sound production. With strings, winds, brass, and percussion, notes are produced by some physical interaction, typically involving a performer's body—and this grounds standard organological categories (such as the chordophones, aerophones, membranophones, and idiophones defined by Erich von Hornbostel and Curt Sachs).[1] Synthesizers do not easily fit into this taxonomic system. They are a kind of "electrophone."[2] But unlike the electric guitar or the microphone, the synthesizer does not merely amplify acoustic events in the world; instead, it *generates* sound through electricity, creating, mixing, and filtering new signals.[3] Both visually and sonically, then, synthesizers are fundamentally technological.

Yet, if you peek inside a piano—at the hammers, dampers, and overlapping strings, at this concatenation of metal, wood, felt, and wire—you will see another complex technological system. It is easy to take this feat of engineering for granted, but the piano emerged from countless technical innovations.[4] For example, the Steinway company has more than a hundred patents, receiving its first in 1857. And though the piano as we know it emerged during the Industrial Revolution, associations between music and technology are much older. Musical machines were publicly exhibited in

DOI: 10.4324/9781003219484-4

18th-century Europe, including musical clocks, orchestrions, and keyboard- or flute-playing androids.[5] Pipe organs found in churches were preceded by the ancient *hydraulis* (water organ), attributed to Ctesibius of Alexandria (3rd century BCE). And archeologists have discovered flutes that are more than 35,000 years old.[6] These Paleolithic instruments were delicate, sophisticated artifacts: some were made by splitting a piece of mammoth tusk, hollowing it out, and rejoining the pieces. These cultural technologies predate Bronze-Age developments such as writing and the wheel by approximately 30,000 years. So, to quote Gary Tomlinson's work on musical prehistory, "musicking was always technological."[7]

Musical instruments, especially, were always technological. Instruments are tools for making sounds. Like other tools, they require certain skills from their users. Playing a bone flute is not easy. It involves the coordination of breath, embouchure, and fingers, and many beginners would struggle to produce a tone on it. This know-how reveals a complementarity of technology and technique that is usefully described by the word "technics," a general term for technical matters.[8] When instrumentalists practice, they develop new skills, new motor habits for producing sounds. The consistent association between sound and action cultivates what I have figuratively called "a *link between the ear and the hand*."[9] Here I draw on neuroscientific research that shows how instrumental practice enhances auditory-motor coactivation in the brain: when an expert instrumentalist hears music that they can play, activity increases in corresponding motor areas; when an expert makes silent performance actions (for example, "playing piano" on a tabletop), activity increases in corresponding auditory regions. In other words, musicians come to "feel" the sound and "hear" the action. Instrumental mediation cultivates a multimodal awareness that is akin to a kind of synesthesia. So, instrumental technique does not only enable musicians to produce new sounds; it also opens up new ways of perceiving, imagining, and knowing music.

This epistemic function is particularly clear with "instruments of music theory." Such instruments are rarely used in musical performance; instead, they are mainly used to investigate and understand sound. Of these, the monochord is likely the best known. As its English name suggests, the monochord often has a single string, though versions with multiple strings have been common since antiquity (when Ptolemy built monochords with eight strings, and then fifteen). The Latin and Greek names, *regula* (rule) and *kanōn* (measuring rod), show it as a kind of musical ruler. It helps to reveal mathematical and acoustical aspects of musical pitches. For example, dividing the string in half produces a pitch that is an octave higher than the open string; this consonant interval, then, corresponds to the numerical proportion 2:1. As David E. Creese argues, the monochord resembles other scientific instruments like the abacus, compass, and ruler, but also mathematical figures and tables. In a sense, the instrument is an interactive, audible geometry diagram.[10] Similarly, Alexander Rehding relates the monochord and other instruments to the "epistemic things" of scientific knowledge production, as defined by Hans-Jörg Rheinberger.[11] Rheinberger explains, "As a rule, instruments enter as *technical things* into an experimental arrangement … , but they can also become *epistemic things*, if in the course of their use they generate unexpected questions."[12] Like more common instruments, then, monochords and other music-theoretical tools have a dual function: they are used to produce sound but also new questions, modes of investigation, and conceptual models.[13]

Synthesizers blur the line between musical and scientific instruments too. They have long been used to study sound, speech, and music.[14] This is true of large mid-20th-century synthesizers such as the RCA Mark II, developed in the 1950s.[15] But arguably, the earliest synthesizer was an "apparatus for the artificial construction of vowels," created in 1858 by the German scientist Hermann von Helmholtz.[16] Helmholtz's device used eight tuning forks, tuned to a harmonic series above a low B-flat. Each tuning fork was activated by an electromagnet, producing a sustained pure tone; its tone, moreover, could be amplified by a hollow resonator when the resonator's lid was opened by pressing a key. "By this means," Helmholtz explained, "it is possible to form, in rapid succession, different combinations of the prime with one or more harmonic upper partials having different degrees of loudness, and thus produce tones of different qualities (*Klänge von verschiedener Klangfarbe*)."[17] By combining different tuning forks at different volumes, Helmholtz imitated vowel sounds and instrumental timbres through additive synthesis. Clearly, this groundbreaking synthesizer was primarily a scientific or music-theoretical instrument, supporting Helmholtz's studies of the physics, physiology, and psychology of sound. In this sense, synthesizers' associations with a mad scientist's lab are apt; they can be understood as scientific instruments that have been co-opted for artistic use.

How can we approach the synthesizer as an epistemic tool? It would be possible to study people who build and use modular synthesizers. What kinds of technical know-how do they develop? How does this affect the ways that they imagine and talk about sound and music? For example, synthesists might be sensitive to subtle variations in tone quality, insofar as those variations are made manipulable by their technology.[18] Yet the present chapter takes a different approach. It juxtaposes three case studies from the 1970s and 80s, which involve thinkers from organology (Herbert Heyde), philosophy (Gilles Deleuze and Félix Guattari), and media theory (Friedrich Kittler). In each case, modular synthesizers stimulated new ideas about music and sound, perspectives on musical ontology and historical musical practices. This investigation approaches the synthesizer as a kind of theoretical metaphor, which can support broader conceptual models.[19]

Considering metaphors as well as physical objects recalls Ludger Schwarte's argument that instruments embody broader processes of instrumentalization. "Instrumentalization," Schwarte writes, "does not only mean a certain making-available of the world, a setting up of possibilities of action, but above all a seeing-as, the intuition of alternative realities, whose foreshadowing is symbolized in the instrument. To grasp what instrumentalization does, we should not consider objects, but rather technical operations … . Instrumentalization is the organization of operative possibilities."[20] In addition to physical tools, then, instrumentalization might incorporate bodies, symbols, languages, and theories.

Heyde's Modular Organology

According to the organologist Herbert Heyde, musical instruments have a "dual nature."[21] Their materials and modes of vibration can be understood in terms of physics and engineering. Yet they also have historical and cultural significance. Because organology examines both technical and cultural aspects of instruments, a blend of scientific and humanistic approaches defines the field. The Hornbostel-Sachs system leans toward science. It categorizes instruments in terms of the physics of sound production, an approach that emerged in the 19th-century organology of the Belgian instrument maker, acoustician, and museum curator Victor-Charles Mahillon.[22] Hornbostel and Sachs wished to ground their classification in science rather than culture because a system that "suits one era or nation may be unsuitable as a foundation for the instrumental armory of all nations and all times."[23] Though the Hornbostel-Sachs system is still well-known today, this taxonomic approach is a product of the 19th century, when acoustics influenced the public imagination and when Europeans increasingly encountered non-Western instruments because of colonialism.[24] Hornbostel and Sachs claimed that their system "is capable of absorbing almost the whole range of ancient and modern, European and extra-European instruments."[25] They were interested in history and culture too, but acoustics offered a kind of objectivity, a way to universalize organology.

In *Grundlagen des natürlichen Systems der Musikinstrumente*, Heyde departed radically from this tradition.[26] The book was published in 1975 by the Central Institute for Music Research in the German Democratic Republic. Instead of acoustics, Heyde's universalizing scientific reference point was cybernetics. Cybernetics developed in the mid-20th century as an interdisciplinary approach to self-governing systems (the term "cybernetics" derives from a Greek word for "steersman").[27] It studied "control and communication in the animal and the machine."[28] That is, it emphasized the systems' abstract structure, applying the same analytical techniques to technological, physiological, psychological, social, or economic systems. In this domain, Heyde's primary source seems to be *Ganzheit und Entwicklung*

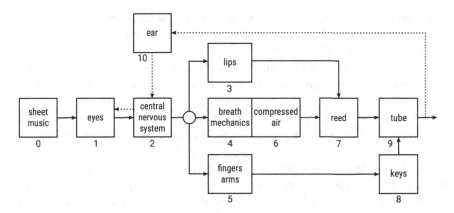

FIGURE 2.1 Heyde's circuit diagram for oboe playing.[36]

in kybernetischer Sicht (1967) by the neo-Marxist economist Oskar Lange.[29] And instead of non-Western instruments, the instruments that catalyzed Heyde's thinking were electrophones. Here he repeatedly cites *Elektrische Klangerzeugung: Elektronische Musik und synthetische Sprache* (1949) by Werner Meyer-Eppler.[30] Combining these perspectives, Heyde developed an original approach—what he called "an organological interpretation of central concepts of systems theory."[31]

For Heyde, every musical instrument system turns energy into sound. In other words, the energy is the system's input, and the sound is its output. Instruments, then, resemble the communication systems theorized by the information theorist Claude Shannon.[32] For Shannon, every communication system transmits some kind of signal (e.g., numbers, letters, sound waves). Each element in the system transforms the signal and passes it along the chain. In a well-functioning system, a relatively intact version of the signal should make it from the source to the destination, despite the interference of noise.

Similarly, with musical instruments, Heyde is interested in "the dependence of the outputs on the inputs" and the overall system's "transmission behavior."[33] And like Shannon, he uses diagrams to represent signal transmission. For example, the diagram reproduced in Figure 2.1 supports his discussion of the transmission behavior of oboe-playing:

> The arrow indicates the course of the signal flow and establishes the structure of the system. Blocks 7, 8, and 9 symbolize the active elements of the instrument itself. All others represent psychophysical body functions.
>
> The reed (7) converts the energy flow—which can be influenced by the breath mechanics (4)—into acoustic energy. The reed (7) in its function can accept several states from the lips of the blower (3) ... . Finally, the

coupled system involves the oboe tube (9), whose effective length is determined through a system of keys (8), which are operated by [the fingers and arms] (5).

All body functions (3), (4), (5) are controlled by the central nervous system (2), which can receive information from a sheet of music (0) by means of the eyes (1).

The whole system can be characterized as an information processing system ... . The system's output is the tonal figure of the sheet music.[34]

A signal encoded in musical notation is transmitted via bodily movements to the instrument, where it is finally output as sound. In keeping with cybernetic principles, feedback loops—e.g., from the sound back to the central nervous system via the ear—allow the system to be self-controlling, to use its output as an input that regulates its behavior. Thanks to this feedback, Heyde explains, "player and instrument represent a control circuit."[35] Both human and technological components can participate in signal transmission, input and output.

In every instrumental system, some energy source or "activator" (Anreger) must be coupled with a "transducer" (Wandler).[37] The energy source might be a player's breath, muscular exertions (as in piano-playing), bellows (as in pipe organs), or an electrical power supply (as in electric organs). But the electrical power supply seems to be the prototypical activator: an unlimited, always ready standing reserve of energy into which a wide range of devices can be plugged.[38] That is, Heyde's cybernetic organology imagines the human body as an energy source, understood in terms of electricity, a technological metaphor. Transducers are also typically understood as electrical devices. They are used in various engineering contexts, though familiar examples are related to music and sound. For example, a microphone is a transducer that converts variations in air pressure into electrical signals.[39] Similarly, an electric-guitar pickup uses magnets to sense the vibrations of metal strings and translate them into electrical signals. With Heyde, the concept of transducer includes many non-electrical elements: "all active elements that have the function of converting mechanical into acoustic energy belong to the class of transducers (strings, membranes, reeds, etc.)."[40] The transducer—as any element that produces audible vibrations—transcends the acoustic differences that define earlier organological classification. Heyde follows a standard cybernetic strategy here, treating strings, drumheads, and reeds as black boxes.[41] He sets aside their physical differences to cut across traditional categories and reveal underlying functional similarities. Moreover, this mode of seeing-as is inspired by electrical instruments, by new musical technologies.

Together the activator and transducer function like an oscillator or noise source. Other elements in the system—what Heyde calls "carrier

TABLE 2.1 Heyde's instrumental modules[45]

Label	German term	English translation	Examples
A	Anreger	Activator	Muscles, lungs, bellows
V	Vermittler	Mediator	Violin bow, guitar pick, piano keys
W	Wandler	Transducer	Strings, membranes, reeds
ZW	Zwischenwandler	Intermediate transducer	Electric guitar pickup
M	Modulator	Modulator	Electronic organ tone filter
Ampl	Amplifikator	Amplifier	Loudspeaker
R	Resonator	Resonator	Violin body, piano soundboard
K	Kopulator	Coupler	Flute tube
[circle]	Kanal	Channel	Violin bridge, flutist's throat
St	Steuerelement	Controller	Fingers, keys, switches

elements"—modify the signal produced by this pair through amplification, modulation, filtering, or control.[42] He defines eight optional carrier elements for a total of ten instrumental functional elements. Heyde discusses each functional element in detail, giving examples and variations. For example, he discusses the modulator as "a device for changing vibrations" received from various transducers. "Its task is the analytic or synthetic formation or alteration of tones," and it may involve either linear or non-linear distortion.[43] For present purposes, there is no need to repeat all of these details, and the summary in Table 2.1 will suffice. Again, Heyde's key move involves modeling non-electric elements (whether instrumental or bodily) as technological processors: the violin bow becomes a mediator, which adapts the activating energy for the instrument, and the flutist's tongue and palate become a control switch.[44] Acoustic instruments and the bodies that play them are reconceptualized via synthesizers and other electrophones.

Ultimately, Heyde presents a diagram that combines all ten instrumental functions, reproduced in Figure 2.2. "The system of all musical instruments may be incorporated into a total system [Ganzsystem] … . Reciprocally, each instrument may be represented as a subsystem of the total system … . The whole instrumentarium can be built from ten different elements."[46] Heyde's Ganzsystem reveals his universalizing goals, but also a commitment to modularity. An instrument's structure is defined by the selection of functional modules and "the energetic, material, and informational couplings" among them.[47] When instruments are categorized in this way—through information theory instead of physics—new organological families emerge. Heyde juxtaposes linear, parallel, and converging systems, which have different abstract, informational structures, different patterns of signal transmission.[48] For example, pianos, violins, and dulcimers are grouped together not

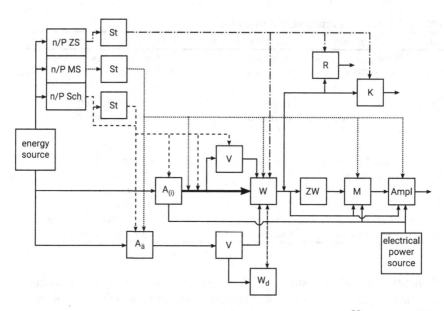

FIGURE 2.2 Heyde's general musical instrument system.[50] For abbrevia-
tions, see Table 2.1. Additionally, Heyde adds subscripts to a
few module labels: A_i stands for internal activator (*Anreger für
innere*), $A_ä$ for external activator (*Anreger für äußere*), and W_d for
double transducer (*Doppelwandler*).

because they all use strings but because they are all "polysystemic linear
systems" with multiple inputs and multiple outputs.[49] For Heyde, then, a
musical instrument is not an organic whole but a system of connected but
theoretically separable modules. So, I would argue that the synthesizers
discussed by Meyer-Eppler served as epistemic things for Heyde, even if he
had little to no experience playing them. Synthesizers stimulated unexpected
organological questions and supported a new conceptual model in which
oboes, drums, flutes, and violins can be seen as networks that, when plugged
into a power source, can generate, modify, and transmit sound—in short, as
"acoustic modular synthesizers."

Deleuze and Guattari's Thought Synthesizer

Synthesizers also ground conceptual models in the writings of Gilles Deleuze
and Félix Guattari. Perhaps their mention of the synthesizer is unsurprising, as
their collaborations—most notably the two volumes of *Capitalism and
Schizophrenia*, *Anti-Oedipus* (1972) and *A Thousand Plateaus* (1980)—feature
an unusually eclectic range of philosophical, psychoanalytic, political, scientific, literary, and artistic references. Nonetheless, it is striking that Deleuze
concluded a 1980 seminar on Leibniz and Kant by stating, "Okay, let's assume

we're really in the age of the synthesizer today … . Isn't there any synthesizer in philosophy, which like a musical synthesizer, is a philosophical synthesizer?"[51] This question explicitly invokes the synthesizer as a model for contemporary thought. But how did the synthesizer inform Deleuze and Guattari's thinking?

According to Martin Scherzinger, their conception of the synthesizer was not primarily based on the actual instrument or its use by new wave bands such as Duran Duran and Spandau Ballet. Instead, Scherzinger argues that their synthesizer is modeled after a "hyperinstrument" imagined by the composer Pierre Boulez, which would involve "electronic sinusoidal sounds" and "conjugations of existent instruments."[52] Just as Boulez's imaginary synthesizer would free sound from its traditional notational and rhythmic limitations, Deleuze and Guattari's abstract machine would open up new possibilities for thought. As such, Deleuze and Guattari's synthesizer would respond to earlier musical modernism, more than contemporaneous popular music; it would reflect an aesthetic fantasy, more than a technological reality.

Though Boulez is an important musical interlocutor in *A Thousand Plateaus*, I would suggest that Deleuze's understanding of the synthesizer was shaped by his student Richard Pinhas. In 1974, Pinhas completed a Ph.D. in philosophy—and his electronic-rock band Heldon released their first album, *Électronique Guerilla*. Heldon's music combined the Moog synthesizer with electric guitar and drums, often on dense, extended prog-style instrumentals. As independent musicians, they did not achieve mainstream success, yet they are widely respected by music critics. For example, when their 1977 album *Interface* was reissued in 2004, a reviewer for *Pitchfork* stated, "To many (including me), Heldon and Richard Pinhas are considered building blocks for whole schools of experimental rock music."[53] From 1976 to 1981, Pinhas "was a session musician, working 300 days a year playing the synthesizer in these big analog studios."[54] During this time, he published an article on synthesis titled "Input, Output," which is cited in *A Thousand Plateaus*.[55] As archival recordings show, Pinhas was also an active participant in Deleuze's seminars at the University of Paris between at least 1977 and 1987.[56] These recordings also illuminate Deleuze's knowledge of synthesizers: in a May 1981 seminar on "Painting and the Question of Concepts," he briefly discussed technical differences between frequency modulation and amplitude modulation and between analog and digital synthesizers. At the same time, Deleuze emphasized that he was covering only "the basics" and deferred to Pinhas's detailed practical knowledge. He started by saying, "see Richard Pinhas for any further comments or corrections," and ended with a direct question to Pinhas: "Do you see something to add here?"[57] This suggests that Deleuze and Pinhas were engaged in ongoing conversations and that, *pace* Scherzinger, the philosopher's interest in synthesizers was inspired, to some degree, by contemporaneous technology and musical practice.[58]

In certain ways, the image of the synthesizer in *A Thousand Plateaus* echoes Heyde's *Grundlagen*. Like Heyde, Deleuze and Guattari highlight connections among distinct elements, abstract diagrams, flows of energy and information, and the cyborg-like interplay of biological, technological, and symbolic elements:

> A synthesizer places all of the parameters in continuous variation, gradually making "fundamentally heterogeneous elements end up turning into each other in some way." The moment this conjunction occurs there is a common matter. It is only at this point that one reaches the abstract machine, or the diagram of the assemblage. The synthesizer has replaced judgment, and matter has replaced the figure or formed substance. It is no longer even appropriate to group biological, physico-chemical, and energetic intensities on the one hand, and mathematical, aesthetic, linguistic, informational, semiotic intensities, etc., on the other.[59]

Yet additional themes emerge here: namely, judgment and continuous variation. And these will, respectively, clarify the synthesizer's philosophical and musical significance for Deleuze and Guattari.

As they state elsewhere in the book, Deleuze and Guattari are thinking about a specific kind of judgment, what Immanuel Kant calls "synthetic a priori judgments."[60] Kant defines synthesis as a kind of cognitive activity, "the action of putting different representations together with each other and comprehending their manifoldness in one cognition."[61] Before Kant, philosophers distinguished between analytic and synthetic judgments. Analytic judgments involve elements that are related by definition: for example, "red is red," "red is a color," or more generally, "A is A." These analytic statements are logically necessary. They are true a priori. By contrast, "the triangle is red" is a synthetic judgment. It puts together—that is, synthesizes—different concepts. This judgment is not necessarily true for all triangles. Instead, it is an empirical observation, made from experience, after the fact (a posteriori). Kant complicated the analytic/synthetic opposition by introducing synthetic a priori judgments, which would be based in experience *yet also universally true* (e.g., "the three angles in a triangle are equal to two right angles"). Deleuze explains this idea in his seminar on Kantian synthesis and time:

> The true a priori synthesis is not between concepts like the empirical synthesis. The true a priori synthesis goes from the concept to the spatio-temporal determination, and vice-versa. That is why there can be a priori syntheses between two concepts, because space and time have woven a network of determinations which can make two concepts, however different they are ... form necessary relations with each other.[62]

So, while synthetic a posteriori judgments combine different concepts directly, synthetic a priori ones connect concepts to or through underlying conditions of space and time, which can make them universally true.[63]

Deleuze appreciates how Kant's synthetic a priori challenged existing philosophical categories but does not aspire to universal truth or logical necessity. For Deleuze and Guattari, this approach to philosophy is outdated, much like 18th-century approaches to painting or music: "Philosophy is no longer synthetic judgment; it is like a thought synthesizer functioning to make thought travel, make it mobile, make it a force of the Cosmos (in the same way as one makes sound travel)."[64] Skeptical readers might feel that Deleuze and Guattari are playing fast and loose here; they might argue that the word synthesis means very different things in 18th-century philosophy and 20th-century music technology. In response, I would point out a historical link, which I have discussed elsewhere, between 19th-century neo-Kantianism and the development of Helmholtz's tuning-fork synthesizer.[65] Moreover, I would emphasize that this terminological coincidence supports a novel cross-domain mapping for Deleuze and Guattari, and such mappings often lead to new conceptual models. In cross-domain mapping, Lawrence Zbikowski notes, "the propositions and structure of a model framed relative to one domain are applied to another domain; if the process is successful (if it allows us to structure effectively our knowledge of some aspect of the target domain), a new conceptual model is born."[66] In both senses of the word, synthesis involves putting together heterogeneous elements. On the Kantian side, this is a cognitive act that produces conceptual blends; on the musical side, a technology that generates strikingly novel sounds. The hybrid "thought synthesizer," then, would be a cognitive technology that generates strikingly novel concepts. And like the style of writing that Deleuze and Guattari cultivate in A Thousand Plateaus, it would be both philosophical and aesthetic, pursuing creative, open-ended play of ideas and affects. The metaphor supports a new conceptual model.

To further understand this new conceptual model, it is important to consider how, for Deleuze and Guattari, the synthesizer opens up distinctive relations to sound:

> By assembling modules, source elements, and elements for treating sound (oscillators, generators, and transformers), by arranging microintervals, the synthesizer makes audible the sound process itself, the production of that process, and puts us in contact with still other elements beyond sound matter. It unites disparate elements in the material, and transposes the parameters from one formula to another.[67]

The synthesizer, then, can generate material by applying abstract formulas to different sonic parameters. This process recalls the RCA Mark II synthesizer, which used punched paper rolls that could be realized in multiple ways, depending on the instrument's settings. For example, in his *Composition for Synthesizer* (1961), which was created on the RCA Mark II, Milton Babbitt set up a correspondence between pitch-class intervals and rhythmic intervals.[68] While this abstract transposition goes beyond sound, the synthesizer also draws attention to sound—not only to timbre but also to the processes of putting sounds together and breaking them apart.

For Deleuze and Guattari, these sonic tendencies are essential to electronic music but also anticipated in 19th-century and early 20th-century composition. "The irruption of brass instruments" in the music of Hector Berlioz and Richard Wagner created "a totally new problem of orchestration, orchestration as a creative dimension, as forming part of the musical composition itself, where the musician, the creator in music becomes an orchestrator."[69] As Deleuze puts it in a 1979 seminar (responding to "a beautiful presentation" from Pinhas), "[Edgard] Varèse straddles the great Berlioz-Wagner tradition of brasses and electronic music."[70] Referencing Odile Vivier's book on Varèse, Deleuze highlights a moment in Varèse's *Ionisation* (1929–30) for percussion ensemble, where the piece suddenly presents only "metallic sonorities."[71] Ideas from the seminar reappear, condensed, in *A Thousand Plateaus*: "Varèse's procedure, at the dawn of this age, is exemplary: a musical machine of consistency, a *sound machine* (not a machine for reproducing sounds), which molecularizes and atomizes, ionizes sound matter, and harnesses a cosmic energy. If this machine must have an assemblage, it is the synthesizer."[72] Here Deleuze and Guattari used synthesizers to reimagine the history of orchestration.

Similarly, they are inspired by the synthesizer's "microintervals." Knobs, sliders, and ribbon controllers allow continuous manipulation of various sonic parameters; they are not limited to the discrete steps of a keyboard.[73] They open up sonic spaces that are smooth, not striated.[74] Again, this supports reinterpretation of the musical past. For Deleuze and Guattari, chromaticism—the move away from diatonic scales, embracing the chromatic scale—is defined by a tendency toward increasingly small microintervals and, ultimately, a smooth pitch continuum. When this tendency extends to all aspects of sound, "a generalized chromaticism" redefines the relation between musical content and form:

> When development subordinates form and spans the whole, as in Beethoven, variation begins to free itself and becomes identified with creation. But when chromaticism is unleashed, becomes a generalized chromaticism, turns back against temperament, affecting not only pitches but all sound components—durations, intensities, timbre, attacks—it becomes impossible to speak of a sound form organizing matter; it is no longer even possible to speak of a continuous development of form.

Rather, it is a question of a highly complex and elaborate material making audible nonsonorous forces. The couple matter-form is replaced by the coupling material-forces. The synthesizer has taken the place of the old "a priori synthetic judgment," and all functions change accordingly. By placing all its components in continuous variation, music itself becomes a superlinear system, a rhizome instead of a tree, and enters the service of a virtual cosmic continuum of which even holes, silences, ruptures, and breaks are a part.[75]

From this perspective, Beethovenian developing variation is creative but essentially linear. It remains oriented toward large-scale tonal and formal *teloi* and recalls the impulse toward universal truth and logical necessity in the work of Kant (Beethoven's older contemporary). By contrast, the synthesizer's continuous variation would be superlinear. Its music would instantiate a distributed, open-ended, multidimensional network. This conceptual model can support novel interpretations of music and music history: for example, they argue that the early 20th-century shift from tonality to atonality is merely a "pseudobreak," insofar as it maintained the hegemony of twelve-tone equal temperament, discrete rather than continuous. Responding to the age of the synthesizer, then, Deleuze and Guattari explore the interplay of music and philosophy, present and past.

Kittler's Wagnerian Machine

The media theorist Friedrich Kittler built an analog synthesizer in the early 1980s, a metallic machine with rows of switches and knobs.[76] Decades later, during a lecture about oscillators, transistors, amplifiers, and related technologies, Kittler discussed the development of the Moog synthesizer. He sketched circuit diagrams and charts of linear and exponential functions on a flipchart and mentioned that he once had "the chance and the honor" of meeting Robert Moog himself.[77] So, synthesizers were clearly important to Kittler. He even suggests that, in popular music since the 1960s, these musical technologies are more important than lyricists or composers. Since lyricists and composers (both authors of musical texts) were irrelevant "in the space of *sound*," "it would be much more fitting to list the circuit diagrams of the facilities and the model numbers of the synthesizers employed (as occurs on the cover of [Pink Floyd's] *Dark Side [of the Moon]*)."[78] Kittler believed that these technical details were important for understanding music in terms of sound and media.

As Rehding puts it, "The basic principle on which Kittler's media theory builds could not be simpler: data goes in, gets processed, and comes out."[79] The basic model recalls Shannon's communication systems.[80] But media, for Kittler, are not limited to gramophones, photographs, telephones, telegraphs, and other modern technologies or even to historical media like print. Instead, he radically extends the concept, arguing, for example, that cities are media.[81] This technology-inspired framework supports Kittler's perspective on sound media. How do different musical media store and transmit sound? Staff notation, he emphasizes, is symbolic. It represents pitches only in terms of discrete letter names. In other words, this medium puts sound through an alphabetical filter, encoding music as text and thereby excluding many aspects of sound.[82] By contrast, recording technology can capture traces of real sound waves: "The phonograph does not hear as do ears that have been trained immediately to filter voices, words, and sounds out of noise; it registers acoustic events as such."[83] Kittler builds on this difference between writing and sound as he distinguishes between the arts and mass media: "The arts ... entertain only symbolic relations with the sensory fields they take for granted. In contrast, media relate to the materiality with—and on—which they operate in the Real itself."[84]

Yet, perhaps surprisingly, Kittler traces the roots of aesthetic media to 19th-century music: "Classical European instruction in musical harmony sought to master the incessant noise all around by means of form and binary coding (major/minor, consonance/dissonance, etc.). Romantic music was—and stayed—a process of decoding such oppositions."[85] Kittler is particularly fascinated by Richard Wagner, particularly in his 1986 essay, "World-Breath: On Wagner's Media Technology." Kittler calls Wagnerian music drama "the first mass medium in the modern sense of the word," and a "monomaniacal anticipation of the gramophone and the movies."[86] Yet Kittler also saw Wagnerian music drama as a monomaniacal anticipation of the synthesizer.[87] Arguably, the synthesizer is an even better model here: audio recording and videography typically require some dynamic input from the world (sound waves, light waves); by contrast, synthesizers and music dramas might be understood to generate their own signals.

Kittler adduces examples from several Wagnerian works. In the prelude to *Das Rheingold* (the opening of Wagner's four-part *Ring* cycle), the orchestra functions like Helmholtz's tuning-fork synthesizer. For around four to five minutes, it gradually builds up a harmonic series above a fixed fundamental. In Kittler's view, "The *Rhinegold* prelude, with its infinite swelling of a single chord, dissolves the E-flat major triad in the first horn melody as if it were not a matter of musical harmony but of demonstrating the physical overtone series. All the harmonics of E-flat appear one after the other, as if in a Fourier analysis; only the seventh is missing, because it cannot be played by European instruments."[88] The prelude neutralizes tonal progression and draws attention to timbre, dynamics, and texture, to a process of additive synthesis. Accordingly, the instruments of the orchestra can be imagined as coupled oscillators, which together produce one expansive, rich, dynamic sonority. Moreover, following Helmholtz's emphasis on the interplay of physics, physiology, and psychology, the sound is understood to affect listeners through physical sensation. As such, Kittler concludes, "Wagner's musico-physiological dream at the outset of the tetralogy sounds like a historical transition from intervals to frequencies, from a logic to a physics of sound."[89] It seems likely that Wagner would have sanctioned Kittler's interpretation: after all, as the composer once wrote to Helmholtz, "for us, harmony turns to sonority."[90] All of this links Wagner's prelude to the historical development of the synthesizer.

While Wagner's orchestra at the beginning of *Das Rheingold* is a source module, it can also serve as a processor module. In Act I of *Lohengrin*, the noblewoman Elsa has been falsely accused of murder and prays for a champion to come to her aid. In Kittler's reading, the orchestra amplifies Elsa's prayer, transmitting it across great distances to call Lohengrin, the mystical knight she has seen in her dreams. Again, this conceptualizes the music drama (and, by extension, the medieval legend) in terms of modern media technology: Elsa generates a signal, an input for the orchestra that processes it and relays it ultimately to Lohengrin. "To make Elsa's barely audible laments into *sound* echoing far away, the orchestra, especially the brass instruments, must take them up."[91] The orchestra as amplifier is part of a modular system and is concerned with sound rather than traditional musical categories.

In addition, the sounds produced in music drama, like those produced by synthesizers, can blur the boundary between tone and noise. For example, at the beginning of Act II of *Tristan und Isolde*, Wagner juxtaposes environmental sounds and horn calls.[92] According to Kittler, "the textual oscillation between the sound of nature and the instruments of the orchestra—between random noise and a hunting signal—corresponds to two equally illiterate horns playing C major and F major at once."[93] By incorporating noise and this supposed illiteracy, "Wagner's new medium, *sound*, exploded six hundred years of literal and literary practice."[94] Music drama, in this view, would be "the first art apparatus capable of reproducing sensory data as such."[95]

A note of caution: Kittler's grandiose claims about Wagner are questionable. For example, thunder effects—which are used in many earlier plays and dramas, including Wolfgang Amadeus Mozart's *The Magic Flute*—already recreate unfiltered, natural noises in the theater. Similarly, Mozart's *Don Giovanni* plays with a certain illiteracy: the finale to Act I features three dances, in three different meters, played simultaneously. In addition to the rhythmic and metric incongruities, Mozart adds intentional harmonic mistakes here, "which cause the musical structure to collapse like a house of cards."[96] Moreover, Kittler's reading of the horn calls in *Tristan* seems to be based on an error. The visual appearance of the score might suggest "C major and F major at once," with the timpani holding a low F, and the horns outlining notated C major and G minor triads. But the horns are transposing instruments: their notated Cs sound as Fs, notated Gs as Cs, and so forth. In this passage, then, the parts create a dominant ninth chord on F, which is entirely consistent with the key signature (B-flat major). There is no C-major sound and no illiterate tonal conflict here. Finally, Gundula Kreuzer argues that Kittler relies on Wagner's ideals, as expressed in the composer's writings, more than the mixed results of their actual material realization. Kreuzer writes that "[Kittler's] Wagner is the PR campaigner of the 1863 preface to the *Ring* libretto and the 1870 'Beethoven' essay (that is, the Wagner of post-Schopenhauerian theorizing), not the exhausted Wagner of 1876 who realized all too late the shortcomings of his dream-turned-brick-and-steel-reality."[97] So, Kittler's take on Wagner and media technology should not be uncritically accepted.

Nonetheless, the goal of the present chapter is neither to agree nor disagree with Kittler but to ask how his views involve conceptual models inspired by synthesizers and other technologies. And that seems clear when Kittler describes Wagnerian music drama as a kind of circuit that connects distinct modules:

> Music drama functioned as a machine that worked on three levels—that is, in three data fields: first, that of verbal information; second, through the invisible orchestra of Bayreuth; and third, with scenic visuality, which involved "tracking shots" and spotlights *avant la lettre*. The text was fed into the singer's throat, the throat's output was fed into an amplifier called "orchestra," and the orchestral output was fed into a light show; finally, all of the above was fed into the nervous system of the audience. Ultimately—when everyone had gone crazy—every last trace of the alphabet had been erased. Data, instead of being encoded in the alphabet of books and scores, were amplified, stored, and reproduced through media.[98]

Obviously, the synthesizer is not Kittler's only model here. For example, "tracking shots" are drawn from film, and the multi-level structure might recall the cinematic division between visuals and soundtrack. Yet, the system's linear chain—which uses analog output from one element as input for the next—does not clearly reflect a cinematic logic. On its own, this structure might seem strange or counterintuitive. Why isn't the orchestral sound an input for the singers, who must constantly listen for cues and who often repeat melodies first introduced by instruments? Shouldn't the singers and orchestra at least be connected in a feedback loop? Why is the lighting driven by the orchestra and not directions in the text? But Kittler's choice of this signal path makes sense when understood in terms of the patching of synthesizer modules, as the media theorist's conception of Wagnerian opera is informed by his own musical hobbies.

Conclusions

This chapter has focused on three case studies from the 1970s and 1980s. Each reflects the authors' disciplinary concerns. While Heyde sought new ways of classifying musical instruments, Deleuze and Guattari developed an experimental philosophy, and Kittler, a technological vision of media. Such differences are also reflected in the authors' key references: for example, Deleuze and Guattari respond to Kant, and Heyde responds to Hornbostel and Sachs. Similarly, they address different audiences. At various levels, then, they have distinct orientations and goals.

Yet all of these thinkers take inspiration from modular synthesizers. In each case, synthesizers guide theoretical ideas and also novel reinterpretations of historical musical practices: Heyde presents acoustic instruments as modular systems; Deleuze and Guattari imagine orchestration in terms of synthesis and chromaticism in terms of continuous variation in multiple parameters; Kittler treats Wagnerian music drama as a chain of inputs and outputs. These retrospective interpretations also align with perspectives on musical ontology. For Heyde, Deleuze and Guattari, and Kittler, music is not simply a form of text. They understand music as sound, as an energetic signal that can include noise. They suggest that music emerges from complex systems of inputs and outputs, which integrate human and technological components.

To some degree, the chapter's analyses remain partial and speculative. Modular synthesizers are not the only technology that inspired these creative, interdisciplinary thinkers. So, the selected case studies could be pursued in greater depth—and others could be added. Nonetheless, Heyde's organology, Deleuze and Guattari's philosophy, and Kittler's media theory suggest ways in which synthesizers have functioned as theoretical instruments, as technologies that support new conceptual models.

Notes

1 Erich M. von Hornbostel and Curt Sachs, "Classification of Musical Instruments," trans. Anthony Baines and Klaus P. Wachsmann, *The Galpin Society Journal* 14 (1961): 3–29. This taxonomy was originally published as Erich M. von Hornbostel and Curt Sachs, "Systematik der Musikinstrumente: Ein Versuch," *Zeitschrift für Ethnologie* 46 (1914): 553–90. Yet, building on the work of Victor-Charles Mahillon, Hornbostel and Sachs effectively reproduce basic organological categories from the Natya Shastra, an ancient Indian treatise on music and other performing arts. See Nazir Ali Jairazbhoy, "An Explication of the Sachs-Hornbostel Instrument Classification System," in *Issues in Organology*, ed. Sue Carole DeVale, Selected Reports in Ethnomusicology 8 (Los Angeles: University of California, 1990), 81–104.
2 Margaret J. Kartomi, *On Concepts and Classifications of Musical Instruments* (Chicago: University of Chicago Press, 1990), 173–74; Curt Sachs, *The History of Musical Instruments* (New York: W.W. Norton, 1940).

3 Jonathan De Souza, "Timbral Thievery: Synthesizers and Sonic Materiality," in *The Oxford Handbook of Timbre*, ed. Emily I. Dolan and Alexander Rehding (New York and Oxford: Oxford University Press, 2021), 348–9.

4 Edwin M. Good, *Giraffes, Black Dragons, and Other Pianos: A Technological History from Cristofori to the Modern Concert Grand* (Stanford, CA: Stanford University Press, 1982).

5 Jonathan De Souza, "Orchestra Machines, Old and New," *Organised Sound* 23, no. 2 (2018): 156–66; Emily I. Dolan, *The Orchestral Revolution: Haydn and the Technologies of Timbre* (Cambridge and New York: Cambridge University Press, 2013); Deirdre Loughridge, *Haydn's Sunrise, Beethoven's Shadow: Audiovisual Culture and the Emergence of Musical Romanticism* (Chicago: University of Chicago Press, 2016); Adelheid Voskuhl, *Androids in the Enlightenment: Mechanics, Artisans, and Cultures of the Self* (Chicago: University of Chicago Press, 2013).

6 Nicholas J. Conard, Maria Malina, and Susanne C. Münzel, "New Flutes Document the Earliest Musical Tradition in Southwestern Germany," *Nature* 460 (2009): 737–40; Jonathan De Souza, "Voice and Instrument at the Origins of Music," *Current Musicology* 97 (2014): 21–36.

7 Gary Tomlinson, *A Million Years of Music: The Emergence of Human Modernity* (New York: Zone Books, 2015), 48.

8 Jonathan De Souza, *Music at Hand: Instruments, Bodies, and Cognition*, Oxford Studies in Music Theory (New York: Oxford University Press, 2017), 2; Lewis Mumford, *Art and Technics* (New York: Columbia University Press, 1952); Bernard Stiegler, *Technics and Time, 1: The Fault of Epimetheus*, trans. Richard Beardsworth and George Collins (Stanford: Stanford University Press, 1998).

9 De Souza, *Music at Hand*, 10.

10 David E. Creese, *The Monochord in Ancient Greek Harmonic Science* (Cambridge: Cambridge University Press, 2010).

11 Alexander Rehding, "Instruments of Music Theory," *Music Theory Online* 22, no. 4 (2016), http://mtosmt.org/issues/mto.16.22.4/mto.16.22.4.rehding.html; Alexander Rehding, "Fine-Tuning a Global History of Music Theory: Divergences, Zhu Zaiyu, and Music-Theoretical Instruments," *Music Theory Spectrum* 44, no. 2 (2022): 260–75, https://doi.org/10.1093/mts/mtac004.

12 Hans-Jörg Rheinberger, "Intersections: Some Thoughts on Instruments and Objects in the Experimental Context of the Life Sciences," in *Instruments in Art and Science: On the Architectonics of Cultural Boundaries in the 17th Century*, ed. Jan Lazardzig, Helmar Schramm, and Ludger Schwarte, English Edition, vol. 2, Theatrum Scientarum (Berlin: Walter de Gruyter, 2014), 3. Emphasis added.

13 See also Thor Magnusson, "Of Epistemic Tools: Musical Instruments as Cognitive Extensions," *Organised Sound* 14 (2009): 168–76, https://doi.org/10.1017/S1355 771809000272.

14 De Souza, "Timbral Thievery."

15 Harry F. Olson and Herbert Belar, "Electronic Music Synthesizer," *Journal of the Acoustical Society of America* 27 (1955): 595; Harry F. Olson, Herbert Belar, and J. Timmens, "Electronic Music Synthesis," *Journal of the Acoustical Society of America* 32 (1960): 311–9.

16 Hermann von Helmholtz, *On the Sensations of Tone as a Physiological Basis for the Theory of Music*, trans. Alexander J. Ellis, Second English Ed. (London: Longmans, Green, and Co., 1885), vi.

17 Ibid., 122–3. For further discussion, see De Souza, "Timbral Thievery," 353–62.

18 Such research would likely build on earlier work on the synthesizer from science and technology studies, such as Trevor Pinch and Frank Trocco, *Analog Days:*

The Invention and Impact of the Moog Synthesizer (Cambridge, MA: Harvard University Press, 2002).

19 As Lawrence Zbikowski discusses, conceptual models "consist of concepts in specified relationships, which pertain to a specific domain of knowledge" (Lawrence M. Zbikowski, *Conceptualizing Music: Cognitive Structure, Theory, and Analysis*, AMS Studies in Music (New York: Oxford University Press, 2002), 15). Conceptual models guide reasoning, supporting ontological claims (e.g., "This interval is a dissonance.") and also conditional statements (e.g., "If this interval is augmented, then it is a dissonance.").

20 Ludger Schwarte, "The Anatomy of the Brain as Instrumentalization of Reason," in *Instruments in Art and Science: On the Architectonics of Cultural Boundaries in the 17th Century*, ed. Helmar Schramm, Ludger Schwarte, and Jan Lazardzig, English edition, Theatrum Scientarum 2 (Berlin and New York: Walter de Gruyter, 2008), 177.

21 Herbert Heyde, "Methods of Organology and Proportions in Brass Wind Instrument Making," *Historic Brass Society Journal* 13 (2001): 4–6.

22 Jonathan De Souza, "Musical Instruments, Bodies, and Cognition" (Ph.D. dissertation, University of Chicago, 2013), 7–14.

23 Hornbostel and Sachs, "Classification of Musical Instruments," 5.

24 See Jairazbhoy, "An Explication of the Sachs-Hornbostel Instrument Classification System"; Jann Pasler, "The Utility of Musical Instruments in the Racial and Colonial Agendas of Late Nineteenth-Century France," *Journal of the Royal Musical Association* 129 (2004): 24–76; Benjamin Steege, *Helmholtz and the Modern Listener* (Cambridge and New York: Cambridge University Press, 2012).

25 Hornbostel and Sachs, "Classification of Musical Instruments," 7.

26 The present discussion focuses on the central second chapter from Heyde's treatise.

27 N. Katherine Hayles, *How We Became Posthuman: Virtual Bodies in Cybernetics, Literature, and Informatics* (Chicago and London: University of Chicago Press, 1999).

28 Norbert Wiener, *Cybernetics; or, Control and Communication in the Animal and the Machine* (New York: John Wiley & Sons, 1948).

29 For an English translation, see Oskar Lange, *Wholes and Parts: A General Theory of System Behaviour*, trans. Eugeniusz Lepa (Oxford and New York: Pergamon Press, 1965).

30 While Meyer-Eppler had also made contributions to information theory (e.g., in his 1959 text *Grundlagen und Anwendungen der Informationstheorie*), he was well known for his involvement in electronic music: he had been involved in founding the historic Studio for Electronic Music of the West German Radio in Cologne, taught acoustics to the composer Karlheinz Stockhausen at the University of Bonn, and contributed to the inaugural issue of the music journal *Die Reihe*.

31 Herbert Heyde, *Grundlagen des natürlichen Systems der Musikinstrumente*, Beiträge zur musikwissenschaftlichen Forschung in der DDR 7 (Leipzig: VEB Deutscher Verlag für Musik, 1975), 22. All translations from Heyde are mine.

32 Claude E. Shannon, "The Mathematical Theory of Communication," in *The Mathematical Theory of Communication*, by Claude E. Shannon and Warren Weaver (Urbana: University of Illinois Press, 1964).

33 Heyde, *Grundlagen*, 22.

34 Ibid., 25.

35 Ibid.

36 Adapted from ibid., 24.

37 Ibid., 60.

38 The concept of "standing reserve" derives from Martin Heidegger, "The Question Concerning Technology," in *The Question Concerning Technology and Other Essays*, trans. William Lovitt (New York: Harper & Row, 1977), 3–35.

39 For a discussion of the microphone, see Carolyn Abbate, "Sound Object Lessons," *Journal of the American Musicological Society* 69, no. 3 (2016): 793–829, https://doi.org/10.1525/jams.2016.69.3.793.

40 Heyde, *Grundlagen*, 27.

41 Ibid., 22.

42 Ibid., 61.

43 Ibid., 46.

44 Ibid., 32, 53.

45 Adapted from De Souza, *Music at Hand*, 34.

46 Heyde, *Grundlagen*, 60.

47 Ibid., 22.

48 Ibid., 62–5.

49 Ibid., 63–4.

50 Adapted from ibid., 62.

51 Gilles Deleuze, "Leibniz: Philosophy and the Creation of Concepts, Lecture 5," trans. Charles J. Stivale, The Deleuze Seminars, 20 May 1980, https://deleuze.cla.purdue.edu/seminars/leibniz-philosophy-and-creation-concepts/lecture-05.https://deleuze.cla.purdue.edu/seminars/leibniz-philosophy-and-creation-concepts/lecture-05Deleuze. Note that Deleuze made this statement three years before the introduction of the Yamaha DX7, the synthesizer that came to define 1980s popular music. For a study of the Yamaha DX7, see Megan Lavengood, "What Makes It Sound '80s? The Yamaha DX7 Electric Piano Sound," *Journal of Popular Music Studies* 31, no. 3 (2019): 73–94, https://doi.org/10.1525/jpms.2019.313009.

52 Martin Scherzinger, "Musical Modernism in the Thought of *Mille Plateaux*, and Its Twofold Politics," *Perspectives of New Music* 46 (2008): 136–7.

53 Dominique Leone, "Heldon: Interface," Pitchfork, 7 April 2004, https://pitchfork.com/reviews/albums/3979-interface/.

54 Aug Stone, "Interview: Richard Pinhas Believes Music Can Fix The World," The Red Bull Music Academy Daily, 25 May 2015, https://daily.redbullmusicacademy.com/2015/05/richard-pinhas-interview.

55 Gilles Deleuze and Félix Guattari, *A Thousand Plateaus: Capitalism and Schizophrenia*, trans. Brian Massumi (Minneapolis: University of Minnesota Press, 1987), 613. Besides Pinhas's article on synthesizers, Deleuze and Guattari cite a 1977 interview with Karlheinz Stockhausen.

56 *The Deleuze Seminars* are currently available online at https://deleuze.cla.purdue.edu/.

57 Gilles Deleuze, "Painting and the Question of Concepts, Lecture 5," trans. Billy Dean Goehring, The Deleuze Seminars, 12 May 1981, https://deleuze.cla.purdue.edu/seminars/painting-and-question-concepts/lecture-05.

58 Now in his 70s, Pinhas continues to create guitar and synthesizer music as a solo artist and collaborator (see https://www.richard-pinhas.com/). He is also the author of a book about Deleuze and music: Richard Pinhas, *Les larmes de Nietzsche: Deleuze et la musique* (Paris: Flammarion, 2001).

59 Deleuze and Guattari, *A Thousand Plateaus*, 121.

60 Ibid., 106, 378–9.

61 Immanuel Kant, *Critique of Pure Reason*, trans. Paul Guyer and Allen W. Wood (Cambridge and New York: Cambridge University Press, 1998), 210.

62 Gilles Deleuze, "Kant: Synthesis and Time, Lecture 1," trans. Melissa McMahon, The Deleuze Seminars, 14 March 1978, https://deleuze.cla.purdue.edu/seminars/kant-synthesis-and-time/lecture-01.

63 In a later lecture in Deleuze's Kant seminar, he discusses how the aesthetic comprehension of rhythm can involve a sense of underlying measure. See Gilles Deleuze, "Kant: Synthesis and Time, Lecture 3," trans. Melissa McMahon, The Deleuze Seminars, 28 March 1978, https://deleuze.cla.purdue.edu/seminars/kant-synthesis-and-time/lecture-03.

64 Deleuze and Guattari, A Thousand Plateaus, 379.

65 De Souza, "Timbral Thievery," 353.

66 Zbikowski, Conceptualizing Music: Cognitive Structure, Theory, and Analysis, 110.

67 Deleuze and Guattari, A Thousand Plateaus, 378.

68 De Souza, "Timbral Thievery," 359.

69 Gilles Deleuze, "A Thousand Plateaus IV: The State Apparatus & War-Machines I, Lecture 1," trans. Timothy S. Murphy, The Deleuze Seminars, 27 February 1979, https://deleuze.cla.purdue.edu/seminars/thousand-plateaus-iv-state-apparatus-war-machines-i/lecture-01. In her work on orchestration, Emily Dolan also highlights the importance of timbre and the formation and breaking-apart of sound. However, though she also discusses Berlioz and Wagner, she traces this back to Joseph Haydn. See Dolan, The Orchestral Revolution.

70 Deleuze, "A Thousand Plateaus IV: The State Apparatus & War-Machines I, Lecture 1."

71 Ibid. See also Odile Vivier, Varèse (Paris: Éditions du Seuil, 1973).

72 Deleuze and Guattari, A Thousand Plateaus, 378.

73 Trevor Pinch and Frank Trocco, "The Social Construction of the Early Electronic Music Synthesizer," in Music and Technology in the Twentieth Century, ed. Hans-Joachim Braun (Baltimore and London: Johns Hopkins University Press, 2002), 73.

74 Deleuze and Guattari, A Thousand Plateaus, 527–8. The opposition of smooth and striated, which Deleuze and Guattari explore at length in A Thousand Plateaus, derives from Boulez. See Pierre Boulez, On Music Today, trans. Susan Bradshaw and Richard Rodney (London: Faber and Faber, 1971), 85–8.

75 Deleuze and Guattari, A Thousand Plateaus, 105–6.

76 A photograph of Kittler's synthesizer is available at https://web.archive.org/web/20220222073643/https://artmap.com/wkvstuttgart/exhibition/jan-peter-e-r-sonntag-2015. A more detailed discussion of the system is included in Döring, Sebastian, and Jan-Peter E.R Sonntag. "Apparatus Operandi1: Anatomy//Friedrich A Kittler's Synthesizer." In Rauschen, 109–46. Leipzig: Merve, 2019.

77 Friedrich A. Kittler, "Non-Linear Oscillators and Computer Motherboards" (Lecture, European Graduate School, 2010), https://youtu.be/CxIHwCnVYIE, 19:39.

78 Friedrich A. Kittler, The Truth of the Technological World: Essays on the Genealogy of Presence, trans. Erik Butler (Stanford: Stanford University Press, 2013), 56.

79 Alexander Rehding, "Discrete/Continuous: Music and Media Theory after Kittler," Journal of the American Musicological Society 70, no. 1 (2017): 223.

80 Kittler, The Truth of the Technological World, 165–77.

81 Ibid., 138–51.

82 Friedrich A. Kittler, Gramophone, Film, Typewriter, trans. Geoffrey Winthrop-Young and Michael Wutz (Stanford: Stanford University Press, 1999), 3–4.

83 Ibid., 23.

84 Kittler, The Truth of the Technological World, 122. See also Kittler, Gramophone, Film, Typewriter, 37.

85 Kittler, The Truth of the Technological World, 53.

86 Ibid., 122. Friedrich A. Kittler, Discourse Networks, 1800/1900, trans. Michael Metteer and Chris Cullens (Stanford: Stanford University Press, 1990), 116; see also, Kittler, Gramophone, Film, Typewriter, 23.

87 Gundula Kreuzer, "Kittler's Wagner and Beyond," *Journal of the American Musicological Society* 70, no. 1 (2017): 232. As Kreuzer notes, Theodor Adorno linked Wagner to the synthesizer before Kittler did.
88 Kittler, *Gramophone, Film, Typewriter*, 24.
89 Ibid.
90 Quoted in Steege, *Helmholtz and the Modern Listener*, 226.
91 Kittler, *The Truth of the Technological World*, 128.
92 Like Helmholtz's synthesizer, valveless hunting horns produce only notes in a given harmonic series. Traditionally, horns required equipment for hunters, who used a conventional system of hunting calls to communicate at a distance. For further discussion, see De Souza, *Music at Hand*, Ch. 6.
93 Kittler, *The Truth of the Technological World*, 129. See also, ibid., 172–3.
94 Ibid., 129.
95 Ibid., 122.
96 Peter Petersen, "Nochmals zum Tanz-Quodlibet im ersten Akt-Finale des Don Giovanni," *Archiv für Musikwissenschaft* 65, no. 1 (2008): 1–30.
97 Kreuzer, "Kittler's Wagner and Beyond," 231. See also, Gundula Kreuzer, *Curtain, Gong, Steam: Wagnerian Technologies of Nineteenth-Century Opera* (Berkeley: University of California Press, 2018).
98 Kittler, *The Truth of the Technological World*, 134.

Bibliography

Abbate, Carolyn. "Sound Object Lessons." *Journal of the American Musicological Society* 69, no. 3 (2016): 793–829. 10.1525/jams.2016.69.3.793.
Boulez, Pierre. *On Music Today*. Translated by Susan Bradshaw and Richard Rodney. London: Faber and Faber, 1971.
Conard, Nicholas J., Maria Malina, and Susanne C. Münzel. "New Flutes Document the Earliest Musical Tradition in Southwestern Germany." *Nature* 460 (2009): 737–740.
Creese, David E. *The Monochord in Ancient Greek Harmonic Science*. Cambridge: Cambridge University Press, 2010.
De Souza, Jonathan. *Music at Hand: Instruments, Bodies, and Cognition*. Oxford Studies in Music Theory. New York: Oxford University Press, 2017.
De Souza, Jonathan. "Musical Instruments, Bodies, and Cognition." Ph.D. dissertation, University of Chicago, 2013.
De Souza, Jonathan. "Orchestra Machines, Old and New." *Organised Sound* 23, no. 2 (2018): 156–166.
De Souza, Jonathan. "Timbral Thievery: Synthesizers and Sonic Materiality." In *The Oxford Handbook of Timbre*, edited by Emily I. Dolan and Alexander Rehding, 347–379. New York and Oxford: Oxford University Press, 2021.
De Souza, Jonathan. "Voice and Instrument at the Origins of Music." *Current Musicology* 97 (2014): 21–36.
Deleuze, Gilles. "A Thousand Plateaus IV: The State Apparatus & War-Machines I, Lecture 1." Translated by Timothy S. Murphy. The Deleuze Seminars, February 27, 1979. https://deleuze.cla.purdue.edu/seminars/thousand-plateaus-iv-state-apparatus-war-machines-i/lecture-01.
Deleuze, Gilles. "Kant: Synthesis and Time, Lecture 1." Translated by Melissa McMahon. The Deleuze Seminars, March 14, 1978. https://deleuze.cla.purdue.edu/seminars/kant-synthesis-and-time/lecture-01.
Deleuze, Gilles. "Kant: Synthesis and Time, Lecture 3." Translated by Melissa McMahon. The Deleuze Seminars, March 28, 1978. https://deleuze.cla.purdue.edu/seminars/kant-synthesis-and-time/lecture-03.

Deleuze, Gilles. "Leibniz: Philosophy and the Creation of Concepts, Lecture 5." Translated by Charles J. Stivale. The Deleuze Seminars, May 20, 1980. https://deleuze.cla.purdue.edu/seminars/leibniz-philosophy-and-creation-concepts/lecture-05.

Deleuze, Gilles. "Painting and the Question of Concepts, Lecture 5." Translated by Billy Dean Goehring. The Deleuze Seminars, May 12, 1981. https://deleuze.cla.purdue.edu/seminars/painting-and-question-concepts/lecture-05.

Deleuze, Gilles, and Félix Guattari. *A Thousand Plateaus: Capitalism and Schizophrenia.* Translated by Brian Massumi. Minneapolis: University of Minnesota Press, 1987.

Dolan, Emily I. *The Orchestral Revolution: Haydn and the Technologies of Timbre.* Cambridge and New York: Cambridge University Press, 2013.

Döring, Sebastian, and Jan-Peter E.R Sonntag. "Apparatus Operandi1: Anatomy// Friedrich A Kittler's Synthesizer." In *Rauschen*, 109–146. Leipzig: Merve, 2019.

Good, Edwin M. *Giraffes, Black Dragons, and Other Pianos: A Technological History from Cristofori to the Modern Concert Grand.* Stanford, CA: Stanford University Press, 1982.

Hayles, N. Katherine. *How We Became Posthuman: Virtual Bodies in Cybernetics, Literature, and Informatics.* Chicago and London: University of Chicago Press, 1999.

Heidegger, Martin. "The Question Concerning Technology." In *The Question Concerning Technology and Other Essays*, translated by William Lovitt, 3–35. New York: Harper & Row, 1977.

Helmholtz, Hermann von. *On the Sensations of Tone as a Physiological Basis for the Theory of Music.* Translated by Alexander J. Ellis. Second English Ed. London: Longmans, Green, and Co., 1885.

Heyde, Herbert. *Grundlagen des natürlichen Systems der Musikinstrumente.* Beiträge zur musikwissenschaftlichen Forschung in der DDR 7. Leipzig: VEB Deutscher Verlag für Musik, 1975.

Heyde, Herbert. "Methods of Organology and Proportions in Brass Wind Instrument Making." *Historic Brass Society Journal* 13 (2001): 1–51.

Hornbostel, Erich M. von, and Curt Sachs. "Classification of Musical Instruments." Translated by Anthony Baines and Klaus P. Wachsmann. *The Galpin Society Journal* 14 (1961): 3–29.

Hornbostel, Erich M. von, and Curt Sachs. "Systematik der Musikinstrumente: Ein Versuch." *Zeitschrift für Ethnologie* 46 (1914): 553–590.

Jairazbhoy, Nazir Ali. "An Explication of the Sachs-Hornbostel Instrument Classification System." In *Issues in Organology*, edited by Sue Carole DeVale, 81–104. Selected Reports in Ethnomusicology 8. Los Angeles: University of California, 1990.

Kant, Immanuel. *Critique of Pure Reason.* Translated by Paul Guyer and Allen W. Wood. Cambridge and New York: Cambridge University Press, 1998.

Kartomi, Margaret J. *On Concepts and Classifications of Musical Instruments.* Chicago: University of Chicago Press, 1990.

Kittler, Friedrich A. *Discourse Networks, 1800/1900.* Translated by Michael Metteer and Chris Cullens. Stanford: Stanford University Press, 1990.

Kittler, Friedrich A. *Gramophone, Film, Typewriter.* Translated by Geoffrey Winthrop-Young and Michael Wutz. Stanford: Stanford University Press, 1999.

Kittler, Friedrich A. "Non-Linear Oscillators and Computer Motherboards." Lecture, European Graduate School, 2010. https://youtu.be/CxIHwCnVYIE.

Kittler, Friedrich A. *The Truth of the Technological World: Essays on the Genealogy of Presence.* Translated by Erik Butler. Stanford: Stanford University Press, 2013.

Kreuzer, Gundula. *Curtain, Gong, Steam: Wagnerian Technologies of Nineteenth-Century Opera.* Berkeley: University of California Press, 2018.

Kreuzer, Gundula. "Kittler's Wagner and Beyond." *Journal of the American Musicological Society* 70, no. 1 (2017): 228–233.

Lange, Oskar. *Wholes and Parts: A General Theory of System Behaviour.* Translated by Eugeniusz Lepa. Oxford and New York: Pergamon Press, 1965.

Lavengood, Megan. "What Makes It Sound '80s? The Yamaha DX7 Electric Piano Sound." *Journal of Popular Music Studies* 31, no. 3 (2019): 73–94. 10.1525/jpms. 2019.313009.

Leone, Dominique. "Heldon: Interface." Pitchfork, April 7, 2004. https://pitchfork. com/reviews/albums/3979-interface/.

Loughridge, Deirdre. *Haydn's Sunrise, Beethoven's Shadow: Audiovisual Culture and the Emergence of Musical Romanticism.* Chicago: University of Chicago Press, 2016.

Magnusson, Thor. "Of Epistemic Tools: Musical Instruments as Cognitive Extensions." *Organised Sound* 14 (2009): 168–176. 10.1017/S1355771809000272.

Mumford, Lewis. *Art and Technics.* New York: Columbia University Press, 1952.

Olson, Harry F., and Herbert Belar. "Electronic Music Synthesizer." *Journal of the Acoustical Society of America* 27 (1955): 595.

Olson, Harry F., Herbert Belar, and J. Timmens. "Electronic Music Synthesis." *Journal of the Acoustical Society of America* 32 (1960): 311–319.

Pasler, Jann. "The Utility of Musical Instruments in the Racial and Colonial Agendas of Late Nineteenth-Century France." *Journal of the Royal Musical Association* 129 (2004): 24–76.

Petersen, Peter. "Nochmals zum Tanz-Quodlibet im ersten Akt-Finale des Don Giovanni." *Archiv für Musikwissenschaft* 65, no. 1 (2008): 1–30.

Pinch, Trevor, and Frank Trocco. *Analog Days: The Invention and Impact of the Moog Synthesizer.* Cambridge, MA: Harvard University Press, 2002.

Pinch, Trevor, and Frank Trocco. "The Social Construction of the Early Electronic Music Synthesizer." In *Music and Technology in the Twentieth Century,* edited by Hans-Joachim Braun, 67–83. Baltimore and London: Johns Hopkins University Press, 2002.

Pinhas, Richard. *Les larmes de Nietzsche: Deleuze et la Musique.* Paris: Flammarion, 2001.

Rehding, Alexander. "Discrete/Continuous: Music and Media Theory after Kittler." *Journal of the American Musicological Society* 70, no. 1 (2017): 221–228.

Rehding, Alexander. "Fine-Tuning a Global History of Music Theory: Divergences, Zhu Zaiyu, and Music-Theoretical Instruments." *Music Theory Spectrum* 44, no. 2 (2022): 260–275. https://doi.org/10.1093/mts/mtac004.

Rehding, Alexander. "Instruments of Music Theory." *Music Theory Online* 22, no. 4 (2016). http://mtosmt.org/issues/mto.16.22.4/mto.16.22.4.rehding.html.

Rheinberger, Hans-Jörg. "Intersections: Some Thoughts on Instruments and Objects in the Experimental Context of the Life Sciences." In *Instruments in Art and Science: On the Architectonics of Cultural Boundaries in the 17th Century,* edited by Jan Lazardzig, Helmar Schramm, and Ludger Schwarte, English Edition, 2:1–19. Theatrum Scientarum. Berlin: Walter de Gruyter, 2014.

Sachs, Curt. *The History of Musical Instruments.* New York: W.W. Norton, 1940.

Scherzinger, Martin. "Musical Modernism in the Thought of *Mille Plateaux,* and Its Twofold Politics." *Perspectives of New Music* 46 (2008): 130–158.

Schwarte, Ludger. "The Anatomy of the Brain as Instrumentalization of Reason." In *Instruments in Art and Science: On the Architectonics of Cultural Boundaries in the 17th Century,* edited by Helmar Schramm, Ludger Schwarte, and Jan Lazardzig, English edition., 176–200. Theatrum Scientarum 2. Berlin and New York: Walter de Gruyter, 2008.

Shannon, Claude E. "The Mathematical Theory of Communication." In *The Mathematical Theory of Communication*, edited by Claude E. Shannon and Warren Weaver. 29–125. Urbana: University of Illinois Press, 1964.

Steege, Benjamin. *Helmholtz and the Modern Listener*. Cambridge and New York: Cambridge University Press, 2012.

Stiegler, Bernard. *Technics and Time, 1: The Fault of Epimetheus*. Translated by Richard Beardsworth and George Collins. Stanford: Stanford University Press, 1998.

Stone, Aug. "Interview: Richard Pinhas Believes Music Can Fix The World." The Red Bull Music Academy Daily, May 25, 2015. https://daily.redbullmusicacademy.com/2015/05/richard-pinhas-interview.

Tomlinson, Gary. *A Million Years of Music: The Emergence of Human Modernity*. New York: Zone Books, 2015.

Vivier, Odile. *Varèse*. Paris: Éditions du Seuil, 1973.

Voskuhl, Adelheid. *Androids in the Enlightenment: Mechanics, Artisans, and Cultures of the Self*. Chicago: University of Chicago Press, 2013.

Wiener, Norbert. *Cybernetics; or, Control and Communication in the Animal and the Machine*. New York: John Wiley & Sons, 1948.

Zbikowski, Lawrence M. *Conceptualizing Music: Cognitive Structure, Theory, and Analysis*. AMS Studies in Music. New York: Oxford University Press, 2002.

3

A TIME-WARPED ASSEMBLAGE AS A MUSICAL INSTRUMENT

Flexibility and *Constauration* of Modular Synthesis in Willem Twee Studio 1

Hannah Bosma

When the editors suggested discussing "modular thinking" in relation to modular synthesis, this intrigued me. Because modular synthesis is associated with creative tinkering, "modular thinking" seemed attractive and interesting. See for example the eulogy in a video on modular synthesis:

> Modular: vast adventurous rewarding
> (& maybe not exactly what you think)
> [...]
> modular allows you to build your own device for your own reasons
> a malleable open instrument that can be different each time you use it
> [...]
> modular is a treat for the imagination
> a place to combine curious bits
> into an evolving, shape-shifting whole[1]

Ruminating on modular thinking, I began to see modularity everywhere. Lego: building blocks to build your own thing. Modular building: a construction technique with prefabricated building modules.[2] Composing your own modular pasta dish: 1) choose pasta, 2) choose sauce, 3) choose toppings, 4) choose dessert, 5) choose drinks.[3] Modules in a university programme. And other associations began to creep in: project-based management and outsourcing, structuring an organization into modules that can be discarded or replaced easily, or that compete against each other, to enhance flexibility and short-term profit, but at the cost of integration and long-term stability and commitment. Is modular thinking perhaps a sign of the times?

DOI: 10.4324/9781003219484-5

Indeed, Blair (1988) suggests that modularity has been characteristic of American culture since the 19th century in such diverse realms as industry, university education, football, poetry, jazz music, furniture and hi-fi equipment.[4] Twentieth-century architecture, art and music dealt with various forms of modularity; for example, Stockhausen's *Klavierstück XI* (1956) consists of 19 modules that may be played in any order, each module determining the tempo, dynamic level and the type of attack of the next one.[5] Manovich (2001) considers modularity as one of the five principles of the (then emerging) new media, that are composed of separate self-sufficient modules, in a computerizing culture.[6] He discerns the pre-computer modularity of mass production, since Ford's assembly line in 1913, from the modularization of culture by computerization.[7] Nevertheless, Kostakis (2019) sees modularity as a "potential of the digital revolution [that] has not been fully realized".[8] While in such accounts modularity is linked to technological, cultural and social developments, the notion of modularity is also used in biology[9] and cognitive science.[10] In each field, the notion of modularity is used in (slightly) different ways, but most have in common that "modular" refers to a system with more or less self-contained, equivalent, interchangeable, recombinable, connectable parts instead of an integrated whole, and goes with decentralization and a non-hierarchical structure. Connectivity and connections of the parts are important. Standardization is another feature that is often related to modularity. In a practical sense, standardization is a prerequisite to connect different modules; historically, the word module comes from the Latin *modulus*, small measurement, and was used in ancient Greek, Latin and Renaissance architecture as a measurement unit to derive the proportions of a building; nowadays the first meaning of "module" is standard unit of measurement.[11] Adaptability, efficiency, ease of customization and handling complexity are among the main benefits of modularity. Callebaut and Rasskin-Gutman (2005) state that "in our world, modular systems, both natural and artificial [...], abound"[12]; nevertheless, it seems that this abundance of modularities may also relate to a particular way of looking at such phenomena—a way of theorizing, structuring, organizing and making that became prominent since the 19th century and is related to the demise of feudalism and the rise of bourgeois culture, (post)modernity, democracy, (late) capitalism, free market economy, (neo-)liberalism, individualism, modernism and electrical, electronic and digital technology.

While the concept of modularity is used so widely, and in different ways, I will focus here on a fringe case of modular synthesis of electronic music: Studio 1 of Willem Twee Studios in 's-Hertogenbosch in the Netherlands. This is a studio with equipment made for other purposes than electronic music, before the first modular synthesizers were developed, and it is functioning as a giant modular synthesizer. Since the equipment under consideration is relatively old, "vintage", this is closely related to the question of its preservation. In what ways is this system modular, and how does that relate to its preservation? As a magnifying glass, Studio 1 highlights some of the attractiveness of modular synthesis in music and reveals a tension at the heart of modular thinking and modular practice.

Willem Twee Studios

Willem Twee Studios (WTS) is part of the Willem Twee music and visual arts center in 's-Hertogenbosch in the Netherlands. WTS is run by Rikkert Brok, Hans Kulk and Armeno Alberts, with the help of some interns and volunteers. The main part of the equipment comes from Hans Kulk's private studio and from the former CEM studio.[13] In 2022, equipment from the musical instrument collection of the Gemeentemuseum Den Haag (Municipal Museum The Hague)[14] was added to the Willem Twee Studios. Musicians, students and anyone interested can book studio time for a price that is considered within the budget of these groups. Additionally, Willem Twee Studios offers workshops and courses, for subscription and for conservatories, co-organizes a yearly electronic music festival, and invites musicians to work in their studios and to give concerts.[15]

Currently, Willem Twee Studios consists of four studios. WTS Studio 1 contains a large collection of analog test & measurement equipment from the 1950s and 1960s as well as analog tape recorders, and is the main focus of this chapter. WTS Studio 2 consists of various synthesizers from the 1970s and 1980s, such as analog modular voltage controlled synthesizers like ARP 2500, two ARP 2600s, Serge Modular, three Oberheim SEMs, Oberheim Mini Sequencer (modified to a modular system by Ernst Bonis, the original owner)[16] and EMS VCS 3 (Putney), and other analog or hybrid synthesizers such as ARP Omni-2, ARP Axxe, Minimoog, PPG1002, Wasp, SCI Pro One, JUNO-60, Yamaha CS-60, SCI Prophet 5, Prophet 600, Korg Polysix, Rhodes Chroma, Oberheim Matrix-6, and some of the first fully digital synthesizers, such as Yamaha DX-7 and DX-7 II FD and Nord Modulars (Classic).[17] Studio 3 is a small space with a variable set up. Studio 4 is the small concert hall Willem Twee Toonzaal, in the same building, that has a Steinway grand piano, a small pipe organ and audio lines to the other studios, so that it can be used for recording reverb or for concert performance as an extension of the studios upstairs. While all activities of Willem Twee Studios are interlinked and WTS Studio 2 is very interesting as well, containing a large, rare collection of functioning vintage modular synthesizers, in this chapter I am focusing on WTS Studio 1 only—triggered by its exceptional setup and its intricate relation with the early history of electronic music and the foundations of modular synthesis.

WTS Studio 1

The genesis of WTS Studio 1 started when, after working with the analog modular synthesizers ARP 2500 and 2600 in the 1980s, in the 1990s Hans Kulk began to collect old test and measurement equipment and analog computers from the 1950s and 1960s, such as sine and pulse oscillators and filters, mostly controlled by hand, some voltage-controlled. His aim was not to form a collection, but to study the old techniques of electronic music with the equipment of that time. This equipment was not developed for making music, but for telephone or radio technology, chemistry research or analog computing. In the 1950s, some of the first electronic music, of Herbert Eimert, Karel Goeyvaerts and Karlheinz Stockhausen for example, was made with such test and measurement equipment. Since the 1960s, similar techniques for sound generation, such as oscillators, ring modulation, filtering, and sample and hold, were integrated in synthesizers like Moog, Buchla, Putney and ARP, making sound generation easier and faster. Whereas the electronic music studios in the 1950s worked with a few test and measurement devices, Hans Kulk collected a large quantity of such equipment, more than 60 devices, and including two analog computers, to use for modular synthesis. While this became too much to keep at home, Hans Kulk's equipment was moved to become a studio at music venue De Toonzaal in 's-Hertogenbosch; with a merger of cultural organizations in 2017, this developed into Willem Twee Studios of the Willem Twee music and visual arts center. With this equipment, Hans Kulk made music and sound mainly for collaborative projects with visual artists like Dineke van Oosten (*24 Hours*, 2012, 's-Hertogenbosch)[18] and José op ten Berg (*VER<F>klanken*,[19] 2016, 's-Hertogenbosch). Subsequently, other musicians, such as Hainbach, Andrea Taeggi and WORM's Dennis Verschoor, were introduced to it at Willem Twee Studios and composed music in Studio 1 or assembled their own collection of test and measurement equipment and analog computers.

The most obvious difference between a regular modular synthesizer and Studio-1-as-synthesizer, is the size and spatiality of WTS Studio 1. Whereas an ARP 2500, consisting of a Studio cabinet, a five-octave keyboard and two Wing cabinets—already a large configuration for a synthesizer—may measure about 2 meters in length and 50 cm in height, and an ARP 2600 about 1 × 0.50 m, the Studio-1-synthesizer consists of an entire studio room. On the right side of the room there is a wall with equipment of ca. 4 × 2 m (nine 19-inch racks), with two analog computers adjacently at the back of the room; on the left side there is a table of ca. 4 m length with a large mixer, several tape recorders, some passive 1950s bandpass filters, and facilities for patching and connecting to the audio interface of a digital computer; plus some additional equipment here and there; with various ways to make

FIGURE 3.1 The right side of Studio 1 (point of view from the back side of the studio). Photo by Lin Houtman, 20 July 2023.

FIGURE 3.2 Close-up of Studio 1; the mixers are used to play Studio 1 as a giant synthesizer. On the left side of the photo the analog computers are partly visible. Photo by Lin Houtman, 20 July 2023.

connections between the units by way of cables, patch panels and mixers (see Figure 3.1 and 3.2).

It is no surprise then that Hans Kulk mentions that, after working on an ARP 2600, one of the positive developments he experienced when working with his configuration of test and measurement equipment was a greater physical involvement. He stresses the spatiality of Studio 1 in various respects. In his account, size and physical appearance are important aspects of the modularity of the equipment: the 19″ norm and the standard U unit, allowing neat placement of the equipment in racks; or the DIN norm of the older German equipment, with a large unit width of 52cm; or the NIM-BIN, a chassis for Nuclear Instrumentation Modules, as another standard. It is also a matter of aesthetics: Hans Kulk likes the uniformity of the cables, dials and knobs of this equipment, and their form and placement, according to Human Factors Engineering—a discipline that takes the capabilities and characteristics of users into account for an effective design. When choosing the devices, he took the aesthetics and uniformity into account and limited the collection to Hewlett Packard and old German equipment. This uniformity and clarity contribute to the experience of Studio 1 as one large instrument. Hans Kulk contrasts this with the abundance of colors and lights, the profusion of different knobs and illuminating cables, that he sees in Eurorack equipment nowadays: "I can't stand that, it would drive me mad!"[20] Although Hans Kulk stopped adding more equipment to Studio 1, he dreams of an endless expansion: "If I would get total freedom and a bag of money, then I would continue, adding more racks and oscillators and … endless! So that I would need a bike to go to all the racks … a hall with islands of racks and equipment, with some nice desks to make large sketches … and with relays to control many signals from one central position … "[21]

After experimenting, as a kid, with guitar, tape echo, cassette tapes and a mixer, and later playing guitar, bass, keyboard and drums in jazz and other music bands, and studying music theory, Hans Kulk bought in the 1980s an ARP 2600 and an ARP 2500 for a low price from respectively a music school[22] and from the national public broadcast organization NOS/NOB,[23] where these modular synthesizers were no longer in use. This was a time when such analog modular synthesizers were considered old fashioned and most musicians were more interested in the then new developments of digital and hybrid synthesizers, such as the Yamaha DX-7, and samplers, MIDI, and the upcoming personal computer technology. To learn more about the technology of the analog synthesizers, Kulk studied the book *The Technique of Electronic Music* by Thomas Wells,[24] which contains bibliographic lists, borrowed more books and ordered scientific articles via the library, and talked and corresponded with experts like Jaap Vink, Stan Tempelaars, Joel Chadabe, Jo Scherpenisse and Ernst Bonis. In *The Development and Practice of Electronic Music*, edited by Jon Appleton and Ronald Perera (1975), which contains a chapter by Joel Chadabe on the voltage-controlled synthesizer, Hans Kulk read that a large modular system is conceptually similar to an analog computer.[25] This raised his interest. And after he found a manual of an analog computer in a secondhand bookstore, he went to a personal computer shop and said that he wanted an analog computer. The salesman

FIGURE 3.3   Close-up of some of the devices of Studio 1, with the Hewlett Packard 3722A noise generator in the middle of the photo (with cables). Photo by Lin Houtman, 20 July 2023.

told him that he had a friend working at Hewlett-Packard who might know where to find one, and indeed, via this contact Hans Kulk was able to buy two analog computers that were stored in the basement of a technical college,[26] for a very low price. By experimenting with the analog computer and connecting it to the ARP, Hans Kulk learned that many concepts from the first modular analog synthesizers come from analog computer technology. He continued exploring analog computer technology and contacted the Faculty of Aerospace Engineering at the Technical University in Delft in the mid-1990s, where they were using large hybrid-analog computers for flight simulation. There, engineer Tak offered him an old noise generator, a Hewlett Packard 3722A (Figure 3.3). Hans Kulk fell in love with its signal quality, its possibilities, its design … he wanted more of this! In the studio where he was teaching, there were large racks that were not used anymore, and he could take them home—by train, one by one … And gradually he found more and more analog test and measurement devices from the 1950s and 1960s via second hand markets of radio amateurs and in sheds of retired engineers.

In the meantime, Hans Kulk became unhappy with his way of working with the ARP synthesizers: it was a pleasure to make complex patches that were running automatically, but when it was running, he missed his role as a musician. He was listening to the first electronic music from the 1950s and noticed that he couldn't generate these sounds with his ARP synthesizers. He then had a dream in which Karlheinz Stockhausen said that he had to use a mixer. The next day he did so: connecting test and measurement equipment, such as sine wave generators and ring modulators, to the mixer and merging these into changing sound textures, moving in space from one loudspeaker to another, by moving the mixer's faders with his hands. He felt like a musician again, making music with his hands and body. For example, he could use the mixer as a hand-controlled envelope generator, making sound contours that are much more free and dynamic than those made with the common voltage-controlled Attack Decay Sustain Release (ADSR) envelope generators that are used in synthesizers since its introduction by Moog and ARP.[27] For Hans Kulk, this hands-on, bodily experience brought back the joy of making music. In 2005 he stopped making electronic music with his ARP synthesizers to concentrate on building his private studio with test and measurement equipment—which became WTS Studio 1. He uses this studio as a musical instrument. Usually, musicians learn to play their instrument well by practicing for years and developing an intimate, embodied relationship with their instrument. Hans Kulk aims to make music with this studio with such an embodied musical expression and "muscle memory".[28] To do so, it is important for him that every dial and knob has one function and that the setup is not changed too often, after he figured out his ideal combination.

Hans Kulk praises the quality and stability of the signals generated by this equipment, in contrast with the (in)famous instability of most analog synthesizers. Such instability may be considered as both an advantage and a disadvantage. Many praise the old analog synthesizers because of their "warm" and "lively" sound, caused by the imperfection and instability of the system; see for example how Pinch and Trocco describe the sound of the Minimoog: "it [...] retained some of the instability and rich peculiarities of the analog world that contributed to its fabled sound" (2002: 214).[29] Such instability is called "analog drift" and may be due to changes in temperature and to interference, strain or other characteristics or imperfections of the internal circuitry.

"Analog drift" is a catch-all phrase used frequently by synthesists as a way of describing why old synths sound better than new synths. Many of these characteristics were considered undesirable when the old synths were created. Many of them are subtle.[30]

My understanding is that "drift" usually is in regard to the effect of component temperature altering pitch slowly over time. All of the subtle variations are usually due to naturally occurring reactions in circuitry that have not been compensated for in the design. The flaw/feature effect. The kind of stuff a software designer is pulling his hair out trying to re-create/ simulate something that in its native environment, just happens naturally.[31]

Tuning, in general, was a significant problem with analog technology. Analog synthesizers were neither precise nor impervious to temperature change.[32]

On the other hand, the instability of the oscillators not only causes the synthesizer to become out of tune, but also hampers additive synthesis, that is, building sound by juxtaposing various sine waves. Hans Kulk explains:

> With the synthesizers, whatever unusual patches I made, it still sounded like "synthesizer". It was not the sound I heard in old electronic music. [...] I want to make sounds with sine waves, because I find these so beautiful ... [he turns on a sine wave sound] Such a wonderful sound ... because the sine waves have no overtones, they are pure. Well, there may be a little harmonic distortion, but with this test and measurement equipment this is so little that you don't hear it. And by adding several sine waves, you can make a new spectrum. However, with a synthesizer, you may read "sine wave" on the front panel, but it is not a pure sine wave at all, it is full of harmonics! You can't do additive synthesis with that ...[33]

For additive synthesis, many stable sine wave generators are required. The problem of doing proper additive synthesis with commercial synthesizers, due to their sine wave oscillators being too few and too unstable, is also described by Peter Manning:

> [A]ny prospect of achieving accurate Fourier synthesis by combining generators was usually overshadowed by problems of frequency stability. Standards of reliability and accuracy were all too frequently sacrificed in favor of commercial design economies, not always serving the interests of more serious composers.[34]

While others found the solution for more accurate additive synthesis in developing digital technology, Hans Kulk went back to the roots of electronic music by using analog devices that are similar to the ones used in old studios like the WDR Studio in Cologne in the 1950s. And instead of the few that were used then, he collected and combined numerous of such original analog devices. Mixing sine waves into more complex sounds became one of the basic techniques of Studio 1 and is also taught to students.[35]

These test and measurement devices are of such a high quality, Hans Kulk argues, because they were originally intended for usage of vital importance, such as radio, telephony and scientific research. For example, ring modulators were used to transpose signals into a higher frequency domain to facilitate their transmission, and later by composers such as Karlheinz Stockhausen to create music[36]; and amplitude modulation (AM) and frequency modulation (FM) are techniques used for radio first and later for synthesizing electronic music sound. The oldest devices are the most durable, in particular the ones that are electronically passive and don't need power, Hans Kulk explains: "they are so fantastically well-built, they are indestructible with normal use".[37] He is delighted how sturdy the old German devices are constructed, and that the Hewlett Packard equipment from the 1960s was built according to "military standards".[38] The modularity of these old systems, and of analog computers, has a similar background: the modularity served the continuous functioning of vital systems—if one module was defective, it could be replaced by another quickly, while the broken one was repaired. Hans Kulk found documentation of such systems since the 1920s, referring enthusiastically to pictures of rooms filled with enormous rack systems.

Modularity

Thus, by assembling Studio 1, Hans Kulk changed his focus from the modular sound synthesis of the ARPs to another type of modularity, that of the test and measurement equipment. This brings us to the question of what this modularity entails and how modularity relates to the practices of Studio 1. The modularity of the systems for which the test and measurement equipment was originally made served two purposes. As mentioned above, quick replacement to allow continuous functioning of systems was one advantage of modularity—when defective, modules or components could be replaced by identical ones, with the same function, to keep the system intact. Another advantage was that it allowed manufacturers to produce basic function modules in series, while various modules could be combined to build a customized system for a specific purpose like a telephone exchange center. What forms of modularity can be discerned in modular synthesis?

Modular sound synthesis is possible because the physics and theory of (electronic) sound are engrained with modularity. The idea that a sound can be considered consisting of an addition of sine waves, closely related to the mathematical procedure of Fourier analysis, is central to the electronic production and analysis of sound; in this respect, the sine wave functions as the fundamental module of sound. In the Helmholtz Sound Synthesizer, invented in 1858, several pairs of tuning forks and resonators function as sine wave sound generator modules.[39] Amplitude modulation and frequency modulation, used before the advent of modular synthesizers in for example radio technology, are also examples of modular conceptions of electronic sound generation. The source–filter model of the human voice[40] is another example of a modular model for the production and analysis of sound.

An obvious form of the modularity of modular synthesis is the practice of patching by connecting the modules of a specific modular synthesizer or rack in such a way that various sounds and sound patterns are formed. This was the way Hans Kulk used the ARP 2500 and ARP 2600, without a keyboard, but

FIGURE 3.4 Hans Kulk (left) and Ernst Bonis at the large table with mixer, computer, tape recorders and various other devices, at the left side of Studio 1. Photo by Hannah Bosma, 12 December 2022.

with a self-built control panel. For a substantial part, such patching may result in automatic processes and patterns of sound. Thus, after patching, the musician becomes more like an observer than a player; the active, creative work consists of patching and listening, and not so much of direct, bodily musical expression as with conventional musical instruments. Missing such bodily musical expression was for Hans Kulk the reason to change gear and leave the ARP synthesizers. In Studio 1, the focus is not so much on patching and changing patches, but on working with a specific patch and changing the sound manually by manipulating faders, dials and knobs. The setup in Studio 1 functions as a large musical instrument that one gets acquainted with through practice and playing; students get lessons, for example, in mixing sine waves and other sounds by hand and ear, and in live improvised controlling of amplitude (loudness) and left-right movement of sound and sound processing with reverb and ring modulator by moving the faders of the mixers.[41] When Ernst Bonis[42] was working with Hans Kulk to implement a patch of Jaap Vink[43] for "Multiplied Feedback", the patching, choosing and connecting the modules, didn't take much time; but then an important part of the work had yet to start: the fine tuning of all levels in the patch is crucial and is done by trying with hands and ear (see Figure 3.4). Or, as Ernst Bonis wrote about this patch: "The parameter settings and modulation options are very decisive for the possible sound output to be formed. [...] In short: with subtle settings, equally nuanced tweaks and listening patience you can explore Multiplied Feedback best".[44] Jaap Vink, a teacher and source of inspiration for Hans Kulk and Ernst Bonis, was indeed using the electronic music studio for improvisatory performance: "To some extent Jaap Vink's pieces are indeed recorded live improvisations, and extending his patches and 'rehearsing' with them was an ongoing process. To see Jaap Vink at work in the studio was to hear the studio coming to life".[45] On the other hand, for many current enthusiasts of modular synthesis patching, making and changing many different patches is the locus of their activity. Instead of stressing continuity—the kind of continuity that allows musicians to develop a deep, musical, visceral and intuitive relation with their instrument—many modular enthusiasts highlight change: "an instrument that can shift function instantly & be different every time / it offers freshness & choice & the unexpected".[46] While Hans Kulk is happy to see the revival of modular synthesis, in the current Eurorack trend he often misses the effort to develop a musical instrument by thinking carefully about setup and purpose and by studying and practicing to learn the instrument well.[47] Such a distinction between continuity and change comes to the fore as well in the different attitudes towards the notation of patches of Hans Kulk versus students and visiting artists. Hans Kulk stresses the importance of notating patches with block diagram notation, including measurements of frequencies, amplitudes, modulation depths, etc., to remember specific patches (see Figures 3.5 and 3.6). "But most young musicians are lazy. If they discover that they cannot 'save' the patch instantly, they are taking a photo of it ... . such nonsense ... . I try to teach them to draw diagrams, but it is difficult".[48]

Next to the modularity of the physics and technology of electronic music and the modularity of the usage and functioning of a specific modular device by way of patching, yet another form of modularity relates to the build of modular systems. In the 1970s, an ARP 2500 was a custom order with a specific combination of modules, cabinets and keyboards.[49] Hans Kulk assembled Studio 1 by carefully collecting and selecting various specific test and measurement devices throughout the years; these devices can be seen as the modules of Studio 1. Contemporary Eurorack enthusiasts tend to expand their systems continuously by buying more and more of all kinds of modules. Thus, each of such modular systems is different, depending on the specific combination of modules. The modularity of the build of the old modular synthesizers (such as the ARP 2500) was concentrated on the phase of acquiring the system, after which it mostly stayed the same. Hans Kulk's Studio 1 has also been stabilized after its formation: although Kulk may dream about endlessly expanding it to the size of a large hall, in fact he decided to stop acquiring more devices. This contrasts with the tendency of Eurorack enthusiasts to keep on expanding and changing their systems by

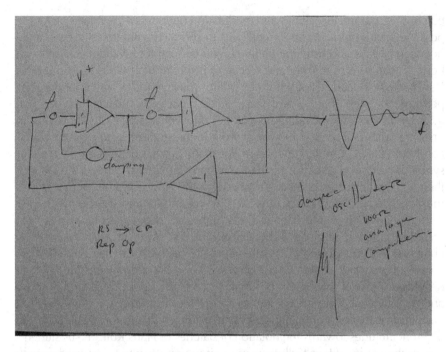

FIGURE 3.5   Example of a block diagram, with a patch for a damped sine oscillation on the analog computer, resulting in a decaying sine tone. Up to six of these circuits can be wired on the analog computer. These are used as part of larger patches, for example controlled rhythmically or in combination with ring modulation. This is used by Hans Kulk, Andrea Taeggi and others. (Telephone conversations with Hans Kulk, 5 and 6 April 2023.) Diagram by Hans Kulk, 31 March 2023.

buying new modules: "modularity allows you to create an instrument that grows with you";[50] and Hans Kulk suggests that a similar dynamic was taking place in the 1990s with the "plugin rage" for computer audio applications. Both Hans Kulk and Eurorack enthusiasts stress that each new module brings an enormous expansion of possibilities for new combinations. "Each new device almost doubles the musical sound palette and options of Studio 1", Hans Kulk elucidates.[51] "Each new addition opens up fresh possibilities for all the others", is the explanation in an introductory video on modular synthesis.[52] Thus, Hans Kulk's decision to not add any more devices to WTS Studio 1 is in line with his emphasis on considering it a musical instrument that needs time and stability for learning and for developing a close and subtle embodied relation. "I don't add any more new devices. But I keep on studying and revisiting all kinds of aspects of the whole thing. And this produces new ideas, for new patches". Moreover, he enjoys the novelties

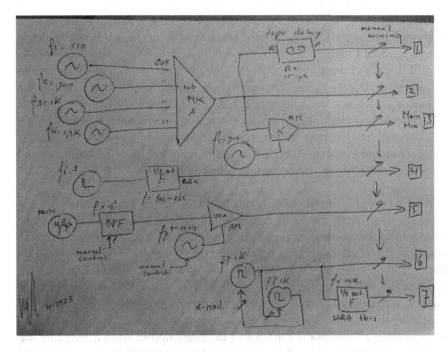

FIGURE 3.6 Example of block diagram notation. Signal source frequencies are given in Hz. Diagonal arrows indicate that signals are varied according to plan or by improvisation. Diagram by Hans Kulk, 11 April 2023. For this patch, various devices of WTS Studio 1 are connected to produce a sound mixture that can be varied manually, playing it as a musical instrument. The patch contains: four sine waves mixed together (2), and parallel processed through tape delay (1) and by a ring modulator (3); a low frequency periodic pulse, filtered for sound color (4); noise filtered for sound color and amplitude modulated for tremolo – both the center frequency of the filter band and the speed of the tremolo are controlled manually (5); two pulse oscillators which are cross-modulated for complex spectra and also filtered (6 and 7).These signals are varied and mixed at the final main mixer while recording. (Email Hans Kulk, 11 April 2023)

that guest musicians introduce by combining analog and digital technologies. However, he is critical about the Eurorack enthusiasts who keep on buying new modules and do not take the time to thoroughly learn the system they have. He considers this a "kind of consumerism".[53] Indeed, complaints of the high financial costs of buying new modules is a common trope on social media. In contrast, some see a positive side to their financial limitations when it restricts their choices, providing continuity in their setup, which enables them to develop a musical relation with their instrument.

I wanted to +1 the notion that I kind of like the restrictions the cost of modular puts on me. The time it takes to save up for each new module ensures that I make a very educated choice about my needs and what to add next, and the gap in between new modules really gives me time to somewhat master those that I already have. As Knobs mentioned in the video: each piece you add not only adds its own functionality but unlocks new potential in all of your existing modules, so with each addition there's always plenty to chew on for a good long time. If I were able to afford a new module every paycheque I probably wouldn't have the intimate level of understanding of my personal system that I do.[54]

Another level of modularity relates to the inside of the modules. Are the modules themselves modular? Basically, all old electronic devices are more or less modular inside, since electronic circuits are made with more or less standard components such as resistors, capacitors and diodes; the introduction of silicon transistors was crucial for the development of the first modular synthesizers of Moog and Buchla.[55] Because of this, old electronic devices can be repaired when the components and a diagram of the circuit are available; this is the case with the test and measurement equipment of Studio 1. However, an electronic circuit can be integrated in such a way that the components are not accessible. For example, in the ARP 2600 the filter was encapsulated in an epoxy filled plastic block so that it couldn't be taken apart, for the protection of the components and to prevent moisture coming in, but also to prevent others figuring out the circuit and thus to prevent reverse engineering and the discovery of ARP's infringement of Moog's filter patent.[56] Electronic microchips are tiny, mass produced integrated circuits in which all the components are integrated. They are small and powerful and the basis of digital technology, but cannot be taken apart and repaired like the old electronic circuits. They have become components or modules in themselves. Moreover, integrating components does not only happen on the level of the microchips; more and more devices are internally glued or melted together—perhaps protecting the components, but also devised as a business model, hindering repair and adaptation by customers or third parties.[57] Newer synthesizers, for example, that are made as inexpensive as possible, are looking slick but are not repairable. Thus, components or modules can become "black boxes", entities with a specific function and an inaccessible, hidden inside.

Preservation and Interpretative Flexibility

The collection of old equipment of the Willem Twee Studios raises the question of its preservation. Collecting and keeping old electronic instruments and devices is the first step of their preservation. However, conservation of these rare instruments is not the main focus of Willem Twee Studios. It is not a museum. The few museums that have such instruments in their collection, mostly display these as objects, without any cables or cords attached and no electric power, and thus without any use function. A few specialized electronic music heritage institutions offer some limited options to play some instruments.[58] In contrast, playing the instruments is the main focus of Willem Twee Studios. One may argue that a musical instrument is not a musical instrument if it is not being used for making music, and that using the electronic music devices is the only way of really preserving the instruments as instruments, instead of conserving non-functioning objects as remnants of the past.

To keep the instruments functioning, they must change. Repair is sometimes needed. Some very skilled and dedicated engineering specialists are associated with Willem Twee Studios, who are able to repair all the old devices, whether these are based on vacuum tubes or microprocessors. Old service documentation is carefully collected for all devices. When devices are repaired, this is documented as well, to keep track of the technical history of the individual devices. There are spare parts and old devices to take parts from to repair the main ones, but sometimes new parts or other solutions have to be found. And sometimes, the time has come that a device is not repairable anymore and that it must be abandoned.

On a more fundamental level, the current environment and use of the equipment of WTS Studio 1 are different from their original habitat, and in that respect, they are not the same instruments as 70 or 50 years ago. The original habitat of these devices was in large racks at laboratories and telephone companies and the like; in Studio 1 they have become modules in a completely different, new network. As discussed above, the test and measurement devices were originally used for telephony, radio, scientific research, etc.—not for music. Around 1950, radio studios started to be used for the production of electroacoustic music; at the WDR studio in Cologne some test and measurement devices were used for the production of *elektronische Musik*. This was already a case of interpretative flexibility of technology, because originally such equipment was not meant for avantgarde music composition. WTS Studio 1 is, as it were, a triple form of interpretative flexibility: it is a re-interpretation of the original interpretative flexibility of electroacoustic music, by taking not just a few devices, but by building, as it were, a modular synthesizer (an instrument that emerged after 1965) with older equipment from the 1950s and 1960s that had very different original purposes. It is a re-interpretation of the original purposes of the test and measurement equipment, a re-interpretation of the original use of such equipment for the early electronic music and a re-interpretation of the concept of the modular synthesizer. Thus, Studio 1 is preserving the old test and measurement equipment, but in a radically new environment, context and network.

Usually, interpretative flexibility, a concept developed within the Social Construction of Technology (SCOT) of Wiebe Bijker, Trevor Pinch and others, is seen in the phases of development or main usage of a technological artifact. The interpretative flexibility of WTS Studio 1 however is taking place after the artifacts have become obsolete. The interpretative flexibility is their preservation (otherwise the devices would have rusted away in some shed), but not a preservation of their original functioning and network—perhaps it is more like a rebirth. And the devices are not only used differently by a different social group (SCOT's interpretative flexibility), they are also functioning differently in/as WTS Studio 1 because of being part of a different network of material objects, technologies and people, in the sense of Actor-Network Theory (ANT). Yet another way of looking at the formation of Studio 1 is that of artistic archiving: making connections between various found objects with a "will to connect" that might be related to the institutive archival art described by Hal Foster.[59]

This raises the question of preservation on yet another level: that of Studio 1 itself, as a unique assemblage of equipment, practices, knowledge and people. Hans Kulk is worried about finding successors, because a combination of various specific kinds of expert knowledge and skills is crucial for the functioning of Willem Twee Studios: technical, musical and social. "Ear training and solfège are a must, for example to tune and set the oscillators. It is not a matter of just pushing a button, it has to be done by ear. But hospitality is another quality that is essential for these Studios. And of course the technical knowledge, especially of the very rare devices".[60] There are a few talented and interested young persons involved in the Willem Twee Studios who provide such a perspective. With one of them, sound editor and artist feferonja, Hans Kulk is making a series of "audio bulletins" , short podcasts of 8–10 minutes each in which he discusses one particular technique or aspect of electronic sound synthesis with sound examples. Initially, he wanted to write a bulletin on each device of Studio 1, about its function, what it was used for and how it is used in Studio 1, what its relations are with other devices, with references, etcetera, but that turned out far too time-consuming.

Instauration, Consolidation, Constauration

Change and flexibility belong to the core features of modular synthesis: "an instrument that can [...] be different every time".[61] As such, it might seem an overdrive of the *instauration* of music – this notion articulates that (seemingly) fixed musical entities, such as musical instruments or musical compositions, are produced, developed and "made to work" again and again in musical practices, such that Johann Sebastian Bach's music is nowadays not the same as it was in Bach's days, while performing and listening practices were, and still are, being developed after the fact of its original composition in such a way that without these practices the music would not exist and appreciated as it is now.[62] To exist, music must be re-made again and again. This notion of instauration, coming from the work of Étienne Souriau, was introduced by Antoine Hennion to overcome the reification of musical phenomena and concepts and to articulate the processes of production, development and renewal that form music.[63]

However, as discussed above, a practice of continuously changing configurations and patches is seen by some as a danger for musical development; likewise, the fast innovation and turning into obsolescence of electronic music technology is a danger for the preservation, the prolonged existence, of electroacoustic music[64]—there is a risk of "destauration", as Hennion suggested.[65] Hans Kulk stresses the importance of not endlessly expanding and changing WTS Studio 1, but keeping the configuration stable for a longer period of time to have the opportunity to develop profound, embodied knowledge of the system, like a musician with their musical instrument. He sees the tendency to change too much and too fast as a flaw of many current modular synthesis and other electronic music practices: "those who go with this flow of fast changing technologies, with what is new and hot—they have to learn to play a new instrument again and again. There are only a very few who are able to develop expert knowledge and skills of various instruments and techniques, they are very rare exceptions".[66] Suzanne Ciani presented a similar argument in her report, written in 1976, of her musical work on a Buchla synthesizer:

> In the practiced performer, a kind of instinct comes into play, and making a transition from one musical material to another is almost a matter of reflex [...] I find that the best performances combine the competence of pre-planned and well-rehearsed playing with the magic of being able to follow one's inspiration when inspired by the audience and the moment. To do the latter, a performer must be familiar with his patch to the point of not having to "think twice" (at least not more than once) about what effect or series of consequences will be produced by a given action.[67]

> [On rhythmic improvisation:] Basically, one must be completely familiar with the rhythmic options of a patch before being able to extemporize. With practice, one can develop the mental and physical reflections to "stay on top" in a performing situation and to play the sound and the space with total control and expressiveness.[68]

Similarly, Michel Waisvisz wanted to develop a musical and virtuosic relation with his electro-musical instrument The Hands, like musicians with their conventional musical instruments, and therefore didn't change

The Hands for long periods.[69] And like many electronic music musicians, Hans Kulk has one computer, containing applications that he likes to use and that work well, that is not connected to the internet and that is not updated or upgraded and thus will not change, thus circumventing the planned obsolescence and forced rapid changes of the contemporary computer industry—changes that hinder users to learn and get used to their applications, their instruments and tools, well.

Whereas classical music tends to be reified into an "imaginary museum of musical works",[70] for electronic music, in an environment of late capitalist rapid technological change and planned obsolescence, the stakes may be different.[71] How to describe the consolidation of a technological configuration, an instrument, a studio, a patch, to allow learning it profoundly, to adjust and get adjusted, to develop an integrated relationship of artist and instrument, to form an attachment?[72] The notion of instauration contains both sides—the process and the stasis, the fluidity and the fixity:

> The word instauration expresses [...] [the] idea of incomplete worlds, made of realities that are then 'calling us' because they need to be sustained to get 'more' existence, as [Souriau] puts it in a radically non-dualistic way.[73]

However, the term instauration may seem to accentuate the renewal more than the need for stabilization. The notion of attachment is referring to the process of artist and instrument forming each other in an intimate bond—but how to refer to the precondition of stability of the instrument? The term consolidation might work: a process of becoming firm together, instrument and artist—without absolute fixity, because what is solid can melt or break down later. But if the term consolidation has too many connotations related to fixity, to power and to corporate capitalism—what about *constauration* to refer to this process of profound and subtle embodied tuning, adjustment and alignment of artist and instrument that requires a relatively stable and restricted environment?

Attachment

WTS Studio 1 is, as it were, a giant synthesizer, with more than 60 modules and the size of a room. In this magnified form, it shows why modular synthesis is so attractive that it revived decades after it was replaced by mass produced integrated synthesizers and computers: the bodily relation with physical objects, connected by hand. Since new Eurorack modules may be (partly) digital inside, it is not "the analog" per se that is characteristic for modular synthesis; more important is the absence of the usual computer interface—screen, keyboard and mouse—and a way of working that is based on bodily movement and hearing instead of seeing.[74] Nevertheless, WTS Studio 1 also reveals tensions and discrepancies in modular synthesis practices. While modularity refers to the possibility to easily change modules and connections between modules and thus the complete system, too much change may hinder musicianship. A particular system needs to be relatively stable and lasting to become a musical instrument: so that the musician may develop a bodily, experiential relation with this particular system. The functional design is very important for this relation as well—related to the importance of the physical, bodily character of such a system. The process of developing such an attachment between musician and modular synthesis instrument, balancing flexibility and stability while fine tuning both musician and apparatus in a reciprocal process, requiring and developing a relatively stable configuration, may be called *constauration*. Modularity makes it possible to efficiently produce options for customization that make it easy to build one's own instrument, but for the configuration to function as a musical instrument, constauration—stability, consolidation and time to develop an attachment—is essential. One may wonder whether such a tension might be inherent to modularity in a wider sense, in relation to organization and production in a neo-liberal, late capitalist society.

Test and measurement equipment was originally not designed and produced for making music. Their use in 1950s electronic music was a form of interpretative flexibility; the development of such technologies for building synthesizers was a further form of interpretative flexibility; and Studio 1 is yet another form. The broader phenomenon of the modularity of the technological and cultural world, and a modular way of thinking and dealing with the world, was a condition to do so. Instead of choosing modules from a specific, limited, predefined collection, such as ARP's modules or Eurorack modules, for Studio 1 a broader, heterogeneous environment formed by rapid technological developments, accessed via older engineers and enthusiasts, functioned as a reservoir or an archive of potential modules, hidden in sheds, discarded by laboratories, dug up from history. The love for these objects and a will to connect opened up their potential to form an assemblage they were not designed for: a giant modular synthesizer, as a musical instrument, consisting of a heterogeneous but carefully selected configuration of devices, developed by musical attachment. While the formation of WTS Studio 1 as an exceptional, giant modular synthesizer was possible because of the modularity of the original technology and of the modular thinking that goes with electronic sound synthesis, it also resists some aspects of contemporary modular technological culture: it goes against the rapid obsolescence of technological devices, instead giving the old test and measurement equipment a second life, and against the ongoing, restless change of technological systems driven by innovation and commercial dynamics, instead allowing a process of attachment, commitment and flexible integration.

Acknowledgments

I thank Willem Twee Studios for welcoming me in their studios and at their events, and Hans Kulk, Armeno Alberts and Ernst Bonis for generously sharing their knowledge and expertise, for their time and for their comments.

This publication is part of the project Preservation as Performance: Liveness, Loss and Viability in Electroacoustic Music (with project number 016.VENI.195.508/6827) of the research programme NWO Talent Programme Veni which is (partly) financed by the Dutch Research Council (NWO) and hosted by the University of Amsterdam.

Notes

1 Video "Why to Modular - 1 – Overview" by Knobs, 10 September 2018, https://youtu.be/rY_7ktKvosY, accessed 10 October 2022. All access dates refer to the date of latest access.

2 See for example the plans for building large numbers of new houses as discussed in Dutch newspapers: Patrick Meershoek, "Bouwen moet sneller: 'Bouwen op een plek waar het huis moet komen is ouderwets'", *Het Parool* 3 September 2020, https://www.parool.nl/amsterdam/bouwen-moet-sneller-bouwen-op-de-plek-waar-het-huis-moet-komen-is-ouderwets~b03c5f10/, accessed 11 October 2022; Ton Voermans, "Een miljoen woningen erbij in tien jaar. Hoe gaan we dat doen?", *Het Parool* 14 March 2022, https://www.parool.nl/nederland/een-miljoen-woningen-erbij-in-tien-jaar-hoe-gaan-we-dat-doen~b9707f10/, accessed 11 October 2022; and Andrew William Lacey, Wensu Chen, Hong Hao, Kaiming Bi. 2018. "Structural response of modular buildings—An overview", *Journal of Building Engineering*, Volume 16, pp. 45–56, https://doi.org/10.1016/j.jobe.2017.12.008.

3 As for example in the restaurant Tucco Real Food Born, Carrer del Consolat de Mar 23, 08003 Barcelona, Spain, https://tuccorealfood.com; restaurant visited October 2019, website accessed 11 October 2022.

4 Blair, John G. 1988. *Modular America: Cross-cultural perspectives on the emergence of an American way*. New York: Greenwood Press.
Wilkinson, Rupert.1990. "John G. Blair, Modular America: Cross-cultural Perspectives on the Emergence of an American Way (Westport: 1988, £33.50). Pp. 165. ISBN 0 313 26317 5". *Journal of American Studies*, 24(3), pp. 452–3. https://doi.org/10.1017/S0021875800034009;
Calhoun, Daniel H. 1990 "'Modular America: Cross-Cultural Perspectives on the Emergence of an American Way' by John G. Blair (Book Review)". *The Journal of Interdisciplinary History* 20(3) p.510–512;
Portes, Jacques 1989 "John G. Blair. — Modular America, cross-cultural perspectives on the emergence of an American way". *Revue française d'études américaines*. 42(1), pp. 472–3.

5 Griffiths 1981: 117–8. Griffiths, Paul. 1981. *Modern music: The avant garde since 1945*. London: J.M. Dent & Sons.

6 Manovich, Lev. 2001. *The Language of New Media*. Cambridge, MA: MIT Press.
7 Manovich 2015: 135–7. Manovich, Lev. 2015. "Remix Strategies in Social Media", in: Eduardo Navas, Owen Gallagher, and xtine burrough. 2015. *The Routledge Companion to Remix Studies*. New York: Routledge, pp. 121–39.
8 Kostakis, Vasilis. 2019. "How to Reap the Benefits of the "Digital Revolution"? Modularity and the Commons". *Halduskultuur: The Estonian Journal of Administrative Culture and Digital Governance* 20(1), pp. 4–19
9 See for example Wagner, Günther P. 1996. "Homologues, Natural Kinds and the Evolution of Modularity". *American Zoologist* 36(1), pp. 36–43; Callebaut, Werner and Diego Rasskin-Gutman. 2005. *Modularity: Understanding the Development and Evolution of Natural Complex Systems*. Cambridge (Massachusetts): The MIT Press.
10 See for an influential account Fodor, Jerry A. 1983. *The modularity of mind: an essay on faculty psychology*. Cambridge, Mass: MIT Press.
11 https://www.merriam-webster.com/dictionary/module, accessed 12 October 2022.
12 Callebaut and Rasskin-Gutman 2005: 3.
13 The Contactorgaan Elektronische Muziek was established in 1956 by the Gaudeamus Foundation in Bilthoven in the Netherlands. During its long history, the studio was located in various places, such as Hilversum, Arnhem and Amsterdam, and since 2006 it collaborated with WORM in Rotterdam after losing its funding. Later, its precious vintage equipment, like the ARP 2500, moved to 's-Hertogenbosch to become part of the Willem Twee Studios. Armeno Alberts was involved with CEM for a long time, in various roles, and was its director. For the early history of CEM, see Tazelaar, Kees. 2013. *On the Threshold of Beauty: Philips and the Origins of Electronic Music in the Netherlands 1925–1965*. Rotterdam: V2_Publishing, and Tazelaar, Kees. 2020. Walter Maas & the Contactorgaan Elektronische Muziek: A lifeline for electronic music in the Netherlands. https://issuu.com/gaudeamusmuziekweek/docs/gmw-jubileumboek-boek4-keestazelaar and https://gaudeamus.nl/en/jubileum/based-on-walter-maas-the-contactorgaan-elektronische--muziek-a-lifeline-for-electronic-music-in-the-netherlands-by-kees-tazelaar/ (accessed 28 January 2024). Armeno Alberts gave an overview of the history of Willem Twee Studios and of CEM in his introduction to the Studios at the expert meeting Modular Synthesis and Preservation in Electronic Music and Sound Art, 24 and 25 January 2024 at Willem Twee Studios, organized by Hannah Bosma.
14 Since 2019, the name of this museum has been changed into Kunstmuseum Den Haag (Art Museum The Hague) (Dollen, Christy. "Naamsverandering museum valt niet bij iedereen in de smaak: 'Zonde!'", Algemeen Dagblad 18 December 2018, https://www.ad.nl/den-haag/naamsverandering-museum-valt-niet-bij-iedereen-in-de-smaak-zonde~a0853a15/, accessed 14 March 2023; "'Onze naam is ongelofelijk onbekend', Haags Gemeentemuseum verder als 'Kunstmuseum'", Omroep West 18 December 2018, https://www.omroepwest.nl/nieuws/3739448/onze-naam-is-ongelofelijk-onbekend-haags-gemeentemuseum-verder-als-kunstmuseum, accessed 28 January 2024).
15 See https://www.willem-twee.nl/agenda/studios and https://youtu.be/AGjRfTupntg?si=Nh4zAkf6bmLdkyOE (accessed 28 January 2024).

   This article is based on Hannah Bosma's contacts with and visits to Willem Twee Studios in 2019–2024, attending to and participating in workshops, presentations, events and concerts, and interviews and conversations with Hans Kulk and Armeno Alberts in person and via internet video conferencing, telephone and email. Due to the Covid-19 crisis, in 2019-2022 contact took place mostly via internet video conferencing. All interviews were in Dutch; translations of quotes by Hannah Bosma.

16 Email Ernst Bonis, 28 Februari 2023.

17 This list is not complete. Information acquired via https://www.willem-twee.nl/agenda/studios#studio2 (accessed 14 March 2023), Ernst Bonis (email 28 Februari 2023) and Hans Kulk (email 3 March 2023).

18 11 November 2012 at Cineac Sonore in the festival November Music in the Verkadefabriek, 's-Hertogenbosch, https://www.novembermusic.net/terugblik (at that time working at Music Center the Netherlands I organized Cineac Sonore in collaboration with November Music); interview with Hans Kulk on this project: https://youtu.be/sJnn_w9gFFA (accessed 11 January 2023); "November Music in the Netherlands". 2013. *Computer Music Journal* 37 (2), p. 9.

19 See Sagasser, Ineke. 2021, "Beeldend kunstenaar José op ten Berg", *De Wijkkrant: Muntel, Vliert en Orthenpoort* 73, June 2021, pp. 12–13, https://buurtkiep.nl/lokale-media/juni-2021/3226; Sagasser, Ineke. 2021. "Elektronische-klankmusicus Hans Kulk", *De Wijkkrant: Muntel, Vliert en Orthenpoort* 75, December 2021, pp. 4–5, https://buurtkiep.nl/lokale-media/december-2021/3456; "Expositie Verfklanken met beeld en geluid bij Dock 5340 in Den Bosch", *Brabants Dagblad* 6 December 2016, https://www.bd.nl/s-hertogenbosch/expositie-verfklanken-met-beeld-en-geluid-bij-dock-5340-in-den-bosch~a6e9e685/; https://www.joseoptenberg.nl/exposities/ and https://www.joseoptenberg.nl/verfklanken/ (accessed 11 January 2023).

20 Interview with Hans Kulk on 30 March 2022.

21 Ibid.

22 The ARP 2600 came from the Tilburgse Dans- en Muziekschool, where Ernst Bonis had a studio to give courses, and is now placed in WTS Studio 2 (email Hans Kulk 12 January 2023, email Ernst Bonis 28 February 2023).

23 This ARP 2500 consisted of two portable wing cabinets and was taken from the basement of the NOB building at the Oude Amersfoortseweg in Hilversum, the Netherlands (email Hans Kulk 12 January 2023).

24 Wells, Thomas H. 1974/1981. *The Technique of Electronic Music*. New York: Schirmer Books.

25 For example: "[M]any analog synthesizer functions and special-purpose digital devices designed for electronic music have parallels in computer systems. The CEMS System, for example, is conceptually akin to an analog computer" (Chadabe 1975: 176). This Coordinated Electronic Music Studio System was developed by Joel Chadabe and Robert Moog and installed at the Electronic Music Studio at the State University of New York at Albany in 1969. It was a voltage-controlled system with extended automation capability, so that even an entire composition could be automated (Chadabe 1975: 168). See also Chadabe 1997: 286–8 and Holmes 2016: 260–61. This automated synthesizer system contained "the world's largest concentration of Moog sequencers under a single roof" (Chadabe 1997: 286). Chadabe, Joel. 1975. The Voltage-controlled Synthesizer. In: Appleton, Jon H. and Ronald C. Perera (eds.) *The Development and Practice of Electronic Music*. Englewood Cliffs, N.J: Prentice-Hall, pp. 138 – 188. Chadabe, Joel. 1997. *Electric Sound: The Past and Promise of Electronic Music*. New Jersey: Prentice-Hall. Holmes, Thom. 2016. *Electronic and experimental music: Technology, music, and culture*. Fifth edition. New York, NY: Routledge.

26 This was a hts (hogere technische school). In the Netherlands, the hts was a form of higher education in engineering with a duration of four years; the grade had the same level as a bachelor grade. Nowadays the name hts is replaced by the umbrella term hogeschool or hbo (hoger beroepsonderwijs), which is often translated as University of Applied Sciences. The analog computers were both a Hitachi model 240. Hans Kulk suspects that this type of computer was specifically

made for the Dutch hts institutes in circa 1970–72, because none of these devices are seen elsewhere; he received this information by talking with and receiving documentation from the Dutch importer Sevanco (email Hans Kulk 12 January 2023).

27 The ADSR envelope generator was specified in a request of Vladimir Ussachevsky to Robert Moog in 1965 for the Columbia-Princeton Electronic Music Center (Holmes 2016: 259–60; Pinch and Trocco 2002: 59).

28 Interview with Hans Kulk in WTS Studio 1, by Hannah Bosma, 28 December 2020, via internet video conferencing.

29 Pinch an Trocco. 2002. *Analog Days: The Invention and Impact of the Moog Synthesizer.* Cambridge, MA: Harvard University Press.

30 Jinsai on 1 March 2010 ,4:41pm, in "What exactly is analog drift?" in the forum on Synthesizers & Samplers / General Synthesizers, of Vintage Synth Explorer, https://forum.vintagesynth.com/viewtopic.php?t=54750, accessed 25 October 2022.

31 Solderman on 1 March 2010, 6:17 am, in "What exactly is analog drift?" in the forum on Synthesizers & Samplers / General Synthesizers, of Vintage Synth Explorer, https://forum.vintagesynth.com/viewtopic.php?t=54750, accessed 25 October 2022.

32 Chadabe 1997: 157.

""Acceptable" drift is highly subjective. By most modern standards, "bad" drift will have you retuning within a couple of hours, if not sooner. Don Buchla considered the Buchla 100 oscillators stable "enough", needing to be tuned about as much as you'd need to tune a violin during a concert. (1.5hrs, roughly?)", Yes Powder on 18 July 2022, 1:31pm, in the forum Modular Synths / Modular Synth General Discussion of ModWiggler, https://modwiggler.com/forum/viewtopic.php?t=263952, accessed 25 October 2022.

See also Pinch and Trocco 2002: 258–261 for the improvement in stability of the oscillators in the ARP 2500 in comparison to the older Moog modular synthesizer.

33 Interview with Hans Kulk in WTS Studio 1, by Hannah Bosma, 28 December 2020, via internet video conferencing.

34 Manning 2013: 116. Manning, Peter. 2013. *Electronic and Computer Music.* Fourth edition. Oxford University Press.

35 Document Willem Twee Studios Synopsis Studio 1 lesson 10 May 2020. Nevertheless, perfect addittive synthesis, in the strict sense, may not be possible in Studio 1 due to distortions introduced by the mixing consoles and other elements of the set-up. (This was discussed in the workshop / expert meeting with Armeno Alberts in WTS Studio 1 on 25 January 2024.)

36 See for example: Davies, Hugh. 1976. "A Simple Ring Modulator", *Musics* 6 (March–April 1976): 3–5.

37 Interview with Hans Kulk in WTS Studio 1, by Hannah Bosma, 28 December 2020, via internet video conferencing.

38 Interview with Hans Kulk on 30 March 2022, by Hannah Bosma, via internet video conferencing.

39 Pantalony, David (2004) "Seeing a voice: Rudolph Koenig's instruments for studying vowel sounds", *American Journal of Psychology,* 117(3), 425–442.

Helmholtz, Hermann von (1859). "Ueber die Klangfarbe der Vocale". *Annalen der Physik und Chemie* 108, 280–90.

Helmholtz, Hermann von (1863). *Die Lehre von den Tonempfindungen als physiologische Grundlage für die Theorie der Musik.* Brunswick, Germany: Friedrich Vieweg und Sohn.

40 Fant, Gunnar. 1960. *Acoustic theory of speech production*. The Hague: Mouton.
41 Document Willem Twee Studios Synopsis Studio 1 lesson 10 May 2020.
  Hans Kulk stresses the importance of pulling the students away from their computer into a tactile relation with the analog equipment, based on bodily movement and focused on doing and hearing, instead of seeing; he also stresses the importance of muscle memory (similar as a musician playing a musical instrument) when working with analog equipment and that this is missing when making music with a computer; Armeno Alberts stresses that this way of working is more related to the practice of musicians and that it may result in a different kind of music than when made with virtual modular synthesizers on a laptop (online video conference interview-conversation with Hans Kulk, Armeno Alberts and Hannah Bosma, 17 June 2020).
42 Specialized in analog electronics, sound synthesis and musical acoustics, Ernst Bonis worked as teacher of electronic music at the Conservatories of Maastricht, Utrecht and Rotterdam, Utrecht School of Music & Technology (HKU, University of the Arts Utrecht) and Netherlands Carillon School, as curator of the musical instruments department at Gemeentemuseum Den Haag, and freelance writer and advisor, inter alia (http://www.ernstbonis.nl, accessed 11 January 2023, and email Ernst Bonis 23 February 2023). He is an important contact and source of information and inspiration for Willem Twee Studios and visits WTS several times a year. I attended his visit at WTS on 12 December 2022.
43 Jaap Vink had an important role in the early history of electronic music in the Netherlands and was staff member, teaching analog studio techniques, at the Institute of Sonology until his retirement in 1993 (Tazelaar, Kees. 2013. *On the Threshold of Beauty: Philips and the Origins of Electronic Music in the Netherlands 1925–1965*. Rotterdam: V2_Publishing.). For Jaap Vink's music see the album (2LP) REGRM 018 EXT (Editions Mego, INA-GRM), http://editionsmego.com/release/regrm018ext, https://www.discogs.com/release/10125640-Jaap-Vink-Jaap-Vink (accessed 11 January 2023), https://recollectiongrm.bandcamp.com/album/s-t (accessed 4 April 2023). Jaap Vink gave his handwritten patches to Ernst Bonis (personal communication with Ernst Bonis 12 December 2022).
44 Bonis, Ernst. 2023. "Jaap Vink's Multiplied Feedback: Remarkable Physical Model-inspired patch from the 1960s", pp. 2–3. My source is the manuscript of this article that I received from Ernst Bonis, in English and Dutch versions, on 23 February and 26 March 2023 by email. This article has been published, in edited and curtailed form, in the Dutch magazine *Interface* 255, February-March 2023, and will be published completely on the website of Willem Twee Studios. Ernst Bonis states: "As far as I know, [Jaap Vink's Multiplied Feedback patch] is the very first musical instrument related analog patch inspired by the operation of a physical model with a time delay as the basic element" (manuscript p. 1 and email 22 March 2023) and argues that it anticipated the MSW and Karplus-Strong models (1983) that were developed in the context of digital sound synthesis (see Roads, Curtis. 1996. *The Computer Music Tutorial*. Cambridge, Massachusetts: The MIT Press, pp. 263–96, for an introduction into physical modeling, MSW and Karplus-Strong sound synthesis). Its exciter-resonator model, in which exciter and resonator are coupled, with its combination of delay and reverb (resonator) and feedback and ring modulation (interactor), gives Vink's patch its organic character: "The hallmark of such a system is the interaction which occurs between excitator and resonator, resulting in a repetitive balancing whole, resoundingly expressed in the dynamic structure of the output signal, both in amplitude and waveform. In perception this means dynamics in loudness and timbre, which is an all-determining characteristic of mechanical and biological acoustic sound sources" (Bonis 2023: 1). (Ernst Bonis uses the term 'excitator' instead of the

more common 'exciter' to avoid confusion, because this last term has another meaning as well in the realm of audio engineering; email 24 March 2023.) Bonis (2023) refers to a video of an instructive demonstration by Kees Tazelaar of an extended variation of the multiplied feedback patch in the analog studio of Sonology at the Royal Conservatoire in The Hague on YouTube: MrSonology, Ring-Modulated Feedback in BEA5, https://youtu.be/X_Bcr_HS9XM (accessed 20 March 2023). Bonis (2023) also refers to Wouter Snoei's composition *Desintegration,* for fixed media and four loudspeakers (1998), that is based on Jaap Vink's MFb patch in the digital domain, see https://www.woutersnoei.nl/en/ compositions/fixed-media/disintegration-2/ (URL accessed 28 March 2023). Ernst Bonis' son Joris Bonis, who was guitarist and sounddesign/soundscape creator of the extreme/black metal band Dodecahedron (2011–20), used complex variations of the MFb patch, amidst many other sound design elements, for the atmospheric sound layers of the album Kwintessens (2017); Jaap Vink's *Screen* (1968) was an inspiration for this band; email Ernst Bonis 2 April 2023; https://www.metal-archives.com/bands/Dodecahedron/3540329941 (accessed 4 April 2023); Béra, Camille. 2019. "De Platon à Dodecahedron: les apports des textures ligétiennes et de la musique électronique chez un groupe de Metal extrême". *Itamar. Revista de investigación musical: territorios para el arte* 5, pp. 143–61, https://ojs.uv.es/ index.php/ITAMAR/article/view/15824/14277 (accessed 4 April 2023), English version: "From Plato to Dodecahedron: The contribution of Ligétian textures and electronic music to the compositions of an extreme metal band", English version: https://www.academia.edu/41362627/From_Plato_to_Dodecahedron_The_con-tribution_of_Lig%C3%A9tian_textures_and_electronic_music_to_the_composi-tions_of_an_extreme_metal_band (accessed 4 April 2023); https://www. facebook.com/ddchdrn (accessed 4 April 2023).

45 Kees Tazelaar on the album REGRM 018 EXT (Editions Mego, INA-GRM), http:// editionsmego.com/release/regrm018ext (accessed 11 January 2023).
46 Video "Why to Modular - 5 – Modularity" by Knobs, posted 11 October 2018, on Youtube, https://youtu.be/YSwv2okfc2E (accessed 2 November 2022).
47 A comparable critique was expressed by Thomas Wells in 1974: "Although equipment designed specifically for electronic music production includes certain features of automation to facilitate sound generation, some composers never progress beyond the superficial capabilities of such time-saving refinements. Yet many early electronic compositions were produced using equipment not originally intended for the production of music; and, in the opinion of the author, some of these pieces remain the most impressive works in the literature of electronic music, both from technical and musical standpoints". Well, Thomas H.. 1974/1981. *The Technique of Electronic Music.* New York: Schirmer Books, p. xvi. Hans Kulk considers this book the best in its kind and with this book he started his learning process into modular synthesis.
48 Interview with Hans Kulk in WTS Studio 1, by Hannah Bosma, 28 December 2020, via internet video conferencing.
49 See "ARP 2500" on Vintage Synth Explorer, https://www.vintagesynth.com/arp/ 2500.php, accessed 3 November 2022.
50 Video "Why to Modular - 5 – Modularity" by Knobs, posted 11 October 2018, on Youtube, https://youtu.be/YSwv2okfc2E, accessed 2 November 2022.
51 Interview with Hans Kulk on 30 March 2022 via internet video conferencing.
52 Video "Why to Modular - 5 – Modularity" by Knobs, posted 11 October 2018, on Youtube, https://youtu.be/YSwv2okfc2E, accessed 2 November 2022.
53 Interview with Hans Kulk in Studio 1, 28 December 2020, via internet video conferencing.

54 Voicedrifter in a discussion in response to the "Why to Modular - 5 – Modularity" video, posted in 2018, https://youtu.be/YSwv2okfc2E, accessed 2 November 2022.

55 Pinch and Trocco 2002.

56 Pinch and Trocco 2002: 263.

57 Such hindering of repairment is not only annoying and expensive for the user, but is also detrimental for the environment, because of the resultant waste. To counter this anti-repair development, various initiatives have been developed, like the Repair Café movement started by Martine Postma in Amsterdam in 2009 (www.repaircafe.org), The Repair Association (www.repair.org), iFixit (www.ifixit.com). See also Postma, Martine. 2015. *Weggooien? Mooi niet! Het succes van het Repair Café*. Oss (NL): Uitgeverij Genoeg.

58 Such as the Swiss Museum for Electronic Music instruments SMEM in Fribourg, Switzerland, www.smemmusic.ch, and the Electronic Music Education and Preservation Project (EMEAPP) in Philadelphia, Pennsylvania, USA, https://emeapp.org/faqs/ (accessed 5 November 2022).

59 Foster, Hal. 2004. "An Archival Impulse" *October*, Vol. 110 (Autumn, 2004), pp. 3–22.

60 Interview with Hans Kulk on 30 March 2022 via internet video conferencing.

61 See above, video "Why to Modular - 5 – Modularity" by Knobs, posted 11 October 2018, on Youtube, https://youtu.be/YSwv2okfc2E (accessed 2 November 2022).

62 Antoine Hennion. "Music and Mediation: Towards a new Sociology of Music". In: Clayton M., Herbert T., Middleton R. *The Cultural Study of Music: A Critical Introduction*, London, Routledge, pp. 80–91, 2003. Antoine Hennion, *The Passion for Music: A Sociology of Mediation*, Surrey (UK), Ashgate, 2015.

63 The concept of instauration was central to a workshop I co-organized with Peter Peters, Denise Petzold and Floris Schuiling at the Lorentz center in Leiden, Music beyond Fixity and Fluidity: Preservation and Performance as Instauration, 12–16 September 2022, where Antoine Hennion participated and presented a keynote presentation and a final reflection.

64 The term "electroacoustic music" is here used as an umbrella term for various forms of musics, sound art and sounding arts where electronic music technology is essential. See the definition by Simon Emmerson and Dennis Smalley in *The New Grove Dictionary of Music and Musicians*: "Music in which electronic technology, now primarily computer-based, is used to access, generate, explore and configure sound materials, and in which loudspeakers are the prime medium of transmission". (Emmerson & Smalley 2001: 59; Emmerson, S. and D. Smalley. 2001. "Electro-acoustic music". In: S. Sadie & J. Tyrrell (eds.) *The New Grove Dictionary of Music and Musicians*. Second Edition. Volume 8. London: Macmillan. pp. 59–67. The preservation of electroacoustic music is a notorious problem, which is addressed in various research projects and publications.

65 Antoine Hennion introduced the term "destauration" in his final comments on 16 September 2022 in the workshop Music beyond Fixity and Fluidity (see above).

66 Interview with Hans Kulk in Studio 1, by Hannah Bosma, 28 December 2020, via internet video conferencing.

67 Ciani, Suzanne. 1976. *Report to National Endowment re: Composer Grant*, p. 1–2. Manuscript available via the shop on the website of Suzanne Ciani, https://www.sevwave.com/product-page/suzanne-ciani-s-report-to-national-endowment, accessed 6 November 2022. The report contains detailed information and discussion of patches and musical techniques and themes: "an outline of a "Basic Performance Patch" which I designed for a Buchla series 200 instrument, and a brief discussion of some of the musical ideas that evolved as a result of working with this patch". (p. 1)

68 Ibid., p. 20. And like Hans Kulk who refers to the importance of Human Factor Design for the devices he collected, discussed above, Ciani stresses the importance of the interface design of the Buchla 200 synthesizer:
"Don Buchla was an expert at designing interfaces. [...] What is the size of a human hand? [...] Don thought, well for a performance instrument, I need to maximize that intersection of the human needs for dimension and the needs of a portable instrument. [...] what he did was he took that position of the hand and designed a touch plate, a keyboard that doesn't look like a traditional keyboard and respects the natural position of the hand. He did this in all of his design. [...] You need to start with the human and then design the technology around that. So some of the new electronic instruments, Eurorack, have a very diminutive size and I'm sad about that because it's very difficult to interact with those modules, because they're so dense and condensed". (Zuegel, Devon. 2022. Suzanne Ciani explains the composition of her sensory career. Podcast and interview with Suzanne Ciani, published 4 January 2022, https://www.notion.so/blog/suzanne-ciani, accessed 6 November 2022.)

69 See Bosma, Hannah. 2017. "Canonisation and Documentation of Interdisciplinary Electroacoustic Music, Exemplified by Three Cases from the Netherlands: Dick Raaijmakers, Michel Waisvisz and Huba de Graaff". *Organised Sound* 22 (2), pp. 228–37, https://doi.org/10.1017/S1355771817000139. Hans Kulk mentions another example of an electronic music artist dedicated to one instrument: Oskar Sala and the (Mixtur) Trautonium (email 4 December 2022).

70 Goehr, Lydia *The Imaginary Museum of Musical Works: An Essay In the Philosophy of Music.* Oxford: Clarendon press, 1992.

71 This resonates with Latour, Bruno. 2004. "Why Has Critique Run Out of Steam? From Matters of Fact to Matters of Concern". *Critical inquiry* 30 (2): 225–48, where Latour worries about the imbalanced focus of (his circle of) academia on "'the lack of scientific certainty' inherent in the construction of facts" in a time of misinformation, conspiracy theories and denial of scientific consensus (227). Likewise, one may question a one-sided appreciation of instability in a time of growing socio-economic, socio-cultural and technological instability.

72 On attachment, see Hennion, Antoine. 2017. "Attachments, you say? ... How a concept collectively emerges in one research group", *Journal of Cultural Economy*, 10:1, pp. 112–21, DOI: 10.1080/17530350.2016.1260629

73 Hennion, Antoine. 2017. "From Valuation to Instauration: On the Double Pluralism of Values". *Valuation Studies* 5(1): 69–81.

74 "One of our main goals is to pull the students away from the computer", according to Hans Kulk on 17 June 2020 in a video conferencing conversation with Hans Kulk, Armeno Alberts and Hannah Bosma.

4

INTERVIEW

Dani Dobkin

Dani Dobkin on Repairing the Computer Music Center Buchla Instrument and Teaching Modular Synthesis

Phone discussion with Ezra J. Teboul and Dani Dobkin, D.M.A. candidate in Music Composition at Columbia University, 18 September 2020.

ET: Ezra Teboul

DD: Dani Dobkin

ET: How did you come to participate in fixing the CMC's Buchla system?

DD: Last summer, Brad Garton[1] became really friendly with his neighbor in Jersey. It turned out his neighbor was the head of the engineering department at Columbia, David Vallancourt.[2] Brad's like, oh, I teach at Columbia too. David was like, oh what a coincidence, I collect guitar pedals and I take them apart in my spare time. And Brad's like, well, we have a synthesizer's that's broken up at the CMC. And then they started talking about the Buchla and David mentioned he had so many students who would love to get the opportunity to work on the machine.

I don't have much of an engineering background. However, I know how to play synths. I've been working with modular for about 10 years now. I started playing the Serge system at Bard College in my second year of undergrad and ... Brad and David were like, you want to get together and try to fix this thing?

So the first thing about the Buchla: it had seen better days, and relatively speaking, it didn't work. Everyone said "well, it worked a

DOI: 10.4324/9781003219484-6

year ago ... " Piece by piece, we took it apart, cleaned everything and started to troubleshoot which areas were most likely to have burnt out.

It was my crash course in circuitry. I had taken a few classes in school, but nothing like opening up a historic synthesizer, following all the hand drawn soldering lines, and testing each individual component. Repairing something, you open it and you're like, well, the soldering seems good here and this looks like it works, and maybe we'll try something else. It was really interesting, a really eye opening experience. There are still some outstanding minor details that we'd still like to fix. We want to rebuild the power system, but that will have to wait until the pandemic dies down.

ET: You've been playing a modular for about a decade ... Had you opened the Serge or had you worked with electronics related to the Serge at all? You were at Bard or taking classes with Bob?[3]

DD: Well, just Bob's musical electronics class, which just gave a false sense of confidence. We had basic schematics and then in my brain, the translation from how a circuit gets built to how this actually works as a mechanism ... kind of just jumped over me.

Bob was at that level of proficiency with these things where it was like asking a mathematician how to do whole number multiplication. Sometimes we'd get answers that make you go "wow. Yeah, okay. wait, what?" But he was one of the most brilliant professors and educators that I ever had the pleasure of working closely with.

I also got a really good introduction to electronics with Douglas Repetto. He was the director of the Sound Art programme at Columbia during my time there. He had a very, "go try it yourself, Dani", kind of a method. He was happy to help you but wasn't going to baby you and do it for you. A real push to get me engaged with the material. And so from there I started reading the Nicolas Collins handmade electronics book.[4] My first year I ended up making a noise circuit and realizing that if you put your hand over the noise circuit a certain way, you became the antenna for a radio receiver. That was one of those pivotal moments. You build it, and the circuit tells you: "try again." And that was a fun process of learning. I'm still not an expert when it comes to it, but I now feel like I have a fundamental knowledge of what it is.

ET: Can you give us the details of troubleshooting either the Buchla's power supply or the filter?

DD: I think the first module that we took out and really tried to examine was that low pass filter. We powered everything up and we tested all the connection points one by one just to make sure that power was going through it and nothing was shorting. The overall system, on an external power supply, was semi-functioning. But when we got to the low pass, we realized it needed more attention. It was funny, funny in a dark way, a grad student and a few undergrads poking around on this historical synthesizer. "Let's see what's wrong with it. How could we go wrong?".

It was also a very paranoid experience! We were all trying our best not to disrupt anything permanently inside the system. Back to the filter- we took it out of the rack and unscrewed the three layer panel. We put the screws from the first panel in a little cup, and we put the little bolts in a different cup. We tried our absolute hardest to make sure that we preserved all the original mechanics and qualities of the way that it was already put in there. I don't remember the whole process, but I remember our feeling when we took it out.

ET: You said you remembered more about the power supply?

DD: So we extracted the power supply. Professor Vallancourt points out that the capacitors could still hold a charge, we should make sure there's no charge in it. And he does his electronic wizardry and discharges the capacitor just so we don't get shocked. Then he opens it up and says, "well, usually the capacitors are the first things to blow. Let's test the rest."

And so he examines the largest capacitor, one of the ones that they don't make anymore. It was the size of your thumb. And he was like, Hmm, maybe this is the one that went wrong- and he wiggles, spins and unsolders it and then takes it out. He tests it, it's fine. And he threw it back in there.

So he gave us confidence, saying go ahead, come on, you can do it. We opened up the power supply and saw that a component had melted and leaked. And so there was something that was shorting two other points of the circuit. I guess it caused it to spark, frying a rectifier.

There were three rectifier components and one had actually exploded. And when we were cleaning out the top of the system, we found the top part of one of them. We really didn't think that much about it- we were like, wow, this kind of looks like a piece of something exploded. So we kept it and then inside of the system was the other part of the rectifier and we were pretty shocked. This thing that we found first was the exploded component that launched itself across the system.

We never found out what was tripping the system. I mean, it could have been the wall voltage, Columbia doesn't have the cleanest power. It could have had something to do with the power surge or the wires that were in the back. The system was one step away from being held up with zip ties. I mean, zip ties would've been wonderful. It was one step away from being held with the old toothpicks, so a power cable sort of swung and hit something and caused a short.

William was the one who suggested "a lot of places sell vintage parts, I think it would do the system justice if we were to replace damaged components with replacements from the exact time that they were installed." William tried to make sure that he did that with the entirety of replaced parts for the Buchla. The sequencer has those little red lights, lights that predate LED's, that blink as the sequence progresses. He found the world's last distributor of those little red lights, and bought the remaining stock in case it ever broke down again. He really wanted to preserve the original magic of the system, which I greatly appreciated.

That's as much detail as I remember. Then we had to move the synthesizer back to Prentis from the engineering building. We called an Uber, and there we were, carrying the priceless 1960s Buchla synthesizer in a rickety old Uber, with one bag of broken components, another bag with the power supply. I don't know what that guy thought we were carrying, but it looked a little suspicious.

I was strapped into the back, holding onto the synth for dear life, it was one of those classic New York drivers, 'I'm going to make a left right here, and I'm going to shift lanes at the last possible second'. It was maybe the longest six blocks of my life.

We were originally going to take it on a cart. I think that's how it got there. I think Brad rolled the synthesizer uphill in an old CMC cart 12 blocks to main campus. Low tech meets high tech I guess.

ET: How has this process of repairing the Buchla changed your synthesis practice, if at all? You perform mostly on a Serge synthesizer.

DD: It's given me a lot of confidence. The Bard College Serge had a loose nut, and a malfunctioning LED. And so I was like, I can open this, I can fix it, and so I did. So that doesn't really inform my practice, but it makes me feel like playing is putting your body in relationship with these machines. Each module is kind of its own ecosystem. You're playing with your ecosystem and you're connecting them together. Whether it is new technology or it's old electronics, it still has a very organic process of communication and connection.

ET: A lot of people are working with digital emulation. You are teaching a class using VCV rack at Columbia?

DD: VCV Rack is a monster in itself. I mean, I was playing with it the other day and I've spent so much money and time building my own physical machine when I could just open the app and play it. Then I remember the reason I started building a modular was so I didn't have to stare at my computer. You can really only easily do one thing at a time. You only have one mouse to click on something with. I have two hands to turn knobs.

But I think VCV, if anything, it's made the modular world a vastly more accessible. I think hardware having a high price range and having a lot of white man history are all forms of barriers. And I love having that kind of an accessible system inside software. Sonically, it's still similar, and although clicking and dragging isn't exactly the same, it means you can teach modular over Zoom during a pandemic.

And now I'm questioning my own mantra. I'm staring at my modular system and asking why it got very heavy, expensive ... But I do enjoy it.

ET: Has this repair process and the confidence you got from it inform your ideas for future instruments you might want to build?

DD: I'm not sure if it informs my work. I think if anything teaching has been more of an influence, I've been doing that for five years now ...

ET: Then how has teaching synthesis techniques and composition on the Serge influenced your work and which parts of your work exactly?

DD: It informs my whole practice as an artist working with live electronics. You're working with this machine, and all of a sudden the room gets warmer and your pitch shifts. I think I've always used working with modular as a kind of a collaboration, in a way that working with VCV Rack isn't because I feel like I'm telling the computer what to do. When I work with the hardware it's a give and take. This has also influenced my visual art practice. I've started working with clay and ceramics. That's a whole part of my work now. And I think the modular and ceramics have a completely similar thought process. It's a lot of brain to hand and object. It has that same scale of ideas.

You're building with your hands and in a direct correlation between your head and your arms and your body and your output. And I think they share that very special bond. And there's also that kind of materiality, if you have clay that behaves in a funny way, or if you leave it out for too long or leave it out for not long enough, they're both such volatile instruments that they just have a similar way of communication. There's also a different relationship to expertise, because of this volatility. You might arrive at some forms through incremental interaction that you can never return to. There's a different approach to control, where expertise and repeatability don't have to be tied together. All together these mediums have been a beautiful study in ephemerality.

Notes

1 Editor's note: Brad Garton is Professor of Composition at Columbia University's Music Department. He was director of the Computer Music Center at Columbia from 1994 to 2022.
2 Editor's note: David G. Vallancourt is a Senior Lecturer in Columbia University's department of Electrical Engineering. For a short article published by Columbia on the topic of repairing Music department's Buchla instrument, see Farabee 2018.
3 Editor's note: Bob Bielecki is Associate Professor of Music at Bard College, where he taught musical electronics classes.
4 Editor's note: see Collins 2020.

References

Collins, N. (2020). *Handmade Electronic Music: The Art of Hardware Hacking*. (Third edition). New York: Routledge.
Farabee, Mindy. (2018). "The Art of Sound." Columbia Engineering Magazine. October 19, 2018. https://magazine.engineering.columbia.edu/fall-2018/art-of-sound.

5

GORDON MUMMA'S SOUND-MODIFIER CONSOLE

Michael Johnsen

Preface

What follows is an analysis of Gordon Mumma's Sound-Modifier Console, an eight-channel processing device he built for the Osaka World Expo in 1970. The author has studied the prototype of this circuit directly and offers perspective on its functioning, down to the circuit level. The device will be viewed in the twin contexts of musical and technological history. It is often helpful to study circuits if one wants to understand the larger field of electronic music, but this becomes essential with experimental composers like Mumma, David Tudor, and others who considered the design of electronics to be an act of composition. Their work often sprung directly from discoveries made within their circuits and is best described in the graphical languages of electronics. The present treatment presupposes a basic understanding of analog synthesis and circuit diagrams. The illustrations presented here are the fullest disclosure; with parts values included to stimulate breadboarding. However, it is hoped that this essay appeals to the interested reader at multiple levels, just as does the best of experimental music.

Context

Gordon Mumma designed the Sound-Modifier Console for the Pepsi Pavilion at the 1970 Osaka World Expo in Osaka, Japan. He'd been recruited by his colleague David Tudor, who was himself part of a burgeoning collaboration between artists and Bell Telephone Labs engineers called Experiments in Art and Technology, or E.A.T.

DOI: 10.4324/9781003219484-7

FIGURE 5.1 David Tudor (left) at the Sound-Modifier Console in the Pepsi Pavilion, Osaka, 1970. Pauline Oliveros in the background (right). Photo Courtesy of Julie Martin.

For its pavilion, the beverage mammoth had entered into a strange partnership with E.A.T, hoping for assistance on the world stage in celebrating "the Pepsi generation". E.A.T was a haven for experimentalists like Deborah Hay, Bob Rauschenberg, Öyvind Fahlström, Max Mathews, Yvonne Rainer, and Elsa Garmire; but PepsiCo wanted a rock concert, preferably by the Beatles (Klüver, Martin, and Rose 1972). The partnership held for 18 months, even as the artists' imaginations fled wildly from Pepsi's tastes and the budget bloated to $2 million. The result was an avant-garde sensory buffet offering very little sugar water. In its main space, a near-perfect spherical Mylar mirror created holographic virtual images and audio ricochets. A slanted tunnel led visitors to a "clam room" which housed a multicolor laser/audio system and localized radio transmissions of insects, fish, and cities emanating from a dozen corresponding floor surfaces. Prospective visitors could grab inductive receivers to hear these transmissions as they moved about the space. From the outside, the building was shrouded in a dense artificial fog environment, which occasionally crept toward neighboring pavilions. A flotilla of human-scale pill-like sculptures roamed around the entrance, their movements slow enough to go unnoticed.

Mutual faith and PepsiCo's pocketbook extended through opening day in mid-March, when the delusional relationship finally fizzled over a chasm of misunderstandings. From their end, the artists and engineers wearied under a patronage they'd thought was a partnership, while PepsiCo was tired of paying more, but getting less. E.A.T's pavilion lasted only a month of its expected six, leaving disappointments but also lessons extending to the present day. In this spirit we turn to Mumma's design itself.

A Design Emerges

The Sound-Modifier Console was just one part of the sound system for which Tudor had been made responsible. In addition to its permanent sensorial atmosphere, Tudor intended to invite a roster of forward-thinking guest musicians to make and perform pieces during the fair. In his mind, the "instrument" he'd offer them would be maximally flexible in distributing and modifying sound. For distribution, he'd envisioned 96 speakers with programmable panning controlled by a paper-tape system. Over time, this was bargained down to 37. Further resources included banks of tape playback machines stocked with a library of hundreds of natural sounds Tudor had solicited from biologists, doctors, and seismologists worldwide. Musicians could bring their own tapes, connect via telephone, or feed microphones and pickups directly into 24 preamplifiers, which mixed, three at a time, into each of eight modifier channels. Mumma would be hired to design the modification circuits. Tudor had proposed 24 channels of modification, reduced eventually to 8 (Nakai 2021, 286).

Mumma was a natural choice for Tudor. Among musical explorers, he was known as a Johnny Appleseed of homebrew electronics. The advent of cheaper transistors had made it much easier for non-engineers on home tinkering budgets to build circuits for music. By the early '60s, integrated circuits (ICs) hit the market, offering complete sub-circuits, integrated on a silicon wafer the size of a single transistor. Homemade mixers, modulators, and transmitters were seemingly within the reach of curious, broke artists. Better still, they dreamt: what kinds of brand new devices might they invent when left to their own imaginations, unhindered by years of engineering curriculum?

As formerly exotic military-industrial parts grew affordable, it became possible to build electronics as a folk-practice—to invent instruments in the kitchen, on one's own terms, with simple tools. In the pre-internet world, practical skills in electronic lutherie spread by word-of-mouth (Mumma 2015, chapter 13). Mumma had the aptitude and a willingness to share what he knew. He'd formed a small company already around 1965 called Cybersonics with the brilliant engineer William Ribbens and he was thoroughly ensconced in the new arts. (Ribbens 2017) Unlike the Bell engineers, Mumma was a musician first, engineer second—a professional non-professional engineer.

His console design would have to balance the contrasting needs of 24 proposals that had been accepted. It should be unprejudiced, sophisticated, and yet friendly to those whose music was not focused on technology, those who "could barely find a light switch in a room", as Mumma reflected (Mumma 2014).

The final design (Figure 5.2), enabled a range of effects from slight to grotesque. Each channel offered as many as three stages of modification, to variously shuffle and generate harmonics, shape loudness, and enhance directionality by filtering. Each stage could be tuned to taste under voltage-control or quickly bypassed. There were 91 knobs and 24 switches on the finished console, implementing a modular approach without resorting to patch cables.

Mumma assembled and soldered the eight-channel device entirely by hand without recourse to printed circuits in the year leading up to the fair. This included power supplies, pots, switches, wiring harness, and finally the modifier boards, which were about the size of a postcard. How these little cards did heavy circuit-lifting with only 3 ICs and 15 transistors will be the focus of the remaining text.

The console itself was literally trashed in Japan by spring of 1970, but the one-channel prototype he'd made survived back home in the States. Around 1973 he gave it to David Tudor, who cherished the box and the work it made possible. After Tudor's death in 1996 it was inherited by Wesleyan University along with the rest of Tudor's instruments and became the subject of the present text. Close study reveals the modifier circuit to be a paradoxical exemplar of '60s DIY hardware—by turns ingenious, quirky and crude.

Circuit Overview and Concepts

The essential source for information on Mumma's Sound-Modifier Console is the concise descriptive chapter he wrote for E.A.T's 1972 book, "Pavilion" by Klüver, Martin and Rose; since reprinted in his own recent text (Mumma and Fillion 2015). It reads like a user's manual, and was perhaps intended as one. Knowing this text thoroughly before encountering the prototype at Wesleyan, the present author expected it to serve equally well as a map to the circuit inside the box, which it does, but only loosely.

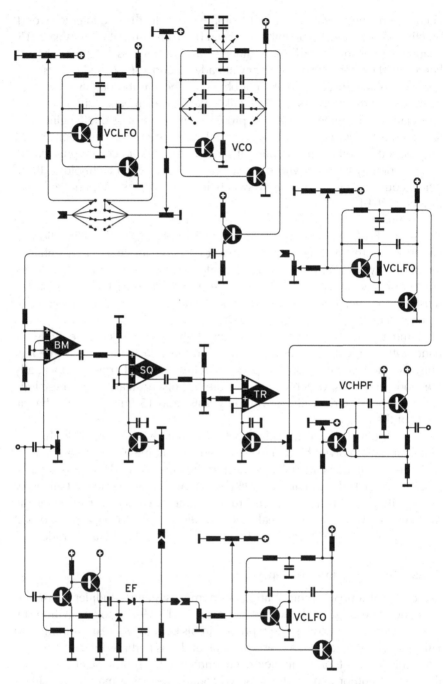

FIGURE 5.2 Sound-Modifier Console: 1 ch. of 8, full schematic, minus bypass switching. Q: 2n2925 IC: mc1545.

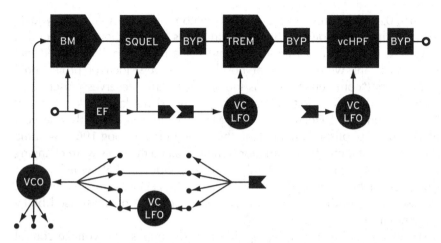

FIGURE 5.3   Block diagram shows signal (top row) and modulation paths. Each of three modification stages may be bypassed, but squelch always follows balanced modulation (BM) stage and is set internally. Input jack also feeds envelope follower (EF), which provides DC control to squelch and three of four voltage-controlled sine oscillators, which themselves modulate the signal path. One oscillator has range- and modulation-switching. The remaining oscillators are sub-audio. Full circuit fits on a postcard-sized pcb.

The console circuit is rich in options despite its humble means, due largely to its use of voltage-control techniques, which were still in their infancy. Voltage-control is a kind of social behavior among circuits: each sub circuit inter-modulates with the others. In the designer's words: "What's significant about the pavilion module is that you interact them. There's so many ways to interact with those parts, to get them cross feeding" (Mumma 2014).

The block diagram of Figure 5.3 provides an overview of the functions. Each channel has a single input and output. The input signal feeds a succession of three modifying sub-circuits in a fixed order. These sub-circuits are a balanced "ring type" modulator, a tremolo, and a high-pass filter; and each stage is easily bypassed by a DPDT switch. Patching is only possible between channels; not between subcircuits. The balanced modulator passes into a squelch circuit to clean up feedthrough and cannot be bypassed. Perpendicular to the main signal path are modulation sources, which act as control signals. Primary among these is an envelope follower, which extracts the loudness contour of the input signal and applies it in five places. This enables subtle envelope-controlled effects. The other four modulation sources are sine oscillators. Three are sub-audio and all of them are controllable by the envelope follower. The oscillators are only tunable over a half-octave or so.

In 1970, high-quality circuits for voltage-control had not yet been widely adopted, though Moog and Buchla instruments did use them. The fact that they were never part of Mumma-Tudor-Cage's regular toolkit, indeed that these musicians were somewhat hostile to the very idea of professional synthesizers ("commercial" seems too strong a term) deserves closer study elsewhere. Though Moog had already published his landmark article on "Voltage-Controlled Electronic Music Modules" in 1965, exactly one year after Mumma published an article in the same journal (Moog 1965, Mumma 1964), it was still much too early for standards since the tools were changing too rapidly. Motorola marketed only 16 op amp types in 1969, which neither Moog nor Buchla used until a few years later. The LED was still unaffordable, so nascent opto-isolators (Vactrol- and Raysistor-type devices) used bulky filament lamps.

Mumma had been developing ideas and circuits for voltage-control himself under his "cybersonics" concept. In his own words, a cybersonic process was one "in which some aspect of the process is fed back as a control mechanism, just as the steersman (kybernḗtēs) controls the boat ... The sound modifies itself. I establish the procedure by which the sound is modified, but the actual modification of the sound is done by the sound itself" (Mumma 1982). In practical terms this works by deriving control signals from some part of the signal—in this case the loudness contour—and applying them elsewhere. Other signal traits, like pitch or spectrum might be also extracted as controls, just as doctors may monitor multiple vital signs. Mumma's circuit method in the console was to use unpowered transistors as voltage-variable resistors. This is examined closely in the oscillator and filter sections of the text.

A separate discussion of each of the sub-circuits follows, slightly out of their order in the channel path. The first two modifier circuits will be treated first, followed by the oscillators, envelope follower, and high-pass filter.

A Balance Modulator (BM) Generates Sidebands

The first and most conspicuous modifier is the balanced modulator stage. Its ability to clangorously denature a signal is familiar to listeners of science fiction and 50s tape music. Better known by the name "ring modulator" (an early species of BM) it was one of the first transforming tools electronic musicians used because it already existed in the radio stations where they experimented after hours. The radio stations were using balanced modulation to do AM broadcasting — a fuller name for BM is doubly-balanced amplitude modulation. Unlike most processors, a balanced modulator requires two signals at its input, which it arithmetically multiplies at its output. In broadcasting, speech would be modulated with a very-high-frequency sine and thrown into the air with a transmitter, then received and de-modulated by the listener. Audio traveled much farther when "carried" by radio frequencies in this arrangement.

Electronic musicians actually want to hear the modulated signal, though, so they choose both input signals from the audio range. What they hear are called modulation sidebands. When a balanced modulator multiplies two sines, the output contains two sidebands only and nothing else. One sideband is the arithmetic sum of the two frequencies and the other is the difference. But sines only have one spectral component. If we replace one of the sines with a voice, then there's a sideband for every spectral component of the voice signal. Recall that the sidebands are determined arithmetically, rather than according to musical intervals. Therefore, any pitch center is moved without regard to musical interval, and the relationships between chord-like components no longer hold. Since a BM generates only sidebands, the original input signal frequencies disappear. This is an extraordinary, ghostly effect, which explains what's meant by "doubly-balanced"—both input frequencies are "balanced" away. Furthermore, it means that the input ports are interchangeable. The console feeds its input to one port, and the other is permanently connected to one of the sine generators, to be discussed in a later section.

Balanced Modulation Circuits

Balanced modulation makes sidebands by working on the amplitude of signals. The astute reader may posit correctly that voltage-controlled amplifiers play a role in this task. But one voltage controlled amplifier isn't quite enough; only a very un-balanced amplitude modulation results. What one needs is a second VCA connected in parallel, but acting in a mirrored sense to the first with its own "un-balancedness" mirroring that of the first. Their parallel connection nulls out the input components and affects the desired balanced condition. The modulator is essentially one signal's voltage-controlled cross fade between another signal and its inverse—one which suppresses both inputs and leaves only sidebands. At a circuit level this can be accomplished in many ways. The oldest method is also the most counter intuitive to modern readers: namely the true ring modulator, which uses only transformers and 4 diodes (Bode 1967). Mumma chose the newest method instead.

Mumma's Design

Mumma took a cutting-edge approach by choosing the newly-minted Motorola mc1545 integrated circuit to generate his sidebands. He'd received the tip from his friend the Bell engineer Fred Waldhauer, "who got me these new military prototype transistors [sic]" (Mumma 2014).

Confusingly, Mumma refers to the chip as "transistors" here, when he can only mean the IC. In the author's discussions with him, he never refers to the chip by name, and requests for clarification have not yet succeeded. The confusion is perhaps explained as follows. The IC came in a metal can, so it looked like a transistor, and was in fact full of many transistors, though it wasn't a discrete one. Motorola was indeed a major military contractor and the chip was both new and costly. In 2021 dollars, an mc1545 cost over $60, and the console used three per channel (Motorola 1969). By contrast, it may be inferred that he is not referring to the discrete transistors in the console, because they were low-cost epoxy-dipped silicon types that had been on the market since at least 1963—neither rare, nor military (Garner 1964). These were the only other active parts in the design (Mumma 1969).

How the MC1545 Does Balanced Modulation and Other Tasks

The Motorola mc1545 had just entered the market in 1969 (MC1545 Datasheet 1969) and was listed—seemingly disguised—as a "gate-controlled two channel-input wideband amplifier". The internal schematic is depicted in Figure 5.4a and may be recognized as a pair of VCAs built around differential transistor pairs. This "pair of pairs" is itself controlled by a third differential pair; completing the kit necessary to do balanced modulation. Figures 5.4b and 5.4c illustrate how various signal port configurations realize this function and others. For instance, using only half of the internal circuit makes a simple VCA. The device symbol looks deceptively like an op amp, but the two internal triangles graphically reinforce the fact that there are actually two pairs of input ports. The "gate" input is a third port, interchangeable with either of the first two. Because the IC is so open, functions like multiplexing two signals under control of a third (FSK, or frequency-shift keying) are also possible, plus it distorts gracefully and operates at video frequencies. Jim E. Thompson, the chip's designer, insisted via email that "the mc1545 isn't a multiplier. It's an RF mux",[1] though it clearly serves that role well enough (Thompson, personal communication). Later IC designs optimized for multiplication-only were less versatile, which may have been what Thompson meant. In fact those later designs overshadowed the 1545 quickly, rendering the device difficult to find today for less that $20, if at all.

Squelch (SQ)

The squelch circuit of Figure 5.4c immediately follows the balanced modulator stage. Its work is performed by the envelope follower, applied to an mc1545 in VCA-mode. This is necessary because balanced modulators

(a)

(b)

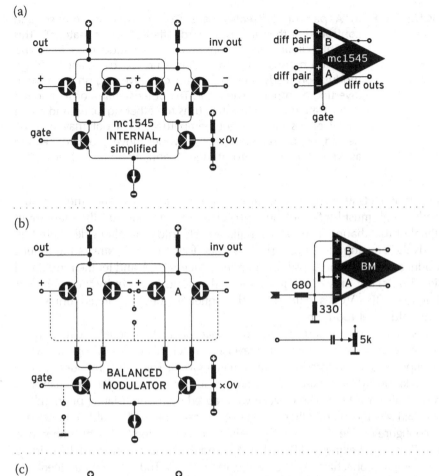

(c)

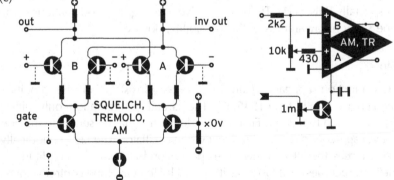

(caption on next page)

FIGURE 5.4 a) A pair of differential pairs (A, B) are cross-coupled and controlled at their tails by a third differential pair (gate,0v). This versatile circuit can perform many amplitude tasks including balanced modulation, crossfading, and fsk. b) Balanced "ring" modulation feeds one signal into one inverting and one non-inverting member of each pair. The other signal is offset and fed to the gate port. Each signal turns the other signal up and down based on its instantaneous amplitude. c) A pot may be connected, enabling feedthrough as wiper moves upward; ceasing as wiper moves toward ground. Modulation signal drives gate.

are imperfect at suppressing their input frequencies and the residual feedthrough must be "squelched". To clean up this murmur, the main input signal of the channel is passed simultaneously into a parallel side-chain that feeds the envelope follower. The envelope follower (EF) extracts the input's loudness contour and uses it to open the squelch VCA and let the modulated signal pass, but only when it's above a threshold level set by the attenuator that feeds the VCA's gate port. The squelch goes unnoticed unless the threshold is set too high.

The author has previously claimed erroneously that the squelch circuit in the prototype modifier is configured in reverse. Namely, that the VCA opened when the input was below threshold, rather than above it. This was subsequently cited by Nakai in 2021 (Nakai 2021), and the author is grateful for the chance to correct his error here. The mistake was caused by mis-translating pin numbers between metal-can and DIP packages. A "reverse squelch" would be quite easy to configure on the IC, by simply feeding the VCA at the "A" ports, rather than the "B" ones (Welling and Russell 1969) and might have enhanced the existing soft-compression characteristic of the modifier in Tudor's mid-70s feedback works, especially Pulsers (1). Tudor's extensive drawings for modifications to the prototype show no awareness of this possibility, however.

Tremolo

The VCA function appears again in the tremolo stage of the modifier, implemented by the last mc1545 IC. The circuit configuration only differs from the squelch in two ways. First, the modulation signal presented to the port is now a voltage-controlled low-frequency sine oscillator. The LFO periodically opens and closes the VCA a few times per second causing the amplitude to "tremble"—sometimes a bit faster, as it's jolted by the envelope control voltage. No sidebands are audible in this case because the LFO is too slow. Finally, the mc1545 is fitted with a potentiometer that allows a controlled amount of signal to feed the opposite-sense input. The resulting wet/dry mix tailors tremolo depth.

Oscillators

Of the four oscillators in the device, only one of them is itself audible, yet they all critically articulate pitch, dynamics, and texture in their modulating roles. The sine oscillator circuits in the console are uniquely instructive in their efficiency and characteristic of their era.

If there is one true miracle in electronics, it is oscillation. What's required is to trick an unmoving DC power supply into vibrating. The shape of that vibration is its waveform, and the rougher its shape, the more harmonics it contains. The sine waveform has no harmonics because its shape is maximally smooth. At the circuit level, it might seem that making clean sines would require finesse and a lot of parts, but the circuit Mumma used was marvelously compact. It is an outgrowth of notch filter design and requires a single transistor. Figures 5.5a through 5.5d illustrate this development in concert with the text.

The Twin-Tee Filter

Understanding the sine oscillator begins with the twin-tee filter of Figure 5.5a. The top and bottom legs of the circuit are the twins: non-identical in this case. The tee formed by 2R-2C-2R is a passive low-pass filter, while the C-R-C is a passive high-pass. They have the same cutoff frequency if the ratio of parts values is strictly obeyed. Cutoff frequency is given by:

$$Fc = 1/(2 \cdot pi \cdot R \cdot C).$$

The frequency response of a filter also has a concomitant effect on phase, which is salient in this case. Namely, at cutoff, the low-pass tee provides a phase shift of −90 degrees, known as phase lag, while the high pass tee provides a phase lead of +90 degrees. By connecting the twins in parallel, the center frequency is shifted 180 degrees (anti-phase) to all others, resulting in its total removal from the output, assuming perfect parts. This is a remarkable feat for a passive filter, making it popular as a hum-removal tool.

The Filter Oscillates at One Frequency

The filter becomes an oscillator in Figure 5.5b, with the addition of inverting gain. Wrapping the twin-tee as feedback from the transistor's collector to base gives gain, but also inverts it 180 degrees. This adds to the filter's 180 degree shift, but only at cutoff, so there's 360 degrees of shift at cutoff and less at every other frequency. It therefore manifests a smooth sine wave because the gain is exactly one (hence regenerative[2]) at the cutoff frequency,

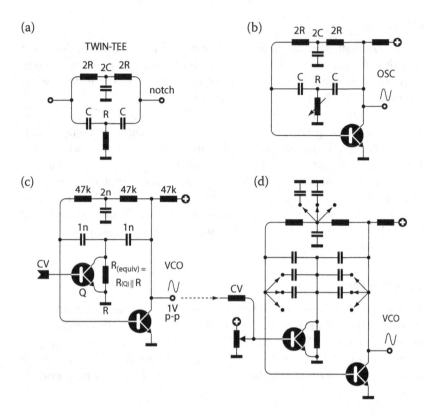

FIGURE 5.5 a) Top leg forms a lowpass filter (–90° lag) with same cutoff as lower leg, which is highpass (+90° lead). In parallel they accumulate 180° phase shift at cutoff; hence making a notch filter of infinite depth. b) From filter to oscillator: Wrapping the twin-tee around an inverting amplifier makes a sine. Pot changes frequency over nearly two octaves, but purity of sine changes as well. c) The transistor as voltage-control element: Q adds a parallel resistance to ground, controlled by positive cv at its base terminal. R(Q) may be varied from megohms down to 10s of ohms. Total parallel resistance is limited by fixed R. d) Range switching adds parallel sets of timing capacitors.

and less than one at all others (hence damping them). The oscillator circuit turns down all frequencies except one, which it amplifies until a sinusoid results. The circuit oscillates equally well at sub audio if large time constants are chosen. However, building this circuit in a world of 10% capacitors and medium-impedance, moderate-gain transistors requires some tweaks to make it sing well.

Tuning

To ensure that the circuit oscillates, the resistor-to-ground R needs to be somewhat less than calculated. Making it a potentiometer allows us to find the edge of oscillations, and to go beyond it. Changing this resistance progressively unbalances the network, (hence the cutoff frequency) which shows at the output—starting with non-oscillation (R too large), then slowly birthing a clean, fragile sine. Adjust further, and the sine grows slightly and stabilizes with R decreasing. Next, the oscillation frequency begins to rise smoothly, and the waveform grows still larger, eventually flattening at the bottom, then halting its glissando. In other words, this resistor controls the purity of the sine, but also converts a fixed oscillator into a variable-frequency one.

Alas, the twin-tee filter is quite dependent on its other Rs and Cs, which limits the oscillator design to a tuning range of about 1.5 to 2 octaves. Though extremely narrow by modern standards, its size and stability convinced Ma Bell[3] to integrate twin-tee oscillators into the DTMF[4] touch tone keypads of its Trimline phones in 1965 (Berry 1966). Mumma solved the range issue in two ways. First, he installed a "register" switch on the audible oscillator, which switched in three sets of matched capacitors (Figure 5.5d). More interestingly, he adapted the basic oscillator circuit to voltage-control.

Adding Voltage-Control

The circuit of Figure 5.5c adds a transistor, which hugs our tuning resistor R, but is conspicuously unconnected to the power supply. When used in this way its collector-to-emitter impedance may be altered (reduced) with the application of positive voltage to its base terminal—in other words, as a kind of voltage-controlled variable resistor—enabling a "programmable" VCO. The apparent resistance of the transistor structure may be varied from megohms to 10s of ohms, so connecting this structure in parallel to the existing R sets a maximum total resistance, namely the value of R (Maynard 1968). It bears repeating however, that the tuning range is still the same, with or without voltage-control. All told, it is an admirable design, and one which proves eminently musical in use.

Mumma had been working on transistor twin-tee sine circuits since at least 1965. We know this because his friend David Behrman's notebooks show one labeled "Gordon's 1965 Oscillator" (Behrman 1969) while Mumma's 1965 letters to the critic Peter Yates mention one "so small that it can be built on the surface of a postcard and no thicker than a piece of corrugated cardboard. I'm considering building more of them to send out as Christmas cards for my friends" (Yates 1966). A series of articles by Fred Maynard in Electronics World from 1963–8 presents numerous twin-tee oscillator circuits, including the voltage-control technique Mumma used (Maynard 1963, 1964, 1968). While it's unclear if he ever saw these articles, his colleague David Tudor did read them just after 1968. They are listed in a note to himself, retrieved from the David Tudor Papers at the Getty Research Institute (Tudor circa 1968). It is tempting to posit a direct connection, but similar voltage-control methods had been already applied widely before Maynard's articles. Mumma's method will re-surface in discussing voltage-control in the high pass filter.

Envelope Follower

The envelope follower (EF) plays an essential role in the modifier's voltage-controlled information system. Its job is to derive an average DC loudness contour from the real-world input signal and send that "intelligence" to govern five circuits: the squelch and all four oscillators. Without this connection to "the world off the circuit board", the modifier channel sounds mechanical and inorganic. As such it plays a "cybersonic" role, in Mumma's parlance. Unlike its synth-world cousin the envelope generator, the follower's job is to follow loudness contours, not invent them.

Mumma's EF circuit may be understood with recourse to Figure 5.6. Recall from the block diagram (Figure 5.2) that the envelope follower shares its input with the main input. As such it forms a side-chain to the signal path. A two-stage transistor amplifier gives a gain of 25 at the second emitter. This ensures a robust envelope will be available to the various CV destinations, which each expect about a volt for maximum swing. The signal is then coupled to a shunt/series pair of diodes, which pass only the positive half of the amplified input. This rapidly charges the large 4u capacitor, rising each time a higher amplitude peak is reached. The final 10k shunt resistor gives the capacitor a slow discharge or decay path to ground, assuming there are no new loudness peaks to recharge it. As such, it follows amplitude increases rapidly, but decays at a

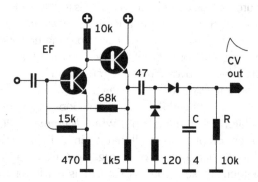

FIGURE 5.6 Envelope follower: Two-stage amplifier with both positive and negative feedback has a gain of 25. Remaining circuit allows positive peaks to rapidly charge capacitor C, while parallel-connected R provides for slow decay to zero. Output may control squelch, filter, and oscillators.

rate determined by C and R. the output has a fast-attack, slow-decay shape, ideal for seamless squelching. The decay can be fruitfully modified by tweaking the timing components R and C. Strictly speaking, Mumma's EF is actually a species of peak follower. Hence, attack is not adjustable in this circuit.

The basic amplifier topology is that of Mumma's "z-amp", itself taken from a 1966 article in Radio Electronics (Sutheim 1966) and used in his own piece Mesa and in the cybersonics splitters he built for Tudor (boxes 0002, 0099) in 1968 (Mumma 1968).

In addition to its role in the prosaic squelching task (see BM/SQ section), the envelope may be applied to all of the VCOs—especially those operating at sub-audio rates. This "emphatic, syncopated heartbeat" effect is characteristic of the modifier's sound.

High-Pass Filter

The filter is the last modifier the signal encounters before reaching the grid of loudspeakers behind the Mylar dome of the pavilion. As such, it is a high-pass type, chosen by Mumma to emphasize the directional characteristics of the acoustically-reflective environment. Like the other modifiers, it may be by-passed, and is subject to voltage control, in this case by a vc-LFO. Mumma notes that "each channel was tuned slightly different … I made it so it was a family of 8, but members of a family are all a little different, right?" (Mumma 2014)

Figure 5.7 depicts the circuit, which is formed around a single transistor and controlled by one of the vc-LFOs discussed previously. Two equal-valued

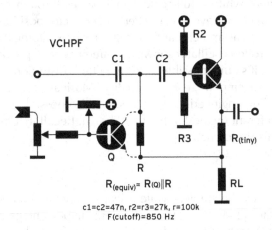

FIGURE 5.7 With dotted lines disconnected, the circuit is a fixed, two-pole Sallen-Key high-pass filter where cutoff is set by C1, C2, R1, and R2/R3. Connecting Q adds a parallel resistance controlled by positive voltages at its base. Trimpot and attenuator tailor the center and depth of response.

capacitors suggest that the response is second-order and closer examination shows it to be a unity-gain second-order high-pass of the Sallen-Key topology.[5] The cutoff frequency is tuned by C1, C2, R1, and the combination of R2/R3. R2 and R3 do double duty as biasing resistors and form the second pole of the filter with C2. The center frequency is set at 850 Hz, and can be tuned upward with the addition of voltage control. Varying any of the fixed elements will change cutoff. Switching capacitors in pairs gives stepped changes, while R2/R3 can't be varied without affecting transistor bias. Therefore, R1 is the best candidate to affect smooth changes in cutoff. Replacing it with a 100k pot gives the desired control, albeit a manual one.

Voltage-Control, Again

Recalling the voltage-control method of the oscillator discussion, it is tempting to transplant what worked there to the filter context, and Mumma does just that. In this case, adding that same transistor control structure gives poor results, even with diligent trimming. Except over the narrowest range, this voltage-controlled tuning distorts the signal asymmetrically, producing strong harmonics.

The trouble here is that the transistor-as-resistor is connected in series, while it is shunted to ground in the oscillator circuit. The structure must behave ohmically in both cases; namely, with the same resistance in either direction. Bipolar transistors can do this only for very small signal levels, unlike their FET relations. In shunt mode, the level at the low-pass tee junction is tiny (about 50mv), while a full-scale signal passes through the series mode element in the filter. By reversing emitter and collector positions to test this, we merely invert the asymmetrical distortion products. Doing the same test in the voltage-controlled oscillator shows no difference, and proves the hypothesis. Alternatively, it's quite possible to see the filter's distortions as a feature, since its goal is to emphasize directionality. Musical aims are sometimes met by disregarding engineering standards. It's also possible that Mumma used an entirely different method in the final design. Recall that the present study is based on his prototype only.

Afterword

Historic circuits shouldn't (only) be studied while asleep—unpowered, and subjected to the dry postulates of circuit analysis. A surprising percentage of retired devices will happily "breathe" under the application of fresh electric potential, affording a direct sensory experience which seemingly elides time. The best way to honor old circuit ideas is to use them.

One consequence of the present study is the production of a set of eight sound-modifier clones in the Eurorack format, to facilitate the performance of key '70s Tudor pieces. Surplus clones from this effort have been deposited in publicly-accessible studios in Japan, Germany, and the United States.

Note

(1) These pieces and their realization are treated exhaustively in Nakai, 2021 458–63.

Acknowledgments

For their smarts and hearts, the author expresses gratitude to: You Nakai, Ron Kuivila, Margaret Cox, Julie Martin, Hallie Blejewski, William Ribbens, and Gordon Mumma.

Please direct corrections and discussions to johnsen.rahbek@gmail.com

Notes

1 Editor's note: multiplexer.
2 Editor's note: regenerative at the cutoff frequency implies that this circuit does not attenuate incoming signal components of that frequency
3 Editor's note: Ma Bell was the nickname of the Bell Telephone company prior to its fragmentation due to antitrust laws in 1983.
4 Editor's note: Dual Tone Multi-Frequency (DTMF) sees two tones assigned to each telephone key. When a key is pressed, these tones are audibly transmitted over the voice channel.
5 Editor's note: Sallen-Key corresponds to one of the most common topologies (circuit configuration) for second-order active filters. For more on this topology, see [add reference for active filter cookbook] or maybe Sallen, R. P. and E. L. Key, 1955. "A Practical Method of Designing RC Active Filters". *IRE Transactions on Circuit Theory*, Vol. CT-2, 74–85.

References

Behrman, David. 1969 December 27. "Gordon's 1965 Oscillator". *Schematic drawing. Stony Point*, NY: Behrman Archive.

Berry, R W. 1966. "Tone-Generating Integrated Circuit." *Bell Laboratories Record.* 1966(10/11): 3l9–323l9.

Bode, Harald. 1967. "The Multiplier-Type Ring Modulator." *Electronic Music Review.* Trumansburg, NY. July, 1967: 9–15.

Garner, Lou. 1964. "Transistor Topics." *Popular Electronics.* 1964(9): 76–103.

Thompson, James E. 2014 April 3. Email with the author.

Klüver, Martin and Rose. 1972. *Pavilion: Experiments in Art and Technology.* New York: Dutton.

MC1545 datasheet. 1969. Motorola Microelectronics Data Book. *Motorola Semiconductor Products Inc*, Phoenix, AZ

Maynard, Fred. 1963. "Twin-T Oscillators Design and Application." *Electronics World* 1963(05): 40–70.

Maynard, Fred. 1964. "Twin-T Oscillators for Electronic Musical Instruments." *Electronics World* 1964(06): 36–79.

Maynard, Fred. 1968 "Twin T's Designs and Applications." *Electronics World*. 1968(08): 35–37, 64.

Moog, Robert. 1965. "Voltage-Controlled Electronic Music Modules." *Journal of the Audio Engineering Society*. 13(3): 200–206.

Motorola Advertisement. "Our New MC1545 Is A Gated, Dual-Channel, Differential Inputs & Outputs, DC Wideband Video Amplifier Integrated Circuit!", Electronic Design 1, January 4, 1969, page 11

Mumma, Gordon. 1964. "An Electronic Music Studio for the Independent Composer." *Journal of the Audio Engineering Society*. 12(2): 240–244.

Mumma, Gordon. 1968. Letters 8 May 1968 and 4 October 1968 to David Tudor. Box 57 folder 3. David Tudor Papers. Getty Research Institute. Los Angeles.

Mumma, Gordon. 1969 October. *Schematic Drawing. Champaign, Ill*. Victoria, BC: Mumma Archive.

Mumma, Gordon. 1982 May 17. Interview with Vincent Plush. https://archives.yale.edu/repositories/7/archival_objects/3183883.

Mumma, Gordon . 2014 October 23. *Interview with the author*.

Mumma, Gordon and Michelle Fillion. 2015. "A Brief Introduction to The Sound-Modifier Console" in *Cybersonic Arts: Adventures in American Music*. Urbana, Ill: University of Illinois.

Nakai, You. 2021. *Reminded by the Instruments: David Tudor's Music*. London: Oxford.

Ribbens, William. 2017 May 25. Interview with the author.

Sutheim, Alan. 1966. "Transistor Line Transformer." *Radio Electronics*. 1966(4): 40–41.

Tudor, David. 1968. circa 1968. Personal Note. Box 40 folder 7. David Tudor Papers. Getty Research Institute. Los Angeles.

Welling, Brent and Russell, Ronald. 1969. "Motorola Application Note AN-475 Using The MC1545." *Motorola Semiconductor Products Inc*, Phoenix, AZ.

Yates, Peter. 1966. "A Ford to Travel." *Art and Architecture*. 1966(2,3): 45–47.

6

ARTIST STATEMENT

Switchboard Modulars - Vacant Levels and Intercept Tones

Lori Napoleon

Escanaba: First Spark of an Idea

While traveling in Michigan's Upper Peninsula, a small town called Escanaba's Historical Museum drew me in. Within its quiet walls stood the town's original telephone exchange. There, alone and in awe, I found that the elaborate banks of switches, patch bays, and weighted cloth cables bore a striking resemblance to modular synthesizers; this moment sparked the desire to breathe new life into this long-silenced apparatus and explore where switchboard operation and musical synthesis overlap (Figure 6.1).

Once a powerful living portal through which all person-to-person messages passed, I created switchboard synthesizers as a way-station for electronic signals once again. The conceptual impetus stems from the interplay of human operators with technology, exploring parallels between the musician, the technician, and their machine as the extension of their work. All instruments were carefully reconstructed by hand—from wiring the audio circuits to cleaning and refinishing the wood. They are interconnected with period cloth cables.

Many developments in audio owe their lineage to these forgotten machines, including the concept of patching in modern plug and jack designs—living fossils of those invented for switchboards in the 19th century. Sonic artifacts from the analog telephone networks leaked into the system, opening up a unique audio frontier that attracted its own subculture of experimenters called "phone phreaks," many of them blind. Recognized among the earliest hackers, I became interested in their obsession with finding, sharing and instigating aberrations in the gaps and silences of calls within their own dedicated communication channels, using the telephone

DOI: 10.4324/9781003219484-8

FIGURE 6.1 Lori Napoleon with her "switchboard modular" synth.

Photo credit: Seze Devres.

network as a giant modular synthesizer and covert chat forum. In this artist's statement I assess my own reflections on modularity in the context of materials, communications history, access and disability, and aesthetics (Figure 6.2).

Built upon combining the salvaged original physical framework with the circuitry and techniques of modular synthesizers, the switchboard's principal structure of the modular patch bay for connecting call circuits is given a new life as a connecting station for sound sources and sound modifiers. The pursuit and acquisition of authentic switchboards has led me to the basements and attics of rural telephone museums, various private collections and media archives, introducing me to a community of vintage equipment suppliers and enthusiasts who, I'm humbled and pleased to say, were as delighted by my endeavors as I was by theirs. These preliminary tools gleaned from telecommunications are specific enough to create a cohesive platform with which to work yet have been malleable and open-ended enough to lead me through a trajectory of exploration, yielding new approaches to performance and composition via the various strands the project has followed since its inception in Escanaba.

Both the switchboard and the synthesizer possess an alluring physical presence punctuated with glowing indicator lights and controls that beckon the user to interact upon first glance but also require some degree of

FIGURE 6.2 Detail of the Switchboard Interface.

Photo credit: Seze Devres.

technical knowledge and training in order to use. While switchboards had rows of input and output jacks manually joined together by patch cords to allow one human voice to hear another human voice, analog synthesizers also contain many inputs and outputs, with manual patch cords connecting the "voice" of one electronically generated sound to another in endless combinations.

Synthesis of Ideas/Searching for Precedents/First Mentors

The spark of a new idea often sends us obsessively in two directions for a period of time: there is the research for precedents and then there is the unabashed and boundless envisioning of future possibilities. This period brought me into the home and studios of artist, inventor and physics enthusiast Leon Dewan, who I had already seen perform at New York City's Issue Project Room as the duo Dewanatron alongside his cousin, artist and furniture maker Brian Dewan. Known especially for their Swarmatron and Hymnotron synthesizers, Leon's home was stuffed to the gills with all manner of bright, colorful and elegantly crafted handmade instruments built with re-appropriated surplus materials, a sculptural wonderland of latches and blinking lights, rotary phone dials and nixie tubes. I soon found myself enthusiastically guided at the helm of the Dual Primate Console, an instrument that directly relates to my fascination with combining telephone and synthesizer technologies. In no time at all I was able to dive into the operations of this marvelous machine, manipulating the decay time of chopped up 8 track recordings, turning telephone dials set up to control counters directed at various sound generators.

This formative, first-hand experience of playing such unique and personal manifestations of the Dewanatron artists' creative propositions gave me the assurance that I could begin a foundation for the platform of switchboard-synthesizers that will in turn guide me on a similar trajectory. Leon became an excellent and very patient advisor when I was in the process of building my new circuits or envisioning a particular mod.

Patchpoint #1: Influential Encounters and Decisions - The End of Naming

My quest towards gaining first-hand knowledge over my specific, irresistible "problem" to solve also led me to a second notably important location—an electro-music festival held at a campground in Bloomingdale, New Jersey in 2009.[1] I recall driving there on my own, with stacks of papers containing schematics in the back seat.

The decisions I would eventually make when first designing my own modular system were inspired by a patch I saw during an introductory workshop on modular synthesis at the conference. A musician named Kevin Kissinger demonstrated a very thorough run-through of basic techniques on his own Aries modular, a system which he populated and has expanded upon since the 1970s. Its formidable presence of sharp, uniform black paneling, sturdy wooden partitions and custom-made backlit green matrix controller was framed by fall leaves in yellow and orange through the window of the rather bland conference room. During a question-and-answer session towards the end, Kevin responded to a delightfully specific request from fellow conference participant and synthesizer virtuoso Don Slepian: a demonstration of the "ARP Newsletter Patch of the Month, January 1974," a dynamically changing, frenetically percussive sample and hold/noise patch that produces random variations of duplets and triplets. As he began with pink noise at the source and described the signal path one by one, I wrote down everything I saw. This patch left a lasting impression on me because it surprised me. This was the first occasion where I would witness the construction of rhythms using a synthesizer rather than a drum machine; my first transparently understood demonstration of how such sounds were modeled from a source. Most noticeable to me as distinguishing characteristics from a dedicated drum machine was a kind of elasticity; not only in the dominant sounds but in the individual nuances between each percussive event—the gaps in between—and also how different the physical body engaged with the instrument as they constructed rhythms within a system outside of the delineated nature of pads and buttons. I later would find that paying attention to, and also bringing attention to the character of in between spaces has held a consistent role over the course of my projects.

The first new PCB module kits I was to acquire, eventually, were chosen with the impressions of this patch and the desire to recreate this experience for myself. Among these were the Noise Cornucopia and Sample and Hold PCBs from Music From Outer Space, two of their ten step sequencers, and the Papareil Synth Labs Warp633 Ring Modulator. Elby Designs based in Australia became a great resource for me as well and helped me to populate my instrument with the basic building blocks such as a module based on a Serge ADSR Envelope Generator and a tribute module to the Steiner-Parker VCF called the Synthacon (Figure 6.3).

Patchpoint #2: Ambiguous Boundaries/Drum Machines/Names

The name of something can influence how you think to use it. A classic Roland 606 drum machine was a gift I'd had for years without doing much at all with it. A battery-operated and thus transportable little machine, an urban legend passed around my hometown was that producer/musician Steve Albini would walk around Chicago programming it, and while I initially embarked upon my own adventures as a roving drum machine programmer, this was short-lived. I find that it speaks to my propensity to construct sounds myself, percussive or otherwise, that it sat dormant until becoming mainly a trigger sequencer in a particularly salient moment. I learned first-hand at the conference that synthesizers can be "drum machines;" and prior to this, even the name put my mind in a certain box about devices having dedicated functions; and thus, in my naïveté (and perhaps no small part a lack of confidence for becoming a musician in the way that I had understood) I had little sense of being able to add anything new or motivation to try. Modular synthesis opened up the box and was the sole motivation for me to feel I might have something personal to add to sound and music. Prior, all of my attempts at music were instruments with clearly circumscribed boundaries between notes, placing events into categories—modular dissolved such categories for me and spoke to my interests, which eventually became clearer to reside in the process of exploring the material of sound itself (Figure 6.4).

Patchpoint #3: Influential Encounters and Decisions - Approaches to Materiality

Despite the many similarities I have drawn between switchboards and synthesizers, the fact that they did, in fact, have some operational require-ments that were very different called for innovation to make the project work.

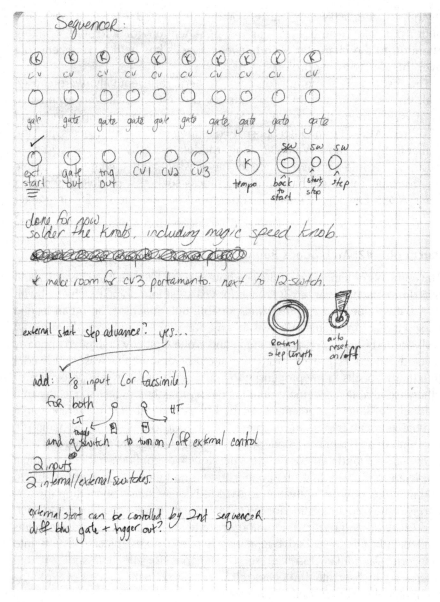

FIGURE 6.3 Initial sketch for switchboard sequencer. Courtesy of the artist.

FIGURE 6.4 Detail of small tabletop switchboard interface.

Photo credit: Seze Devres.

Several technical innovations were required to solve functionality discrepancies in this project such as the need to compensate for the significantly
lesser amount of continuous, dialable parameters in the original switchboard,
(which was much more generously populated with discrete controls such as
switches).

To resolve this, one of the most important technical modifications was to use
a number of the phone jacks in a rather unusual way: as a purely physical
connecting interface for potentiometers/knobs. All electronic musical instruments contain significantly more reasons to have variable voltage control via
knobs (volume control, pitch changes, degrees of effects/modulations of the
sounds, etc.) than switchboards did. To become a legitimate instrument, I
needed to make way for knobs on an interface that was essentially rows of
patchboards. While the first thought might be to get out the drill, one of the
most important suggestions made for my project prevented me from mutilating
my coveted switchboard with holes on the peripherals of the wood paneling.
Coming from Howard Moscowitz at the electro-music conference: why not
use the three connectors intrinsic in potentiometers (the wiper and the two
contacts) as connecting liaisons into the three electrical contacts in the
already-existing phone plugs (the tip, ring and sleeve?)? A kind of hybrid
connector; a potentiometer that I can plug directly into a switchboard jack that
is wired appropriately to the PCB?

Accordingly, I spent a day sawing TRS plugs in half, soldering leads connecting each of the three conductors to the three terminals on each potentiometer followed by another several days of protecting the fragile electrical contacts in resin, rounding their edges with plumber's epoxy and coating the exterior with liquid rubber.

Simple, elegant, and something I would never have thought of, for this thought requires a mindset that only then was introduced to me. Soldering potentiometers into plugs that can go directly into a jack also made them modifiable and interchangeable: some could be rock-solid; others could be more loosely soldered and allow for the user/operator to control a given parameter with their hands by squeezing the contacts together manually. The most important lesson was to solve a problem by looking at the essence of how a piece of machinery works in its physical, material form, and not be limited by the function any given object is conventionally used for. The electrons have no regard for what the intended purpose was. They "see" the metal, a conductor, and they travel. Erase these barriers, and new functions, uses, and innovations of any given tool emerge, thereby revealing a strong interplay between what is possible with the circuits/modules themselves and the discovery of what's possible within the repurposing of pre-existing controls or technical equipment of any kind (Figure 6.5).

Patchpoint #4: The Sandbox

"I am finding out what's possible with these controls; entrusted to the process ... this brings about the ability to learn something new, not only limited to my imagination—to be surprised."

These thoughts, scrawled across an old journal by my newly forged commitment to modular synthesis, are also reflective of my mind's eye when I first made the decision to attend a two-year graduate program Interactive Telecommunications Program at NYU, beginning in 2007. A sandbox where you are encouraged to try and to fail many times, to venture outside of your previous fields of expertise, to be unafraid of pursuing projects whose ideas exceed prior technical training and approach obstacles with curiosity, collaboration and mentorship, perseverance, and sheer belief in the possibilities instead.

I knew nearly nothing about synthesis of any kind before attending NYU, and spent the first year trying and failing quite a bit at a number of projects related to sound, light, and electricity. When the "idea I hoped to stumble upon and fall in love with" which brought me there eluded me, I focused on correctly wiring servo motors to stop nervously quivering as if they had consumed too much coffee and other technical quandaries with the tools at hand. I never did stumble upon the idea I would fall in love with in New York

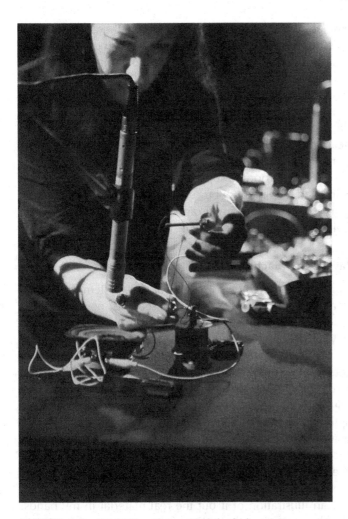

FIGURE 6.5 Exploring the material of electricity.

Photo credit: Jason Isolini, courtesy of ISSUE Project Room.

City as I'd envisioned; it happened far away from the buzzing excitement of that 4th floor of the Tisch building on Broadway, but, fittingly, within the spaciousness of nature; a quiet lighthouse museum of Escanaba with the gentle waves of my familiar Lake Michigan lapping along the shore; an intimate campground setting in New Jersey with a self-selected, niche community of synth enthusiasts; it was following these experiences that I knew that modular (and DIY) was going to be how I would approach my remaining time there.

Learning an unfamiliar medium will introduce a whole spectrum of feelings in order to see it through—from wonderful curiosity and the triumph of discovery to the dread of being "stuck" via having something to express in the moment that has become thwarted for days by an unknown technical issue. This is why I have found it is good to try and adopt certain constraints, spend an ample amount of time on less, but intentionally selected pieces of gear and processes—and then just explore and explore until you feel the most right, and the most "you." One aspect that took longer than anticipated was the intense cleaning of the switchboard in order to remove corrosion that was interrupting clean audio signals. Not a difficult task, but extremely time-consuming. Another technical challenge was figuring out how to power multiple modules from the appropriate supply. I had never worked with full wall-power before and had to learn about fuses, bipolar supplies, connecting multiple modules parsimoniously ... Circuits and concepts had to briefly wait for a period, in service of the sheer necessity of the correct power.

Patchpoint #5: Playing the Machine

Yet, as the main and predominantly solo user of this system (my main collaborator being a novelty and special-interest seeking, neurodivergent brain) this handmade process motivates my drive towards composing with the raw transparency of analog electronics.

Modular synthesis became my sole entry into music composition, perform-ance and instrument building. Despite prior exposure to non-modular analog synths with keyboards, drum machines and samplers, the freeing up of my hands via patching and automatic, generative processes was not only reminiscent of sculpture in which one seems to touch the sound, but it served as a multi-sensory and multi-modal extension of listening, as opposed to constantly "doing" ("playing" an instrument).

I am exploring the physics of sound and this material which is electricity. The interactions of these elements are what produce the sounds. Not an illusion of a thing or an illustration of it but the real material in my hands. Awareness of a real phenomenon—the entity of the medium—the interac-tions of these elements are what produce the sounds.

Patchpoint #6: Overlaps and Portals

In a broader sense, there is a transformation into an operator that happens upon anyone who makes use of modular or any complex, semi-predictable electronic equipment; often, this act of being a technician, a programmer, an operator behind the other, collaborative performer (the instrument) extends our senses and physical limitations. Operators of machines and operators of instruments overlap, manually connecting circuits, responding to feedback, making adjustments when necessary.

FIGURE 6.6 Voice changer and headset synth.

Photo credit: Seze Devres.

Telephone operators were often portrayed working the switchboard, cable in hand; these daring and socially ambiguous women represented the anonymity of the disembodied, yet powerful, faceless voices on the other end and the public intrigue towards the then-most prominent occupation of a woman in technology. The operator is a living portal through which all messages and information pass.

The job at the switchboard was one of the first occupations for women (let alone one directly associated with cutting-edge technology and a certain omnipresence over the goings-on of entire communities). As this occupation paved the way for many women to make their own income and have a sense of the community beyond their own domestic life, they were truly pioneers (Figure 6.6).

The telephone switchboard operator held a strong mystique for me, as an introvert whose machines "give me wings" in the manner that Suzanne Ciani has described in her own sentiment over her relationship with the Buchla: a photograph of a sole woman in a room full of large machines as she connected liaisons with her patch cables, the illustration of her weaving networks of power lines in her hands. The magazine ads of the period of the telephone operator (known also as the similarly pseudonym-like handle "The Exchange") presented her as a lone, confident worker, her expression serene and focused as she and her counterparts operated an inexorably complicated monolith of a machine that stayed running at all hours of the day and night, through thunderstorms, floods and snowfall.

An initial story that captured my heart was paraphrased from an article in the 1886 issue of "Electrical World" in Carolyn Marvin's (1990) book *When Old Technologies Were New*, which chronicled the tales of the brave operator women of the telegraph lines, stationed in remote desert outposts along railroad tracks, "whose only glimpse of the world she has left behind her … in the trains which pass or repass two or three times during the day." (27)

Patchpoint #8: Systems for Encoding Space

Where is information located? You need appropriate means to decode it. You cannot hear the information recorded on a cassette tape without an electromagnetic coil to replay it. The information is not always easily locatable. "EERHT OWT NOW" I wrote this backwards on a piece of paper, as I tried to wrap my head around where the sound was as the tape passed along the coil, where to splice.

What the machines evoke is important to me, especially their transparency in atmospheric conditions: Telephone exchanges were often struck by lightning causing the metal drops, the electromechanical indicator that preceded burnished glass-lit lamps, to fall in unison. Robust, the intention to withstand decades of use results in their own signature sounds, the "click" of a toggle, the pulses of the dial soundtracking the rhythms of communication (Figure 6.7).

Incessant static, blinking lights, behemoth machines, voices and connection through the severest weather—Crosstalk, trans-continental echo, and

FIGURE 6.7 Detail of patch bay.

Photo credit: Seze Devres.

encryption are examples of sounds to be dealt with physically and conceptually. The noises in the networks, these signature sounds were not meaningless; even the "unwanted" sounds in the space between communications contained information about adjacent technologies, geographical locations, and atmospheric conditions. Telegraph operators were called "Sparks."

Perhaps this is what I should call myself.

Note

1 See https://electro-music.com/ for further information and context.

Reference

Marvin, Carolyn. *When Old Technologies Were New: Thinking about Electric Communication in the Late Nineteenth Century*. Oxford: Oxford UP, 1990.

Websites

http://www.meridian7.net/
http://antenes.net/
http://www.fairradio.com
http://www.nhtelephonemuseum.com/home.html
http://www.telephonearchive.com/dmworkshop/items/kellogg_wall_switchboard.
    html
http://www.1911encyclopedia.org/Telephone
http://www.cgs.synth.net/
http://www.musicfromouterspace.com/
http://www.angelfire.com/music2/theanalogcottage/
http://elby-designs.com/asm-2/asm2.htm

7

EURORACK TO VCV RACK

Modular Synthesis as Compositional Performance

Justin Randell and Hillegonda C. Rietveld

Eurorack is a hybrid, post-digital system for modular synthesizers produced by both experimental developers and established brands. With "post-digital," we mean that the module designs and modular patches include both analogue and digital processes; each modular system has the potential to be a unique combination of oscillators, filters and modulators that draw on a heritage of vintage analogue electronics as well as modern digital signal processing techniques. It offers a standard set of parameters for modular configuration, which is an attractive feature that aids its popularity, and which enables shared knowledge with a thriving scene of developers, music makers and enthusiasts. Designers develop modules, some of which are open source, which musicians and sound designers can assemble into bespoke creative systems. Within each modular patch, a specific combination of instruments and audio filter modules is interrelated; in other words, a network of embedded relations affects how each module operates. Starting from a relatively minimal setup, unexpected feedback loops with a generative module can produce an intricate sonic system. Interfacing with human creativity, the modular system produces musical sound through both its patch design and its serendipity; as modular performer Suzanne Ciani (2018) observes, composition is hereby constantly in process as a continuously developing performance. The attraction of working in this way was encapsulated, back in 1968, by minimalist composer Steve Reich:

> I am interested in perceptible processes. I want to be able to hear the process happening throughout the music. To facilitate closely detailed listening, a musical process should happen extremely gradually. (Reich 2002:34)

DOI: 10.4324/9781003219484-9

Although initiated by the music maker, we argue that a generative patch seems to take on its own form of agency, enabling an explorative creative dynamic. Here we understand generative music as a configuration of modules that allows a certain amount of randomization and that is, in effect, "an open dynamical system" (Bown 2011: 73). In this, we follow Waters (2021) who argues that music instruments are assemblages, dynamic processes within a performance ecosystem in which "the non-standard instrument can be seen to be typical of human/instrument entanglements" (138). Such an ecosystem integrates the performer and instrument in a multiform environment. A modular generative patch has its own unique affordances and may therefore be understood as a performer within the ensemble; as a patch, the instrument becomes a performing partner with the musician in a creative relationship where boundaries between human and machine are symbiotic. Such an immersive musical relationship between human and machine produces a specific techno-aesthetic. Simondon (2012: 3), proposes that an embodied aesthetic is possible through *techne*, the craft and its related knowledge produced through making, suggesting in an unsent letter to Derrida back, in 1982, that,

> … contemplation is not techno-aesthetics' primary category. It's in usage, in action, that it becomes something orgasmic, a tactile means and motor of stimulation.

> Aesthetics is … also … the set of sensations … of the artists themselves: it's about a certain contact with matter that is being transformed through work.

Applying this concept to working with Eurorack hardware modules, a sonic techno-aesthetic develops within an active creative relationship with modular music technology. In this context, our research addresses interactive modular compositional performance, whereby engagement between human creativity and machine-generated serendipity leads to an improvisatory music performance and composition.

Investigating how we engage with the techno-aesthetic of a generative modular practice, we take an auto-ethnographic approach. Our case study is of a patch within a hybrid setup consisting of Eurorack hardware modules and VCV Rack, a screen-based virtual open-source version (VCV Rack 2021). We thereby reflect on forms of creative practice in which the distinction between analogue and digital sonic processing is demonstratively blurred. We investigate generative musical processes as well as the distinction between physical and virtual musical interfaces. Our methodology consists of the preparation of a hybrid generative modular patch and noting the experience of flow in improvisation,[1] supported by the descriptive practice of recording, and of reflecting on these dynamic processes. In this case, the attention is focused on the pleasure in making music with a generative network of machines that at times seems to have a mind of its own. This is because a generative process offers a form of musical improvisation, in which a musician creates material based on a set of rules. In doing so, our research addresses creative practices that are shaped by the intersection between the performer and analogue-digital hybrid modular systems. In the discussion that follows, we will first address the instrument, and next move the discussion to the case study to illustrate the interrelationship between musician and instrument, and finally expand the discussion to the wider environment, within which the creative practice with a generative modular patch can be understood.

Through a comparative exploration of creating music with Eurorack's physical modular system and VCV Rack as its on-screen manifestation the aim of our observations is to understand how the physical and virtual interfaces of a generative patch shape the process of musical interaction. In our observations, we are particularly interested in how the multi-touch and kinetic interaction with a modular system stimulates the flow of compositional improvisation, or *comprovisation*, as computer music composer Dudas (2010) names it. Hands-on interactivity puts an emphasis on listening, which is of particular importance in an era where the arguably reductive, yet visually seductive, interface of the screen-based DAW (digital audio workstation) dominates in electronic music-making. With the latter, we refer to DAWs that favor a linear timeline approach to the arrangement, which is currently ubiquitous in music production practices (Strachan 2017). Whilst VCV Rack is a DAW, it operates according to the principles of interlinking and re/combination of Eurorack's modular system that similarly suggest a mycelium-like[2] underlying structure from which sonic structures spring forth and contract as the modular network of patch cords evolves. With such an instrument, the exploration of sonic textures and micro-tonalities is foregrounded over quantized timings and tunings.

The Eurorack standard was initially designed by Dieter Doepfer in 1996 (Doepfer 1998), who defined the physical and electronic characteristics of this modular system (Groves *et al.*, 2020). It provides an open, yet standardized, framework that encourages manufacturers, musicians, and sound designers to experiment with potential formats in sound creation. On a practical level, a standard format resolves issues in integrating equipment with different voltage to pitch standards, module sizes, and power requirements, thus providing a platform for designers to create modules that fit and function within a single environment. More significantly, the Eurorack standard facilitates the cross-fertilization of established analogue designs with digital counterparts. The ubiquity of self-contained DAW software in electronic music production can make the cost of investing in music hardware difficult to justify, instead enticing young musicians and producers to start out on relatively affordable and even free computer-based alternatives (Strachan 2017). The change from physical to virtual modes of music-making has had a profound effect on the creative process; and yet, it has arguably provided a catalyst for the resurgence of interest in modular music making.

The techno-aesthetics that emerge from the affordances of modular music-making emphasize sonic textures and intensities over linear structures. In this way, modular synthesis can enhance an affective relationship with electronica. In a brief historical discussion of the Moog synthesizer, Kristen Gallerneaux (2018: 21–2) describes the experience of working with modular hardware as a "retreat from compressed sound and the closed system of purely automated and digital music-making," further stating that:

> Using patch cords, wires, and keys to shape sound is at once frenetic, random, and meditative. There is a sort of ritualistic power in commanding the signal, and then determining how it lives and expands or decays into noise.

Moog synthesizers nevertheless offer an inbuilt keyboard, while our practice-led case study is based on a generative modular system, which purposely veers away from the use of a traditional-music interface. Within a generative patch, the artist is not in complete command, as the sonic outcomes are surrendered within the serendipity of a dynamic process between the human musician and the electronic instrument.

The first step in designing the modular patch is to research modules that can work together as a live performance system. For our case study, the initial prototypes were explored in VCV Rack as it is easier to try out different configurations of modules and connections in a virtual environment before investing in modular hardware and spending time on optimizing the physical layout of hardware modular patch. The main elements required for this patch are synthesis (sound generators) and (sound) sequencing modules; further utility modules are required to create a control schema, or concept, based on algorithmic processes. The concept considers how such a hybrid setup should work, both sonically and gesturally; and as Doornbusch (2002: p. 155) observes on composition practices in algorithmic patching, "it seems that there is no set method for mapping data from the domain of the conceptual, gestural or structural to the musical domain." In our case, the preference was to use hardware as the hands-on performance component, and VCV Rack as an algorithmic extension. The selection is based on a consideration of how modules will be patched together which is mainly gleaned from working with them—much like any instrument, this is mostly a tacit form of knowledge, derived through *techne*, within the practice of making. Once a particular configuration of modules has been established, it provides a platform to explore the system by creating different connections between the modules. This phase is where the creative potential of a system is defined. It can take multiple rehearsals to establish various ways of patching the system to fully understand how each module can be interacted with. Engaging with the gestural concept of a generative system, a set of controls are combined to enable an improvised performance using both predictable and unpredictable sequences and sounds.

The generative modular patch developed for the case study (see Figure 7.1) is based on Mutable Instruments' sequencing module *Marbles* as the main algorithmic generator—a virtual version of which can be found in VCV Rack as Random Sampler. This module generates random CV[3] data based on an algorithm defined by Émilie Gillet and draws inspiration from Buchla's Source of Uncertainty module and linear feedback shift-registers as a way of generating CVs. The outputs of the modules are grouped into three types labeled T, X and Y:

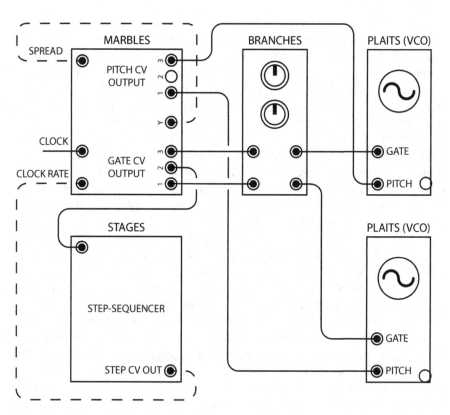

FIGURE 7.1 Inputs are on the left and outputs are on the right of each module.

- T 1, 2, 3 produce gate signals (for example: Note On/Off) depending on the position of the Bias control.
- Y produces a random CV, which is typically used as a modulation signal.
- X 1, 2, 3 produce random CV that can be looped into short sequences and typically used as pitch CV data.

There are three controls for the main compositional parameters of the generated CVs. The "Rate" parameter controls the timing of gate events on the T outputs as well as the random pitch CV of the X outputs. The "Spread" parameter controls the divergence of the X CV outputs, which in simple terms means it can control the melodic contour from small jumps to larger leaps (assuming this is how it is used). The "Deja Vu" parameter controls the probability of repetitions, from fully random to what is also referred to as a locking sequencer where random data repeats into a sequence. This is a high-level overview of what could be broadly defined as the compositional parameters that control the underlying arrangement of sounds in a generative patch.

With *Marbles* as the main generative component, the step sequencing module *Stages* is connected in a recursive feedback loop, providing CVs that modulate the clock rate parameter in Marbles. This results in a dynamic interaction between the two modules, where one sequencer (*Marbles*) generates pseudo-random sequences and the other (*Stages*) modulates the rhythmic structure. The outputs of Marbles are processed by the filtering module Branches, which introduces an algorithmic filtering technique based on Bernoulli gates.[3] This module controls the density of musical events (Rowe 1992) and can be further modulated by CV signals from either *Marbles* or *Stages*.

As in any modular patch, it requires at least one sound synthesis module, such as an oscillator, to produce a sound which is referred to as a voice; each voice can be a particular sound. In the context of a performance-oriented patch, a range of voices are selected and manipulated that can work across different registers. For this case study, the Plaits module, a macro-oscillator, has been selected to produce the voice. *Plaits* can synthesize a wide range of sounds such as conventional sawtooth and pulse-width modulated waveforms, percussive sounds, and more unusual granular textures and tones. As a self-contained voice module, it places some limitations on the sonic possibilities, but this simplicity makes it easier to use in a performance context. Oscillators can be combined to create further dynamic interactions at the point where the sounds are synthesized.

Our practice of patching embraces a symbiotic network of relations between the sequencing and synthesis elements. Similarly, modular musician Richard Devine observes that,

Within this architecture, you then discover these interesting interactions between different modules. It's been an incredibly intuitive tool for creating abstract textures and sonic spaces. (cited in Bjørn and Meyer 2018: p.146)

The creative practice illustrates how humans are entangled in a music-making network in which the instrument is, ultimately, a process. Through the curation, consideration and selection of modules, integrated complexity is built from simplicity. An element of compositional performance becomes apparent as parameters trigger a range of changes across both the arrangement and the sounds themselves. In this sense, our understanding of a music instrument differs from the ecological approach of Waters (2007, 2021), in which players remain human, as we are concerned with the material process of an immersive sonic-bodily-architecture in which, we argue, the human and machine are both performers, both agents in the act of *musicking*.[4] The unique design of a modular ensemble contributes to the performative compositional process by interacting and responding. Such a relationship reminds of Latour's Actor Network Theory (ANT), which Strachan (2017: 8) explains in relation to making music with electronic music technologies:

(E)ach network of creativity ... is distinct and relational according to the make-up of a given network. ... (T)echnologies are active in the production of experience for the human actor.

In the ecosystem of the generative patch, there is a creative dynamic between instrument control as performance and the response to machine-generated material. As such, the generative process enhances the fuzzy space in which the human participant is partly in control in a network of machines.

Working with a generative patch using *Marbles*, highlights the intersection between the human performer's creativity and the characteristic serendipity of the performing modular patch as instrument. Its brand name, Mutable Instruments, suggests the malleability in the affordances of its modules. Having previously developed DIY synthesizers, this was one of the companies that had been at the forefront of hybrid analogue and digital modular innovation between 2012 (Bjørn and Meyer 2018) and 2022 (Synthead 2022). Such a blend of technologies is intrinsic to the way these devices are designed and manufactured, with each device offering unique sonic affordances that are subsequently manipulated in interaction with other modules and the performing musician-producer. Like many module designers, Mutable Instruments' Émilie Gillet places emphasis on the design of the module faceplate, reinforcing the importance of how the layout of the controls shapes the mode of interaction and thus the process of music and sound creation:

One thing I didn't truly anticipate is that people can make music on systems without external MIDI controllers or sequencers. I come from this background which was really focused on having a MIDI sound generator, which you hook up to a sequencer or computer, in order to make it play music. (Why We Bleep, 2019: podcast)

The way in which the modular sound is produced fundamentally reshapes the creative process. This is unlike a software interface, for which the mode of visual interaction is screen-based, and the interface is predominantly via a mouse and keyboard.

Modules that work with randomization can be traced to Don Buchla's modular synthesizer principles, developed over half a century ago in collaboration with composer and musician Morton Subotnick, who was looking for "an electronic music easel" (Redbull, 2015: t 40:30), to sculpt "sound in a canvas of time and space" (t 1.22) and create a "visceral" electronic music for the future. In the words of music writer Harry Sword (2021: 180), Subotnick was "obsessed with the machine itself … describing the process as one of constant invention, bewitched by possibilities." In his quest to develop an ecstatic aesthetic, Subotnick saw himself as a conductor while remaining in the role of composer, laying out his musical plans in advance, and working with Buchla to design modules with microtonal affordances required for his purpose. With generative synthesis, though, one is additionally enthralled by the unexpected, submerging oneself within the flow of sonic affordances while combining, connecting, rupturing, recombining, and overall tweaking electronic sound over a period of time. As electronic musician Alessandro Cortini puts it during an enthusiastic demonstration of his Buchla synthesizer: "I embrace that these machines … aren't completely controllable … it's a different chess game every time" (Sonic State, 2015: t 6:24).

A trend in music software design is to recreate analogue devices through skeuomorphic software interfaces. Bourbon (2019) shows that, on the one hand, there is the practical aspect of creating a visual interface that mimics real-world controllers that have been established by hardware equivalents, down to a recognizable comfort-inducing wood paneling of a (sometimes fictitious) vintage synthesizer. On the other hand, there is the (arguably contentious) issue of recreating the 'inaccuracies' of temperamental analogue circuitry in which oscillators drift in and out of tune, and amplifiers introduce subtle harmonic distortion. Overall, it appears there is a dominant pre-occupation with reinforcing the notion that software is a true representation of physical hardware. This tendency can also be found in the creation of VCV rack, which the online electronic music instrument shop Synthtopia (2017) describes as being "designed to be used as a complete DAW for creating modular synthesizer compositions or as an extension of hardware modular systems." However, there are issues with the assumption that virtual versions have exactly the same affordances - they do not. For example, VCV Rack can save and recall patches while hardware patches are more challenging to note and reproduce. The physical interface of a Eurorack module enables an interaction between performer and instrument, whereas the generalist interface combination of mouse and screen with VCV Rack presents a barrier to a more direct hands-on creative approach.

Certain companies have tried to overcome interface issues through the introduction of bespoke MIDI controllers, such as Ableton's *Push* and Native Instruments' *Maschine*. Whilst MIDI controllers can create a similar tactile experience to hardware, it is not quite the same. MIDI parameters can be customized to different mappings which means the performer has to remember the position of various controls; consequently, it offers a polymorphous interface that differs from the fixed layout of a hardware synth. Furthermore, the aging 7-bit resolution doesn't offer the granularity required for more expressive and subtle timbral changes whereas analogue circuitry operates with continuous control voltages. Eurorack may, in part, be understood as a response to the software-driven virtualization of music-making, and to the subsequent successes and failures of MIDI devices in bridging the barrier of simulation. Eurorack modules have come to fuse analogue and digital technologies into a new paradigm that is primarily experienced through a physical interface, which re-shapes the creative process of making electronic music and sounds.

VCV Rack provides a useful way to prototype ideas for modular patches that a musician may implement with hardware modules. The advantage of VCV Rack is that it also opens possibilities that are difficult and even impossible to achieve in hardware, such as the ability to save and recall whole modular systems, or the use of polyphonic modules and connections, simplifying a complex task in the hardware realm. After spending some time experimenting, and discovering modules that only exist in VCV Rack, it makes sense to explore the potential of combining both software and hardware into a hybrid instrument. The main consideration is in deciding how to perform with the visually dominant screen-based VCV Rack, in contrast to Eurorack modules that we have observed enhance hands-on interaction. A bridge between hardware and software may be achieved with, for example, the *ES-9* sound card module by Expert Sleepers, as used in our set-up. This was designed to receive audio signals as well as control voltages. It is similar to a standard sound card, other than it provides connections that are compatible with Eurorack modules with switchable DC filters.[5] The latter are required to transmit low-frequency CV signals that would otherwise be filtered out on many sound cards. Using this approach enables the combination of both Eurorack and VCV Rack modules in a hybrid instrument consisting of both hardware and software. Part of the compositional practice is in the selection of physical modules as various haptic interfaces, and in the combination of the software elements on screen. This helps to learn the resultant instrument and to create a common performance interface that can be used for a range of different patches and compositions.

In the environments of Eurorack and VCV Rack, the dialogue between software and hardware is an ongoing process shaped by a complex web of commercial companies, smaller outfits, and enthusiasts. It is within this scene, or social ecosystem, that new modules are developed which can lead to entirely new approaches to the creative process. The open-source ethos leads to a community of modular designers and practitioners who exchange knowledge at music and showcase events (such as *Superbooth* in Berlin, Germany), as well as via web forums (such as *Modwiggler* at modwiggler.com). There are unintended uses of the modules, with unexpected outcomes, spin-offs. The design mutations, and an open-source ethos lead to new affordances in a fusion of old and new, whilst also stimulating dialogue between designers and end-users. For example, Émilie Gillet, the module designer of Mutable Instruments, states that:

I was surprised that the hardware and plug-in companies were promoting their products as if they contained magical ingredients. I wanted to dispel the anxiety about such a way of revealing the contents. We wanted the user to be able to customize the module to get around those restrictions. (Clockface Modular 2017)

This open and fluid approach to design enables modular designers to respond to the creative needs of the Eurorack community. In the selection of modular parameters there is no analogue versus digital, and in the context of their post-digital affordances, patching could even be understood as being analogous to coding. By providing the infrastructural affordances of a module, its designer can be perceived as an absent performer within the creative network of modular patch.

Reflecting on the creative practice of generating pseudo-random musical structures, different processes become interconnected into larger networks. Interacting with the modular system, the patch responds, creating further mutations. In this dynamic setting, parameters are explored and evaluated on their sonic qualities in the context of a performance. Serendipity plays a role throughout a modular performance until a desired musical sound emerges; when this is achieved, the main connections branch and solidify into a patch that forms the basis for a composition. As with any new instrument, it takes some practice to learn how to perform with the patch, and so it helps to record as much as possible to be able to review and iterate. To some extent the process can feel like jamming with a band, an experience that may be comparable to jazz improvisation as described in the ethnomusicological research of Elina Hytönen-Ng (2013), during which musicians riff on each other's responses within the flow of music making. The best take is used as a whole piece, or the most interesting parts are edited together from several takes, while avoiding over-editing to maintain a sense of continuity. In effect, compositional performance becomes a recorded practice.

In summary, the experience of both the tactile and kinetic aspects in performing with hardware modules is of particular importance in working with Eurorack. Module designers, such as Mutable Instruments' Émilie Gillet, place much emphasis on how parameters shape the interaction during the design process, including the haptic considerations of the panel layout. During the performance, or music session, there is constant interplay between the affordances of the modules and their networked links on the one hand, and the active listening and flow of haptic responses by the performing artist on the other hand, that together produce an entangled form of musicking. There is a sense of intimacy in working with sound in this manner. As architect Juhani Pallasmaa (2012: 53) observes, "(s)ight isolates, whereas sound incorporates; vision is directional, whereas sound is omni-directional. The sense of sight implies exteriority, but sound approaches me ..." Working with the haptic and sonic architecture of a hardware modular system, the division can blur between human and machine, between self and other, together as performers in a compositional network that articulates a dynamic creative relationship.

"Look at the network, and it starts to look at you," Merlin Sheldrake (2020: 186) observes in the context of mycelial research, as he reflects on the use of metaphors by scientists. In asking the question of how we hear, see, perceive, and engage with the techno-aesthetic of making sound and music with Eurorack and VCV Rack, we were able to draw comparisons as well as indulge in metaphors that help to familiarize a musical communication with what seems, at times, like a close encounter of the third kind. In modular systems such as Eurorack, there is a return to Subotnick's and Buchla's 1960s quest to paint and sculpt electronic music in a way that is unique for such instrumentation. Like Reich in the opening statement from 1968, the modular musician longs for "perceptible processes," for "closely detailed listening," and for "a musical process (that) should happen extremely gradually" (2002:34). While human subjectivity is increasingly intertwined and entangled via the internet, interconnected modular patches offer a sense of private yet social space. The patch is like a musical friend, interacting and responding, allowing the composing performer to test the affordances of the equipment as it offers up randomized variations as though suggestions from a co-creator. And the entanglement extends further, across the very digital communication networks that the hardware modular seems to offer an escape from, as friendships, module designs, and patch ideas are forged, fused, branched, broken, and abandoned, across a dispersed Eurorack social network that implicitly exists within its post-digital patches.

Notes

1 With the idea of a flow experience, we refer to Csikszentmihalyi's (2002) notion of an autotelic condition, in which the creative process is about enjoyment in the improvisatory activity without prescribed external goals.
2 We take our metaphorical inspiration here from the *Mycelium* module synthesis conference held in Spring 2021 (Cohen 2021).
3 CV: Control Voltage.
4 Small (1998: 9) describes "musicking" as: "to take part, in any capacity, in a musical performance, whether by performing, by listening, by rehearsing or practising, by providing material for performance (what is called composing), or by dancing."
5 DC means Direct Current, the opposite of Alternating Current (AC). Most digital soundcards tend to filter frequencies below 20 Hz as they are not deemed necessary for conventional listening purposes. These would typically have a DC filter on the output.

List of References

Text

Bjørn, K. and Meyer, C., with Nagle, P. as editor (2018) *Patch & Tweak: Exploring Modular Synthesis*. Copenhagen: Bjooks.

Bourbon, A. (2019) 'Plugging in: Exploring Innovation in Plugin Design and Utilization.' Hepworth-Sawyer, R. et al. (Eds.) *Innovation in Music: Performance, Production, Technology and Business* (1st ed.). London and New York: Routledge.

Bown, O. (2011) 'Experiments in Modular Design for the Creative Composition of Live Algorithms.' *Computer Music Journal*, 35(3), 73–85. 10.1162/COMJ_a_00070

Ciani, S. (2018) Foreword. In Bjørn, K. and Meyer, C., with Nagle, P. (2018) *Patch & Tweak: Exploring Modular Synthesis*. Copenhagen: Bjooks.

Clockface Modular (20 May 2017) 'Interview with Designers: Émilie Gillet (Mutable Instruments). Approaching the Unique Module Design Philosophy of a Popular Manufacturer.' *Clockface Modular*. https://en.clockfacemodular.com/blogs/waveguide/interview-with-designers-emilie-gillet-mutable-instruments

Csikszentmihalyi, M. (2002) *Flow: The Classic Work on How to Achieve Happiness*. London: Rider.

Doepfer, D. (1998) 'Zeit-Tabelle.' *Doepfer*. https://doepfer.de/time.htm

Doornbusch, P. (2002) 'Composers' views on mapping in algorithmic composition.' *Organised Sound*, 7(2), 145–156. 10.1017/S1355771802002066

Dudas, R. (2010) '"Comprovisation": The Various Facets of Composed Improvisation within Interactive Performance Systems.' *Leonardo Music Journal*, 20, Improvisation, 29–31. http://www.jstor.org/stable/40926370

Gallerneaux, K. (2018) *High Static, Dead Lines: Sonic Spectres and the Object Hereafter*. London: Strange Attractor Press.

Groves, W., Rivas, T., Humiston, T., Winship, G. and Handley, J. (14 Feb 2020) 'Beginner's Guide to Eurorack: Case Basics, Power Supplies, and Your First Modules.' *Reverb*. https://reverb.com/news/beginners-guide-to-eurorack-case-basics-oscillators-filters

Hytönen-Ng, E. (2013) *Experiencing 'Flow' in Jazz Performance*. Burlington VT and Farnham: Ashgate.

Pallasmaa, J. (2012) *The Eyes of the Skin: Architecture and the Senses* [Third Edition]. Chichester: Wiley.

Reich S. (2002) *Music as a Gradual Process. Writings on Music*. Oxford: Oxford University Press.

Rowe, R. (1992) *Interactive Music Systems: Machine Listening and Composing*. Cambridge, MA: MIT Press.

Sheldrake, M. (2020) *Entangled Life*. London: Penguin.

Small, C. (1998) *Musicking: The Meaning of Performing and Listening*. Middletown CO: Wesleyan University Press.

Simondon, G. (2012) 'On Techno-Aesthetics' (trans. De Boever, A.). *Parrhesia*, No. 14, 1–8.

Strachan R. (2017) *Sonic Technologies: Popular Music, Digital Culture and the Creative Process*. New York and London: Bloomsbury.

Sword, H. (2021) *Monolithic Undertow: In Search of Oblivion*. London: White Rabbit.

Synthhead (8 August 2017) 'VCV, a New Open-Source Virtual Modular System, to Debut at Knobcon.' *Synthopia*. https://www.synthtopia.com/content/2017/08/08/vcv-a-new-open-source-virtual-modular-system-to-debut-at-knobcon/

Synthead (11 December 2022) 'Mutable Instruments, R.I.P.' Synthopia. https://www.synthtopia.com/content/2022/12/11/mutable-instruments-r-i-p/

VCV Rack (2021) 'About.' *VCV Rack*. https://vcvrack.com/manual/About

Waters, S. (2021) 'The Entanglements Which Make Instruments Musical: Rediscovering Sociality.' *Journal of New Music Research*, 50(2), 133–146. 10.1080/09298215.2021.1899247

Waters, S. (2007) 'Performance Ecosystems: Ecological Approaches to Musical Interaction.' Paper presented at *Electroacoustic Music Studies Network EMS-07* Proceedings, Leicester, United Kingdom. http://www.ems-network.org/spip.php?article278

Video

Cohen, O. (20 March 2021) *Mycelium Symposium March 2021 - Modular/Virtual Modular Performances, Workshops, and Presentations*. https://youtu.be/nnipBoTOQ2g

Redbull (2015) Morton Subotnick Talks Silver Apples, Wild Bull and San Francisco. *Red Bull Music Academy*. YouTube. 15 May 2015. https://youtu.be/fmA7mJzh9os

Sonic State (14 May 2015) Alessandro Cortini - Fire Up The Buchla. *SonicState*. YouTube https://youtu.be/_o4yFxumAuA

Audio

Why We Bleep (2019) Podcast Interview with Gillet. https://audiojunkie.co/podcasts/why-we-bleep/episodes/why-we-bleep-005-the-mutable-instruments-emilie-gillet-interview-podcast-remastered-in-writing-edition

8

STRANGE PLAY

Parametric Design and Modular Learning

Kurt Thumlert, Jason Nolan, Melanie McBride, and Heidi Chan

Introduction

Critical work in music education has begun to challenge dominant Eurocentric curriculum and educational practices, as well as normative constructions of musicianship and musical literacy grounded in Western 'talent regimes' (Lubet 2009a; Lubet 2009b) and in 'performativity'-oriented music education (Kanellopoulos 2015). This body of literature draws our attention to the many and varied systemic barriers that children and young people face in public schools, as well as the need for critical perspectives attuned to challenge the ableist, racialized, and gendered realities of many students' music education (see Benedict, Schmidt, Spruce and Woodford 2015; Thumlert, Harley and Nolan 2020; Thumlert and Nolan 2020). Since the mid-1990s, journals associated with the MayDay Group have interrogated oppressive practices in music education by mobilizing critical theory, and by forging more culturally responsive and participatory pedagogies. Building on more recent calls to confront social justice obstacles in music education (Hess 2017), mainstream publications like *Music Educators Journal* (NAfME) are now examining taken for granted aesthetic and pedagogical suppositions in music education, as well as related institutional practices that reproduce cultural hierarchies and social inequalities. What is more, communities and researchers are coming together to rethink music learning and technology based on principles of equity and inclusion (see Drake Music, est. 1997; Canadian Accessible Music Instruments Network, est. 2021) and through initiatives supporting disability-led design of musical instruments (bell *et al.* 2020). Much work needs to be done, however, as dominant forms of music education in public schools remain moored to anachronistic curricular forms and inherited aesthetic coordinates, as well as

DOI: 10.4324/9781003219484-10

fundamentally ableist principles that foreclose—sometimes in advance, but also through instructional processes—possibilities for sustained participation and self-inclusion in sound-making worlds and practices.

Eurorack modular synthesis is a format that represents a dynamic site, practice, and sociotechnical model for rethinking learning with sound, and by implication, for revisioning music learning in formal and informal contexts, within and beyond music classrooms. Since the mid-1990s, the phenomenon of Eurorack synthesis has provided authentic contexts for engaging sound, technology, and music production differently—illuminating alternative means and pathways for inquiry, learning, making, and community participation. Given that there are very few spaces where Eurorack modular synthesis is formally taught, modular synthesis provides a unique site and figure for understanding self-directed engagement and inquiry, as well as emerging sociotechnical contexts for learning with sound, making, and developing a skilled practice (Thumlert *et al.* 2021).

In this chapter, we focus directly on material-centric (McBride 2018) engagements with the tools and sounds in play, examining what and how people learn with modular synths and through inquiry-driven processes of 'parametric design' (Strange 2022). Specifically, what can we learn from modular synthesizers through tangible, interactive engagement with these tools, and through our experiments, improvisations, and compositional play? We are curious to know how the aesthetic and technological properties of modular synthesizers disrupt normative genre boundaries and frustrate traditional judgements of taste embedded in Eurocentric aesthetic and educational norms. We then unpack how modular synthesis (and its resurgence in popularity in the Eurorack format) may provide alternative learning pathways and practices for self-inclusion in music making worlds, bringing into play much of which has been nullified in dominant music curricula.

Drawing upon technology studies and critical discourses in music education, we start with a critical examination of traditional forms of music education. We then draw upon Allan Strange's landmark guide for learning modular synthesis, *Electronic Music: Systems, Techniques and Controls* (2022) to help us examine how learners engage the properties of sound and the elements of music in modular environments. Allen Strange (1943–2008) was a composer and professor of electronic music at San Jose State University, a close associate of Don Buchla and other members of the San Francisco Tape Music Centre, and author of the Buchla Music Easel manual. First published in 1972, *Electronic Music* stands, arguably, as the first comprehensive work on engaging the practice of modular synthesis. We suggest that, in many ways, Strange's text anticipates the work of Tim Ingold (2013) and other contemporary researchers examining the so-called 'materiality turn' in the social sciences, providing an alternative and more inclusive framework for learning with sound-based materials and tools.

Distributions of the Sensible: 'Talent Regimes' and the Null Curriculum

Recent critical work has identified exclusionary structures in music education. This body of research examines hegemonic Western/Eurocentric discourses, values and hierarchies in music curriculum (Benedict, Schmidt, Spruce, and Woodford 2015); ableist assumptions about tool use that mediate inclusion and create barriers for disabled and neurodiverse learners (bell 2017; Darrow 2015; Matthews 2018); and longstanding patterns of privilege and marginalization that remain unchallenged in music education settings (Schmidt 2020). While much critical work interrogates standardized music literacies and/or seeks to decolonize the explicit curriculum (Hess 2015; Kaschub 2020; Kivijärvi and Väkevä, 2020), we examine music education's *hidden curriculum* (Jackson 1968) and *null curriculum* (Eisner 1979): the underlying suppositions and structures that normalize forms of 'proper' tool use, inclusion and exclusion, and related pedagogical processes that may lead to learner (self)disqualification and eventual disengagement from sound-making worlds.

Where the explicit curriculum is comprised of manifest educational aims, policy, content (repertoire), ensemble systems, and skills-based performance objectives, the hidden curriculum is defined as all that is learned, indirectly or incidentally, through embodied experience, everyday action, and routinized doing in schools. The hidden curriculum is comprised of our unstated presuppositions. It is, in short, our pedagogical and aesthetic 'common sense': our embodied understandings of music and musicianship, and the normative means—the corpus of rules, principles, routines, gestures, and procedures—for transmitting and assessing practical knowledge and skill. Alongside the hidden curriculum, Eisner (1979) draws our attention to the null curriculum: all that is tacitly excluded or rendered invisible, inaudible, intangible—all the representations, experiences, and varieties of play, inquiry, tool interaction, and modes of sound making and doing that are effectively 'nullified' in and through the normative interworking of the explicit and hidden curriculum (Nolan and McBride 2014).

By way of analogy, the conjunction of explicit, hidden and null curricula enacts what Jacques Rancière (2004) calls a distribution of the sensible: a self-evident regime of (unequal) roles and relations, including educational power/knowledge relations (Rancière 1991), which are pregiven in a form of community. Insofar as a distribution of sensible (*partage du sensible*) establishes a horizon for community (being together), it also delimits the 'respective parts and positions' of that community, as well as self-evident boundaries for those who 'naturally' have no participatory 'part' at all (Rancière 2004: 12–14). For Rancière, the stakes of the aesthetic are always interwoven in matters of equality. As an aesthetic figure, any distribution of the sensible delimits and normalizes how sensory/aesthetic experience is perceived, composed, generically classified, and differentially valued (or rendered invisible or inaudible). In musical worlds, a distribution of the sensible defines the normative field of what counts as music and what is perceived as 'noise', as well as horizons of aesthetic possibility for compositional doing or performance within a particular institution or historical context. If Pinch and Trocco remind us that what counts as music is always 'contested territory' (2004: 10), Rancière signals how relations of (in)equality—and tacit forms of inclusion and exclusion—are always at stake in these aesthetic 'contests'.

In music education, the possibilities of aesthetic experience and pedagogical action are largely mediated through the dominant curricular 'common sense' of inherited institutional routines and cultural-aesthetic priorities, along with largely *uncontested* pedagogical means for obtaining related curricular goals. If, for example, the *explicit* curriculum imposes a taxonomy of learning objectives that conjoin Western art music, discrete technical skills, notational decoding, and sequentialized benchmarks for acquiring 'musical literacy', the hidden curriculum also communicates cultural suppositions about what music is, what it means to be a musician; about competitive individualism, rank and aesthetic hierarchies; as well as ableist norms that associate musicianship with technical alacrity, 'talent', virtuosity, or professionalized roles. Under these conditions, authentic musicianship is narrowly defined in advance of learning, and is located not in immediate tangible relations, explorations and making, but in Western aesthetic ideals or specialized cultural identities to which we might collectively aspire to or teach *toward*. At the same time, authentic meaningful learning opportunities like improvising, designing sound, and composing music are (indefinitely) deferred until after other 'requisite' forms of musical literacy have been meted out.

Dominant forms of music education thus prefigure the very terms of participation—and the tacit forms of exclusion or the means of self-disqualification over time. Lubet (2009a; 2009b) asserts that the hidden curriculum of music education is shaped by culturally normative 'talent regimes' grounded in Western music traditions. Talent regimes and talent discourses define 'right ways of being talented' (Gaztambide-Fernández *et al.* 2013: 130), conceive of talent, ability, and 'giftedness' as attributes of hyper-abled individuals and frame musical competences and learning goals in terms of technical competences toward which learners are gradually trained and competitively socialized. Sequenced curricular expectations thus police the work and play of sound-making, increasingly narrowing the ambit for participation as students are progressively graduated 'up' through the system (Thumlert and Nolan 2020). Students may procedurally come to understand, through the linear-developmentalist process of instruction itself, that they can't do (yet), can't participate authentically (yet), are not 'good enough' (yet), and will likely never be 'good enough' in relation to imposed cultural ideals about music/musicianship (see Lubet 2011).

Indeed, these norms, ideals and instructional pathways are ratified at the earliest ages through doctrinaire methods associated with Orff, Suzuki, Dalcroze and Kodály, which inaugurate children into musical worlds characterized by bourgeois cultural norms and replicable, uniform (group) mechanics. In schools, music and learning music is de-composed, compartmentalized and then sequenced into sets of skills to be gradually acquired, assessed, and evaluated on the way to some future condition of musical competency. In order to progress, however, students are locked within the amber of heteronomous curricular mediations over which they have no meaningful agency or authentic embodied competence (Thumlert 2015). This pedagogical distribution of the sensible disqualifies learners with disabilities from the outset and enacts a procedural 'winnowing out', over time, of the majority of music learners in schools (Lubet 2009a: 730; see Rancière 1991).

For those who are initially 'able' and invited to play, ensemble instruments—woodwinds, brass, strings, percussion, etc.—are generically presented as static technologies defined in terms of their own intrinsic characteristics and normalized aesthetic functions. Ableist expectations for what an instrument is, can do, and how it fits in the spectral whole are built into the dominant Western curriculum and repertoire. While conventional instruments are relatively 'fixed' in terms of material structure and range of sonic outputs, the tautology that a musical 'tool is what it is' according to inherited cultural meaning, an ensemble function, or sets of proper motor skills coordinated with an idealized performance body (bell 2017: 123; Honisch 2019), serves to stabilize and de-historicize the tool, insulating it, and those who play it, from further vibrotactile inquiries, inclusive adaptations, or extended techniques that might challenge barriers to participation in sound making worlds.

Insofar as the hidden curriculum of music education instills particular values and normalizes pedagogical forms, notational literacies and ways of talking about music, it also excludes or nullifies other modes of engagement and sonic exploration: interactive experimentation with diverse vibrotactile materials and sound-making tools; learning through open-ended inquiry, experimentation and sound design; and authentic opportunities to learn, play, and compose outside of abstract notional systems and Eurocentric music theory frames. In formal contexts, diverse students are rarely permitted to follow intrinsic interests, idiosyncratic pathways, and modes of autonomous inquiry and self-directed practice where embodied competences might be enacted and advanced ongoingly through authentic practice (Howell 2017; Thumlert and Nolan 2020; Schmidt 2020). Even in early years education, these alternative approaches to working with sound and music-making technologies are, for the most part, consigned to the null curriculum in favor of promoting culturally hegemonic, ableist and adultist orientations to music learning (Nolan 2020).

In the next section, we explore modular synthesis as a dynamic location for activating much that has been nullified in music education's dominant distribution of the sensible. We draw upon the work of Allan Strange (2022) to consider a material-centric (McBride 2018) approach to working with tools, for exploring the properties of sounds, and for engaging the elements of music in ways that emerge dynamically through play with modular synthesizers and the unique environmental affordances they provide.

'Forgotten Basics': Sound Learning from the Inside Out

In his introduction to *Electronic Music: Systems, Techniques and Controls*, Allen Strange (2022) examines how 'musical consciousness' is always contingent upon aesthetic and historical structures where evolutions in technology, discourses, and artistic inquiry/practice co-determine what is perceptible as noise or music, and under what contexts and conditions sounds are or are not, or may ultimately become, hearable as 'music'. Along with Don Buchla and other members of the San Francisco Tape Music Centre, Strange was interested in shifts in 'musical consciousness', particularly where novel musical tools, interfaces, and techniques were in play, and where forms of 'noise' might *become* perceptible as music. In this respect, Strange's *Electronic Music* challenged, and continues to disrupt, a normative distribution of the sensible (Rancière 2004) governing what can be perceived as music, a music technology, or a proper interface for artistic inquiry, doing and making with sound.

Electronic Music: Systems, Techniques and Controls is described as a guide for learning techniques: procedures for producing 'events', organizing forces, and making 'things happen' (2022: 3). As such, it offers an alternative orientation to sound-based learning, one which proceeds by estranging us from normative assumptions about musical culture by first reminding us of certain 'forgotten basic ideas' (1). Leaving genre hierarchies and aesthetic categories behind, Strange brings us back to the principle that a musician or composer 'is ultimately concerned with different manners of vibration' (7), that all musical processes come down to 'temporal pressure variations' on the body (8), that is, 'variations in air pressure at various rates', and that, further, we can 'make music from anything we choose to make music from' (2). Throughout the text, Strange draws analogies between electronic tools and acoustic ones, giving us fresh insights into the fundamental properties of vibrotactile materials across material contexts and aesthetic environments. While Strange asserts outright that his book is 'not a text of musical theory' (3), *Electronic Music* nevertheless re-orients readers to fundamental principles of sound and music, and to that extent helps us rethink music learning through inquiry with materials, and in ways that contest a dominant distribution of the sensible in formal music education.

What is endorsed throughout the book are open-ended explorations of the *parametric* relationships among synthesis tools, and of the many entangled variables in play when interacting with or patching modules, routing signals and voltages, and generating/controlling energies, timbres, patterns, movement, and shapes. Strange's approach, rather than a prescriptive one that decides in advance how the materials might be uniformly controlled to obtain predicted outcomes, invites the player to patch, listen and respond to the character of the tools and how they work (alone and together) toward the users own (co)emergent purposes. Strange outlines inquiry-driven procedures based in tool interaction, deep listening and response, and experimenting and improvising with materials, to learn with and from them. Boundaries between making sounds and composing music are at the same time challenged: the player or composer is not following traditional notation systems or external models, but is invited to learn, immersed in continuous interactions with and alongside the tools and their outputs, and to expand upon and beyond the text rather than be constrained by it (Strange 2022: 10).

If Strange is interested in, and attuned to, shifts in 'what counts' as music, a musical tool/interface, or skilled practice, and how one goes about learning a practice, educational institutions, by contrast, tend to conserve practical and aesthetic boundaries, predetermining what is 'sensible'—what is hearable, aesthetically, as a valid musical gesture or event. As noted above, curricular forms organize how sounds/music are classified, along with direct and indirect means of determining how we should differentially value, 'feel' or be affected by various musical forms, scale systems, modes and idioms. With predetermined cultural precepts for learner percepts, this distribution of the sensible extends to coordinating what is hearable as a valid musical gesture, genre or competent sound-making event (Thumlert and Nolan 2020). Insofar as students are fitted to conventional instruments in relation to standardized literacies and ensemble roles, extrinsic patterns of practice are imposed and reproduced. Historically, these educational endeavors, as Ingold puts it, are in their purposes and means of instruction *hylomorphic*: that is, they enact a gradualist and linear mode of 'training [that] molds the raw material of immature humans to pre-existent designs' (2017: 5) in order to programmatically replicate the given design, or at least some rudimentary features of it.

If Ingold critiques dominant forms of hylomorphic instruction (where known goals and standardized outcomes direct learning processes and organize environmental contingencies so that students might ultimately obtain or embody those predicted ends), Strange's book offers a point of departure for rethinking learning by situating the practitioner within the 'meshwork' (Ingold 2009) of a modular environment and tool ecology where actors make and learn from the 'inside out'. Throughout the text,

Strange dramatizes the forgotten basics of sound-making, analogically, by making equivalences between how electronic sound-making tools and non-electronic materials work. Strange makes connections back and forth between electronic music systems and other sound-making materials (bows, reeds, strings, hair, metal cylinders, fingerboards, 'quantizing' frets or tone holes, muting materials, membranes and even fluorescent lights and television tubes) to explore their common auditory and vibrotactile properties in relation to the bodily forces in play when we interact with various instrument and their respective resonant structures and sonic outputs. Strange's point of departure enables learners to begin with tangibly real and grounded reference points for practice and inquiry, such as vibration and experimentation with multiple materials and sound sources, as well as various means of modulating voltages and attendant outputs/sounds.

The orientation to sound and making modeled by Strange at once helps us demystify musical aesthetics while also energizing sound-based learning with the spirit of tangible inquiry, play, experimentation and improvisation. Like any other form of energy, electricity can be controlled to generate and complicate vibrations (waveforms), their pitches, fluctuations, amplitudes, shapes and patterns over time. As Strange states, 'electronic sound is only "electronic" in terms of generation and control … The electronic musician relates to electronic instruments in precisely the same manner [as a cellist]. Both are concerned with how fast, how hard and in what pattern [the materials] are vibrating' (11).

The cello player attentively exerts force to press a string to the fretboard and pull/push the bow, the 'bow-hairs force the string and sound board to vibrate and cause the air pressure in the string's environment to fluctuate' (11). The player of a modular synthesizer attentively organizes voltages to control the pitch of an oscillator or the rate of an LFO, the cut-off frequency and resonance point of a filter, or the periodic force of an amplifier. This not much different than placing a finger on a cello fingerboard to select the pitch, plucking or bowing the string to control the attack, decay and amplitude (dynamics), and moving the finger on the fretboard to create vibrato (LFO) or to glide between intervals (Thumlert et al. 2021).

The point being: the electronic musician and player of acoustical instruments share common concerns and are fundamentally doing the same things. Strange's text enables us to re attend to, and focus on, the materials in play, the techniques for generating particular sounds and timbres, and in so doing we see what is shared in common across electronic and acoustic materials. Using similar analogies, electronic music pioneer Daphne Oram (2016) referred to the dance of capacitors and resistors where, in producing musical events, energies are held in tension or variously released. For a sound to 'happen', the gesture or event can take many forms: magnitudes can be embodied as 'a poised drumstick, a controlled bow, a flexed finger, a held breath. Release the tension and the result is a flow of sound … waves of compression alternating with rarefaction which beat against our eardrums' (2016: 19). As Oram continues, 'by pursuing further analogies between electronic circuits and the composing of music, we will be able to gain a little insight into what lies between and beyond the notes' (20). Here, Oram in turn hopes to 'unbend' ossified conventions surrounding sound and music in order to excite both scientific and artistic possibilities (21–6).

While Strange highlights what is common across different music making environments, he also highlights the unique characteristics and properties of modular synths, whose 'unfixed structures' provide near limitless variation in relation to what is actualizable sonically and compositionally. Materially, a modular synthesizer is an assemblage of discrete sound-making devices, or modules. Voltage-control is the 'operational technique' of modular synthesis, the physical practice and artistic means by which, using patch cables, users interconnect the different components of their system to route audio signals and steady or fluctuating control voltages (Strange 2022 : 4).

Whereas a traditional instrument is relatively fixed or structurally consistent—'its structure being the result of its technical evolution tempered by the demands of its evolving literature', modular synths are 'not fixed' (3): their outputs are mutable and dynamic, affording sound makers a polyphonic frequency palette that can articulate or glide through the Western 12-tone array, while offering an expanded range of pitches with an equally pliable horizon of timbres, both acoustically organic and synthetically otherworldly (Thumlert *et al*. 2021). That is, a modular synthesis system, by virtue of its vey modularity, has 'no pre-defined structure', but is better initially understood as 'a collection of possibilities—a set of musical variables or parameters such as pitch, loudness, space, timbre, etc., that exist in an undedicated state' as a 'yet to be produced musical event' (3). With a multiplicity of different techniques and interactive means for controlling relationships among sound sources, signals and controls, the 'art of electronic music' thus requires the organization of variables into 'desired structural relationships' that actualizes musical gestures out of a near infinite horizon of possibilities (3).

At the site of material interaction, the synthesizer, Strange suggests, is best understood as both the component tools and the user/player who (by way of patching) organizes and routes relationships among the different modules in play. By the same token, a patch, then, is more than a set of cables physically connecting different components to create a signal path leading to an amplifier/speaker; it is also a specific configuration of malleable relations and interwoven variables through which users *and* modules interact with, affect, and respond to one another (Thumlert *et al*. 2021). Each patch, gesture or work(flow) can take a different 'shape' or evolving architecture, uniquely reflecting, according to Ingold's (2013) view of learning, the fluid relations between the materials and the learner where students are co-transforming themselves with and through their tools/materials, through ever increasing levels of complexity (Nolan and McBride 2014).

Due to the unfixed structures of modular synthesizers, the very building blocks of sound design disrupt the ways we think about music and learning instruments in formal educational spaces. The affordances of modular synthesizers enable users to break down the basic elements of music into discrete unit relationships and sound-making operations, providing alternative pathways for sound-based inquiry and learning from the inside out. Strange coined the term 'parametric design' as the key means for learning and making within a modular environment, defining it as an analytic process in which one navigates or envisions 'the individual and corporal influences of all the parameters of an existing or imagined sound' (2022: 10). For Strange, any musical gesture is already implicitly 'parametric'. One cannot acoustically perceive pitch without perceiving a sensation of loudness, tone quality, duration, and so on. However, the implicit parametric variables we encounter when we listen to music or hear a sound need to be explicitly attended to and coordinated when shaping sounds with Eurorack synthesizers, using voltages to control and modulate the parameters of other tools.

For example, a single physical gesture on an acoustic instrument (a press on a piano key) is, in modular synthesis, distributed across multiple modules and unit sub-operations that are patched together to produce what Strange calls a *composite event*: these elements may include the use of a signal source (e.g., an oscillator) and respective wave form(s) (sine, sawtooth, triangle, square); the control and modulation of frequency, timbre, and amplitude; and the control of envelop properties that shape the sound gesture over time or limit it to a clipped attack. Filters can be used to subtract frequencies and taper the harmonic character of a sound, and/or the gesture can be augmented by other modules to multiply harmonics and increase timbral complexity (i.e., subtractive synthesis, in the traditional Moog format, or additive synthesis in the Buchla tradition). Subaudible waves from low-frequency oscillators (LFOs) can be patched to further modulate the tonal and timbral character of a sound, 'sweep' through a frequency spectrum, or create dynamic ornamentation (trills, vibrato, tremolo, etc.). LFOs and other voltage sources can trigger, gate, sequence, and even randomize voltages in ways that yield near infinite possibilities for sound modulation, harmonic spectra, texture, dynamics, rhythm, space, and so on. As the player of a prepared piano may introduce new timbres (by attaching objects to strings), the player of a modular synth 'folds waves' to complicate sound coloration.

The accomplishment of one gesture or patching technique feeds forward into emerging questions and new challenges that may arise through interaction with tools: organizing chords, polyphony and counterpoint; working with stepped/arpeggiated sequences or glissando, articulated or slurred notes; controlling the rise/fall and differential slew of wave shapes/ functions; gauging the sonic effects of different triggers, pulses and gate durations; exploring quantizing functions, microtonality, or sound grains; hearing/seeing what happens when you attenuate (or attenuvert) signals, or mix multiple signals together; 'pinging' a filter to sculpt percussive notes or resonant sound artifacts; modulating one oscillator signal (sine wave) with the wave output of another oscillator (FM synthesis) to build complex timbres without the use of filter. By way of particular circuit designs, some modules enable you to introduce elements of chaos, indeterminacy and chance into parametric play. Other modules emulate the properties and movement of magnetic tape and afford modes of recording, splicing, and looping associated with *musique concrète* and early electronic music pioneers like Buchla, Subotnick, Oram and Strange himself. In short, the learner can dwell within the shaping of a gesture, can loop recursively to hone a technique or refine a pattern, or unspool in unpredictable, non-linear directions (where talent regimes and normative judgements of taste no longer contain learning opportunities).

What modular synths invite players to do is build sound gestures—and learn—from the ground up, through additive and subtractive processes, hit and miss operations, where the learning is inseparable from the idiosyncratic movement of actualizing sound/music from the 'undedicated state' (Strange 2022) of virtual (Deleuze 1988) parametric relations and composite possibilities. While Strange provides meticulously detailed guidelines for 'control' and frequently elucidates principles by drawing attention to the mathematics and physics of musical and electroacoustic phenomena, a recurrent message of the text is that there is no single 'right way' to use a modular synth, no uniform starting point or workflow. Here, Oram asserts that one 'controls the craft' in as much as one is 'taken along with it, while responding to 'unforeseen circumstances' as they evolve or emerge (2016: 27). Ingold echoes, inverting theory-to-practice views of learning, that it is *through* practice and making—*through* our engagement with authentic materials, processes and improvisational inquiries in the world—that theoretical and practical knowledge emerge and skillfully expand (Ingold 2013).

Just as there is no one fixed or 'right' path for patching signals, there is no one sequential or developmental trajectory for learning the practice. As novices, we can begin with a single gesture using a minimal number of modules and shape a single sound, explore a timbral color wheel, and excite overtones; we can start with musical sequence and then explore melody and harmony using conventional tempered tunings or by exploring the untempered spaces between intervals; we can embroider relations between musical figures and textural grounds; or we can design a generative work that evolves over time, bringing into play the more entangled parametric relationships among different modules and control voltages. Through situated exploration and doing, learning actors enact, as Gee puts it, an 'embodied empathy for complex systems' (2009: 4), as well as gain increasingly deeper understandings of the rules of the game, and how to creatively bend or critically modify them to a user's own purposes. That is, with immediate interactional feedback, where meaning is situationally immanent in the contexts of actors' practical engagement with materials, tools, and their affordances (Ingold 2017), a novice learner can begin to authentically design, compose, and make music with modular tools through open-ended exploration—without *their* understanding and explorations needing to conform to any external plan, curricular script, or imposed (hylomorphic) expectation for knowing (Thumlert *et al.* 2021).

Anticipating by decades the work of more recent 'new materialist' theory, Strange urges readers/students to attend to materials and forces and their relations, to not be limited by external references or hylomorphic exemplars, whether they be the manuals, the front panel graphics of the modules themselves, or the many exploratory etudes and schematics provided in the book. By contrast, the learner is enjoined to learn with and from the materials, to make and learn from mistakes, and to:

> 'Try everything! Never wonder "*should* I do this?" Instead, "I wonder what will happen *when* I do this?" may lead to an expansion or development of a unique technique' (Strange 2022: 10, italics in original).

To adopt this attitude of inquiry is to affirm an experimental relation to yet unknown and possibly unknowable outcomes. 'I wonder what happens *when* I do this?' is, for Strange, a question that may lead to new insights in relation to the open-ended exploration and the possibilities parametric organization, the perceiving and coming to understanding of the affordances of tools (Norman 1988) and environments (Gibson 1979) in developing a skilled practice.

Patching, in this vein, is not process of imposing theory or applying abstract rules on objects or people, but an action-oriented means of coming to understand, through perception and response, what different modules might do under different conditions, in different parametric relations with one another or even themselves, for example, when self-patching a module. Rather than the imposition of ideal forms (i.e., a predetermined aesthetic outcome), there is continuous movement where goals and intentions themselves are modulated through the processes of tool interaction and attending to feedback.

Drawing on Ingold's work, Van der Kamp et al., speak to a similar mode of learning where attention is directed and renewed through the process of interaction and inquiry: student 'searching is deliberate … [but] what is revealed—if anything—is not known beforehand. It is the active, inquiring, *expanding* experiencing that counts, not the particular content or outcome of the experience, [and this] entails a degree of uncertainty for students in their encounters' (2019: 5). Fittingly, Ingold describes this mode of learning a skill as 'an unbroken, contrapuntal coupling of a gestural dance with the modulation of the materials' (2009: 434).

From Null Curriculum to Modular Practice

The mode of learning described above activates much of what has been suppressed or excluded through the interworking of music education's hidden and null curriculum. Attending to the 'forgotten basics' of sound learning and making with materials provides an alternative pathway for 'ground up' self-inclusion in sound/music making worlds. Indeed, attending to the 'forgotten basics' is the antithesis of regimented, decontextualizing and fundamentally alienating 'back-to-basics' educational ideologies. Much of what we have described above in relation to modular synthesis practice—interaction with tools, the shaping of sound gestures, and constructive exploration of properties of sound and the elements of music—implicitly or explicitly speaks to the opportunities of learning through open-ended play and authentic compositional making: immersive interaction in feedback-rich environments that rewards experimentation and renews attention and involvement; contexts that invite players to take risks ('try everything'), fail, and persist ('try again') through increasingly complex and dynamic problem spaces. Practice feeds forward into new inquiries and involvements, where struggle and doing hard things are welcomed as a necessary part of the learning process, and are inseparable from ongoing parametric play: the idiosyncratic movement of trying, listening, abductively figuring out, making and getting better. These modes of inquiry and learning are indifferent to institutionally-assigned learner 'deficits' or 'lacks' to be endlessly remediated, and there are no distant, extrinsic objectives to conform to.

If music education's dominant distribution of the sensible predetermines the forms of community, the ensemble roles/positions and (hylomorphic) means of obtaining outcomes, as well as the cultural boundaries and modes of address (cultural, ableist, racialized, gendered) for who can participate and how, then the multiplicity of sounds, purposes and contexts associated with modular synthesis can be seen to enact a *redistribution* of the sensible, along with expansions in 'musical consciousness'—in what counts or is perceivable as music, and who can make it or do it. In learning about and shaping sound and music, the practice is largely indifferent to exclusionary genre boundaries, as well as disrupts traditional metrics of 'talent' and comparative/competitive ranking that are often explicit or tacit aspects of—and indeed barriers to—learning an instrument in formal institutions (Bolden 2010; Lubet 2011).

By contrast, modular synthesizers require us to attend to elemental building blocks of sound. To make music, we coordinate and assemble gestures and compositive events 'from scratch'. In doing so, we also attend to the elemental materiality of our listening and sound-based learning environments. This shift in attention and possibility in turn enables us to re attend to the parametric properties of music education itself, its dominant distribution of the sensible, and to rethink the possible forms and functions of sound making technologies and who can use them. This approach to tool-use also allows us to consider various feedback loops between informal and formal learning spaces, and to empower students to 'take charge of' and 'explore their particular interests and idiosyncratic ways of learning' (Howell 2017: 251). Technology, in this context, is evaluated in terms of how tools support 'autonomy and independence ... qualities [that] are invariably compromised, if not negated entirely, within the restrictions and rigidity of most formal music education institutions' (Howell 2017: 251).

Music making tools need to be re-evaluated, too, based on how learners can adapt tools to their own purposes and interests, rather than forcing people to adapt to, or be 'configured by' (Woolgar 1990), the instrument, or to fail based on external determinations of 'progress', 'success', 'talent' or 'beauty'. Strange's forgotten basics help us attend, as well, to instruments as historically situated and evolving (adaptable, hackable) tools. Instead of accepting instruments and ensemble roles as static or historically settled, as is often the point of departure in schools, we might look to technological change, process, and adaptation. Hugh Le Caine, for example, invented the first voltage-controlled synthesizer, the Electronic Sackbut, in the mid-1940s based on a Renaissance-era brass instrument (the sackbut) which was itself developed from trumpet-like instruments utilizing a then novel (ca. 15th century) trombone-like slide to control pitch: the affordance of a telescoping slide allowed the brass player to glide between notes. Analogically, Le Caine's electronic sackbut, using voltage control, provided a new—and wider—range of glide and expressiveness, and with 'extended timbral opportunities afforded by electronics' (Vail 2014: 4). The point being, what we call a 'trombone' is less an ontologically stable or 'settled' ensemble instrument than a knot or node in an ongoing process of variation in, or modulation of, the materials we use to make things vibrate and resound, and in the new forms of community that may co-emerge with these modulations.

Indeed, modular synthesis entangles sundry communities and domains of interest/inquiry—from conventional aesthetic/arts-based approaches to music making to electrical engineering, the sciences and DIY electronics tinkering and so on. Even prior to the era of Eurorack, synth manufactures were advertising analog synthesizers as an educational means for demonstrating scientific principles of music and acoustics. More dramatically, electronic music establishes common meeting grounds for artistic, technological and scientific inquiries, where one field of interest might be engaged through the lens of another, or where a problem can be seen from two (or more) directions at the same time. Indeed, parametric design, as Strange defines the term, is a mode of modular thinking that cuts across disciplines and domains, and even anticipates recent calls to integrate computational thinking (Grover and Pea 2013) and algorithmic literacies (Thumlert et al. 2022) into 21st-century learning environments. Again, transdisciplinarity is rare in formal music education, where cultural/aesthetic norms foreclose productive boundary crossings and converging access points, standardize ways of musical knowing, and stabilize our relations to instruments and tools.

All of this, of course, has material implications for inclusion, for co-creating new positions of capacity and participation. Beyond the multiple access points and boundary-blurring modes of play, modular interfaces can be modified in multiple ways by particular users. The electronic properties and 'unfixed structures' of modular synths afford customizable interfaces for controlling voltages, including touch (knobs, buttons, sliders, grids, joysticks, etc.) and conductive surfaces (for controlling voltages and adjusting resistances), and more recently computer-based virtual systems (from Nord Modular to VCV Rack). Emblematically, Don Buchla's first synths were constructed based on his own interest in creating new interfaces between musical instruments and humans, and for addressing the limitations of working with magnetic tape (Bernstein 2008). Just as modular synths offer an 'undedicated state' of sonic possibilities for 'yet to be produced musical events' (Strange 2022: 3), so too they offer materials, precedents and processes for imagining yet to be designed musical interfaces and instruments, created by and with (rather than for) diverse learners (bell et al. 2022; Hamraie 2017; Matthews 2018). To disrupt marginalizing and ableist orientations to music education, this kind of imaginative work aligns with recent calls to 'hack' music education based on disability-led design initiatives (bell et al. 2020; bell 2017; Matthews 2018). As bell et al., state, this line of inquiry may 'lead to new instrument designs and modifications to include a broader spectrum of people, especially as it relates to bodily differences' (2020: 13). Significantly, Eurorack synthesis also broadens how we think about the possibilities of sound, technology and related shifts in musical consciousness, and how we might also, then, rethink culturally hegemonic constructions of music and musicianship in learning spaces. Any act of self-inclusion or unexpected 'part-taking' signals a redistribution of the sensible: a reordering in normative positions of participation 'where the palette of sanctioned sensibilities shifts ground' (Toila-Kelly 2019: 124), and where domains of possible experience, practice, and participation are challenged and extended by the excluded themselves (Rancière 2004; Rancière 1998).

Conclusion: Parametric Play and Material-Centric Learning

At the turn of the century, Lev Manovich's (2001) *The Language of New Media* offered a comprehensive theorization of emerging media and new artistic practices. In this work, Manovich asserts that the synthesizer was the first instrument to embody 'the logic of all new media—selection from a menu of choices' (2001: 126). Grounded in a symptomatic reading of technoculture, the synthesizer, for Manovich, is crystalized as a kind of representative metaphor, a prodrome for a logic of creativity associated with the computer age and techniques of remix (selection and recombination). Displacing the romantic genius of the modern era, the paradigmatic artist of the electronic age is bound to a different principle of production: 'modification of an already existing signal' (126). The synthesizer player is then recast, somewhat deterministically, as 'technician turning a knob here, pressing a switch there—an accessory to the machine' (Manovich 2001: 126). In Manovich's reading, the synthesizer announces the era of remix, where musical artists create using tools with menus of presets, banks of plug-ins, and ready-made patches, a collage of quotes sampled from the existing 'database of culture' (Manovich 2007: 5).

Manovich's reading, however, obscures the variegated history of the synthesizer (Bernstein 2008; Collins, Schedel and Wilson 2013; Pinch and Trocco 1998; Pinch and Trocco 2004; Vail 2014) and thus signally mischaracterizes the tools/users, their practices and modes of inquiry, and their contexts. In short, Manovich conflates voltage-control modular and semi-modular synthesis—in the Buchla, Moog, EMS, Serge, or Doepfer/Eurorack traditions—with mainstream, post-70s digital keyboard synthesizers and, later, MIDI, MPCs, digital laptop systems, software synths with prefabricated digital patches, collage-making 'remix apps' and the like. This new ideology of creativity in turn misrepresents what and how people are learning and making when engaging with Eurorack synths, as well as through processes of parametric play, as described by Strange. The conflation of all 'synths' into a single category, based on the logic of cultural selection and digital re/combination, occludes, as well, vital analogies between electronically and acoustically generated sounds, and how these analogies create common spaces for exploring vibrotactile materials and learning from 'the inside out'.

Very few Eurorack synth users would see themselves as 'accessories' to a machine. That said, few likely would argue that the machine is an accessory to them. Strange's text speaks to the uniquely entangled and enmeshed relationships (Ingold 2009) between tools and users, contexts and environments, where the agency and identity of the various synthesizing actors in play are constituted and co-defined by the patchwork of their various interactions, responses and actualizations. Parametric play, in this regard, challenges the romantic ideal of the composer as a gifted autonomous genius, as perpetuated in Western 'talent regimes', as well as moves beyond newer ideologies of creativity based on the selection and recombination of ready-made elements, digital presets, and pre-set conventions: the digital 'remix' view of creativity that has become fashionable in recent educational theory, especially where new media are in play.

By contrast, this chapter signals the opportunities of modular thought and practice for challenging the hidden curriculum of mainstream music educa-tion, that both nullifies possibilities while rationalizing inequalities as a 'natural' part of our everyday learning environments. We highlighted opportunities for open-ended, process-oriented and material-centric modes of inquiry over the execution of prescribed curricular skills and performative displays of 'talent' which are, too often, programmatically scripted around exclusionary tools, standardized benchmarks, and adult-centered cultural expectations and professionalized identities that are, paradoxically, not even actualizable in institutional spaces like schools. What Eurorack modular synthesis thus offers and models, and what we are interested in further exploring, are approaches to learning with sound that support idiosyncratic points of departure, with self-determined learning purposes and modes of doing that feed forward into ever-expanding domains of playful wayfinding, making, practice and skill.

Acknowledgements

This research was supported by the Social Sciences and Humanities Research Council (SSHRC) and York University's Catalyzing Interdisciplinary Clusters (CIRC) program (Designing Sound Futures). The authors have repurposed portions of this chapter for a professional audience in music education.

References

bell, adam patrick. (2017), '(dis)Ability and music education: Paralympian Patrick Anderson and the experience of disability in music', *Action, Criticism & Theory for Music Education*, 16:3, pp. 108–128.

bell, adam patrick, Bonin, David, Pethrick, Helen, Antwi-Nsiah, Amanda, and Matterson, Brent (2020), 'Hacking, disability and music education', *International Journal of Music Education*, 38:4, pp. 1–16.

bell, adam patrick, Dasent, Jason, and Tshuma, Gift (2022), 'Disabled and racialized musicians: Experiences and epistemologies', *Action, Criticism, and Theory for Music Education*, 21:2: pp. 17–56.

Benedict, Cathy, Schmidt, Patrick, Spruce, Gary, and Woodford, Paul (2015), 'Why social justice and music education?', in C. Benedict, P. Schmidt, G. Spruce, and P. Woodford (Eds.), *The Oxford Handbook of Social Justice in Music Education*, Oxford: Oxford University Press, pp. xi–xvi.

Bernstein, David (2008), 'The San Francisco Tape Music Center: Emerging art forms and the American counterculture', in D. Bernstein (Ed.), *The San Francisco Tape*

Music Center: 1960s Counterculture and the Avant-Garde, Berkeley: University of California Press, pp. 5–41.

Bolden, Benjamin (2010), 'Talent', *Canadian Music Educator*, 51:4, pp. 4–6.

Collins, Nick, Schedel, Margaret, and Wilson, Scott (2013), *Electronic Music*, Cambridge: Cambridge University Press.

Deleuze, Gilles (1988), *Bergsonism*, New York: Zone Books. https://www.zonebooks.org/books/24-bergsonism

Darrow, Alice-Ann (2015), 'Ableism and social justice: Rethinking disability in music education', in C. Benedict, P. Schmidt, G. Spruce, and P. Woodford (Eds.), *The Oxford Handbook of Social Justice in Music Education*, Oxford: Oxford University Press, pp. 204–220.

Eisner, Elliot (1979), *The Educational Imagination: On the Design and Evaluation of School Programs*, New York: Macmillan.

Gaztambide-Fernández, Rubén A., Saifer, Adam, and Desai, Chandni (2013), '"Talent" and the misrecognition of social advantage in specialized arts education', *Roeper Review*, 35:2, pp. 124–135.

Gee, James Paul (2009), 'Games, learning, and 21st century survival skills', *Journal of Virtual Worlds Research*, 2:1, pp. 3–9.

Gibson, James J. (1979), *The Ecological Approach to Visual Perception*, Boston: Houghton Mifflin.

Grover, Shuchi and Pea, Roy (2013), 'Computational thinking in K–12: A review of the state of the field', *Educational Researcher*, 42:2, pp. 59–69.

Hamraie, Aimi (2017), *Building Access: Universal Design and the Politics of Disability*, Minneapolis Minnesota: University of Minnesota Press.

Hess, Juliet (2015), 'Decolonizing music education: Moving beyond tokenism', *International Journal of Music Education*, 33:3, pp. 336–347.

Hess, Juliet (2017), 'Equity in music education: Why equity and social justice in music education?' *Music Educators Journal*, 104:1, pp. 71–73.

Honisch, Stefan Sunandan (2019), 'Virtuosities of deafness and blindness: Musical performance and the prized body', in Y. Kim and S. L. Gilman (Eds.), *The Oxford Handbook of Music and the Body*, New York: Oxford University Press.

Howell, Gillian (2017), 'Getting in the way? Limitations of technology in community music', in A. Ruthmann and R. Mantie (Eds.), *The Oxford Handbook of Technology and Music Education*, New York: Oxford University Press, pp. 449–463.

Ingold, Tim (2009), 'The textility of making', *Cambridge Journal of Economics*, 34:1, pp. 91–102.

Ingold, Tim (2013), *Making: Anthropology, Archaeology, Art and Architecture*, London and New York: Routledge.

Ingold, Tim (2017), *Anthropology and/as Education*, New York: Routledge.

Jackson, Phillip W. (1968), *Life in Classrooms*, New York: Holt, Rinehart and Winston.

Kaschub, Michelle (2020), 'Are we ... mindful of the critical relationship between repertoire and curriculum', *Music Educators Journal* 106:3 pp. 8–9.

Kanellopoulos, Panagiotis A. (2015), 'Musical creativity and "the Police": Troubling core music education certainties', in C. Benedict, P. Schmidt, G. Spruce, and P. Woodford (Eds.), *The Oxford Handbook of Social Justice in Music Education*, Oxford: Oxford University Press, pp. 318–339.

Kivijärvi, Sanna and Lauri Väkevä (2020), 'Considering equity in applying Western Standard Music Notation from a social justice standpoint: Against the notation argument', *Action Criticism and Theory for Music Education*, 19:1, pp. 153–173.

Lubet, Alex (2009a), 'The inclusion of music/the music of inclusion', *International Journal of Inclusive Education*, 13:7, pp. 727–739.

Lubet, Alex (2009b), 'Disability, music education and the epistemology of interdisciplinarity', *International Journal of Qualitative Studies in Education*, 22:1, pp. 119–132.

Lubet, Alex (2011), *Music, Disability, and Society*, Philadelphia: Temple University Press.

Manovich, Lev (2001), *The Language of New Media*, Cambridge Massachusetts: MIT Press.

Manovich, Lev (2007), 'What comes after remix?' Manovich.net, winter 2007, http://manovich.net/index.php/projects/what-comes-after-remix. Accessed April 28, 2021.

Matthews, Charles (2018), 'The social model of disability from a music technology (and ADHD) perspective', [Blog Post] http://ardisson.net/a/?p=363. Accessed 15 February 2021.

McBride, Melanie (2018), 'Tangible inquiries: A study of aroma materials and sources in the built and botanical Environments of Grasse, France', Ph.D. thesis, Toronto: York University.

Nolan, Jason and McBride, Melanie (2014), 'Beyond gamification: Reconceptualizing game-based learning in early childhood environments', *Information, Communication & Society*, 17:5, pp. 594–608.

Nolan, Jason (2020), '(Self)Interview with an autistic: Intrinsic interest and learning with and about music and the missing modality of sound', *Canadian Music Educators*, 62:1, pp. 7–14.

Norman, Donald (1988), *The Design of Everyday Things*, New York: Doubleday.

Oram, Daphne (2016), *An Individual Note of Music, Sound and Electronics*, London: Anomie.

Pinch, Trevor and Trocco, Frank (2004), *Analog Days: The Invention and Impact of the Moog Synthesizer*, Cambridge Mass: Harvard University Press.

Pinch, Trevor and Trocco, Frank (1998), 'The social construction of the early electronic music synthesizer', *International Committee for the History of Technology (ICOHTEC)*, 4, pp. 9–31.

Rancière, Jacques (1991), *The Ignorant Schoolmaster: Five Lessons in Intellectual Emancipation*, Stanford: Stanford University Press.

Rancière, Jacques (1998), *Dissensus: On Politics and Aesthetics*, London: Continuum.

Rancière, Jacques (2004), *The Politics of Aesthetics: The Distribution of the Sensible*. London: Continuum.

Schmidt, Patrick (2020), 'Doing away with music: Reimagining a new order of music education practice', *Journal of Curriculum Theorizing*, 35:3, pp. 44–53.

Strange, Allen (2022), *Electronic Music: Systems, Techniques, and Controls*, Toronto: Responsive Ecologies Lab.

Toila-Kelly, Divya P. (2019), 'Rancière and the re-distribution of the sensible: The artist Rosanna Raymond, dissensus and postcolonial sensibilities within the spaces of the museum', *Progress in Human Geography*, 43:1, pp. 123–140.

Thumlert, Kurt (2015), 'Affordances of equality: Rancière, emerging media, and the new amateur', *Studies in Art Education*, 56:2, pp. 114–126.

Thumlert, Kurt and Nolan, Jason (2020), 'Angry noise: Recomposing music pedagogies in indisciplinary modes', in P. Trifonas (Ed.), *The Handbook of Theory and Research in Cultural Studies and Education*, Berlin: Springer, pp. 1–23.

Thumlert, Kurt, Harley, Daniel, and Nolan, Jason (2020), 'Sound beginnings: Learning, communicating and making sense with sound', *Music Educators Journal*, 107:2, pp. 66–69.

Thumlert, Kurt, Nolan, Jason, Chan, Heidi, and Kitzmann, Andreas (2021), 'Together, apart: Modular sound communities in the age of COVID-19', *The Journal of Music, Health, and Wellbeing*, Special Issue, pp. 1–15.

Thumlert, Kurt, McBride, Melanie, Tomin, Brittany, Nolan, Jason, Lotherington, Heather, and Boreland, Taylor (2022), 'Algorithmic literacies: Identifying educational models and heuristics for engaging the challenge of algorithmic culture', *Digital Culture & Education*, 14:4, pp. 19–35.

Vail, Mark (2014), *The Synthesizer: A Comprehensive Guide to Understanding, Programming, Playing and Recording the Ultimate Electronic Instrument*, Oxford: Oxford University Press.

van der Kamp, John, Withagen, Rob, and Orth, Dominic (2019), 'On the education about/of radical embodied cognition', *Frontiers in Psychology*, 10, pp. 1–9.

Woolgar, Steve (1990), 'Configuring the user: The case of usability trials', *Sociological Review*, 38:1, pp. 58–99.

9

GRID CULTURE

Arseni Troitski and Eliot Bates

My impression is that whenever I just make something for myself or for the sake of itself I'm doing it wrong or somehow pursuing a dead end. As if there is a cultural endgame to all creative pursuits that should result in 1000-fold commodification. This to me makes it very difficult to sort the conventions and expectations of doing a thing from the value it has to me, not recording music, not taking commissions, not spoiling leisure with obligation.

> Imminent_Gloom, Dec. 25th 2020, https://llllllll.co/t/on-singularity-creation-repair-and-commodity/39710/2

I think once the Lines community hit a tipping point, we realized that we needed to have a set of ground rules that folks could look at and just be like, "How am I going to be treated coming into this space and how, what expectations will the space have of me walking in?" … The real message of the code of conduct is to help people who may not have felt comfortable publicly contributing to be like, "No, we have your back." These are the things that we hold, that hold value for us. And we will protect those things so that when you come in here, your experience is positive and if it isn't positive, we will help.

> Dan Derks, interviewed in Grosse 2019

For me, Monome is about unlimited possibilities for open-ended exploration. Grids and arcs appeal to me for this reason in the same way that modular synthesizers do. There's nothing telling me I have to do things in a certain way. I can define the device in my own terms and make things as simple or complex as I need or want.

> Sandy, Feb. 27 2017, https://llllllll.co/t/why-monome/6725/6

DOI: 10.4324/9781003219484-11

These three statements all pertain to a subset of modular synthesis centered around the hardware ecosystem developed by Monome and software developed largely by end-users. They simultaneously pertain to individuals and/or a community of users of a message forum platform called lines (https://llllllll.co/). In all three statements we find descriptions of values, and evidence of processes of valuation, although precisely what is being valued in each statement would appear to differ: the commodifiable value of doing specific activities, the protection of the social values of a community, and the use-value of "open-ended" technological objects for an individual user.

Rather than treating values, economic or moral, as pseudo-objects or discrete states in the world, we are interested in "how valuations are made" (Hennion 2017: 70). In Graeber's "ethnographic theory of value," production produces not just commodities but social relations, and "by extension, human beings, who recreate themselves and each other in the very process of acting on the world" (Graeber 2013: 223). As Meyer and Wilbanks note, "valuation is produced through distributed and heterogeneous processes: products, practices, principles, and places are valued, each interacting dynamically with the others." Beyond just "technical and production aspects", this also constitutes "a valuation of social links and of specific forms of organization and/or marketization" (Meyer and Wilbanks 2020:117). That said, what constitutes social links within a milieu that is singularly structured around the production, reconfiguration, and use of a set of technical objects? What kinds of sociability and valuation are intrinsic to a "gear forum"?

In this essay, we provide our initial ethnographic research about lines, which at a superficial level can be considered a *platform*, but which is also synonymous with a *community* of 8800 users, and with 606k unique posts authored by these users that in aggregate constitute a *discourse network* (*Aufschreibesysteme*, see Kittler 1990).[1] lines—as platform, community, and discourse network—is primarily designed to support the development, documentation and user-generated code efforts specific to a set of hardware synthesis-related objects (the two we will focus on are called Norns and Grid; we'll discuss what they are and what they do through the subsequent sections). Extending from the value/valuation questions, how do the vibrant materialities and the interface/design/reconfiguration aspects of these objects relate to the sociocultural norms and values found within lines? How does the concept of "grid culture" emerge out of this particular imagined community in response to pre-existing and competing places of exchange and connection, and what are its various distributed meanings to the numerous lines users?

Our research methodology for this project is rooted in the methodological questions posed by critical organology (Bates 2012), but in an expanded form that accounts for the material semiotics (Law 2009) of digital materialities (Pink *et al.,* 2016). One of us (Arseni) has been an active lines user since 2017 and plays a Norns as part of his own modular synthesis environment;

the other (Eliot) "lurked" on lines until 2020 but has been an active participant in other audio technology message fora since 2001. That said, we chose not to employ an interventionist participant-observational approach in studying lines. The community is heavily invested in documenting itself (including considerable podcast and video content hosted outside of the platform per se), and within some lines threads we found ample "limit cases" and "critical cases" that resulted in rich discussion and self-reflection on the part of lines users. In this sense, this chapter documents how a music technology community comes to terms with its own capacity for critical discourse and knowledge production. Our wider forum analysis data set included a contrasting set of "general" threads totaling over 10,000 posts and additional threads pertaining to specific code objects—approximately 2.5% of the post history of the lines forum—a subset of which we analyzed with a variety of discourse analysis techniques. Our primary focus was on 1) the vocabularies used to describe interface, community, values, and the qualities of technical objects; and 2) the ways that the domains of technical objects, interface, community, music and values were discursively linked, which we examined within threads ranging from everyday discussions that featured statements of personal value, to routine types of "category errors," to utopian manifestos about lines as a *grid culture*.[2]

Lines as a Gear Culture

While all active message fora such as lines blur the distinction between platform, community, and discourse network, lines is more specifically a space for the discussion, showing and audition of audio technologies, and therefore best conceptualized as being a *gear culture* (Bates and Bennett 2022). The gear culture concept is useful for understanding how and why groups of people socially organize around a certain set of technological objects, rather than, for example, a shared taste in musical styles (which would typically be described as subcultures or scenes). In the case of lines, many of the objects in question are part of a Monome ecosystem that began with user-customizable hardware controllers—a class of objects that do not produce sound by themselves but rather provide tactile control of software running on general-purpose computers or synthesizers, samplers and other electric and electronic sound-making objects. The best known of them is simply known as Grid, which on the surface would appear to be a grid of buttons illuminated by LED lights: the *grid culture* moniker derives from the sociotechnical activity around this specific object. Monome later developed objects that facilitated the connection of computers and computation to modular synthesizers. Most recently, they expanded into providing their own dedicated computing platform. However, the technological objects at the center of the lines gear culture are not limited to just these hardware objects, but include the many code objects that were authored either by the principal Monome developers or by users. Additionally, by 2018 lines became one of the most significant places for socializing around hardware synthesis in general, especially for Eurorack-format modular synthesizers and the conceptual creations of niche designers such as Peter Blasser (Ciat-Lonbarde).

While, as is the case with all gear cultures, not all discussion on lines *must* relate to one or more Monome hardware objects, or to the code objects that are shared amongst users, the lines' instantiation of the Discourse forum platform has been configured overwhelmingly around technological objects (Monome, and more widely, modular synthesis hardware, field recording equipment, recording gear), and users maintain this forum emphasis.[3] For example, in the week leading up to 28 February 2021, we found that in the 193 active threads:

- 45 were about code objects for Monome-designed hardware,
- 44 were about electronic musical instruments (unrelated to Monome),
- 31 were about Monome-ecosystem hardware objects,
- 16 were about audio technological objects (other than instruments),
- 16 were about music/sound/art/photography/film-making processes,
- 17 were general / off-topic,
- 15 were about music, podcast, and synthesis-related film releases,
- 6 were about upcoming or recent events,
- 2 were about the Disquiet Junto thematic collaborative recording series,
- 1 was about acoustic drums.

Thus, 71% of participated threads were organized around some audio technology-specific object—and similar object-oriented discussion often slopped over into the process and release-specific threads, too.

While some core community members in 2021 formulated goals to deemphasize the gear-centric tendency of discussions,[4] without Monome's hardware objects, user-produced code objects, non-Monome synthesis objects, and the forum platform software itself there would be no social formation we could call "lines." Lines as gear culture, therefore, is a significant part of the co-constructed *associated milieu* (Simondon 2017: 59) of these objects understood as *technical individuals* (ibid.: 21), with their specific materialities and *digital materialities* (Pink et al., 2016). Simondon discursively conflates the individuation/associated milieu relations between technical objects (for a steam engine-as-technical-individual, water and air would be key elements in the associated milieu) with the individuation/associated milieu relations for living organisms (where the associated milieu would be other organisms and the natural world that sustain the living organism, which of course also include water and air) (Mitchell 2012: 79). In gear cultures, individuals routinely make category errors, confusing their relations with technological objects with their relations to other people. This commonly manifests through the anthropomorphizing and personification of technological objects, and attribution of agency to them. Simondon's conflation, therefore, is a productive starting point for charting the *kinds* of relations between the lines platform-community-discourse network, a set of technical objects that are routinely remade through recoding, other music-related technical objects, and individual users. That said, if we subscribe to Simondon's dictum that our alienation from technologies stems from not understanding them as cultural actors and "technical individuals," these category errors *could* be regarded as productive towards the goal of "domesticating" technologies, albeit with caveats: Carolyn Marvin critiques this very notion with regards to late 19th-century "electric houses," which at the time were idealized and self-described as a "feudal fortress against the world" (1988: 78). These category errors, however, are destructive of human-social relations when objects or abstract notions of economic value become regarded as equivalent to the value of human life (Hornborg 2019).

While in some regards lines' sociocultural norms and practices differ from those in other technology-specific gear cultures, we also find notable parallels. Evangelos Chrysiagis wrote of the "intrinsic value" in the "sharing" of resources within DiY music communities (2020: 744) in the context of Glasgow's 2010s DiY music scenes; similar values are regularly proclaimed at lines, too. Just as was found the live-coding and algo-rave community studied by Christopher Haworth, we find widely shared ideologies about open-source (as a positive ethics) and community-produced tools, which on lines led some object designers to author visionary, manifesto-style texts that gloss "materials, programming languages … and software ethics" (2018: 568). However, as was the case with another utopian-minded experimental music technology platform, the now defunct Res Rocket, alongside the expansion of the user base exists an increasing "capacity of the anonymous online community to depart from the utopian model and exert their own agency" (Haworth 2020).

In March 2015 Brian Crabtree, who co-founded Monome with Kelli Cain, stated that in the course of transitioning the old monome.org fora (running the forum software Vanilla) to the current Discourse platform, the Monome founders were "structuring for inclusivity beyond grid culture" and at the same time aiming at preserving modes of discussion previously established by monome.org community.[5] Three years later, when reflecting on this transition in an interview with Darwin Grosse, Crabtree said:

> I just realized that having a platform and a community to speak about things that weren't just grids was what I really wanted, that it was clear that the Monome community forum was not just a kind of company support forum. It was a culture that had emerged really really quickly. It was brought together by a mutual interest, but it really extended much further.

> So [in] the move to basically de-brand the Monome community and make it not about Monome anymore […] we were trying to find a way to, this is funny realizing my nervousness around trying to make broad generalizations about the forum is really clear, like it only reinforces the fact that it is not mine in any way (Grosse 2018).

So what *are* these objects that require so much care, and a whole networked community surrounding them? Monome's inventions began with the Grid, which at first glance might look similar to other button-based control surfaces on the market, but which is intended to be designed, redesigned, configured and reconfigured by its users so that it can become a veritable "open, interactive instrument." The Arc is similar in concept to the Grid, but brings "openness" (flexible reconfigurability) to a knob-based controller instead. Grids and Arcs can readily be connected to a computer where they control software-based instruments, for example, a Max/MSP patch. But because of the considerable expansion of interest in computerless hardware synthesis in the 2010s, Monome subsequently developed a set of Grid-connected Eurorack sequencer modules (White Whale, Earthsea, Meadowphysics, Ansible) and Eurorack scripting platforms Teletype and Crow. Their most recent non-Eurorack object, Norns, is a purpose-built sound computer used to provide convenient access to the kinds of synthesis, audio processing, and sequencing-related applications, running as SuperCollider and Lua scripts, for which users would customarily have needed a general-purpose computer. For many of these objects, both the hardware designs and code used to operate them are open source. Notably, each of these objects *requires* scripts and code in order to do anything, and different scripts change the kind of interfacing of which Monome hardware objects are capable. Therefore, what might initially seem to be a technical problem (getting code to work), or a matter of musical instrument design (coming up with the ideal interface), becomes an opportunity for social interaction around code objects.

Interfacing the Community

One of the key shifts in 21st-century modular synthesis has been a transition away from the historical "analog versus digital" debate (Scott 2016) and towards the careful and deliberate design of instrument interfaces. Much of the "problem" with 1980s-90s digital synthesizers pertained to the limitations of the tactile controllers of such instruments (Théberge 1997: 75–7), and to the limitations of MIDI—the 7-bit digital standard used for inter-device operability. While these problems persist with some instruments today, within the Eurorack format, Scott Jaeger (designer of the Harvestman and Industrial Music Electronics modules) was the first to experiment with coupling a computation core with the continuous control made possible through routing "analog" control voltages (rather than MIDI commands).

We mention this, since much of the backlash against "the digital," it turns out, was more a problem with interfaces and their implementation than with the mode of synthesis or the underlying technology used for waveform generation. As Thor Magnusson, an active participant in the NIME (New Interfaces for Musical Expression) conference and designer of computer-based instruments notes, "the digital instrument *has* an interface, whereas the acoustic instrument *is* an interface ... In digital instruments, their computational nature and arbitrarily mapped control elements result in technologies that feel thin, yet powerful" (Magnusson 2019: 44). If and when control elements are arbitrarily mapped, then they can be readily *remapped* by changing code.

When thinking about interfaces and computation-based modules, we should distinguish between several phenomena and therefore several *kinds* of interfacing at play (Bates 2021). Within the Monome ecosystem, the Crow and Ansible are both Eurorack-format modules that on the surface lack any obvious means of user control and consist only of jacks that you plug cables into. They are not objects that you buy to directly touch and to play music: they interface *between other hardware objects,* providing sequencing and control voltage processing capabilities in modular systems. In contrast, the Arc presents the user with two or four large knobs and LED circles, and the Grid presents the user with 64 or 128 illuminated buttons: on the surface, both are controller-type tactile user interfaces. The Norns, which internally consists primarily of a small computer, is ontologically more "complex," in that it contains audio inputs and outputs, USB ports, a screen, tactile controls, and Wi-Fi: it mixes user control with hardware interfacing to dedicate its embedded computing potential to audio signal processing and musical performance.

All of these objects have two additional interfacial registers. Beyond the nominal operability concerns, all of Monome's products share a minimalist aesthetic sensibility with regards to their look and feel,[6] what Simondon terms techno-aesthetics (2012: 1). Those objects that do involve controllers are especially prized for their haptic feedback: the ludic potential contained on the surfaces of the hardware objects matters considerably to some users (Bates 2023). And, perhaps most important of all, they are intended to be redesigned by end-users, either by loading existing user-written applications, or by creating new ones. The design and configuration of applications introduces the whole set of interfacing modes found within any computing environment, including the graphical user interfaces of integrated development environments (IDEs), and the class interfaces (especially the application program interfaces, APIs) that allow code chunks to interface with others. In the case of Monome's products, many of them, in turn, benefit from the class libraries provided by Open Sound Protocol (OSC). An open standard developed by CNMAT (the Center for New Music and Audio Technologies at UC Berkeley), OSC facilitates communication between objects via serial interface devices (USB being the best-known today). By using OSC, Monome's developers and end-users could address their own interfacing needs via a relatively easy set of human readable instructions that, having existed since 2002, was already familiar to many music software coders.

With this in mind, what are the stakes of an interface, and why do we care about them? For Jan Distelmeyer, "interfaces perform conduction," which means not only the conducting of electricity, but "the social, educational, and political meaning of leadership and guidance like in 'to conduct somebody/groups' and 'conduct politics'" (2019: 85). Changing modes of human-user interfaces, as in the case of Monome gear that expands user interface domain onto scripting platforms accessible to non-expert coders, affects user-instrument interaction models and redefines both their performative and social aspects. To what extent do Distelmeyer's findings apply to lines and Monome art worlds? What might "conduct politics" mean within a grid culture?

For Grid, the original and most visually iconic object in the Monome ecosystem, one of the fundamental interface ideas was to present 64 to 128 illuminated buttons with reprogrammable actions "decoupled" from their LEDs. This essentially breaks the interfacial convention found within mass-produced instruments, where LED states are obligatorily tied to their respective buttons—or in other words, where the visual and the haptic/tactile would be indexical in a reduced sense. Programming the Monome Grid's LEDs is entirely handled by whatever application or script that communicates with the Grid at the moment, allowing coders and end-users to configure the Grid as a data display, state / condition / progress indicator, alphanumeric generator, or canvas for digital art (Grosse 2014).

Norns can be immediately ascertained as an aesthetic object that was designed with minimalist aesthetics, a lack of any labeling, and a minimum of control elements. At one level, interfacing with Norns involves the correlation between the look/feel and haptic response from engaging with this box. At a second level, Norns was given audio input/output, USB ports and WiFi to allow it to interface with audio technology and computing technologies. At a third level, and the one we are most concerned with in this section, Norns is a platform that hosts scripts: interfacing, here, involves customizing what Norns does, how Norns interacts with other hardware objects, and how Norns responds to human users.

Today, in addition to Norns, other developers make synthesis-specific microcontroller-based platforms that permit "easy" user modifications of the firmware. These include Teensy/Teensyduino (the system on which ornament & crime was built), the Electrosmith Daisy, and Lich platform by Befaco/

Rebel Technologies. Some module developers such as Mutable Instruments build on widespread embedded computing devices (e.g., STM32) and publish the firmware code, which simplifies the process for end-users to create their own custom firmwares. Dozens of products for the Eurorack market are built on the above, all of which have resulted in at least one user-authored firmware that circulated and gained popularity. But with 271 user-generated libraries for the Norms, each of which turns the Norms into a "different instrument," Norns serves as the platform with the largest community of enthusiast end-users generating and sharing script libraries. Conduct, here, is predicated on contributing to openly shared code libraries.

One explanation might be the complexity of the code: Mutable Instruments' main code, for example, is written in C++, and only a small number of prospective enthusiasts both know that language and how to do DSP coding in it. Norns' interface code is written in Lua, a simpler "higher level" language that assumes less prior knowledge of object-oriented programming conventions and is expressly designed for embedded applications. If a user wanted to build a new on-screen graphical user interface, they can start with the set of auxiliary code objects provided by the Norns development team. If they wanted to do a different kind of digital signal processing or sound generation, they can draw upon the extensive library of DSP classes written for SuperCollider—another high level, cross-platform, well-documented language that is Norns' audio engine. Similarly, many sample-manipulation scripts can be built on a Norns-specific sub-engine called softcut (developed by Ezra Buchla). If a user wanted to work with notes, scales and chords in a language more familiar to musicians, they can use the provided Lua library called musicutil.

What has been the result of Norns' open-ended scripting environment in terms of new interface development, and the social interaction around the creation and modification of scripts? Norns user scripts are published in separate forum threads where end-users share their experiences and use scenarios, express gratitude, ask for help with the scripts and provide feedback in the form of bug reports and feature requests. New scripts become not just opportunities to experiment with creating new interfaces, they are social opportunities for co-working on code. The script author acts as a peer or collaborator rather than an omniscient superior entity: end users are commonly requested to modify code, to expand the feature set, or to test scripts and their specific features. Other times, however, users fork a script, modify it to suit their own specific requirements, and then share that modified code with the lines community.[7] Whenever a script reuses code fragments or relies upon libraries from other coders, in the original post of the thread the author explicitly enumerates the code sources and credits the authors. For instance, in a representative opening post to one of the scripts' threads the author credits two users for contributing to documentation of the script. They also thanked fourteen end-users and "everybody else" who "tested, contributed feedback, gave encouragement, and shared artifacts while this script was coming together"—and credited two core developers for creating Norns, its codebase and inspiring "many artists" to "build, deconstruct, and share."[8] Finally, scripts often attract interest and social engagement when end-users share performance videos featuring the script that highlight specific ludic aspects of the interface that the user finds appealing.

Before Norns was released, some prospective users claimed to be "more excited for norns to run other people's code than my own" while others expressed concerns about being a "'consumer' of other people's code" or being able to "contribute much to the growth of the platform."[9] Dan Derks, the author of the cheat codes script (one of the more popular code objects for Norns), comments that he perceived the intended goals for the platform as being "closer to enabling artists to build their own tools, to modify code to meet their particular performance needs" and refers to the community interaction model, where end-users would employ community-contributed scripts without modifying or creating code, as "unexpected.[10]" However, the social life of Norns scripts seems to accommodate both interfacial paradigms: one where end-users interact with scripts as concretized objects, and one where users treat scripts as sources of programmatic ideas and opportunities for customization.

Discourses and Community

When beginning to examine the relations between lines-specific socio-cultural aspects and the concerns of objects and object-interface design, we mapped out the rates at which keywords occurred. While these word frequencies do not provide detailed accounts of meanings or usage, nor do they inform us about individual user attitudes towards those words, they do provide basic information about hegemonic discursive formations and the key terms around which thread topics are framed (Simon and Xenos 2004). Usefully, they expose the *absence* of other terms. Moreover, for prospective users browsing the forum to determine if they vibe with the space, or for casual users considering whether they want to participate in a thread, a cursory and informal word incidence/absence assessment will be as significant, if not more so, than a careful or nuanced reading of posts.

We first analyzed the most significant Norns development thread—where the project was first announced, user input was solicited, and development charted—which contained 2629 posts. Beyond the obvious terminology of products related to Monome/Norns, or directly related languages/platforms/protocols (e.g., MIDI, Lua, USB, SuperCollider, Instagram, OSC, Linux, Wifi, Organelle), the most frequent vocabulary relates to objects or pseudo-

FIGURE 9.1 High incidence words in the first major thread about Norns.

objects. These include (from most to least frequent): *scripts, thing, code, device, sound, audio, thread, engine, interface, stuff, computer, music, modular, screen, instrument, video.* Fewer recurring keywords pertained to humans, either individually or collectively: *people, community, everyone, mine, user, others.* The most frequent verbs/nominalized verbs include: *use, get, work, make, think, want, see, need, know, look, love, control, running, start, learn.* We also get a sense of the "desired qualities" of both Norns objects *and* Norns discourses in the adjective frequency list: *new, different, possible, excited, fun, power, cool, simple, amazing, hard, beautiful, together*[11] (Figure 9.1).

In a general thread about the "UX of Music Instruments & Tools," the 124 posts are overloaded with terminology specific to objects and pseudo-objects, verbs that (typically) pertain to these or attitudes about these, and perceived qualities of the objects. Regarding the most frequently mentioned kinds of objects, we find (in order of decreasing frequency): *instrument,*

FIGURE 9.2 High incidence words in a general thread about user interfaces.

things, interface, music, piano, keyboards, tool, software, thing, layout. The most frequently used verbs are: *think, design, use, get, make, play, work, feel, need, want, look, sense, find.* Frequently mentioned qualities include: *different, easy, musical, isomorphic, new, visual, advanced, mastery, flat, abstract, potential, traditional.* As was the case with the Norns development thread, few keywords pertained to people and human matters, either individually or collectively; those that did were: *people, user, designers, team, others, brain, mind, human, muscle, players*[12] (Figure 9.2).

In comparison to other music-technology themed fora, lines moderators have a greater tolerance for user-generated threads that suggest changes to moderation policies, code of conduct, and community values. One such thread that we analyzed, "Code of Conduct?", generated 138 posts from its inception in 2017 to the last post two years later. These types of threads, while less common than single-object threads, are nonetheless useful for understanding the vocabularies that are leveraged when dealing with a problematic situation that is not specifically about one of the Monome-ecosystem products. Unlike the previous threads, users leveraged a wider vocabulary to discuss matters pertaining to individuals or community. The most frequent terms included: *community, moderators(mods), members, people, identity, others, ideology, someone, ideologies, culture, communities, gender,* and *sex/sexual* (the terms *class, race, disability* and *ethnicity* did appear, but considerably less frequently). In terms of the qualities (of things, communities, or people), we find: *good, political, welcome, committed, clear, inclusive, open, equitable, new, respectful, right.* That said, the verbs and nominalizations don't differ as much as might be expected: *think, make, flag, see, want, discuss, feel, treatment, know, say, moderation, conduct, need, help, express, find, communicate, get* (Figure 9.3).[13]

FIGURE 9.3 High incidence words in a thread about a proposed lines code of conduct.

Even without a more nuanced contextual discourse analysis, it should be clear that regardless of the topic, certain discursive tropes are consistent across even ostensibly unrelated thread topics. In all three of the above threads the verbs *get, make, think, want,* and *need* figured prominently; in the code of conduct thread and one of the technical threads the verbs *see, know,* and *find* are also significant. Reading for context, we find that lines posts (here and elsewhere) overwhelmingly foreground discourses of individual agency and experience by nearly exclusively employing state-experiential verbs and action-process verbs (Cook 1979). On the other hand, when confronted with the problem of what a code of conduct could or should govern, although two users provided salient and informed discussions of prior successful implementations in feminist-organized spaces, the terms *inclusive, open, equitable, respectful* mainly appeared in the many rewrites to the code of conduct done by a lines moderator and not in regular user vocabularies (except for a few users who noted that they were offended by these keywords). Although regular users often note that lines is a "nice" or "gentle" space (in contrast with reddits or other gear-focused sites like gearspace.com and modwiggler.com),[14] we nevertheless see a mismatch between the individual action/experience-focused emphasis (on the part of general users) and the discursive performance around "inclusiveness" (on the part of moderators).

Additionally, excepting a few words indicating ubiquitous musical instruments, we note that none of these threads prominently discusses any specific aspects about music or art, whether their theories or formal aspects, their sensuous, aesthetic, or affective aspects, or music's many social meanings. A few threads on lines do do this, but they're bracketed off from other threads. In fact, only in one recurring thread do users specifically discuss recent recording releases. This is not to imply that lines users are not interested in music at all, but rather that it is not what brings the community together.

To understand why this may be, we applied a more nuanced mode of contextual discourse analysis to a 2021 thread entitled "Why is this forum less about music and more about stuff." Here lines users ranging from forum moderators and product developers to "power users" to occasional posters debated the topical focus of the forum. The original poster (initiator of the thread, almost always shortened to "OP") was curious why a thread that they had authored about a new music release had been deleted and the original post merged into the sprawling "new release" thread (meaning it was unlikely to generate any ongoing, meaningful user discussion). The responses to the OP were illuminating. Many users noted, wholly positively, that lines was about "documentation and gear setup," that gear was "objective" and easier to talk about than art, that gear discussion was practical and had a "nominal value" (whereas music chatter had an unclear value), and that gear lacked "translation issues." The moderator who had deleted the OP's earlier thread noted that threads about music needed to lead to conversation, and suggested ideas for how that might be achieved in the future. But other users suggested that lines shouldn't be about music at all but about art (not surprising considering the art-school origins of the Monome ecosystem at the California Institute for the Arts), and that although lines typically had more "openness" to the discussion of emotional/spiritual spaces, that philosophical/artistic discussions were typically not "useful." One regular user, summarizing the themes that we explore in this essay, suggested that the main purpose of lines was for users to explore "how you interface with the interface": this post received 18 likes from other lines members.[15]

In another 2021 thread entitled "How Did You Get Here," the original poster hoped to solicit the personal scenarios that led to users joining the lines community.[16] While some respondents mentioned attributes of the human-social conventions of lines being part of the reason that they *stayed*, 75% claimed that they found lines and began participating due to an interest in Monome's hardware objects, in music device-related coding (i.e., code objects), or in synthesis gear in general. For the 25% of users who did *not* explicitly mention that gear attracted them to the community, their reasons to join included "community values" or escaping the "toxic atmosphere" of other gear-centric music fora (five users), word-of-mouth from personally knowing other community members (three users) or from hearing about lines from a well-known influencer-type (three users), and the recurring Disquiet Junto recording collaboration that is hosted at lines (three users).

Most crucial for us, however, is articulating the links between the dominant object-centered draw of lines, and the secondary interest in human-social matters. Similar to other gear cultures, in this thread (and elsewhere) there is plenty of technological fetishization and alienation evident, whether users indicate their aesthetic attraction towards other users' Norns ("a little box that caught my eye"), or frame Monome objects' *functionality* as intriguing by being unclear (one poster had long been fascinated with Norns whilst trying "to work out what it was actually doing"). Some users' narratives suggest an unwritten assumption that owning/using Monome gear might be a requirement to fully participate, or that their own activity is contingent upon new gear object "rollouts" and the ensuing community participation in creating code for these new things. Other users, in comparing lines to other gear-centric fora, write posts that confuse or conflate the boundaries between technological objects, social values, and gatekeeping discursive moves—kinds of "category errors" that are frequent on other gear forums. Music or art is only occasionally discussed, and even there, the user who was attracted to Monome by seeing an early Grid performance of Brian Crabtree that "completely blew me away & made me think differently about chopping samples & performance interfaces" elides an artist, an instrument and its interface, a musical space/activity, and a personal reaction in a noteworthy way.

Conclusion

When we were formulating our research, we were curious to explore what it meant that the founder of Monome defined lines as a "grid culture." Was lines-as-grid-culture similar to gear cultures such as muffwiggler.com and gearslutz.com—both notorious not only for their sexist naming schema but for the ubiquity of toxic masculine discourse and for the shared social value placed on "gear acquisition syndrome"? (Bates and Bennett 2022) Did the encouragement for lines users to create and share their own code and to remake the interfaces of their instruments contribute to a radically different kind of sociability? The answer to both is complicated. In podcasts and in some forum posts, some moderators repeatedly note that they do not view lines as a gear forum, emphasizing the community, design, and utopian aspects instead. That said, many active users disagree *in practice*: considerably more than half of participated threads are framed around specific technological objects or kinds of objects. The implementation of the Discourse platform, the presence of a standard convention where posters suggest to other posters to voluntarily remove or edit threads, and the social value placed on liking posts, all corroborate the statements made in the "Code of Conduct?" thread that lines is in practice a "nice place." But a nice place *for whom*?

While there has been no demographic analysis of professional synthesists, in 2021, women continue to number fewer than 10% of professional audio producers/engineers, many electronic music festivals and labels continue to feature exclusively male (and largely if not exclusively white) rosters, and the accomplishments of the extremely significant women and trans* electronic music performers that *are* working in this space are continuously undermined by their "silencing in electronic music histories" (Rodgers 2010: 6). Research has shown that the problem begins with the "leaky pipeline" of high school and university-level music technology degree programs (Brereton *et al.*, 2020), and further attrition results from the pervasive microaggressions against women and trans*/non-binary engineers and producers (Brooks *et al.*, 2021). The problem is not just studio-sited; in John Richards' experience of teaching the *dirty electronics* series of DiY instrument-making workshops, gender parity was never achieved in music department-hosted events but only in ones hosted in art departments (2016).

Since 2018 many users have "migrated" from toxic masculine sites such as muffwiggler and gearslutz to lines, precipitating a "context collapse" (Marwick and boyd 2011) due to the work these new users did to recenter conversations around acquisition and hoarding discourse—and around normalizing technological exhibitionism (pics of your gear) as a practice for engendering social prestige (Bates and Bennett 2022). This move was met with *some* resistance, however, when some longstanding Monome/lines users strategically used the affordances of the Discourse platform to discourage the context collapse.[17] The value placed on users re/making the interfaces of their instruments and tools, when acted upon, does reduce technological alienation and is potentially compatible with the values of DiY, "do-it-together" makerspaces, and feminist/genderqueer organizing writ large (Kori and Novak 2020).

On the one hand, our analysis found that even in threads where such topics would be appropriate, there was a widespread avoidance of discussing questions of social identity, most notably any matters pertaining to race, ethnicity, class, and nationality; gender and sexuality have only been discussed within a handful of threads, and there, the focus has been on nonbinary and trans* participation in electronic music. While some of this may masquerade as being supportive of an "inclusive community" (a word pairing that does appear on lines), when taken alongside the vocabularies we discussed above, this "silent" identity is a foil for white masculinity. To the extent that lines is a "safe space" (The Roestone Collective 2014), as of 2023 it is nevertheless most successful in being a "nice place" for white men who are reasonably affluent—and secondarily for those able (and willing) to most of the time subjugate their social identity so they can pass as cisgendered white men. On the other hand, to the extent that the moderators' stated values—inclusion, decentering commercial posting, and creating open-ended (underdetermined) tools that can be used for creating art—are widely held amongst users and inform their social interactions, then lines continues to have the *potential* to be a *grid culture that is not a gear culture*.

Notes

1 "About." https://lllllllll.co/about [access date: 17 August 2023].
2 http://archive.monome.org/community/discussion/18527/lines-monome-forum-migration-begins [accessed 28 February 2021]. This is the original instance of the term on lines.
3 Discourse is the name of an open-source message forum software introduced in 2014, developed by Civilized Discourse Construction Kit, Inc.
4 "We value the decentralized emergence of information, not simply reacting to a single voice. We appreciate an environment that does not encourage hyper-consumption. we have no intention of being product-focused." https://lllllllll.co/t/why-we-made-this/30 [accessed 28 February 2021].
5 http://archive.monome.org/community/discussion/18527/lines-monome-forum-migration-begins [accessed 28 February 2021].
6 https://lllllllll.co/t/approaching-norns/13236 [accessed 28 February 2021].
7 https://lllllllll.co/t/mouse-updated-with-usb-keyboard-mouse/41562/35 [accessed 5 March 2021].

8 https://lllllllll.co/t/cheat-codes-2-rev-210303-threshold-live-input-recording/38414 [accessed 7 March 2021].
9 https://lllllllll.co/t/approaching-norns/13236/565 [accessed 3 March 2021].
10 https://lllllllll.co/t/community-funded-bounties-for-norns-development/39641/25 [accessed 26 February 2021].
11 https://lllllllll.co/t/approaching-norns/13236 [accessed 27 February 2021].
12 https://lllllllll.co/t/ux-of-music-instruments-tools/2772 [accessed 27 February 2021].
13 https://lllllllll.co/t/code-of-conduct/9727 [accessed 27 February 2021].
14 Gearspace.com, from 2002–21, was named Gearslutz, and regular users routinely referred to each other as "sluts" and used other sexualized vocabulary to discuss audio technologies (Bates and Bennett 2022). Modwiggler.com, also until 2021, was named Muffwiggler. Although both sites changed their names ostensibly to foster more "inclusion," both continue to be environments featuring locally specific hegemonic masculinities that many outsiders (and some insiders) regard to be especially toxic.
15 https://lllllllll.co/t/why-is-this-forum-less-about-music-and-more-about-stuff/41949 [accessed 28 February 2021].
16 https://lllllllll.co/t/how-did-you-get-here/41504 [accessed 3 March 2021].
17 This included a mix of strategic group use of the "heart" (like) feature for lines-appropriate posts, the use of the post slowdown feature to prevent users from commenting more than once a day on contentious posts, quote-responses followed by a dissection of the problematic post, and open discussion of adding problematic users to one's "ignore" list.

References

Bates, Eliot. 2012. "The Social Life of Musical Instruments." *Ethnomusicology* 56 (3): 363–395. 10.5406/ethnomusicology.56.3.0363.

Bates, Eliot. 2021. "The Interface and Instrumentality of Eurorack Modular Synthesis." In *Rethinking Music through Science and Technology Studies*, edited by Christophe Levaux and Antoine Hennion. London: Routledge. pp. 170–188.

Bates, Eliot. 2023. "Feeling Analog: Using Modular Synthesisers, Designing Synthesis Communities." In *Shaping Sound and Society: The Cultural Study of Musical Instruments*, edited by Steven Cottrell. New York: Routledge.

Bates, Eliot, and Samantha Bennett. 2022. "Look at All Those Big Knobs! Online Audio Technology Discourse and Sexy Gear Fetishes." *Convergence: The International Journal of Research into New Media Technologies* 28 (5): 1241–1259. 10.1177/1354 8565221104445.

Brereton, Jude, Helena Daffern, Kat Young, and Michael Lovedee-Turner. 2020. "Addressing Gender Equality in Music Production: Current Challenges, Opportunities for Change, and Recommendations." In *Gender in Music Production*, edited by Russ Hepworth-Sawyer, Jay Hodgson, Liesl King, and Mark Marrington, 219–251. New York: Routledge.

Brooks, Grace, Amandine Pras, Athena Elafros, and Monica Lockett. 2021. "Do We Really Want to Keep the Gate Threshold That High?" *Journal of the Audio Engineering Society* 69 (4): 238–260.

Chrysagis, Evangelos. 2020. "When Means and Ends Coincide: On the Value of DiY." *Journal of Cultural Economy* 13 (6): 743–757.

Cook, Walter A. 1979. *Case Grammar: Development of the Matrix Model (1970–1978)*. Washington: Georgetown University Press.

Distelmeyer, Jan. 2019. "From Object to Process. Interface Politics of Networked Computerization." *Artnodes* 24: 83–90. 10.7238/a.v0i24.3300.

Graeber, David. 2013. "It Is Value That Brings Universes into Being." *HAU: Journal of Ethnographic Theory* 3 (2): 219–243.

Grosse, Darwin host. 2014. "Brian Crabtree." *Art + Music + Technology (podcast)*, Sep 21. Accessed Feb 1, 2021. https://artmusictech.libsyn.com/podcast-047-brian-crabtree

Grosse, Darwin host. 2018. "Brian Crabtree (monome.org)." *Art + Music + Technology (podcast)*, Sep 30. Accessed Feb 1, 2021. https://artmusictech.libsyn.com/podcast-247-brian-crabtree-monomeorg

Grosse, Darwin host. 2019. "Dan Derks." *Art + Music + Technology (podcast)*, Dec 15. Accessed Feb 1, 2021. https://artmusictech.libsyn.com/podcast-306-dan-derks

Haworth, Christopher. 2018. "Technology, Creativity, and the Social in Algorithmic Music." In *The Oxford Handbook of Algorithmic Music*, edited by Roger T. Dean and Alex McLean, 557–581. New York: Oxford University Press.

Haworth, Christopher. 2020. "Network Music and Digital Utopianism: The Rise and Fall of the Res Rocket Surfer Project, 1994–2003." In *Finding Democracy in Music*, edited by Robert Adlington and Esteban Buch, 144–163. London: Routledge.

Hennion, Antoine. 2017. "From Valuation to Instauration: On the Double Pluralism of Values." *Valuation Studies* 5 (1): 69–81.

Hornborg, Alf. 2019. *Nature, Society, and Justice in the Anthropocene: Unravelling the Money-Energy-Technology Complex*. Cambridge: Cambridge University Press.

Kittler, Friedrich A. 1990. *Discourse Networks 1800/1900*. Translated by Michael Metteer and Chris Cullens. Stanford: Stanford University Press.

Kori, Lisa, and David Novak. 2020. "Handmade Sound Communities." In *Handmade Electronic Music: The Art of Hardware Hacking*, edited by Nicolas Collins, 3rd ed., 393–405. New York: Routledge.

Law, John. 2009. "Actor Network Theory and Material Semiotics." In *The New Blackwell Companion to Social Theory*, edited by Bryan S. Turner, 141–158. Chichester: Blackwell Publishing.

Magnusson, Thor. 2019. *Sonic Writing: Technologies of Material, Symbolic, and Signal Inscriptions*. New York: Bloomsbury Academic.

Marvin, Carolyn. 1988. *When Old Technologies Were New: Thinking About Electrical Communication in the Late Nineteenth Century*. New York: Oxford University Press.

Marwick, Alice E., and danah boyd. 2011. "I Tweet Honestly, I Tweet Passionately: Twitter Users, Context Collapse, and the Imagined Audience." *New Media & Society* 13 (1): 114–133. 10.1177/1461444810365313.

Meyer, Morgan, and Rebecca Wilbanks. 2020. "Valuating Practices, Principles and Products in DIY Biology: The Case of Biological Ink and Vegan Cheese." *Valuation Studies* 7 (1): 101–122.

Mitchell, Robert. 2012. "Simondon, Bioart and the Milieu of Biotechnology." *Inflexions* 5 (March): 68–110.

Pink, Sarah, Elisenda Ardèvol, and Débora Lanzeni. 2016. "Digital Materiality." In *Digital Materialities: Anthropology and Design*, edited by Sarah Pink, Elisenda Ardèvol, and Débora Lanzeni, 1–26. London: Bloomsbury.

Richards, John. 2016. "Shifting Gender in Electronic Music: DIY and Maker Communities." *Contemporary Music Review* 35 (1): 40–52.

Rodgers, Tara. 2010. *Pink Noises: Women on Electronic Music and Sound*. Durham: Duke University Press.

Scott, Richard. 2016. "Back to the Future: On Misunderstanding Modular Synthesizers." *EContact!* 17 (4). https://econtact.ca/17_4/scott_misunderstanding.html.

Simon, Adam F., and Michael Xenos. 2004. "Dimensional Reduction of Word-Frequency Data as a Substitute for Intersubjective Content Analysis." *Political Analysis* 12 (1): 63–75.

Simondon, Gilbert. 2012. "On Techno-Aesthetics." Translated by Arne De Boever. *Parrhesia* 14: 1–8.

Simondon, Gilbert. 2017. *On the Mode of Existence of Technical Objects*. Translated by Cécile Malaspina and John Rogove. Minneapolis: Univocal Publishing.

The Roestone Collective. 2014. "Safe Space: Towards a Reconceptualization." *Antipode* 46 (5): 1346–1365.

Théberge, Paul. 1997. *Any Sound You Can Imagine: Making Music/ Consuming Technology*. Hanover: Wesleyan University Press.

10

MODULAR ECOLOGIES

Bana Haffar

How does modular synthesis relate to personal transformation, picnics, a community center, and the desire to translate? Without the advantage time and age give to see the bigger picture of things, I will attempt to correlate these nested spheres from my present vantage point.

As you read on, consider the sections as prototype modules, with various speculative patch points connecting them throughout; a module being a self-contained piece of a set that can be combined through format to build something larger. This approach has been a practical way to start building connections between the diverse activities I've found myself involved in since my journey in synthesis began.

Breaking Apart in Order to Put Back Together Differently

Modular synthesizers are not like traditional instruments. Created in the 1960s, they are still in their infancy relative to the piano, guitar, violin, etc. There is no standardized notation or systems for learning, no practicing of scales, no repertoire, they don't fit neatly into orchestras and bands, and the point of entry for players is as varied as their capacity for sound creation.

For the musician, they offer an opportunity to break music down into its basic sonic components and rebuild it in a way that reflects their unique voice. For the non-musician, they offer a chance to participate in the world of sound, free from the need to understand music theory. They are teachers of sound and music is the byproduct of the breaking down and rebuilding of sound into new forms and structures.

Coming from a traditional schooling in classical and contemporary music, I was drawn to modular synthesizers because they prompted me to question

DOI: 10.4324/9781003219484-12

my institutional conditioning in systems of music and sound. I had to begin again, one element at a time. Music began to change from something that I de facto learned, practiced, recorded, and performed in narrowly prescribed ways to a participatory, creative, and autonomous activity.

As my perception of music shifted into atomized experimental sound units, a modular effect began to take root in my thoughts on performance structures and later into community and ways of learning.

The modular effect is, as I see it, a state of constant questioning and rearranging of the status quo by breaking it down again and again into smaller units of inquiry. Through this process, new forms of organization, and ultimately expression, can be created out of alternate combinations of previous arrangements. It's not about the erasure of earlier structures, it's about re-combining the pieces in service of possibility, which is to me, the essence of synthesis.

Formations

In a massive abandoned shipyard in a small town in the Pacific Northwest, rusted beams, trusses, and piles of scrap metal rattled in glorious cacophony. The same type of surprise and exhilaration babies have as they discover their screaming voices for the first time came over me, mixed with the fear of getting caught for trespassing. I slid up the volume fader on the mixer anyway.

Modular on the Spot co-founder, Eric Cheslak and I were headed to Portland from Los Angeles in my Toyota Matrix hatchback loaded with a pair of QSC K10 speakers, a Honda EU2200i generator, cable bags, four cases of modular synthesizers, camping gear, and blankets. We were searching for unique environments to set up and experience our modular synthesizers along the drive, scratching every itch—"what if we set up there?"

We found ourselves setting up in coastal Redwoods, bouncing modular beats off smooth poured concrete at the bottom of various skate bowls, synthesizing bird song from a mossy clearing in a fern forest, and agitating architecture in the dockyard mentioned above.

In hindsight, we were operating in the spirit of 70s UK sound system culture; stealthily setting up and tearing down our modular sound system as quickly as possible without permits, before getting caught—generator, speakers, systems—in and out.

The precariousness of the location dictated how much time we had to build our patches. Once the patches were sufficiently generative, we would stand back and listen as the sounds pervaded the surroundings, adding a magical dimensionality.

Moving out of the solitary stasis of the studio, into the dynamic outdoors was liberating. What I had always imagined to be an indoor white-glove studio instrument came to life in a totally new way when placed outdoors and given the space to interact with the incidental sounds of its surroundings. The synthesizer became an extension of the ecology and vice versa.

It was these formative excursions that led Eric and I to try and recreate what we had experienced on our road trips with friends back home in Los Angeles.

Modular on the Spot

The first gathering happened in July of 2015, in a small park overlooking the LA river (Figure 10.1). The parameters were simple, friends were invited to bring their modular synthesizers and a blanket, plug into the mixer, and share a patch with the human and non-human environment. We were exploring together, rather than performing for each other. The stakes were low and the experience was rich. This was not a studio or a club, it was a collaboration between player, community, and ecology, free and open to anyone.

Eventually, we moved out of the park and down into the river, at the base of a concrete embankment under a transmission tower, away from hovering park rangers and permit requests. Each month, we would mount a hand-drawn sign next to a small opening in the fence, marking the entrance to our hidden sonic oasis in the middle of Los Angeles.

FIGURE 10.1 Modular on the Spot, two-year anniversary, Los Angeles River (Photo: Pablo Perez).

The new spot became *The spot*—a spontaneous urban amphitheater of lumpy concrete, boulders, tunnels, running water, herons, mallards, geese, sagebrush, willows, giant reeds, and scattered homeless encampments. We were making music *with* the environment and the city was participating—helicopters, cars, trucks, sirens, birds, tumbleweed, wind, and running water, merged with the modulars and fused with the luscious natural reverberation of the concrete embankment.

This was an expanded way for us to experience and experiment with our instruments, together, outside of the isolation of our studios and expectation-laden music venues. Listeners sat, laid, stood, danced, and moved wherever they wanted on the river bank, translating into comfort, openness, and receptivity. There were no set times or lineups and playing order was decided on the spot. We would begin in the late afternoon, with music transitioning us into the evening. At the end of the night, the "heads" would stay back and huddle around the glowing systems, enthusiastically exchanging praise and patch techniques, encouraging newcomers to play at the next event.

Unbeknownst to us, a sound and a scene were born. The idea spread and continues to root and sprout itself rhizomatically across the country and overseas with local initiatives adopting the identical format—generator/ speakers/ modular synthesizers/ outdoor/ free.

The Beirut Synthesizer Center

I found the words in the image above spray painted on a wall in Beirut just days after meeting with future center co-founders Elyse Tabet, Ziad Moukarzel and Hany Manja, to discuss the idea of opening a community space for synthesis in Beirut (Figure 10.2). The timing was providential and confirmed what we had been discussing—the resource we need most is dedicated space.

Each of us had been loosely cultivating the idea of a synthesizer center in our minds, but that night our visions gathered over black tea and chocolate Depression cake. We decided we were going to pool our equipment and open an artist-led community space in response to the atomized, isolated, and cost-prohibitive era of modular synthesis.

We wanted to create a central physical space that would connect people with resources, to enrich our creative practices, and facilitate the learning of others. The space would provide free access to equipment, weekly open hours with facilitators, a small reference library, and a variety of workshops, small-scale performances, screenings, and lectures. With the help of some seed money, we found a space and opened our doors in June of 2021.

Since its opening, the center has become a new kind of sanctuary—a focused environment dedicated to the study of synthesis and connection with fellow learners of all backgrounds. A space where the formation of a new community is possible, one rooted in inquiry and mutual investigation.

FIGURE 10.2 Grafitti on a wall in Beirut (Photo by author).

There's also the reality of Lebanon in all of its multidimensional turmoil, and why spaces like these are especially needed in places suffering unfathomable despair, but I've chosen to omit that part of the story, because it would need an entire book, or chapter at the very least, and focus instead on the space itself and its relevance to any city.

Monday, 23 August 2021
Beirut

Gratitude
for this small portal of sound space we've created
a space we can come to work and think
with or without power
learning how to share space and energies

(personal journal entry)

The Beirut Synthesizer Center is not a school in the traditional sense. It is predicated on the notion that self-directed learning *occurs* when people have access to equipment and those who can answer questions in a dedicated space, activated by fellow learners.

We Need Any Kind of Center

Modular on the Spot and The Beirut Synthesizer Center were both formed in response to the problem of space, both physical and metaphorical.

As we move into new ways of making music with new instruments, it behooves us to explore alternative ways to make, learn, and share work, beyond the club and the school, beyond industry and academia.

Modular on the Spot creates a spontaneous and movable gathering space, where people with a common interest can briefly meet, share work in a free, outdoor, low-stakes environment, and disperse. The Beirut Synthesizer Center is similar but operates from a fixed location, not only connecting those with a shared interest in synthesis but in the specific case of Lebanon, it connects people with social, economic, even catastrophic, states in common.

Autonomy and accessibility are equally important aspects of both spaces. Autonomy implies a freedom to experiment, fail, modify the structure as needed, and resist pressures of expansion. Accessibility implies a low barrier to entry, financially and skill-wise, and the space to ask questions and seek answers in person. Autonomy is not to be conflated with lack of structural integrity. In many ways, the quality of study, work, and play is richer when intrinsically motivated.

The concept of self-directed arts-based learning is nothing new. There exists a rich lineage of artist-led experimental pedagogical models worth revisiting from Black Mountain College to The San Francisco Tape Music Center, and more recently The Synth Libraries in Portland and Prague, to name a few. All of these shaped the formation of the Beirut Synthesizer Center.

I'm not suggesting that we run out and shut down conventional universities and performance venues. I'm advocating for a post-institutional re-thinking of the ways knowledge is shared through the parallel development of alternative spaces for making, learning, and sharing. I'm uplifting the idea of flexible, free (or almost free), dedicated, micro spaces that provide access to equipment, space to learn, and the opportunity to connect.

Translation

As we began gathering materials for the Center in Beirut, we thought it might be useful to create a glossary of translated synthesis terms from English to Arabic, starting with the word synthesizer.

Because of the complexity of the word, a direct translation turned out to be a difficult task, and was quickly eclipsed by the logistics of opening the space so we proceeded with a transliteration, a long, inelegant transliteration that roughly read "seeenthaasaizr" in Arabic.

Meanwhile, the search for a viable translation continued. WhatsApp messages bounced back and forth as each of us searched for and proposed synthesizer-adjacent Arabic words, but nothing gelled.

The word felt irreducible, uncontainable. The more I dissected it, the more slippery it became. The quest for translation started folding in on itself … What is synthesis? Are we combining, mixing, adding, subtracting, transforming? How can we possibly describe this process in a single word? Does the word even exist in Arabic, or do we need to invent one?

Extending the idea of synthesis to translation (at the risk of belaboring the word) what we were attempting to do was *synthesize* a term; transform from one form (English) to another (Arabic). But this was more than a word, it was a concept that would require a deeper level of understanding to translate, opening the door to larger questions about language as a way of seeing the world.

Arabic is a vast language of millions of words, compared to the thousands in the English language. It is a florid language with oral and poetic roots, not well suited for technical writing. It is complex and highly nuanced. Many words name the different kinds of love, friendship, rain, stages between sunset and sunrise … the list goes on.

The nuance of the language is mirrored in the micro tonal structure of the maqams (modes) used in Arabic music. An important characteristic of the quarter tones in the maqams is that they are shifting targets. Their pitch waivers infinitesimally depending on the mode, the location of the note in the melody, the region the music is being played, and the mood of the player.

In recent years, synthesizer companies have added microtonal functionality by allowing users to program a quarter tone on any key, equally spaced between the two adjacent notes, flattening centuries of musical nuance. Sequencers that simply state "Arabic scale" in the menu of scale quantizations are even more reductionist.

The soul of expression in Arabic language and music lies in the in-between places, the infinitesimal variations that are honored by name.

Surrender

Something is inevitably lost in the process of translation. Trying to shoehorn the word synthesizer into a single Arabic word feels like a surrendering of language.

The idea of creating a new dictionary of Arabic synthesis terms has the capacity to shift the focus from translation to participation. Participation by finding ways to articulate processes in synthesis through the nuanced capacity of the Arabic language, matching it to the complexity of electronic sound, rather than artificially flattening for the sake of convenience.

How can the inherent richness and capacity of Arabic music and language be composted into new words that encompass new technologies, such as synthesizers. What if the compositional logic behind the many words used in Arabic to express nuanced variations could be extended to different types of oscillators and filters. A single, evocative word that alludes to "low pass filter" for example, or a new word that evokes the energy of a looping envelope. What if a word in Arabic is named before the sound, creating a kind of speculative language of synthesis. Could this lead to the development of new techniques and sounds? How can the language of synthesis evolve through the consciousness of Arabic.

The Arabic language has the capacity to introduce poetics into technical language. Instead of simply translating the word synthesizer, why not create a parallel language of technical expression?

The world of synthesis is especially fertile ground for semantic experimentation because of its newness and the fact that it does not belong to any one culture. It is fair game for all and ideas on technique, notation, expression, and terminology have not been codified to the extent they have been for other instruments.

As an Arab woman living between East and West, I am constantly grappling with my diasporic heritage and questions of increased generational cultural erasure. Synthesizers have become a part of my imaginative toolkit, helping unlock the possibility of alternate futures. Futures where active participation overrides mere translation and consumption.

View from a Height

"People bring each other into activity."
 - M.C. Richards

Much like a patch ends up being an evolved mess of cables, dancing voltages, dead ends, and revelations, so is this writing. Certain connections can be traced back to an initial idea, others left to mystery and chaos.

Modular synthesis affected my personal relationship to music, expanded into spontaneous community, expanded further into the founding of a fixed space, and made its way into language. Each stage prompting a re-thinking of existing structures and proposing an experimental alternative. A re-thinking of the process of music making on a personal level, a re-thinking of where and how musical performance happens, how learning occurs, and a re-examination of my connection to Arabic language and music, embracing it as one with participatory capacities I had not previously considered.

Imagine synthesizers and synthesists as an emergent micro-species. One that can benefit from a specific ecosystem in order to discover their transformative capacities. An ecosystem grounded in community access, experimentation, autonomy, and participation through networks of micro-zones for self-directed experimental outliers wanting to learn how to do things together.

This type of environment is a reflection of the format itself. Modular synthesizers are tools of re-imagining. They encourage alternative ways of creating sound, re-arrangement, flexibility, and honor what is small, from the tiniest sound to the tiniest knob turn, every gesture holds the potential to affect the entire patch.

Connected in format, each module is an independently operable unit that can be connected to a larger structure in infinite ways. Much like the ideas presented in this book. Each a unique vessel, flexible in its formation, connected to the same source of modular thought. One that encourages questioning, loosening, and reimagining externally imposed systems of self, music, hardware, and culture. They are instruments that show us what it means to be independently interdependent. Each module serves a unique function, but also needs to be connected to the rest of the patch in order to make sound. Modules don't make sound on their own, they need to be connected. And so do we.

11

OURORACK

Altered States of Consciousness and Auto-Experimentation with Electronic Sound

William J. Turkel

If modular synthesists generally honor one prohibition, they avoid plugging an output into an output, lest they overload their electronics and fry something. Here we wire up cultural and occultural currents, patching our way into territory somewhere between the New Ordinary (Stewart 2007; Berlant & Stewart 2019) and *True Hallucinations* (McKenna 1994) … and hope the whole thing doesn't go up in smoke.

The Palimpsest

The manual for the ERD/BREATH Eurorack module states that it "allows for controlled summoning and nourishing of (artificial and benevolent) demons, for the offering of sacrificial vapours, and for any modular use of multiple incenses or other smoke-producing resins, herbs and natural substances" (Howse 2019). To call it a manual is perhaps misleading. It is a single page, apparently a bad photocopy or scan, easily dismissed as postmodern pastiche. But where would be the fun in that? If you have gotten so far as considering the purchase of a limited-edition module for making burnt offerings, you already take a hermeneutic approach to texts, artifacts, and experiences. Blow away the smoke and read on.

It is evident that the text specific to the ERD/BREATH module has been pasted into a two-page spread from the Betz edition of *The Greek Magical Papyri in Translation*. The first text on the verso is from *PGM* VII. 540–78 on lamp divination[1] (Betz 1986, 134–5). The supplicant is directed to light a lamp and censer, burn an offering of frankincense on grapevine wood, and make an invocation. Translated, the formula to be spoken begins "Come to me, spirit that flies in the air, / called with secret codes and unutterable names ..." This

DOI: 10.4324/9781003219484-13

will draw the spirit of Hermes Trismegistos down into the body of an "uncorrupt, pure" boy who serves as a medium. Or perhaps not, as the text of the formula is mutilated, and rituals have been known to fail with a variety of consequences ranging from the imperceptible to the catastrophic. Scholars who study ritual failure argue that "in many cases participants and spectators alike learn more about the 'correct' performance of a ritual by deviating from, rather than by adhering to the rules" (Hüsken 2007, 337). And the *Oblique Strategies* tell us to honor errors as hidden intention, but the question remains: *whose* hidden intention? (Eno and Schmidt 1979)

During a talk and demonstration given at Noise Kitchen in Prague, the designer Martin Howse describes the origins of ERD/BREATH and a companion model called ERD/LICHT which responds to flickering light (say a candle). "Maybe two years ago I was invited to take part in a festival on an archaeological site in 'ancient Greece'". He makes the air quotes with his right hand.

> "I became interested in temperature change and in smoke production and how within this whole kind of like ancient Greek world like what the purposes of ... sacrifice, which involved producing large amounts of smoke or burning things, these were somehow, I don't know. I was interested in the idea that these were somehow feeding ... feeding demons or something like this or invoking demons or invoking dreams and there were various arguments set against sacrifice, that it was this way of nourishing demons and I started thinking about how this somehow could relate to more modern concerns with smoke and pollution and so on and how these could also be some kind of technological nourishment for demons." (Noise Kitchen 2020, 43:36)

Howse pinches some sage from an envelope and drops it on the heating element of the ERD/BREATH module. Smoke curls upward into his bearded face, becoming a thick plume. He manages to position a green laser so it shines through the sage smoke, causing the ERD/LICHT to respond with staticky bursts of noise and resonant clunks (Noise Kitchen 2020, 47:00). (The ERD/BREATH manual notes that "a small sample of sage is included for preliminary cleansing of the modular environ".)

So, *The Greek Magical Papyri*, having been suitably emended, might serve as a manual of sorts for this module after all. The next spell in the original (*PGM* VII. 579–90) describes an ouroboric phylactery marked on metal leaf or hieratic papyrus that can serve as "a bodyguard against daimons, against phantasms". In the ERD/BREATH manual, however, that spell has been effaced and overwritten with two quotes. The first is a fragment from the 3rd-century alchemist Zosimos of Panopolis, describing Nature's transformation of itself through the method of "breathing in and breathing out" (Taylor 1937 cited in Cheak 2019). This is followed by a quote from *On Nature* by Empedocles: "They behold but a brief span of a life that is no life, and, doomed to swift death, are borne up and fly off like smoke" (McInery 1963).[2] Each of us experiences so little in a lifetime and takes it for the whole. And speaking of smoke dispersing, recall *Job* 5:7, "Yet man is born unto trouble, as the sparks fly upward".[3]

The figure of a snake biting its tail, from the original spell, remains in the ERD/BREATH manual. Esoteric glyphs are inscribed within the circle made by the snake's body, representing the names of power of the great god. If you squint at these glyphs—squiggly lines connecting little nodes—they look a lot like the diagrams that represent modular synthesis patches. And to many modular synthesists, the image of ouroboros will call to mind the teachings of Allen Strange on the "programming and meta-programming of the electro organism" (Strange 2013). Strange's classic text on modular synthesis was first published in 1972, and, despite being difficult to obtain now, remains one of the foundational documents of the modular tradition (Strange 2006).[4] The text serves as a kind of book of spells in its own right, both in the verbal sense of describing ways to influence, control or compel modules to do your bidding, and in the nominal sense of leading to states of enchantment.

Here is one example. In a discussion of the fluctuating random voltages created by the Buchla 266 Source of Uncertainty module, Strange presents a patch correlating these voltages. The Buchla 266 has a controllable parameter n, which ranges from 1 to 6, and a pair of outputs, labelled $n + 1$ and 2^n. As n increases, the $n + 1$ output provides a linearly increasing number of quantized random voltages. At the 2^n output, on the other hand, the number of quantized random voltages increases geometrically. In the patch that Strange depicts, a clock module generates a regular pulse which triggers the remote voltage source. Each time it is triggered, the $n+1$ output is multed and used to control the pitch of a voltage-controlled oscillator and the n value of the random source. At the same time, the 2^n output is multed to control the cutoff of a filter on the oscillator's output and the period of the clock. Strange writes "bright timbre is accompanied by longer events, longer events are accompanied by greater range probabilities for pitch, and the number of range probabilities for pitch selection is correlated geometrically with the number of spectral choices! This tail-chasing configuration can really consume many hours in the studio" (Strange 2006, Figure 6.71 and 85). Implementing such patches provides something of an initiation for modular synthesists, who usually encounter this particular idea first in variations of the Krell patch of Todd Barton (Barton 2012a). The more you experiment with such systems and come under their spell, the more it seems like they are trying to communicate with you.

Reminded of What Dreams Sound Like

Barton's comment upon discovering the Krell patch was "this patch almost seems to breath" (Barton 2012b). That's right: a patch breathing in and breathing out. Barton named it for the music of the long extinct alien species in the film and novel adaptation *Forbidden Planet* (1956). The Krell were able to manifest and manipulate real world phenomena through thought alone, and the story involves astronauts exploring a massive cubic machine—nuclear powered and self-repairing—that the godlike alien race embedded in the titular planet. That is before the Krell "were wiped from existence" "in one night of unknown, unimaginable disaster" maybe 200,000 years before (Stuart 1956).[5] While exploring the planetary machine, the astronauts themselves come under attack from an invisible assailant. This is from the novel:

> I said, "For God's sake, man, what was it you heard?" and he said, "Well—it was like—like something breathing, sir."

> That jolted me; and it seemed to make him more nervous still, just remembering. He said, "Something awful big—" His face was white now. "But—but there wasn't anything there, sir! There wasn't anything any-place!" (Stuart 1956)

The novel does not mention the music of the Krell. That was imagined and created by Bebe and Louis Barron, whose "electronic tonalities" for the film served to introduce a mass audience to avant-garde electronic music composition for the first time. Inspired by Norbert Wiener, the Barrons "created individual cybernetic circuits for particular themes and leit motifs" and recorded them to tape, noting that "each circuit has a characteristic activity pattern as well as a 'voice'. Most remarkable is that the sounds which emanate from these electronic nervous systems seem to convey strong emotional meaning to listeners". One of the tracks on the 1976 *Forbidden Planet* LP (track 13, on Side A) is entitled "Ancient Krell Music" and this is presumably what inspired Barton in naming his own compelling version of a tail-chasing patch (Barron and Barron 1976).

The agencies of *Forbidden Planet*, the planetary forces and invisible assailants, the unseen things that breathe in and breathe out—and eradicate—turn out to be "monsters from the id", unleashed by Krell technology and responsible for killing both Krell and humans alike. In Douglas Kahn's apt description, the story is "a pre-*Sputnik* space-age rendering of Shakespeare's Tempest seen through the highball lens of cocktail-party Freudianism" (Kahn 2013, 97) In their liner notes, the Barrons claimed that people told them their electronic tonalities "remind[ed] them of what their dreams sound like" (Barron and Barron 1976). But with the technology of the Krell, these are such dreams as stuff is made on. They have at least as much to do with the occult cosmology of Shakespeare's world as with the bourgeois psychology of the 1950s. For in the Renaissance, "[s]pirits, demons, and unseen active effluvia comprised the invisible technology of nature's marvels" (Floyd-Wilson 2013, 1).

For some thinkers of an esoteric bent, this is as true in the age of electronics as it was in the age of rebirth. Writing about the occult in the early 1970s, for example, Colin Wilson suggested that the novel *Forbidden Planet* taught more about phenomenological psychology than Freud: "The subconscious mind is not simply a kind of deep-seat repository of sunken memories and atavistic desires, but of forces that can, under certain circumstances, manifest themselves in the physical world with a force that goes beyond anything the conscious mind could command" (Wilson 1971, 516). These sorts of powers give rise, on the readings of some occultists, to what Patrick Harpur calls 'daimonic reality': profound experiences of angels and aliens, cryptids and crop circles, in imaginal rather than literal space (Harpur 2003).[6]

The Prospero of *Forbidden Planet* is a character named Morbius, a doctor from the previous expedition who had been stranded for two decades with his daughter after the rest of his crew mates were killed and his wife died of natural causes. Now Morbius may be a real surname. The Ancestry website reports a single family with that name who lived in Texas in 1920, attested by a meagre 49 historical documents.[7] I prefer instead to think of the character's name as an approximation to the British pronunciation of Möbius, an homage to the German mathematician who was one of the discoverers of the strip that bears his name. Or really re-discoverers since the figure was known in antiquity. Like the ouroboros, the head of the Möbius strip is connected to the tail, but only after it has been half-twisted. This results in a surface with only one side, and a single curve around its boundary. It has weird properties. A *Flatland*-style creature living in the strip is indistinguishable from its own mirror image, for example (Weisstein, n.d.). Perhaps when you close your eyes, you see red ants crawling along the surface of a Möbius strip in the Escher engraving, chasing one another's tails (Escher 1963).

Cybernetic Circuits

Electric and electronic circuits that sing, breathe and talk have been encountered in a wide variety of settings over more than a century. Whether the circuits are drawing down agencies from airy nothing or are themselves somehow alive depends on who you ask. In folklore, electronic devices have been subject to haunting and possession since their inception, the "technologies serving as either uncanny electronic agents or as gateways to electronic otherworlds" (Sconce 2000, 4). One context for these paranormal affordances of the electronic can be found in the murky prehistory of electricity.

Although we tend to forget it now, the origins of humankind's manipulation of electricity lie in the handling of a few species of strongly electric fish. These include the African electric catfish, a salient inhabitant of the environments where our species evolved; the marine torpedo, a ray commonly encountered on Mediterranean shores; and the so-called electric eel (really a kind of fish) of South American rivers. For most of the human career there were two ways to experience substantial electric shock, either by touching one of these fish or being struck by lightning, and for the most part those were understood to be completely different phenomena. Beginning in antiquity, we have records of strongly electric fish being used in magical and medicinal contexts, although the practice almost certainly predates the development of writing. Humans, electric fish, and other animals formed the circuits by which people came to understand how electricity worked, gradually disenchanting what had been an occult force. Experimental electrical apparatus expanded to involve inorganic materials like metal, glass and water, and the organs and tissues of recently killed animals. (Think of Luigi Galvani's frog preparations or gruesome electrical experiments with human cadavers). Over time, the inorganic component of our technology has increased to the point where electricity seems naturally to belong to the domain of physics, and bioelectricity has been relegated to the status of subfield. But people still put themselves and other animals into circuits of many kinds, and the stories we tell about electricity often ascribe life and agency to forces that science now assures us are inanimate (Turkel 2013).

The transduction of sound into electrical signal and vice versa, the audification of electrical signals, dates to the 19th century and the development of electric microphones and loudspeakers. Emil Du Bois-Reymond, an inveterate builder of circuits made from the bodies and tissues of frogs and electric fish, wired a telephone to a galvanoscopic frog so that when he shouted "Jerk!" into the mouthpiece the leg twitched on command. The choice of magic word was crucial here, as sounds with deeper overtones turned out to be more effective. When he called out "lie still" the frog's leg would not react (Dombois 2008; Turkel 2013). As far as I know, no one is using modular synthesis with a microphone and e-stim to make their own bodies or those of freshly-killed animals twitch and dance on command (updating Stelarc's *Ping Body*, the robotized dog mummies of Survival Research Labs, or the grotesque faces electrified by Duchenne du Boulonge) ... or at least they haven't gone public with it yet (Stelarc 1996; Edmondson 1983; Vale and Juno 1983; Duchenne de Boulonge 1876). But the legacy of hybrid organic/inorganic apparatus continues with projects like *cellF*, a synthesizer controlled by human neurons grown on an electrode array. Rather than responding to voice commands, the neurons were given a hit of dopamine before a show: "it really pepped them up," Andrew Fitch reports (Fitch 2015). Recall that Weiner's original definition of cybernetics stressed control, as well as communication (Weiner 1961).

Even circuits composed without living tissues seem to live. In 2017, Minimalist composer Caterina Barbieri told Dennis DeSantis of Ableton that she likes using modular synthesizers in her practice because she can leave every piece she plays "as a real, living organism somehow" (DeSantis 2017, 5:12). Or there is modular artist Richard Devine, speaking animatedly at a Make Noise panel in March 2013 while passing his hands suggestively over a patch:

> "I will set up a patch that's self-sequencing and mutating and changing over time and doing all these things and then I just leave it … you basically create what I call this floating, a little electrical organism, that's moving around this little neural network you've made and it just floats in that space for that one time, like this little ghost that you have … it's a fascinating sort of feeling you get that I've not really experienced with any other instrument I've ever played. No matter how hard I was tripping [laughs]." (Make Noise 2015, 1:37)

Speaking in another interview with *Future Music* the same year, Devine noted "and then when you pull the patch cables out it's lost forever" (Isaza 2013).

Manipulating a modular synthesizer is intimate and immersive: one is almost continually engaged in relations of communication and control with a perceived other, and the lines of force go in both directions. For those in search of deeper communion, there are many ways to incorporate the practice of circuit bending in modular—perturbing a circuit with one's body and listening for a response. These include patch points that use the conductivity of human or other animal bodies (Landscape AllFlesh),[8] a biofeedback sensor and random voltage generator (Instruo Scíon),[9] an assortment of capacitive sensors including Mister Grassi, Peter Blasser's touch rungler module (Blasser 2016), theremin-style antennae (Doepfer A-178 Theremin Voltage Source),[10] modules that starve others of power (like the ADDAC 300),[11] and a wide variety that have bent circuits in their design.[12] Reed Ghazala, one of the originators of the practice, describes circuit bending as "send[ing] probes to alien worlds: We don't know what's there. But we want to" (Ghazala 2005, 4).

Perhaps the best known of these bent circuits is the Benjolin of Rob Hordijk, available both as a Eurorack module and a standalone instrument, and its precursor, the Blippoo Box. In a widely circulated paper from *Leonardo Music Journal*, Hordijk described the process of creating an instrument that was already 'bent by design', which he noted paradoxically contradicted the definition of circuit bending (Hordijk 2009a). In search of the instrument's sonic character, which he wanted to be a generalized 'electronic sound' rather than a synthesized sound per se, Hordijk found inspiration on the Barron's *Forbidden Planet*. The chaotic core of the device is the 'rungler'. If you're asking yourself, "What the $#%$ is a rungler?" it turns out that Hordijk himself has answered that very question: "The purpose of the rungler is to create short stepped patterns of variable length and speed … It needs two frequency sources to work and basically creates a complex interference pattern that can be fed back into the frequency parameters of the driving oscillators to create an unlimited amount of havoc" (Hordijk 2009b). Note how the serpent bites its own tail. The complexities of the circuitry make it possible for an experienced interlocutor to anticipate the direction the instrument will take, if not to predict it. Just like a conversation.

Called with Secret Codes and Unutterable Names

One anthropological analysis of divination is that the diviner makes sense of "random chance—defined culturally as god, gods, spirits, nature, the dead—[which] produces a complex sign". This sign is then interpreted "using a culturally determined canon of interpretive techniques, texts, images, and myths" (Lehrich 2003, 171). In 1967, the ethnomethodologist Harold Garfinkel reduced this idea to perhaps its most skeletal form. He gathered a group of undergraduates in search of advice for their personal problems by telling them that they could participate in an alternative form of psychotherapy. The students were encouraged to describe their issues to a counsellor via an intercom. Along the way, they could pose 10 questions and the counsellor would provide a simple "yes" or "no" answer for each question. Unbeknownst to the students, these answers were simply chosen from a table of random numbers with equal probabilities for each. The flip of a coin, in other words. What Garfinkel showed, however, was that the students themselves narrated these random outcomes into a meaningful interaction "by searching for and determining pattern, by treating the adviser's answers as motivated by the intended sense of the question, by waiting for later answers to clarify the sense of previous ones, by finding answers to unasked questions—the perceivedly normal values of what was being advised were established, tested, reviewed, retained, restored; in a word, managed" (Garfinkel 1967, 94).

If it is possible to have a meaningful dialogue with a stream of random bits while cold sober in a sociology lab, imagine the effects of set and setting—not to mention state of consciousness—when communing with "(artificial and benevolent) demons" via a complex patch with a life of its own (cf Zeitlyn 1995). The unpredictable outcomes created by contemporary modules are far more sophisticated than random bit streams, and thus provide much more grist for the interpretive mill. For example: the many heirs of the Buchla 266 Source of Uncertainty spit out various kinds of constrained randomness; there is the Rungler, fascinating and mercurial; there are a wide variety of chaotic modules that trace fast or slow orbits around strange attractors (Andrew Fitch of Nonlinearcircuits has created many of these, the most popular probably being his Sloths); there are modules that deviate probabilistically from fixed rhythmic grids; and there is even the ADDAC 405,[13] which is intended to generate the *relabi*, or self-erasing pulse, of John Berndt. Relabi is an "experiential gestalt," a "palpably coordinated plurality of events that appears to be cyclical but simultaneously suspends identification of a uniform pattern". According to Berndt, relabi "seems to have an unusual ability to induce odd states of mind" and "sustained exposure … can space you out in a strange way" (Berndt 2009). But really, something similar might be said of prolonged exposure to the outputs of any of these options, singly or in combination.

In 1971, the psychopharmacologist Roland Fischer published a useful "cartography of ecstatic and meditative states" in *Science*, describing a continuum from highly aroused 'ergotropic' states of ecstasy or mystical rapture to low-energy 'trophotropic' states like yogic samadhi. He chose to represent this graphically with a lemniscate (the infinity sign) putting the normal "I" in the middle and the modes of hyper- and hypo-arousal on the two lobes of the figure (Fischer 1971). As Jonathan Weinel and others have argued, this scheme maps rather neatly onto a wide range of musical genres from techno (ergotropic) to ambient (trophotropic) (Weinel 2018). Both ends of this arousal spectrum are particularly well served by modular synthesis, in the sense that it is much easier to create fast hypnotic drums for raving, or dreamy Rings-into-Clouds for navel gazing, than it is to create something musically sophisticated that appeals to a critical listener in full possession of their faculties.

Speaking of possession, there are few things quite as uncanny (the familiar made strange) as other-than-human voice. Listening to hundreds of crows massing in the sky and trees outside his Berlin studio in the autumn, Martin Howse began musing about simulating their calls. He realized that crows mimic other sounds, like telephones and human voices, which led him deep into the history of speech synthesis and vocoders. When this research became extensive enough that it threatened to overwhelm his original idea, he distilled it out from the crow project. The eventual result was a speech module called ERD/WORM designed "to imagine what could be produced by a worm of varying lengths, if it could produce a voice" (Noise Kitchen 2020, 35:00-37:30). What he imagined can be pretty unnerving. Once I was zoned out listening to ERD/WORM say "miracle, miracle, miracle" over and over to the point of semantic satiation, when the module suddenly and clearly said "bad worm".[14] In fact, a number of the vocal fragments that ERD/WORM 'composts' were spoken by people now dead. "That's the challenge," Howse wrote in an e-mail interview with Ezra Teboul, "how to say insert the influence of Edgar Allen Poe directly into the software and hardware" (Teboul 2015, 197).

One might take a disability studies perspective on ERD/WORM (and other modules like the Synthesis Technology E950 Circuit Bent VCO, which licensed Texas Instrument's 'Speak and Spell' technology), and interpret speech modules in the context of echolalia, palilalia, and other speech impairments. Howse's manual for ERD/WORM supports this reading to some extent, straying toward the trope of the disabled body as cyborg (Blume 1997; Kafer 2013):

"communication becomes the business of circuit-bent human-worms, opening human and animal speech to the cut-up, to stammerings, stutterings and tics; to an extra-human outer-word terrain of numbers burrowed and nibbled by the earth worms. ERD/WORM feasts on this electronic legacy" (Howse 2017, 6–7)

Instead of cripping the module here, however, I will focus on it as a generator of something akin to Enochian, the language reportedly spoken by angels to John Dee and Edward Kelley in the Elizabethan era, and, according to them, last understood by Enoch, biblical patriarch and great-grandfather of Noah. I don't mean to focus on angels specifically or on Dee's interpretation of his and Kelley's experiences. Rather, following Egil Asprem we might see Enochiana as "part of a perennial system, which stems from much higher authorities, whether these be the angels themselves, a selection of Atlantean or Rosicrucian secret chiefs, or other esoteric sources of knowledge. Perhaps Dee and Kelley failed to understand the significance of their revelations, or perhaps they were even deliberately misled by 'the angels,' whoever they were?" (Asprem 2012, 3). In other words, the ERD/WORM module can be used as a generator of an 'artificial and benevolent' glossolalia, however we want to take that. I think such imaginal uses are clearly signaled in Howse's manual, which has an epigraph from Michael Maier: "The WORM was for a long time desirous to speake, but the rule and order of the Court enjoyned him silence, but now strutting and swelling, and impatient, of further delay, he broke out thus ..." Howse's description of the module's function is framed by vermicular and serpentine graphics reproduced from the canon of Western esotericism, including two separate depictions of ouroboros, one from the Chrysopoeia of Cleopatra in the Codex Marcianus and the other resembling the drawing that appears in Theodoros Pelecanos' 1478 copy of the lost tract attributed to Synosius of Cyrene.[15]

The uncanny utterances of ERD/WORM better lend themselves to what Frederic Bartlett called the 'effort after meaning' than other (pseudo)randomly produced sounds (Bartlett 1995). For one might imagine some dim communicative intent behind unexpected rhythmic, timbral, or harmonic events, but it is quite hard *not* to hear it in urgently repeated words, guttural syllabic clusters, vocalic buzzing, or wordless moaning. The rational or skeptical mind can always dismiss such experiences as apophenia. An occultist, however, might hear Enochian revelations or find theurgical uses for the module in the manner of Iamblichus. "Language that to us has no sensible significance," Dale B. Martin writes, "is nevertheless valuable because it is sensible to the gods and even to the 'divine intellect' that is in us. Furthermore, by means of strange, barbaric words the soul is elevated to the gods and joined together with their power" (Martin 1995, 100). Or, moving ever further from the cultural frameworks of Near Eastern antiquity, a psychonaut might use it to receive a xenolinguistic download, "The alien Other find[ing] symbolic means—language—to transmit its message from the hyperdimensional realm … into baseline reality" (Slattery 2015, 48–9).

Such secret codes and unutterable names seem to play a particularly significant role in the work of 'synthesynthesist' Peter Blasser, who designs Eurorack modules under the pseudonym Ieaskul F. Mobenthey (Roe 2011). Blasser's schematic circuit diagrams are filled with non-standard icons with equally non-standard names. One, which looks like a cross between an hourglass and the alchemical symbol for tutty, rotated 45 degrees, is called "STABB COUNCILLOR". Another, which resembles a Geneva stop mechanism, is the "SPESAL CUCK", not to be confused with the same symbol when it is encased in an aura of sorts, and thus becomes the "TRANSOB POST AMPLIFIER INPUT" (Blasser n.d.) These symbols and dozens of others are haphazardly laid out in little electronic villages and connected with meandering cow paths to create larger units with names like "Tarp", "Arpserge" and "Tarpterge". If you squint at Blasser's schematics, they look a bit like the names of power of the great god. These esoteric designs are transmuted into equally warped physical circuit board layouts, and in use the IFM modules themselves seem to inherit a quirky unpredictable magic from their circuit-as-sigil construction method. As above, so below.

Let's go meet Blasser—where else? —in a grotto. In a video documentary about his "gonzo" circuits, we see the young man warily approach a cave in a large rock beneath an overhead power line, trying to see if there is anyone inside before getting too close. Much of the graffiti sprayed on the rock near the entrance has been effaced, but not the word WORMIAN, which is uppermost in block caps. Blasser is carrying a cloth bag full of Tocantes, an instrument that he says he "designed to play in a cave". Entering the cave and sitting down, he pulls out the Tocantes and sets them on the ground, lights a small round purple smoke bomb and throws it into the depths and then begins playing. He lights some more smoke bombs and resumes playing. The colored smoke becomes pretty thick. Billowing around, it gets the better of him and he begins to cough. Cut to a scene outside the cave: smoke is pouring from the opening and drifting into the sky. Roll credits (Synth Docs 2016, 10:46-13:13).

Gateways to Electronic Otherworlds

Victoria Nelson reminds us that our word 'grotesque' comes from the Renaissance Italian *grotte*, "from *crypta*, a Latin borrowing from Greek meaning 'hidden pit' or 'cave'". The grotto was a "'place of birth and death, passing away and rebirth, descent and resurrection,' a highly charged microcosmic container of selected physical objects that drew down the arcane energy of counterpart Forms in the superior world" (Nelson 2003, 2). Tocantes notwithstanding, caves and grottos don't really lend themselves to most forms of modular synthesis and the challenge becomes finding some other way to tap into their telluric potential to form portals. Martin Howse's ERD/ERD Earth Return Distortion module, for example, passes audio and control signals through a little sample of soil in the module itself, but also allows the possibility of routing the signal outside of the modular and through nearby earth. When ERD/ERD was still available for purchase, Howse offered discounts in return for soil samples such as "Edgar Allen Poe earth from Baltimore, vampire earth from Transylvania or Whitby".[16]

Another alternative is suggested by practices of radiesthesia and ghost hunting. Vlad Kreimer, Ukrainian/Russian mystic, musician, and practitioner of 'esoteric engineering' at SOMA Laboratories, makes devices that are well suited to both spiritual exploration and sonic expeditions into high strangeness.[17] Kreimer's handheld Ether is an 'anti-radio', a wide-band RF and magnetic receiver which captures "all the interference and radiation that a traditional radio tries to eliminate [with tuned filters]". It works particularly well to soundtrack an urban infiltration (Ninjalicious 2005) or for psychogeographic wandering in desolate (post)industrial spaces, transforming them into the Chernobyl Exclusion Zone or the *zona* of *Roadside Picnic*, by immersing the user in a cacophony of alien electronic voices that are always present but usually inaudible. This can be a strange experience, not least because it simultaneously partakes of both of Mark Fisher's categories of the strange: such spaces are already *eerie* because they cause us to ask, "Why is there nothing here when there should be something?" and they are *weird* because of the juxtaposition of "that which does not belong", namely creepy acousmatic sounds. "The weird brings to the familiar something which ordinarily lies beyond it … " Fisher tells us. If you search for the source of the sounds that you are hearing, you veer back toward the eerie, which "is fundamentally tied up with questions of agency. What kind of agent is acting here? Is there an agent at all?" (Fisher 2016, 13–7).

Perhaps if you are sensitive enough you don't need the Ether or any other kind of ghostbox to hear these sounds. As Colin Dickey wryly remarked in an article on the broken technology of ghost hunting in the *Atlantic*, "the best tools for tracking down spirits have always been the ones fallible enough to find something" (Dickey 2016). A longstanding idea in esoteric thought is that the human mind itself might be a kind of filtered receiver, tuned by evolution to focus on the here and now and to eliminate perception of anything irrelevant to survival. Change the tuning and you open the door to ghosts or spirits, hyperdimensional entities, aliens, the Akashic Records, the implicate order, cosmic oneness, or what have you (Kripal 2017). Of course, there is also a longstanding notion that such sensitivity is peculiarly feminine. Writing about the "feminization of channeling" in the late-19th and early-20th centuries, Jill Galvan notes that women were considered to be well suited to "the work of mediating others' transmissions ... a single vocation that took many forms—typing, telegraph operating, and occult mediumship". Complementing the idea of electronics as prone to haunting, "those interested in paranormal contacts took new communication devices as their models and aids" (Galvan 2010, 2, 8).

In a semi-autobiographical vignette Kristen Gallerneaux writes of eaves-dropping on her grandmother's séances as a child:

> "The smell of rose perfume floating through the sulfurous tendrils of extinguished candles. Cherry tobacco smoked in a pipe. The sound of thick pencils. A noise like 'sssssssssshhck.' The belly of a snake sliding across rough paper. She could hear that the line was continuous. The pencil never lifted from the page, and so there was no tapping of lead on paper on top of wood. ... Soft mumbling. Questions. Have you ever heard a ghost walk? The shapes were coming down the stairs, behind her, as they always did, every second Saturday night of the month. She had learned, over time, to not be afraid of them. They were people once. 'It's not the dead that will hurt you,' her grandmother would tell her. 'It's the living'" (Gallerneaux 2018, 58-9 epub)

Gallerneaux is now an artist and curator of communications and information technology at the Henry Ford Museum. Her explorations with haunted media include a "dirt synth",[18] a Eurorack modular setup that uses a custom module somewhat like Howse's ERD/ERD to amplify the sonic qualities of samples of soil she collected from various sites of poltergeist activity. Patch cords directly from the soil samples feed from the interface into off-the-shelf modules like the 4MS Spectral Multiband Resonator and Mutable Instruments Clouds. If you are familiar with those modules, you would expect the output to be quite pretty, and it is, with a bit of a rough edge. The dirt synth includes a ribbon-circuit module so the person playing it can put themselves into the circuit (Gallerneaux 2018).

The easiest gateways to locate are not to be found in grottos or other "hyperlocal landscapes" as Gallerneaux puts it. One needs instead to turn inward. Vlad Kreimer's Quantum Ocean is an "electro-shamanistic project" that pairs a binaural noise generator with a series of psycho technical exercises for achieving new mental states and perceptions. The QO is a small black metal box with headphone jack, a pair of pads that can transmit the signal through your fingertips, and another pair of sensitive pads that can be used to tune the instrument to an object, another person or to your own internal states. The extensive manual for the QO reveals Kreimer's serious intent: to put the user into direct contact with the zero-point fluctuations of quantum fields, via the activity of analog semiconductors. "[Y]ou hold in your hands a piece of the deepest secret of the Universe, maybe even to the doors outside" (Kreimer 2019, 13).

His video announcing the device, however, is straight out of *Psychic Discoveries Behind the Iron Curtain* (Ostrander and Schroeder 1970) or a Victor Pelevin novel. It begins with a Ken Burns-style slow zoom into the first image, which is of Buddha seated in a lotus position surrounded by a globe of energy, with a red heart behind, in what I take to be a torsion field, floating in front of Earth and Moon, and 3D molecules, and a diagram of the Zodiac, and a chromosomal karyogram, and there's more but the image is already being replaced; it is followed by a pair of stylized, severely rendered, red-eyed Lenin heads merging into one another; then a painting of a young woman in a peasant blouse soldering electrical apparatus with vacuum tubes; then a schematic diagram of a vacuum engine generating 'free energy' ... OK, you're fifteen seconds into a video that is three and half minutes long (Kreimer 2020).[19] You've learned that the QO comes from Kreimer's "research into very unusual phenomena, the underground movement of post-Soviet science called Second Physics ... in an attempt to find an alternate reality outside the traditional laws of physics" (Kreimer 2020).[20] Whether you take it seriously or not, magic begins in imagination.

Scriptio Inferior

The underwriting of a palimpsest can be revealing in its own right. Comparing the ERD/BREATH manual with the *Greek Magical Papyri in Translation*, we can recover the passages on the recto that were occluded by text specific to the module.[21] The first is an invocation that calls the ear of heaven, ear of the air, ear of the earth, and so on. The second is a long "fetching charm for an unmanageable [woman]". In this spell, the supplicant invokes seven gods, accusing the woman in question of stating that each of these gods is deficient in some way: one does not have ribs, one was castrated, two are "by nature a hermaphrodite", and so on. If the spell is successful, a daimon will seize the woman and bring her to the supplicant, "inflamed with passion, submissive". The third is a spell "from the Diadem of Moses". If you sleep with the plant snapdragon under your tongue, rise early and recite the names before you speak to anyone "you will be invisible to everyone". A variant of it can be used as a love spell (Betz 1986, 135). By its absence, each of these texts fleshes out the story presented here.

Judging by the discussion on online forums such as *Mod Wiggler*, most modular synthesists are preoccupied with musical productivity and with calling ears to their work, recording performances or recording tracks for distribution. Given the cost of the equipment, this is very much a space of privilege and posturing, and feminist critics have noted that "persistent militaristic technology and aesthetic priorities of rationalistic precision and control epitomize notions of male technical competence and 'hard' mastery in electronic music production. These have produced and been constituted by their opposite: non-technical or 'soft' knowledges and practices that are coded as female". Complicating discourses of sound reproduction is the idea that women are "always already entwined with a logic of *reproduction* ... that ties women to age-old notions of passivity, receptivity, and maternality" (Rodgers 2010, 7, 12). Such discourses tend to be heterosexist as well as gendered. As an alternative to 'hard' mastery or virtuosity, we might gesture toward Jack Halberstam's idea of the "queer art of failure" (Halberstam 2011). Rather than focusing almost exclusively on the production and reproduction of sound for various imagined futures, we might see this preoccupation as a heteronormative construct, following Lee Edelman or the xenofeminist thinkers he inspired (Edelman 2004; Hester 2018). The possibility of creating an ephemeral sonic work/ing to serve esoteric, ritualistic, or paranormal ends is hinted at from time to time but receives very little sustained discussion.

The suggestion that one can gain other-than-human powers by sleeping with a plant under one's tongue raises both the question of drugs and the limitations and potentials of the liminal body. There is, of course, an extensive literature on altered states of consciousness and psychedelic or entheogenic drug use, and sonic phenomena are often encountered in this context (Davis 2019). One thinks, to take a single example, of the McKenna brothers tripping in the heart of the Putumayo in the early 70s, when Dennis started channeling very faint transmissions from somewhere, "experimenting with a kind of humming, buzzing vocal sound" deep in his throat, and ended up buzzing like a giant insect when he locked on to a "snapping, popping, gurgling, cracking electrical sound". And then, you'll recall, he began to rupture 3D space with his voice and bend it into higher dimensions (McKenna 1994, 68). The famous experiment at La Chorrera, Graham St John writes, "was a dog's breakfast of high esoterica: ayahuasca analoguing, telepathy, telekinesis, channeling, and a presumed transdimensional alteration whereby one of them was to be transformed into 'a DNA radio transmitting the collective knowledge of all earthly life, all the time'" (St John 2015, 90–1). But the McKennas didn't have a modular synth, and at this remove it's difficult to tell what difference one might have made to their explorations.

Locking into synesthetic resonance with loud buzzing noises is also reported in some UFO abduction and astral travel experiences. One experiencer told ethnographer Susan Lepselter of awakening paralyzed in the middle of the night with non-human 'objects' beside his bed. One put a wand to the top of his head, and a tingle began moving slowly down his body. "And my brain going wn-wn-wn-wn-wn-wn-wn-wn—Have you ever heard a dynamo?" he asked. "It's loud. It's mighty loud—dynamo—it's a *power*-producing unit. ... And it produces a lot of noise wnwn**wnwnwn** that's what—inside of my brain. ... But my body tingled—outside my body. And I couldn't move" (Lepselter 2016, 35).

In his classic *Journeys Out of the Body*, Robert Monroe described one of his earliest experiences as follows

> "As I lay there the 'feeling' surged into my head and swept over my entire body. It was not a shaking, but more of a 'vibration', steady and unvarying in frequency. It felt much like an electric shock running through the entire body without the pain involved. ... Frightened, I stayed with it, trying to remain calm. I could still see the room around me, but could hear little above the roaring sound caused by the vibrations." (Monroe 1977, 24)

As Monroe learned to work with liminal and hypnogogic states, he discovered a wonderland of altered consciousness in waking sleep paralysis, now often captured by the motto "mind awake, body asleep". It is possible, rewarding even, to experiment with modular or semi-modular synthesizers in these states, but the sensory, haptic, and somatic potentials of your body have been drastically altered. You are blind and paralyzed, but you can use capacitive, electromyographic or theremin-style interfaces to capture myoclonic jerks or eye movements and convert them to control voltages. You can also hack inexpensive electroencephalogram devices like the Neurosky MindWave, to capture brainwave data to use in the same way that you would use a low frequency oscillator.[22] In addition to capturing your brainwaves, you can also entrain them by generating binaural beats or isochronic tones. In liminal states you become aware of your hearing cutting in and out, but you can supplement audio with various kinds of tactile feedback using vibrating motors and solenoids. The sounds you hear, whether external or endogenous, can trigger visual displays, kinesthetic jolts, smells or tastes. Your memory and awareness will fluctuate. If you decide to set out on this path, you will find that the experience and knowledge of disabled synthesists is key: many modules that are perfect for someone who is temporarily able bodied just aren't useful in this setting. One place to start is with the Koma Elektronik Field Kit and Field Kit FX, which are Eurorack compatible, potentially battery powered, and provide a wide range of ways to incorporate unconventional sensors and transducers into your rig.[23]

One of the joys of modular synthesis is the incredible variety of modules available, and the number of different uses that 'patch programmability' permits them to be put to. Many synthesists are in search of the perfect live case for performance, a distinctive sound, a streamlined and productive music-making workflow, FX to process their acoustic instruments, a sound design arsenal, a spendy hobby, or the satisfaction of any number of other goals. Modular can accommodate them all. For a few, however, the processes of patching and repatching, arranging and rearranging racks or cases, can become a different kind of *solve et coagula*, a practice leading toward ineffable experience, gnosis or transmutation. For them, there is ourorack.[24]

Notes

1 PGM = *Papyri graecae magicae.*
2 The quote has been changed by Howse from "span of life" to "span of a life".
3 King James Version.
4 Starting in July 2018, Jason Nolan of Ryerson University began spearheading an effort to get the book legally reprinted for the hundreds of modular enthusiasts who have expressed interest in owning a copy. See https://www.modwiggler. com/forum/viewtopic.php?f=4&t=203573
5 Stuart is the pen name of English author Phillip MacDonald who adapted the screenplay by Cyril Hume. MacDonald was better known for writing thrillers and detective novels.
6 Harpur's formulation draws on Jung and especially Henri Corbin. He is also the author of *Mercurius* (1990), one of the most detailed alchemical novels I know of.
7 https://www.ancestry.com/name-origin?surname=morbius
8 https://www.landscape.fm/allflesh
9 https://www.instruomodular.com/product/scion/
10 http://www.doepfer.de/a178.htm
11 https://www.addacsystem.com/en/products/modules/addac300-series/addac300
12 e.g., Industrial Music Electronics A Sound of Thunder Mk II http://www. industrialmusicelectronics.com/products/8, Synthesis Technology E950 https:// synthtech.com/eurorack/E950/, MengQi Karp https://static1.squarespace.com/static/ 56122b94e4b01402b90cce28/t/57c4ff8737c5815e27dec06a/1472528274146/Karp +KRPLS+Manual.pdf, QU-Bit Data Bender https://www.qubitelectronix.com/shop/ data-bender
13 https://www.addacsystem.com/en/news/introducing-addac405-vc-relabi-generator
14 6 February 2018. https://www.modwiggler.com/forum/viewtopic.php?f=16&t= 146386&p=2752020
15 The former is Codex Marcianus graecus 299 (Venice); the latter is Fol. 279 of Codex Parisinus graecus 2327.
16 https://www.1010.co.uk/org/ERD.html
17 https://somasynths.com/
18 http://kristengallerneaux.com/dirt-synth/
19 When I first saw the video, I paused it at this point so I could order a QO from Moscow for myself. It is scheduled to arrive the day I'm writing this.
20 The Second Physics website covers "non-local interactions, the influence of consciousness on physical reality, new sources of energy" (at least according to Google Translate) at http://www.second-physics.ru/

21 *PGM* VII. 591-92, *PGM* VII. 593-619 and *PGM* VII. 619-27.
22 https://learn.sparkfun.com/tutorials/hackers-in-residence---hacking-mindwave-mobile/all
23 https://koma-elektronik.com/?product_cat=field-kit-series
24 Just before the winter solstice in 2020, while I was plugging away on this article, I saw a new module appear on Modular Grid, the ERD/All the Colours of the Noise. It was explicitly billed as "an Ourorack noise voice," which seemed to me to nicely sum up Howse's design approach and which provided a wider context and a title for what I was trying to accomplish with my essay.

References

Asprem, Egil. 2012. *Arguing with Angels: Enochian Magic and Modern Occulture.* Albany, NY: SUNY Press.

Barron, Louis, and Bebe Barron. 1976. "Music Notes," *Metro-Goldwyn-Mayer's Forbidden Planet LP*. Planet Records.

Bartlett, Frederic C. 1995. *Remembering: A Study in Experimental and Social Psychology.* Cambridge: Cambridge University Press.

Barton, Todd. "Building the Krell Muzak Patch." *Vimeo video*, 5:14. August 29, 2012a. https://vimeo.com/48466272

Barton, Todd. "Krell Music by Todd Barton on Buchla." *MatrixSynth*. August 28, 2012b. https://www.matrixsynth.com/2012/08/krell-muzak-by-todd-barton-on-buchla.html

Berlant, Lauren, and Kathleen Stewart. 2019. *The Hundreds*. Durham, NC: Duke University Press.

Berndt, John. "'Relabi': Patterns of the Self-Erasing Pulse." *JohnBerndt.org (blog)*, 2009. http://www.johnberndt.org/relabi/Relabi_essay.htm

Betz, Hans Dieter, ed. 1986. *The Greek Magical Papyri in Translation*. Chicago: The University of Chicago Press.

Blasser, Peter. "Ieaskul F. Mobenthey --- Mister Grassi Manual." *Synthmall.com*, 2016. http://synthmall.com/ifm/ifmGRASSI.pdf

Blasser, Peter. [Fourses PortDock] *Ciat-Lonbarde.net*, n.d. https://www.ciat-lonbarde.net/paper/fourses.pdf

Blume, Stuart S. 1997. "The Rhetoric and Counter-Rhetoric of a 'Bionic' Technology." *Science, Technology and Human Values* 22, no. 1 (Winter): 31–56.

Cheak, Aaron. 2019. "Circumambulating the Alchemical Mysterium" *from Alchemical Traditions: From Antiquity to the Avant-Garde*. Auckland: Rubedo Press. http://www.aaroncheak.com/circumambulating

Davis, Erik. 2019. "Resonance." In *Unsound: Undead*, edited by Steve Goodman, Toby Heys and Eleni Ikoniadou. Falmouth, UK: Urbanomic.

DeSantis, Dennis. 2017. "Interview with Caterina Barbieri." *YouTube video* 52:07. https://youtu.be/nxECAD3NwQE

Dickey, Colin. "The Broken Technology of Ghost Hunting." *The Atlantic*, November 14, 2016. https://www.theatlantic.com/science/archive/2016/11/the-broken-technology-of-ghost-hunting/506627/

Dombois, Florian. 2008. "The 'Muscle Telephone': The Undiscovered Start of Audification in the 1870s." In *Sounds of Science—Schall Im Labor (1800–1930)*, edited by Julia Kursell. Preprint 346. Berlin: Max Planck Institute for the History of Science. https://www.mpiwg-berlin.mpg.de/sites/default/files/Preprints/P346.pdf

Duchenne (de Boulonge), G.-B. 1876. *Mecanisme de la Physionomie Humaine*, 2nd ed. Paris: Libraire J.-B. Bailliere et Fils. https://archive.org/details/mcanismedelaphy00duchgoog

Edelman, Lee. 2004. *No Future: Queer Theory and the Death Drive*. Durham, NC: Duke University Press.

Edmondson, Bill. "Mark Pauline," *Bomb* 6, Summer 1983. https://bombmagazine.org/articles/mark-pauline/

Eno, Brian, and Peter Schmidt. 1979. *Oblique Strategies*, 3rd ed.

Escher, M. C. "Möbius Strip II." February 1963. https://www.gallery.ca/collection/artwork/mobius-strip-ii

Fischer, Ronald L. 1971. "A Cartography of the Ecstatic and Meditative States." *Science* 174 (4012): 897–904.

Fisher, Mark. 2016. *The Weird and the Eerie*. London: Repeater Books.

Fitch, Andrew. "1st cellF Concert." *Nonlinearcircuits (blog)*, October 5, 2015. https://nonlinearcircuits.blogspot.com/2015/10/1st-cellf-concert.html

Floyd-Wilson, Mary. 2013. *Occult Knowledge, Science, and Gender on the Shakespearean Stage*. Cambridge, UK: Cambridge University Press.

Gallerneaux, Kristen. 2018. *High Static Dead Lines: Sonic Spectres and the Object Hereafter*. London: Strange Attractor Press.

Galvan, Jill. 2010. *The Sympathetic Medium: Feminine Channeling, the Occult, and Communication Technologies, 1859–1919*. Ithaca, NY: Cornell University Press.

Garfinkel, Harold. 1967. *Studies in Ethnomethodology*. Englewood Cliffs, NJ: Prentice-Hall.

Ghazala, Reed. 2005. *Circuit Bending: Build Your Own Alien Instruments*. Indianapolis, IN: Wiley Publishing.

Halberstam, Judith. 2011. *The Queer Art of Failure*. Durham, NC: Duke University Press.

Harpur, Patrick. 2003. *Daimonic Reality: A Field Guide to the Otherworld*. Enumclaw, WA: Pine Winds Press.

Hester, Helen. 2018. *Xenofeminism*. Cambridge, UK: Polity.

Hordijk, Rob. 2009a. "The Blippoo Box: A Chaotic Electronic Music Instrument, Bent by Design." *Leonardo Music Journal* 19: 35–43.

Hordijk, Rob. "What the $#%$ is a rungler?" *Electro-Music Forum*. November 14, 2009b. https://electro-music.com/forum/topic-38081.html

Howse, Martin. ERD/WORM Manual. 2017. https://1010.co.uk/org/ErdWormOnlinewhitemanual.pdf

Howse, Martin. *ERD/BREATH Manual*. 2019. http://1010.co.uk/org/breath_manual.pdf

Hüsken, Ute. 2007. "Ritual Dynamics and Ritual Failure," in Ute Hüsken ed., *When Rituals Go Wrong: Mistakes, Failure, and the Dynamics of Ritual*, 337–366. Leiden: Brill.

Isaza, Miguel. "Richard Devine Interview." *Future Music*. June 18, 2013. https://www.musicradar.com/news/tech/richard-devine-interview-578399

Kafer, Alison. 2013. *Feminist Queer Crip*. Bloomington, IN: Indiana University Press.

Kahn, Douglas. 2013. *Earth Sound Earth Signal: Energies and Earth Magnitude in the Arts*. Berkeley: University of California.

Kreimer, Vlad. 2019. *QO, SOMA Laboratory*.

Kreimer, Vlad. 2020. "QUANTUM OCEAN Demo, SOMA Laboratory." *YouTube video*, 3:33. https://www.youtube.com/watch?v=-zfV11aX6s0

Kripal, Jeffrey J. 2017. *Secret Body: Erotic and Esoteric Currents in the History of Religions*. Chicago: University of Chicago.

Lehrich, Christopher I. 2003. *The Language of Demons and Angels: Cornelius Agrippa's Occult Philosophy*. Leiden: Brill.

Lepselter, Susan. 2016. *The Resonance of Unseen Things: Poetics, Power, Captivity, and UFOs in the American Uncanny*. Ann Arbor, MI: University of Michigan.

Make Noise. May 6, 2015. "Panel including Richard Devine." *YouTube video*, 16: 26. (Event date March 24, 2013) https://www.youtube.com/watch?v=HelYLzUZkKY

Martin, Dale B. 1995. *The Corinthian Body*. New Haven, CT: Yale University Press.

McInery, Ralph. 1963. *A History of Western Philosophy, Volume 1: Beginnings to Plotinus*. Notre Dame, IN: University of Notre Dame.

McKenna, Terence. 1994. *True Hallucinations: Being an Account of the Author's Extraordinary Adventures in the Devil's Paradise*. San Francisco: HarperSanFrancisco.

Monroe, Robert A. 1977. *Journeys Out of the Body*. New York: Harmony Books.

Nelson, Victoria. 2003. *The Secret Life of Puppets*. Cambridge, MA: Harvard University Press.

Ninjalicious. 2005. *Access All Areas: A User's Guide to the Art of Urban Exploration*, 2nd ed. Toronto, ON: Infiltration.

Noise Kitchen. November 11, 2020. "Noise Kitchen Invites Martin Howse." *YouTube video*, 1(17): 36. https://www.youtube.com/watch?v=sVU_ovF-Ufg

Ostrander, Sheila, and Lynn Schroeder. 1970. *Psychic Discoveries Behind the Iron Curtain*. Englewood Cliffs, NJ: Prentice Hall.

Rodgers, Tara. 2010. "Introduction" in Tara Rodgers, ed. *Pink Noises: Women on Electronic Music and Sound*. Durham, NC: Duke University Press.

Roe, Nat. "Psychic Circuits: Peter Blasser of Ciat-Lonbarde." *Rhizome*, February 2, 2011. https://rhizome.org/editorial/2011/feb/2/psychic-circuits-peter-blasser-ciat-lonbarde/

Sconce, Jeffrey. 2000. *Haunted Media: Electronic Presence from Telegraphy to Television*. Durham, NC: Duke University Press.

Slattery, Diana Reed. 2015. *Xenolinguistics: Psychedelics, Language, and the Evolution of Consciousness*. Berkeley, CA: Evolver Editions.

Stelarc. 1996. "Ping Body," http://www.medienkunstnetz.de/works/ping-body/

Stewart, Kathleen. 2007. *Ordinary Affects*. Durham, NC: Duke University Press.

St John, Graham. 2015. *Mystery School in Hyperspace: A Cultural History of DMT*. Berkeley, CA: Evolver Editions.

Strange, Allen. 2006. *Electronic Music: Systems, Techniques, and Controls*. New York: McGraw-Hill.

Strange, Allen. 2013. *Programming and Meta-programming in the Electro Organism*, 2nd ed. Buchla Electronic Musical Instruments. Originally printed in 1974.

Stuart, W. J. 1956. *Forbidden Planet*. New York: Farrar, Straus and Cudahy.

Synth Docs. November 22, 2016. Gonzo Circuits: The Synths of Peter B / Ciat-Lonbarde. *YouTube video*, 13:45. https://www.youtube.com/watch?v=f9oZqTS0-HM

Taylor, F. 1937. "The Visions of Zosimos," *Ambix* 1: 88–92.

Teboul, Ezra. 2015. *Silicon Luthiers: Contemporary Practices in Electronic Music Hardware*. MA thesis. Hanover, NH: Dartmouth College.

Turkel, William J. 2013. *Spark from the Deep: How Shocking Experiments with Strongly Electric Fish Powered Scientific Discovery*. Baltimore: Johns Hopkins University Press.

Vale, V., and Andrea Juno, eds. 1983. *RE/Search: Industrial Culture Handbook, #6/7.*

Weinel, Jonathan. 2018. *Inner Sound: Altered States of Consciousness in Electronic Music and Audio-Visual Media.* New York: Oxford University Press.

Weiner, Norbert. 1961. *Cybernetics or Control and Communication in the Animal and the Machine,* 2nd ed. Cambridge, MA: MIT Press.

Weisstein, Eric W. "Möbius Strip." From MathWorld--A Wolfram Web Resource. n.d. https://mathworld.wolfram.com/MoebiusStrip.html

Wilson, Colin. 1971. *The Occult: A History.* London: Hodder & Stoughton.

Zeitlyn, David. 1995. "Divination as Dialogue: Negotiation of Meaning with Random Responses," in Esther Goody, ed., *Social Intelligence and Interaction: Expressions and Implications of the Social Bias in Human Intelligence,* 189–205. Cambridge: Cambridge University Press.

12

PATCHING POSSIBILITIES

Resisting Normative Logics in Modular Interfaces

Asha Tamirisa

Introduction

In the last few years, there has been a flurry of name and terminology changes in the music technology and "tech" communities that demonstrate a concern for how language reflects gendered and racialized biases in culture. "GearSlutz," the pro-audio message board changed its name to "GearSpace" following a petition signed by more than 5,000 people asking for a name that " … more appropriately represents the gear community" (Ran 2021). Soon after, "MuffWiggler," a popular message board for modular synthesizer enthusiasts, whose name is a double entendre that refers to the names of two Electro-Harmonix guitar pedals, in addition to making a vague sexual suggestion, changed its name to "ModWiggler." Around the same time, the Internet Engineering Task Force announced they were removing the technical terminology "master" and "slave"—terms used to denote the relationship between main and peripheral devices—explaining that "oppressive or exclusionary language is harmful" (Conger 2021). The music software company Ableton also removed this terminology from their software interface. Such changes reflect an acknowledgement of, and desire to shift, the privileging of a "… perspective of an archetypal Western, white, and male subject" in technical and audio-technical practices (Rodgers 2010, v).

This chapter moves from this point— that language, metaphors, and logics in technology articulate beliefs about subjectivity, bodies, and power. Analyzing language and logic in technical design is then critical to the larger project of understanding and shifting assumptions of a normative masculine subject in electronic music. In this chapter, I focus on the metaphor of "male" and "female" in patching interfaces as well as the organizational logic of

DOI: 10.4324/9781003219484-14

modularity, demonstrating that technical language and design is indicative of historical, social and cultural logics. With a generative spirit, this chapter offers how we might see existing technologies in dialogue with this analysis—in other words, how this critique may also shed an affirming light on historical and contemporary technical designs. I also offer an example of a self-designed interface that is motivated by the framework of this analysis.

With the revisions in technical language described above, a common response is that mere shifts in language don't necessarily produce material change. Surely, such gestures should not be understood as the cadence of systemic change. However, this critique separates language from the material when language and metaphor and its instantiation in technical objects generate senses of relation: to ourselves, to sound, to technology, and to each other. The analysis of metaphor is deeply revealing of gendered and racialized logics from a historical and cultural perspective, and I argue that shifts in language *can* produce new relationships and expressive possibilities—that interrogating and responding to the gendered, racialized, and normative nature of the technology is linked to the possibilities for systemic change (Wajcman 2004, 12).

Performing and playing with modular synthesizers is about imagining, exploring, and finding connections, and being curious about what can't yet be predicted. In this spirit, this chapter contributes to ongoing connections being made between feminist science and technology studies, media archeology, sound studies, and sonic practice. Beginning with a description of modular[1] patching, I continue with an analysis of how patching interfaces insinuate a gendered relationship between human and machine. I then discuss how the logic of modularity mediates constructions of difference. Following a discussion of the role of metaphors in interface design, I analyze modular patching interfaces that utilize novel metaphors and logics, analyzing the way these novel interfaces operate beyond the conventions of standard modular syntehsizer systems. Presupposing that modular patching interfaces are reflective of a particular historical and cultural moment, I ask: what kinds of patching interfaces might we imagine at *this* particular moment?

Modular Patching Interfaces

The interface design of modular synthesizers emerged out of analog computing and telephone switchboard technology, both of which came into prominence as electronic tools in the later part of the first half of the 20th century (Neal 2016, Meyer 2020). In modular patching interfaces, the circuitry of signal generating and signal processing modules lies behind a flat surface covered with jacks. Signals are routed within this system using "patch cables," cables that are plugged and unplugged to move and process audio and control signals in the machine. These systems contain many separate modules—hence the name modular synthesizers—which are regions that are separated based on functionality.

On a conventional modular synthesizer, there are signal generating modules (oscillators that produce different wave types at varying frequencies/pitches) signal processing modules (filters that shape the harmonics of a signal, or amplifiers that boost or attenuate a signal's amplitude), as well as logic modules (logic operators that can shape timing and probabilities). Conventionally, the signals of one module, for instance the output of the oscillator, would be connected to the input of a signal processing module, such as a filter, via a patch cable that connects to the jacks of the respective output and input. On the interface, there may also be toggle switches, pushbuttons, and potentiometers on the modular interface that allows the user to manually change parameters like the frequency of an oscillator, or the "amount" that a signal may be affecting a particular parameter. Via patch connections, as well as manual adjustments, sound generating and processing systems are set up and composed.

Mating Connections, Gendering Interaction

Patching interfaces on modular synthesizers rely on a binary biological sex metaphor. On modular synthesizers, the panels on the machine's cabinet house "female" input jacks. Inputs and outputs on each module are accessed via these ports (the "female" jacks) through a patch cable connection. The patch cable has "male" ends on either side, which are plugged into the modules to connect them to play the instrument. Following this metaphor, there is an implicit alignment of the feminine with the machine and the user as masculine.

Modular interfaces are far from alone in articulating a binary biological metaphors through design principles. In many forms of hardware components, including other media cables and peripherals, plumbing pipes, nuts and bolts, and even spaceship docking mechanisms, two components with complementary but different shapes that fit together are called "mating connections." In these "mating connections," "female" ends on components imply inward sockets, whereas "male" ends imply a protrusion. According to these conventions, "female" and "male" are technical terms that describe the type of "gender" of the component (Eveleth 2015, Wikipedia 2022).

In architecture, biological sex and "gender" metaphors denote not only topographical features of a structure, but also its quality. Architect Simon Unwin describes that "female" designs emphasize domes and concave shapes, and are seen as nurturing and accommodating forms. "Male" structures, such as columns and skyscrapers, emphasize power, exclusiveness, and penetration of the environment. These associations reveal two layers of correspondence: these topological forms to specific human sexual biologies, and these biologies to conventional gendered qualities related to power and connection (Unwin 2019, 37). In other words, "female" designs correspond to spherical shapes and are associated with the conventional "feminine" qualities of being receptive and compliant whereas "male" designs correspond to defined, protruding shapes and are associated with conventional "masculine" qualities of dominance and superiority.

With electronics cable components, there is a similar correlation of conventional gendered qualities to connector topology. The "female" connector, or the "jack" is generally stable and fixed on the technical object. It is designed to receive and hold the "male" connector (Wikipedia 2022). The "male" end or "plug" is usually located on a mobile cable end. Using an electrical outlet and electric power cord as an example, the outlet would be described as "female" and is fixed on the wall, and the power cord would have a "male" end that fits into the socket. While there are exceptions (in the case of extension or adapter cables and components, which extend the reach of a "male" connector via a cable with a "female" end on one side and a "male" end on the other), the general sense with electrical cables is that the jacks, or the "female" ports, are fixed and passive, while the plugs, or "male" ends, are mobile and active. Other mating connections, such as inter-flight spacecraft docking, follow this correlation where the spacecraft with an active "probe" docks onto the passive spacecraft with the socket-shaped "drogue." This design and language align with normative understandings of biological sex and the attendant correlation with the gender binary, where the masculine entity is active and agential, and the feminine entity is fixed and passive.

The conflation of the biological metaphor "male" or "female" to the "gender" of the component, the normative "gendered" qualities ascribed to such objects, and the implied sexuality and heterosexuality of the "mating" connection is revealing of a historical tendency to conflate biological sex and gender. Rose Eveleth (2015) rightfully laments in *A Modest Proposal for Re-Naming Connectors and Fasteners*, "I really see no reason why my electronics need to not only confuse sex and gender[...], but then use those confused terms to describe pieces of hardware."

The "difference" generated through this language and design underscores other dichotomies that conventionally align with masculine and feminine gendered qualities: human and machine, intellect and instrument, rational and complicated, self and other, etc. ... This oppositional framework "... ranks the two polarized terms so that one becomes the privileged term and the other its suppressed, subordinated negative counterpart" (Grosz 1994, 3). This arrangement obscures notions of misogynistic control in the encounter as we are culturally comfortable, and in fact comforted by, the idea of human control over machines.

While one might argue that this language and design is purely functional and benign, it encourages an exclusionary type of visual and social language and imagery, within audio cultures. This is visible, for example, in the way feminine bodies are often portrayed in and aligned with music technology in advertising and other circulated imagery.[2]

Correlations between technical design and misogyny reify problematic notions of gender, sex, and power, reflecting not only who developed these technological designs, languages, and logics, but also who the presumed user may be (Akrich 1997, 207–8). In a blog post about the Professional Audio Manufacturers Alliance (PAMA)'s suggestion to nix "male" and "female" from technical terminology, Peter Kirn says "… in the audio world, which *has* been hostile traditionally to women and trans students, producers, and professionals, you do have an actual problem. That's a material problem, not a philosophical or political one, even. It is an obstacle to teaching, inclusion, and growing the industry" (2021). The distilling of normative cultural and social logics in technical design is a reification of bias and difference, and, as Kirn states, there are stakes in their continued presentation.

It's not new knowledge that the use of electronic and digital technology has been a key feature and source of patriarchy and global order in Western modernity (Wajcman 1991, 6). This ranges from the global economics of electronics manufacturing to the minutiae of cultural interactions in the field.[3] As an example of the latter, Bob Moog is quoted saying: "And the fact that it's all this goddamn hardware, you know, that made it a guy thing too. It was halfway between being a musician and hot-rodding your car" (Pinch & Trocco 2002, 155). Despite a strong lineage of women and non-binary artists and engineers contributing to the development of sound, music, and music technology (including the development of Moog's synthesizers), normative masculinity is often recapitulated as the default (Rodgers 2010, 2).

In several discussions of gender in modular synthesis, I've heard the question raised about the relationship of modular synthesizers to telephone switchboards and analog computing—of which the labor force was thoroughly feminized in the early and mid-20th century (Chun 2011, 37 and Rodgers 2015, 20). Such comments suggest that this history may present an intervention in the sense of default masculinity in computing and music technology. While absolutely true that this history demonstrates that the capacity for technical skill has no gender, a fundamental difference exists in how men and women were organized and perceived in technical fields, which aligns with conventional logics of the gender binary: men as "producers" and women as "reproducers." As an example, in early computer programming, it was understood that "planners," who were mostly men, decided on the operations that the women "coders" carried out—a separation of intellect and physical work that reified hierarchies of gendered labor (Chun 2011, 35). Rather than being seen as expert users of the machines, women were conceived of as *part* of the machine, receiving and executing instructions.[4] Eventually the work performed by women in these roles was absorbed by the machinic processes, leaving men in their role as expert "programmers." Computers—both the machine and the women workers—may have been feminized, but that needs to be distinguished from what might be feminist (Chun 2011, 33). As Judy Wajcman describes, the absorption of women's technical labor into a binary logic is an example of "… the way hierarchies of sexual difference profoundly affect the design, development, diffusion, and use of technologies" (Wajcman 2004, vii). Thus, binary gendered and hierarchical organizational logics can continue to recapitulate and haunt interventions in technical fields, even as they may appear otherwise.[5] As such, the manifestation of cultural logics in technical design and praxis can be understood as mutually constitutive.

Modularity & Cybernetics in Synthesis Systems

Modularity has become a default way of thinking about the generation of electronic sound—as a tool that allows for the building of sound from "scratch," composing sounds through the combination and multiplication of various oscillators, shaping timbre through filtering, and creating shapes in time through sequencing, enveloping, and amplitude modulating. In modular systems, these functions are separated into the components that comprise the machine. As such, constructions of difference are fundamental to modular design and the interactive possibilities within the system. Modular logic, and its relationship to cybernetics, represents a very particular way of imagining difference, organizing the components of a system, and mediating interaction between them.

The evolution of the term "module" offers some scaffolding for this organizational logic. In the 18th and 19th centuries, it was used to describe a standard unit of measurement that could be multiplied or repeated depending on the desired size or ratio. In other words, like a "model," the module was a small-scale unit that could be duplicated to create a larger form (Blair 1988, 2). In the early 1940s, Le Corbusier expanded on this definition when he envisioned a system of measurement based on human bodily proportions. A module in this understanding was a component that related to others proportionally. Le Corbusier's use of the term indicates a shift from thinking of modules as duplicates of the same component to thinking of modules as distinctly different from one another (Blair 1988, 2).

In the United States, the definition of "module" changed in the post-war period (roughly 1945–68) to describe a unit with a specific function and feature, other than size, that differentiated it from other units in a configurable system. It was no longer a standard replicable component, or one that had likeness to the others, but one that was self-contained and unique. Its inherent and essential difference gave the module its identity within the system and legitimized its separateness from other modules. Just as important as the module's difference was the expectation that it would be compatible and configurable with all other elements of the system. With significance given to re-configurability, modular design shifted emphasis away from a known "whole" and instead prioritized the individuality of each component and the possibilities of reconfigurability (Blair 1988, 3). John Blair argues in *Modular America* that, while this organizational logic is immaterial, it can nevertheless be identified as a distinctive feature of American mid-century design and culture (4). This compartmentalization, he argues, demonstrates a particular way of governing the relations between parts and wholes (5).[6]

Modular logic is entangled with cybernetics, a framework developed by Norbert Wiener in 1948, which used metaphors from animal physiology to conceptualize interactions in complex systems (Dunbar-Hester 2010, 114). Wiener held out that electronic circuits in computing machines were "precise analogues" to "neuronic circuits and systems"—thus implying that there was something "essentially" similar about biological systems and electronic systems (Weiner 1948, 7, 22). The bodily metaphors in cybernetics were alluring: the unknown aspects of technology and electronic communication felt knowable through the tangibility of human behavior and the human body. Moreover, if humans and machines could be understood on similar terms, then one could more easily imagine them working in concert.

Cybernetic systems rely on notions of difference. Components must maintain their separate identities in order to have meaningful exchanges (Manovich 2012, 36). In other words, for communication between two entities to have significance, the entities need to be perceived as separate and essentially different. Cybernetics highlights the role of "feedback" in systems interaction and regulation, where information from one area of a system is sent and received by another, and has a causal effect on the second entity (1948, 114–5). Theories and applications of cybernetics conceptualized this response in different ways, either emphasizing the capacity for systems to produce "homeostasis" or stability, *or* complex "autopoietic" interactions.

N. Katherine Hayles usefully organizes this variety within cybernetic discourses into "waves," of which the first two are applicable to this discussion. The first focuses on self-regulating, equilibrium-seeking systems ("homeostasis") while the second oriented around complexity arising out of the interaction of components within a system ("autopoiesis") (1999, 131). In "Listening to Cybernetics: Music, Machines and Nervous Systems 1950–1980," author Christina Dunbar-Hester (2010) describes how experimental composers and sonic artists in this time period navigated these two "waves." While generally interested in "change over constancy, evolution over equilibrium, complexity over predictability," (2010, 116) it was generally the juxtaposition of control and indeterminacy that sound artists were exploring (126–8).

In a modular synthesizer, the creation of these relations and systemic interaction between inputs and outputs allows for "play" within the system. Playing the instrument is about exploring the flexibility within each module and the connections that can be made between them. However, as

Dunbar-Hester explains, modular systems—while suggesting a "second-wave" cybernetic approach—are "haunted" by first-wave ideals (Dunbar-Hester 2010, 116). While allowing for numerous connections between modules to be made, modular systems mediate and reduce these relations to specific transactions. The localism of interactions inhibits a sense of awareness for the functioning of the entire system. In the process of being integrated into a single system, modules are individuated based on notions of difference. This quality of differentiation suggests an essentialism of each component in a modular system. The components of modular systems are configurable, but unaware of their global context. Moreover, the modulation of an entity via its inputs does not change the module; the module is stable, and always returns to its original state (McPherson 2011, 27).

Interfaces & Metaphors

Interfaces, while instantiated in technology, are a figuration of relation. Interfaces are the logic of how the internal ("inter") is expressed and understood through the external appearance of an object ("face"). Interfaces attempt to present the unrepresentable, and make visible and palpable what can't be seen (Hookway 2014, 1). The interface is a "… zone of interaction … " that both generates connection and codifies difference (Galloway 2012, vii). The interface produces a language and means of relating, and is the site of exchange (Galloway 2012, 31). Interfaces often draw on metaphors to construct this means of relation. The metaphors and logics chosen are not arbitrary. They are chosen and based on our experiences and understandings of the physical world.

> "Interfaces themselves are effects, in that they bring about transformations in material states. But at the same time interfaces are themselves the effects of other things, and thus tell the story of the larger forces that engender them." (Galloway 2012, vii)

In the attempt to represent that which does not present itself as readily engageable, interfaces draw on other knowledges that are more familiar. For instance, while one could imagine a seemingly infinite number of ways to engage with frequency or pitch of sound, many synthesizers use a keyboard interface, given that the keyboard is a prevalent, domestic instrument and offers a quick means of relating to the production of sound. In this instance, the piano keyboard is a metaphor—the keys do not connect with a lever and hammer that strikes a physical string. Instead, the gesture of the depressed key is connected with circuitry that produces the corresponding pitch.

Thus, through the piano keyboard interface, the user is "… understanding and experiencing one kind of thing in terms of another"—the crux of a metaphor (Johnson and Lakoff 1980, 5). Metaphors encourage a perceptual transfer, where correspondences are made and mapped between entities. Metaphors do not convey "truth"—it's understood, for instance, that the piano keyboard on the synthesizer is not connected to mechanical elements, or that hardware components aren't actually biological (Unwin 2019, 2). Metaphors require a suspension of fact. Still, while not "truth," metaphors convey a "sense," a way of relating to an object that transforms the perception of the object (Unwin 2019, 3). The ease and plausibility through which a metaphor "works" is reflective of the social and cultural conditioning and consensus around the metaphor's operator (Unwin 2019, 21).

"For while the interface might seem to be a form of technology, it is more properly a form of relating to technology, and so constitutes a relation that is already given …" (Hookway 2014, 1)

As Wendy Hui Kyong Chun (2011) illustrates in *Programmed Visions*, metaphors are all over electronics and computation. They are, as Chun states, part of what makes machines feel "user friendly," turning what is otherwise cold and abstract into something knowable. Computers and technology "depend on and perpetuate metaphors" (55). Metaphors convey a spirit of interaction. For instance, the computer "desktop" and file-cabinet metaphors used in "file" (even the word "file" is a metaphor for an object with data) organization offer the sense of the computer as a place for work and productivity. Metaphors allow for more than understanding, they are a "… transfer that transforms," a logic through which an object or process is animated (56).

Interfaces allow for the creation of physical memory with the material that they represent. Thus, interfaces with their attendant metaphors produce an epistemology, a particular way of knowing what lies beneath the surface. It is also a way of coming to know the self: the subject is defined in the articulation of the boundary between human and machine (Hookway 2014, 1). As Madeleine Akrich describes, "Once technical objects are stabilized, they become instruments of knowledge," (221) figuring and "measuring a set of relations between heterogeneous elements" (205).

In the section that follows, I offer a set of examples that demonstrate how technical language and design manifest and materialize these relations, and how they can generate new kinds of performative, expressive, and relational experiences of self, machine, and sound.

Patching Possibility

In Zach Blas's speculative design organization, *QT* ("Queer Technologies"), Blas "critiques the heteronormative, capitalist, militarized underpinnings of technological architectures, design, and functionality" through an array of tools, applications, and interventions Included in this array is *ENgendering Gender Changers*, a package of cable adapters presented as a consumer product that Blas describes as "a 'solution' to Gender Adapters' male / female binary ..." (Blas 2022). Blas's use of "solution" can be interpreted as a recognition of the problem of gender binaries as a logic of design, whether technical or social/cultural. In an accompanying "manual" for how to use these technologies, Blas describes the *ENgendering Gender Changers*, as "linking forbidden spaces" and describing them as hardware that produces "pure possibility" (Blas 2008, 87). Blas's speculative design offers a spirit through which modular patching interfaces can be analyzed as critical objects in dialogue with technical design history, and as presenting new modalities for sonic understanding and expression.

Androgynous Nodes

Peter Blasser is a synthesizer builder who has built instruments under the name Ciat-Lonbarde for over ten years. In his instrument design, Blasser pays close attention to language and metaphor, recognizing that new forms of relation are created in the process: "In naming a concept, much is done towards taming it." (2015, 17)

His sonic epistemologies verge on myth, as he develops imaginary paradigms and invents language to describe his instruments and their behaviors. Some of his design concepts include: *Nabra*, an analog brain, or an assemblage of modules connected together; *Toucharce*, a musical instrument responding to touch; *Shinth*, a synthesizer meant to be played by touch for circuit-benders who find their materials in the trash; *Rolz*, a circuit formed by simple sub-components in a closed loop.

One such novel design concept that responds directly to the "male" and "female" logic of inputs and outputs on conventional modular synthesizers is Blasser's "*Sandrode*," a circuit node formed by an intersection of inputs and outputs. He describes Sandrodes as being "androgynous," "... both input and output or neither" (Blasser 2015, 32–3). He cites Burrough's mouth-anus as well as radio transmitters, which can switch from transmitter to receiver with a difference in relative amplitude, as sources of inspiration (70). *Sandrodes* sometimes appear on his instruments as brass rods and sometimes as capacitive touch surfaces. While Blasser's description of the Sandrode as "androgynous" keeps the metaphor in a biological register, the commentary sheds light on the directional, sexualized and gendered implications of conventional patching interfaces that he is responding to in his design. The idea of accessing and interfacing with a circuit that both responds to and supplies an electrical signal is a great example of how the language and logics of modulars can expand and produce new ways of thinking about signal flow, patching, and patching interfaces.

littleBits

Ayah Bdeir's "littleBits Synthesizer Kit," produced in collaboration with Korg between 2013 and approximately 2018, allows for modular patching by moving Lego-sized modules and connecting them via magnetic symmetrical connection points. The size of the kit, and the way the modules can easily be connected and disconnected, allows for the whole system to feel completely reconfigurable, as opposed to the monolithic feeling that many electronic instruments present. As Lyn Goeringer astutely described in a discussion on the new media listserve "empyre," "… the system is no longer reliant on patch cables, but looks towards a system that can be reshaped and reconfigured by moving the basic components … The motion and adaptability of the system is limited not by a physical object that was created by another person … where I must understand their logic and design …" (2014).

Magnetic connections such as the ones on the littleBits kits would not suit sturdier and larger electronic instruments. The advantage produced, however, is a modular system that is highly portable, playable for younger children, and more accessible in terms of price-point. The littleBits synthesizer kit exhibits a different design logic than its modular predecessors and garners a completely different audience owed to its pricing, pricing, non-age or gender specific aesthetics with which it is made available. It also sheds the sexualized language of standard patch interfaces through its novel design. Thus, littleBits is another example of how modular logic can be adapted to produce new ways of interfacing that change the social landscape of who sees themselves as sonic producers and synthesists.

Leaky Matrices

My own motivation for thinking through the implications of language and design in patching interfaces arose out of working with the ARP2500 synthesizer. Produced in 1970, the ARP2500 utilizes a switch matrix system that is decidedly different from the conventional jack and plug design of analog synthesizers. Patching with this system involves connecting inputs and outputs on each module by sliding a vertical switch to assign them to one of the 20 numbered buses that lie on the horizontal channels of the switching system. Then, the signal travels along that horizontal channel and can be distributed to any number of destinations on that channel, and picked up anywhere on that channel on the machine. The switch matrix interface offers an alternative to standard jacks and plugs, sidestepping the conventional "male/female" language that they insinuate, and additionally offering a greater ease of creating connecting points within the modular system.

As I used the ARP2500, I began to think about how inputs and outputs lose their identity once assigned to a particular bus—inputs and outputs might become a sort of conglomerate of voltage fluctuations, or a sort of voltage "blob." Additionally, the interface can encourage less linear and directional signal processing logic than jack/plug patching interfaces: multiple outputs can be assigned to the same bus, and an output can be picked up by any number of inputs. While conventional modulars might also allow for this type of distribution and connection through banana cables (which can be stacked to send one signal to multiple places) or through signal multipliers ("mults") which duplicate a signal so it can be sent to a number of destinations, it is more the rule rather than the exception with the ARP2500's interface. One of the production critiques of the ARP2500 was "crosstalk"—that signals might "bleed" to other buses due to inadequate isolation between channels. While undesirable from an objective and technical perspective, it does create a sense of "leakiness" in the interface, pointing towards the "possibility" mentioned by Zach Blas, where signals move into "forbidden" zones outside of where they have been instructed to move. The contrast between the switch matrix system and patch cable-patching encouraged me to think about how an interface might create a different set of possibilities for interfacing and for signal flow through its design.

Intermodulations & Sympathetic Connections

> Now, in the 20th century, when science is so prominent, can it link more closely with imagination? Can science unbend sufficiently to present scientific facts in such a way that they excite artists? Can artistic creations equally excite the scientist? Do both the scientist and the artist need a new range of metaphor, verging on mythology: a new set of analogies …? (Daphne Oram 1972, 6)

In 2015, I designed an interface in response to some of the ideas that arose out of my critique, thinking, and playing with modular interfaces. The result, the "Matrix-Harp," is a hardware interface that utilizes capacitive sensors to make patch connections, and a grid of interwoven stretch sensors to modulate parameters. The interface is a digital interface—it connects to a computer which performs the sensor mapping and audio synthesis. This interface explores plug-less patching, inspired by the ARP2500 and motivated by the desire to move beyond the sexual implications of a standard plug and jack system (Figure 12.1).

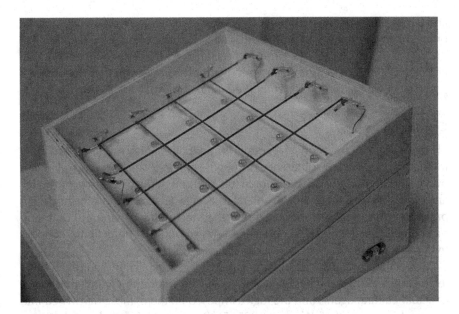

FIGURE 12.1 Photo courtesy of the author.

In addition to the standard plug and jack system indicating a binary gendered logic, it also does not lend itself well to performance gestures. As I imagined the Matrix-Harp, I wondered about a softer gesture and material for patching. As an electronic musician, I very much enjoy thinking about building systems out of signals and logic, but I find the action of patching to be very jarring. The discrete and sudden gesture can take me out of the flow of exploring the fluidity of signal processing. The action of plugging and unplugging cables can itself generate a clacking sound, which needn't be considered "noise," but can feel intrusive in a performance practice.

As a response, I used capacitive nodes on a matrix grid to designate connections between inputs and outputs. Since they are capacitive, I imagined that connections could only be made if I am physically connecting them—a design choice that asks that I be physically a part of the patching process at all times.[7] As such, my patching decisions are informed by the reach of my fingers and my previous patching choices, as much as any sound I wish to make.

The system also creates a sense of "interconnectedness" between "modules" or functions. In standard modular systems, each module is distinct in its function, and connections are localized. For example, connections between modules A and B are otherwise unknown by module C, despite the fact that they exist within the same system. Because the capacitive sensors are located below the stretch sensors in the Matrix Harp, and because stretch sensors are arranged in a woven interlocking grid, patch connections and parameter changes are "felt" throughout the sonic system. Stretching sensors to connect A and B may also alter the parameter C. Thus, all modules are somewhat "aware" or impacted by changes in the system. Given the lack of numeric feedback on the system (i.e., there are no notched knobs), the interface encourages engagement through listening rather than a more cartesian measuring of parameters, again, asking for embodied interfacing with the system.

Conclusion

This chapter focuses on the design implications of modular patching interfaces in asserting difference through gendered and sexualized metaphors, and in modular organizational logic. Through the discussion of "male" and "female" as a metaphor in interface design and modularity as a design principle, this analysis demonstrates that technologies are cultural and historical artifacts that relay understandings of bodies, subjectivity, connectivity, and power.

Given the array of possibilities of engagement with sound through technology "… mismatches inevitably occur between the engineering logics of actual cultural agents and the logics imposed on them, those hardwired into the machine from afar" (Greene 2005, 5). By encouraging dialogue between cultural critique and design logics, this chapter demonstrates the expressive and performative potential of materializing alternative metaphors and logics in technology.

Notes

1 I am indebted to "The Poetics of Signal Processing" by Jonathan Sterne and Tara Rodgers (2011), which animated the notion of signal processing as a historical and cultural concept, shapeshifting what I had understood to be a solely technical and creative process. I am also indebted to "U.S. Operating Systems at Mid-Century: The Intertwining of Race and UNIX," by Tara McPherson's (2011), which corroborated the notion that technical design is cultural and ideological.
2 As one example, this Korg MS-10 advertisement, that reads "Jack your patchcable into my body"—it is unclear whether this was an original advertisement or a caption later added, but the sentiment is nevertheless explicit https://www.matrixsynth.com/2012/05/korg-ms-10-jack-your-patchable-into.html
3 For more information, see Lisa Nakamura's "Indigenous Circuits: Navajo Women and Early Electronics [3] Manufacturing" (2014), Lucie Vagnerova's "Nimble Fingers in Electronic Music: Rethinking Sound Through Neo Colonial Labor" (2017);
4 This is not to say that the women "coders" did not also make critical decisions in computation, but that their decisions were uncredited given these labor structures.
5 For further reference: Bliss, Abi. 2013. "Invisible Women." The Wire Magazine.; Rodgers, Tara. 2015. "Cultivating Activist Lives in Sound." Leonardo Music Journal. Vol 25; Tamirisa, Asha. 2021. "Sonic Activism in the Integrated Circuit." *Feminist Review*.
6 For further discussion on modularity, society, and computation, see Tara McPherson's (2011), in "U.S. Operating Systems at Mid- Century: The Intertwining of Race and UNIX," in which McPherson describes this type of compartmentalization as exemplary of the "separate but equal" logic that took hold in the US at mid-century

(25). On the heels of the Civil Rights Movement, oppressive racial logics were disguised through systems of difference that perpetuated (and in some ways, amplified) segregation and racism. McPherson says that these social structurings and developments in modular computing are very much interdependent, commenting that the resistance of both parties to consider the relevance of the other as part of modular logic itself. McPherson is not arguing that there is a direct causal relationship between both, but that they both represent a distinct move toward modular logic.

7 This design is akin to the EMS Synthi.

References

Akrich, Madeleine. 1997. "The De-Scription of Technical Objects." *Shaping Technology, Building Society: Studies in Sociotechnical Change*. Cambridge: MIT Press.

Blair, John G. 1988. *Modular America: Cross-Cultural Perspectives on the Emergence of an American Way*. Westport: Greenwood Press.

Blas, Zach. 2022. "Queer Technologies 2008–12". https://zachblas.info/works/queertechnologies/

Blas, Zach. 2008. *Gay Bombs: User Manual*. Queer Technologies Inc. Self-published.

Blasser, Peter. 2015. *Stores at the Synth Mall*. MA Thesis, Wesleyan University. Middletown: Wesleyan University.

Conger, Kate. 2021. "Master,' 'Slave' and the Fight Over Offensive Terms in Computing." https://www.nytimes.com/2021/04/13/technology/racist-computer-engineering-terms-ietf.html

Chun, Wendy Hui Kyong. 2011. *Programmed Visions: Software and Memory*. Cambridge: MIT Press.

Dunbar-Hester, Christina. 2010. "Listening to Cybernetics: Music, Machines, and Nervous Systems, 1950–1980" *Science, Technology, & Human Values*, Vol. 35, No. 1, pp. 113–139.

Eveleth, Rose. 2015. "A Modest Proposal for Re-Naming Connectors and Fasteners" The Last Word on Nothing. https://www.lastwordonnothing.com/2015/11/27/a-modest-proposal-for-renaming-connectors-and-fasteners/

Galloway, Alexander R. 2012. *The Interface Effect*. Cambridge: Polity Press. 14

Greene, Paul D. 2005. "Introduction: Wired Sound and Sonic Cultures." *Wired for Sound: Engineering and Technologies in Sonic Cultures*, Middletown: Wesleyan University Press.

Grosz, Elizabeth. 1994. *Volatile Bodies: Toward a Corporeal Feminism*. Bloomington: Indiana University Press.

Hayles, N Katherine. 1999. *How We Became Posthuman*. Chicago: University of Chicago Press.

Hookway, Branden. 2014. *Interface*. Cambridge: MIT Press.

Kirn, Peter. 2021. "So Yeah, Let's Just Use Plug and Socket — Industry Group Recommends Obvious Change in Terminology." Create Digital Music. https://cdm.link/2021/07/so-yeah-letsjust-use-plug-and-socket-industry-group-recommends-obvious-change-in-terminology/#:~:text=So%20yeah%2C%20let's%20just%20use%20plug%20and%20socket%20%E2%80%9 3%20industry%20group,recommends%20obvious%20change%20in%20terminology&text=The%20Professional%20Audio%20Manufacturers%20Alliance,it%20with%20something%20not%20horrible

Lakoff, George and Mark Johnson. 1980. *Metaphors We Live By*. Chicago: University of Chicago Press.

Manovich, Lev. 2012. *The Language of New Media*. Cambridge: MIT Press.

Meyer, Chris. 2020. "Analog Computers and Modular Synthesis." Alan Pearlman Foundation. https://alanrpearlmanfoundation.org/analog-computers-and-modular-synthesizers/

McPherson, Tara. 2011. "U.S. Operating Systems at Mid-Century: The Intertwining of Race and UNIX." *Race after the Internet*. Ed. Nakamura, Lisa, and Peter Chow-White. New York: Routledge, 21–37.

Neal, Meghan. 2016. "This Is What Synths Made of Repurposed Telephone Switchboards Sound Like." Vice Magazine. https://www.vice.com/en/article/yp33em/listen-to-the-ambient-music-ofrepurposed-telephone-switchboards

Oram, Daphne. 2016 (reprint from 1972). *An Individual Note of Music, Sound, and Electronics*. London: The Daphne Oram Trust & Anomie Academic.

Pinch, T. J., and Frank Trocco. 2002. *Analog Days: The Invention and Impact of the Moog Synthesizer*. Cambridge: Harvard University Press.

Ran, Cam. 2021. "Gearslutz, Please Change Your Name." change.org. https://www.change.org/p/gearslutz-gearslutz-please-change-your-name

Rodgers, Tara. 2010. *Pink Noises: Women on Electronic Music and Sound*. Durham: Duke University Press.

Rodgers, Tara. 2010. "Synthesizing Sound: Metaphor in Audio-Technical Discourse and Synthesizer History." Montreal: McGill University.

Rodgers, Tara. 2015. "Tinkering with Cultural Memory." *Feminist Media Histories*. Vol. 1, No. 4, pp. 5–30. 10.1525/fmh.2015.1.4.5

Sterne, Jonathan, and Tara Rodgers. 2011. "The Poetics of Signal Processing." *Differences: A Journal of Feminist Cultural Studies*. Vol. 22, No. 2-3.

Unwin, Simon. 2019. *Metaphor: An Exploration of the Metaphorical Dimensions and Potential of Architecture*. New York: Routledge.

Wajcman, Judy. 1991. *Feminism Confronts Technology*. University Park: Pennsylvania State University Press.

Wajcman, Judy. 2004. *Technofeminism*. Malden: Polity Press.

Weiner, Norbert. 1948. *Cybernetics: or, Control and Communication in the Animal and the Machine*. 2nd Ed. Cambridge: MIT Press.

Wikipedia. 2022. "Gender of connectors and fasteners." Wikipedia Foundation. Last modified 29 June 2022. https://en.wikipedia.org/wiki/Gender_of_connectors_and_fasteners

13

DRAFT/PATCH/WEAVE

Interfacing the Modular Synthesizer with the Floor Loom

Jacob Weinberg and Anna Bockrath

Draft/patch/weave is an ongoing exploration into the expressive possibilities that arise from making meaningful connections between the contemporary modular synthesizer and floor loom. This project began with an intention to create textile and sound-based works that explored the intersection of music and weaving, and we narrowed our focus towards developing a process that emphasized the common mechanics found within these machines.

We settled on weaving drafts as a starting place in developing our technical and creative processes. Though weaving drafts are commonly used for various kinds of hand weaving, and their use in synthesis and electronic music is rare, when taken out of their functional context, they become a set of instructions that can be used and interpreted towards different ends. Mechanizations of the loom and synthesizer are similar in that they can interpret a binary system—1s and 0s/ rising and falling/ over and under—into analog texture and movement. The weaving draft is composed of binary data, and so can be applied to both handweaving and modular synthesis.

This project also participates in a tradition of alternative musical interfaces found in modular synthesis and explores ways in which both of these machines make digital information concrete.

Weaving Basics

A weaving draft provides instructions for preparing *warp* threads, vertically running threads, on a loom for weaving (also known as *dressing* the loom), and a sequence for weaving *weft* threads, horizontally running threads. The configuration of chosen warp threads influences the textile's pattern as a weaver moves weft through the warp. On a floor loom, a weaver raises warp threads by

DOI: 10.4324/9781003219484-15

pushing down on a pedal or *treadle* with their feet, which raises all warp threads on the loom's corresponding *shaft,* a frame on the loom that holds the warp. The warp is held in place by small brackets called *heddles.*

A common weaving draft shows the configuration of the warp threads on a grid across the top of the draft (the *threading*), and the treadling sequence on the right. The small grid on the top right corner—the *tie-up* configuration—shows how each treadle is to be connected to a particular shaft; treadles can access more than one shaft at once. Usually, a marked cell in the threading indicates that the warp is to be put through a heddle on the shaft at that position, and a marked cell in the treadling matrix indicates that that pedal should be depressed at that stage of the weaving sequence.

Using the information provided in the threading, treadling, and tie-up matrices, one can create a weaving pattern by filling in which threads are marked in the threading at each step of the treadling, line by line. Once the entire treadling sequence is iterated through, a single instance of the pattern should be displayed in the large, central area of the draft called the *drawdown.* A weaver then takes the pattern and multiplies it vertically and horizontally in a textile. For weavers today, the entire process of rendering a pattern using the method above can be automated through computer software, which can also emulate how a pattern might look by assigning colors to the warp and weft.

In some ways, the weaving draft acts as an abstraction of the mechanics of the floor loom, since each main part of the loom is represented in the diagram. However, exact visual congruity between the pattern described on the draft and the textile itself is not guaranteed, as material factors such as thickness, texture, and color can influence how legible the pattern is in the finished textile.

Drafting Patterns in Max/MSP

We wanted the movement of the modular synthesizer's control voltages to reflect the floor loom's ability to build intricate patterns and textures through routing. To do this, we decided to reconstruct a weaving pattern generator in Max/MSP (Max), as that environment is well-suited to processing sequences of information, digital signal processing, and generating MIDI and control voltages. We identified the weaving draft as the focal point for this interface because it provides the core information from which a textile can be woven, making it a meaningful basis for determining control data. Creating a weaving program allows us to preview and design patterns thread-by-thread, step-by-step, and affords us the greatest number of possibilities in routing control information to the synthesizer. We input the threading pattern and the order of treadling in the interface, and then a sequencer generates the pattern as it advances through each row of the treadling sequence. The sequencer can be triggered by an internal clock, or via an external trigger. By emulating and automating the mechanics of weaving drafts, the interface built in Max also emulates the mechanics of the floor loom itself.

In the drawdown window, the pattern is displayed as a series of bright or dim, or, on or off pixels. Brightness here corresponds to the filled-in elements of a draft done by hand, which in turn, correspond to those parts of the weaving where the weft crosses under the warp (on) or over (off). A high state of 1 in the program corresponds to the warp being physically raised on the loom, and a low state of 0 not being raised at all.

The weaving program can be "tapped" for data at any point in the system, giving maximal choice over synthesizer control. We typically have generated the following types of signals from weaving patterns:

- Treadle number being used, which can be used to distribute gate or trigger signals
- Which shafts are being accessed, which can serve as a kind of switch or multiplexer
- Lists of values generated as multiple cv outputs (up to 16) by each shaft, or combination of shafts, which can be interpolated between their high and low states, creating varying modulation envelopes
- Values of individual points in the generated weaving pattern, accessed by coordinates, which can be read in a number of different directions across the drawdown.

As the data is generated from the weaving draft, it is sent out to the modular system. All control voltage is outputted using 14-bit MIDI-to-CV converters (allowing 16383 voltage increments). On a hardware modular synthesizer, we have access to 16 control voltage outputs in total, where a weaving pattern may use as many as 72 heddles at one time. Because of this, we either instruct the program to compress the values of each row of the draft into a 'resolution' that can scale to the 16-output limitation, or we can have the program select a range of the pattern that may be read at a time, for instance, setting the program to output warp threads 1–16, or 2–17, etc.

We opted for using MIDI-to-CV over DC-coupled audio outputs because it was simpler to scale numbers generated in Max to voltage levels, because we had a greater number of outputs at our disposal than we would have using audio.

Hardware and Software

We should note here that, although we focused primarily on hardware modular synthesis as the instrument to be controlled by the weaving program in Max, we have also used software versions of modular synthesis as well.

While there are software programs that directly emulate modules and manufacturers of hardware synthesizers, the Max/MSP platform was suited well enough to the task, being a modular coding environment itself, and allowing more precise customization over sound control.

There are advantages to software, primarily because we can have an unconstrained number of inputs and outputs. This becomes particularly useful when employing synthesis or processing techniques that require an array of inputs and outputs all performing similar functions, such as fixed filter banks (including Buchla-style spectral processing), and additive oscillators. Aside from the often prohibitive cost of their hardware implementations, their software versions let us scale the number of controls, inputs, and outputs to the dimensions of a given weaving pattern.

We felt that implementing some combination of synthesis that includes both hardware and software was most successful, as software implementation affords more flexibility and scalability, but hardware affords tactility, concreteness, and helps define the technical implementation of our approach as more symmetrical, because it provides a physical counterpart to the floor loom.

Even with the numerous advantages to software, we found that having both a physical floor loom and a hardware synthesizer together when showing the project has advantages in engaging and educating viewers. Folks who are unfamiliar with either weaving or synthesis tend to have an easier time learning if they can physically interact and see the process unfolding. The hardware modular synth makes the modular aspect more legible, as folks can easily differentiate each module and learn what functions it performs.

Performing Weaving Drafts

This system built with Max runs in real time, immediately outputting control data as it is generated. In this way, we have used the program to not only create synthesizer patches, but also to perform our drafts in tandem with the floor loom as it weaves. To do this, we trigger the sequencer externally from motions on the loom. We have preferred using triggers generated by the audio events created by the loom operating, which are processed through an envelope follower and window comparator. The loom has a distinct sonic character, with long, metallic rattling from the heddles, and sharp knocks by the beater as it is returned to its resting position after it compresses each line of weft, signaling the end of one step in a sequence. It seemed more appropriate to center this aspect of floor loom operation as an integral part of the project when playing the weaving in real time. We typically have 1–2 microphones recording the movement of the shafts and have a contact mic positioned to pick up the beater as it is put back into its resting position.

We opted out of using sensors to track the activity of the loom because we felt that the loom was its own instrument, and that by using sensors as external inputs to the sequencer clock, we would be abstracting the loom into an alternative musical interface or controller in service of to the rest of the system, where instead, we center the weaving draft itself in this role as a hub or common ground between our two independent machines. When running in real time, the entire system evokes a feeling of the patch or program as a kind of performance.

Creative Process

Our creative process was directly informed by the tools and techniques we created, meaning that much of the work was centered on the practices of synthesis and weaving, and their historical and methodological overlaps. The project has been worked on as a series of iterations, with each iteration being comprised of a weaving pattern and a completed textile, modular patch, and audio recording. Within each iteration, we followed the steps of weaving draft selection and programming first, then textural choices, and then the creation of the textile and recording of the sound piece.

Our decision making around weaving draft selection has evolved with our process. Patterns were chosen as much for their visual possibilities as they were for the potential rhythm modulations that could be generated using our framework. Most of our drafts were variations, or sometimes direct uses of traditional weaving patterns found in Western textile arts that have existed for hundreds of years if not more. We felt it more significant to program older patterns into Max, rather than create new ones from within the program itself. The patterns themselves are more durable, are more legible to the listener/ viewer experiencing the work, and usually carry a rich set of meanings, contexts, and associations.

We then program the patch in Max, the synthesizer, and the loom. Just as we must set the treadling and threading pattern correctly on the computer, the loom must also be dressed or 'programmed' to the draft's specifications, only in a physical format. We then choose what kinds of textures and materials the draft will embody the draft. This is a time when the immediacy of Max has been most useful because we have been able to preview the pattern both in terms of the modulations it can provide, and what it might look like as part of a full textile by tiling the pattern in the drawdown window.

Texture applies both to sound and tactility here: the pattern comes to life only when made concrete through our aesthetic choices of sound and thread. It is the aspect of our process that some may see as most 'creative' because it is the point of our project where we are most left to our own aesthetic preferences for how the pattern may look or sound. We have sought to arrive at a place where the sound is reflective of the material choices and vice versa, and both sound and textile are doing justice to the design of the weaving pattern itself. In the end, we have a textile, a sound piece, and the draft that was used to create them.

For live performances of this process, we wanted to emphasize the weaving draft as a kind of composition or choreography that can be played. In addition to syncing the program in Max/MSP with the pace of the loom itself, we took the sonic elements of weaving and presented them alongside the sound generated by the synthesizer. In a recent show, we used exciters to play back the audio from the synthesizer by vibrating against the walls next to the loom so that the piece was heard in the room alongside the sound of weaving. The weaving draft becomes a score or choreography.

Reflections

We have viewed this collaborative process primarily as a means of exploration towards connecting two disparate machines and disciplines that are more alike than they may seem. With that in mind, the project was not conceived as having an overarching theme, message, or conceptual aim. If anything, it was started with an ethic that is rooted in process, questioning, exploration, education, and clarity and demystification around each of these machines. Even so, we continued to encounter certain ideas, thoughts, and wider concepts that inform our work, which we will discuss below.

We felt that the overlap between modular synthesizers and textiles and their shared history in computation was a motivating factor, but not an end. If anything, acknowledging their relationship to computation was the very first step. Despite our mentions in our technical discussions, we have said relatively little here and spoke little through our collaboration about the specific ways that looms and synthesizers relate to computers in terms of their specific technologies, or their involvement in the history of technology. That discussion can narrow the project into a focus on computation in the 20th and 21st centuries, where we seek to understand computation and digitality as a much older, broader, and more fundamental human practice.

If anything, our project conveys something more about how the floor loom and the modular synthesizer relate in how they juggle their analog and digital aspects simultaneously than how they fit neatly into a lineage or technological history. This is best illustrated by returning to the core functionality that Max had in this process, which was generating arrays and matrices of binary data and acting as a hub between the two machines by dictating the high and the low—in other words, it communicated through a shared reliance on quantity that is reflected analogically in the loom and synthesizer's machinery. The loom shafts are raised and lowered, and the voltages of the synthesizer rise and fall. While they operate through different physical or qualitative means, they can be controlled together across a shared digital framework, a common quantitative language that can be scaled in proportion between each machine.

Digitality or 'the digital' is often spoken about in relation to (or tension with) analogicity or 'the analog;' notably in the recent world of modular synthesis today, there are endless discussions and internet squabbles over the virtues and vices of analog modular synthesis, which for some, is said to have an infallible authenticity and quality of sound. For many others, this notion of analogicity represents a fetishization of a group of technologies that has made the world of electronic musical instruments elitist and unobtainable in their resurgence and resurrection. Another more common and less charged distinction between the digital and the analog may be found in the differences between discrete or continuous values, which is most often spoken about in terms of the quality or resolution of digital-to-analog and analog-to-digital converters in the context of modular synthesizers. We have already visited this concept here in talking about the bit resolution of our MIDI-to-CV conversion.

However, we want to evoke a broader sense of the analog and digital aspects of these machines, which are best conveyed through Alexander Galloway's explorations of the analog and the digital. On the digital he writes,

> First, the digital or *logos* relies on a *homogeneous substrate* of elements that are differentiated quantitatively. Those famous 'zeros and ones' get the most attention, but the rest of the integers are just as digital, as are the natural numbers overall and the rational number line as a whole. (Galloway 2019.)

For Galloway, distinctions of the analog and the digital are separated from associations with consumer electronics and made plain as complementary ways of approaching the world. Earlier in the same blog entry, Galloway writes,

> ... *analogos* [referring to the analog] is not the negation or inversion of *logos* [referring to the digital] --and thus, by extrapolation, the analog is not the opposite of the digital--but rather, in some fundamental sense, its twin or echo. (Galloway 2019.)

If we briefly allow ourselves to extrapolate on these remarks and speak of the analog as being of material and the digital of concepts, logic, and order, then we can locate the fascination of computation between these machines—not necessarily in their perfect relationship with one another with machines that came before or after, but rather in the way they are able to bring tactility, timbre, and aesthetics to quantity and logic. We see the sound pieces and textiles as echoes of each other, just as they are echoes of the weaving drafts we used.

Finding a clearer relationship between the analog and digital aspects of this process feels significant here. Both modular synthesizers and hand weaving have been enjoying a resurgence among other older forms of media as an assertion of the primacy of the physical world, not unlike those who profess the virtues of all analog circuitry. This resurgence is all happening within a world that is increasingly mediated through screens and operating systems. We have cultivated a tense battle between the analog and the digital.

The analog here is merely a stand-in for tactility, or perhaps the feeling of having greater access to the means through which something is produced. Indeed, modular synthesizers are touted for their tactility, their handmade quality, and their DIY electronics aesthetic. Similarly, when we weave with a floor loom, we can see the textile take shape from start to finish and can touch the threads at every step of the way. Handmade textiles are seen as an answer to the toxic world of fast fashion. In both disciplines, we are engaging with an aura of the handmade, of care towards the material, and perhaps even the hope or feeling of escaping alienated labor.

But we know that through these tools, the materials are made beautiful by their mediation of the digital; thread is ordered to make forms and electric signals are modulated into compositions. The physical stuff, the moving wooden parts of the loom and the electrons running through the circuits of the synthesizer are made sense of and made special within a digital framework. We scale up this thought and recognize that the aura that these tools have in the 21st century is very much dependent upon the scaled-up digital backdrop we live in today. Certainly, this flavor of infatuation with materiality would not have existed when Buchla and Moog invented their respective systems.

In describing our decision making around choosing weaving patterns, we mentioned that we have generally favored older, established designs over newer or drafts improvised by us. Aside from the reasons mentioned earlier, we also acknowledge the broad impact that the history of weaving had on this project. In much of the older writing on weaving, we encountered lots of metaphorical language comparing weaving on a loom to playing an instrument and comparing a weaving draft to sheet music. One writer, Eliza Calvert Hall remarked,

> Often I find myself thinking and writing as if weaving and music were sister arts. A draft is like a long bar of music and the figures or marks on it are the notes. When I see a weaver at his loom I think of an organist seated before a great organ, and the treadles of the looms are like the pedals and stops of a musical instrument. (Hall 1912.)

Hall's words remind us of our real-time use of Max when we set the loom in sync with the modular synth. When placed specifically in the context of the modular synthesizer, her words resonate with the idea that modular patching itself could be placed in conversation with weaving and music composition, because all three can be performed or can be understood within a performer-composer relationship. After working through several iterations of making textiles and audio recordings, we recognized that placing older weaving patterns into Max amplified the continuity that exists between old and the new technologies and brought greater depth to our project. Hall's thoughts, which predate most of the innovations used to facilitate our work, have only strengthened our preferences in our weaving draft choices.

When discussing the similarities between weaving and computers specifically, many point to the Jacquard loom as the center of this relationship, because of the direct influence Jacquard weaving has had on early computer programming with punch cards. Although we do not deny the significance that Jacquard weaving has in the history of textiles and computers, we see it as less significant for our purposes. The complexity of these machines and the patterns they can produce can escape the legibility of the weaving itself, meaning it is hard to understand the weaving patterns and the textile structures at first glance. For us, the drawback of Jacquard is like the problem of teaching someone a basic logical circuit by means of examining the circuitry of an iPhone; the technology in focus is not simple enough to maintain clarity throughout our creative process and to be easily understood by folks who lack technical knowledge. For us, clarity around our technical approach was a high priority.

We especially felt the need to bring clarity to our process when showing the work in spaces like art galleries, where our intention and process were conveyed by literally bringing the loom and synthesizer into the space and demonstrating it through live performance and visitor participation. Visitors were able to create weaving patterns using an extended version of our Max patch on an iPad and use the floor loom, all of which was hooked up for real-time use. In recent years, increased interest has also brought a kind of esotericism or mysteriousness to these machines, which we felt we wanted to dispel. Both are deceptively simple to use and operate, and we have both seen them leveraged to create an aura around a particular product or music project without adding anything to the final work. We felt that there was a kind of literacy being built among those who were able to interact with the work, akin to the ethos of makerspaces and DIY electronics groups, which cultivate a kind of reclamation of technology. We were pleased that instead of obfuscating the process or the functionality of these tools, the Max patch also helped to illuminate them by giving immediate feedback to pattern changes.

Reflecting on our process, we thought it important to note where the logic of the weaving draft itself stops and where we had to step in as artists.

Weaving drafts convey the underlying structure of a textile while determining very little about what materials are being used. Similarly, signal flow across a modular synthesizer may provide a general idea or expectation of the resulting sound, but it does not determine the kind of modules being used across a signal chain, for instance, the sound source, filter type, or envelope shape. There are aesthetic and interpretive choices applied to the initial structure by an artist or musician. In this way, our textiles and sonic works are determined only up to a point by the system we have developed and our choices in how the pattern is connected to the modular synthesizer.

As said earlier, we believe that our technical and creative processes were designed with the intention to produce works using two machines that were meaningfully related to each other, but the reflections described above have been helpful reminders to us that the world we live in has produced a certain lens through which we understand technology, and the kernel of the matter is that much of what we consider advanced technology today is quite old but just performed much faster or more efficient. When examined within the context of our approach to digitization and quantification, it becomes increasingly harder to define where one kind of technology qualitatively stops and where another begins. We find this especially true today, where computers seem to convey a promise of being able to connect everything to everything else.

At the time of writing this, our project is still ongoing, and we are leaving it open-ended for the moment. We are thinking about ways we could show the work in the future, how our process could be made more legible in a performance or gallery space, and ways it could expand across different media.

Work that we have completed in this project can be found at draftpatchweave.xyz.

References

Galloway, Alexander R. "General Formula for the Digital and the Analog." February 20, 2019. http://cultureandcommunication.org/galloway/general-formula-for-the-digital-and-the-analog.

Hall, Eliza Calvert. *A Book of Hand-Woven Coverlets*. Boston, MA: Little, Brown, and Company, 1912.

14

COMPOSING AUTONOMY IN THRESHOLDS AND FRAGILE STATES

David Dunn and David Kant

Introduction

One of the more interesting questions to emerge from electronic music practice in the later half of the 20th century is that of autonomy. Artists and instrument makers, such as David Tudor, Don Buchla, Eliane Radigue, and Bebe and Louis Barron, built circuit systems capable of producing complex and unpredictable sonic behaviors, systems that were less akin to traditional instruments but rather exhibited intrinsic tendencies, character, and agency with which the performer had to contend, systems that were more akin to life than any a traditional musical instrument.

Recently, the question of autonomy has been taken up en masse by hobbyists, professional artists, instrument builders, and electronic musicians, with a resurgence in analog style modular synthesis and proliferation of modules designed to produce complex, self-organizing patches. What is autonomy? How do we design it? How do we recognize it? How do we interact with it? Most importantly, why has it captured the infatuation of artists and electronic music practitioners, causing a shift away from music composition as the specification of moment-to-moment detail to the role of systems design—that is, the creation of a generative system from which music emerges?

In this chapter, we revisit a 2010 article, *Thresholds and Fragile States*, by composer David Dunn, which describes the design and implementation of a custom-built, autonomous circuit network inspired by the self-organizing behavior of acoustic ecosystems. Dunn explains how the circuitry, a modular synthesis feedback network of coupled chaotic oscillators, produces a seemingly endless variety of ever-changing electronic soundscapes and describes the underlying design philosophy, which is based on the theory of autopoiesis.

DOI: 10.4324/9781003219484-16

This chapter contains a reprint of the original 2010 publication by Dunn, expanded with documentation of Dunn's approach to performing on the instrument, together with a new section detailing recent work by David Kant to study how autonomy arises from it. One of the central questions posited by Dunn is whether or not the behavior exhibited by the network is specific to its instantiation in analog form. In the final section, Kant describes a digital realization of the oscillator network and details two computational experiments intended to study how autonomy arises from it.

Part 1: *Thresholds and Fragile States* (Reprint)

A Descriptive Score for a Generative Sound System
Using Networked Autonomous Chaotic Oscillators
David Dunn
2010–11

Introduction

Several years ago I spent time living on a houseboat deep into the Atchafalaya Basin of Louisiana. Nights were spent making forays into the swamp to record continuous night sounds without interruption. One of the most striking features of this sound world was the abrupt transition between distinct collectives of sound makers. One group would hold center stage for hours and then suddenly fade to silence. Within minutes a whole new cast of sonic actors replaced them. The dynamic quality of these dense soundscapes, with their fantastic spatial motion, impressed upon me a sense that—beyond the communicative agenda of individual living sound generators—there was some underlying emergent logic at work to drive them into a global patterning. It was as if there were multiple chains of communication linking together a fractally delineated field of interlaced non-linear sources. These communicative chains not only extended outward in all directions but also up and down levels within a potentially infinite array of organizational hierarchies.

No scientific study of such global interactions and feedback between diverse sound makers within an ecological network has ever been attempted. This project is a step towards modeling the assumption of such an underlying emergent dynamical force. The circuits herein described give rise to autonomous sound behaviors that aestheticize mathematical pattern formation and are also a tool for the exploration of dynamics that help weave sounds together in the natural world. In many ways this work is an attempt at understanding pattern formation in natural sound systems. Comparisons and interactions between these natural and artificial systems might shed light on how similar dynamical properties might be operating at their generative levels.

As an artwork, these autonomous circuits are intended to either stand-alone as a conceptual entity that embodies dynamical behaviors or as a system of closure that can be structurally coupled to its surrounding auditory world. Under such circumstances, their behavior can be perturbed and constrained by the environment's behavior but should also be understood as a metaphoric machine expression of the autonomy of the living rather than as merely information processing devices.

The autonomous audio devices articulate an underlying assumption of biological autonomy through a basic design implementation. Two identical analog feedback circuits exist as closed autonomous units that can be structurally coupled through the simple connection of a shared resistance network. After initial conditions are established through the setting of a few potentiometers, the coupled circuits are allowed to behave autonomously in a self-organizing manner. Any sounds produced by the circuits emerge as a type of "conversation" that is allowed to continuously drift through novel behavioral domains that exhibit repetitive action at a local level but tremendous global diversity over extended time periods. In this sense the circuits resemble the closed nervous systems of living unities that are under constant perturbation from other similar closed nervous systems. The intention is not to simulate the high-level functioning of biological organisms and their cognitive capacities but rather to take this question down to its most primary level of autonomous-closure machines where self-organization is more obviously inseparable from behavior (Figure 14.1).

There are several famous examples of autonomous analog computation within the history of electroacoustic music practice and most likely many more instances that exceed the well known by an exponential factor. The

Machine Modeling of Biological Autonomy

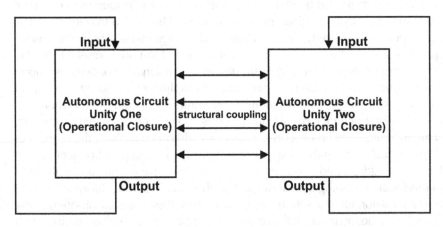

FIGURE 14.1 Block diagram of the instrument's double oscillator setup.

various forays into this area of exploration can largely be divided into two categories: 1) those that were derived by the "patching" of recursive non-linear feedback networks within commercially available analog synthesis instruments (Moog, Buchla, Arp, Serge, etc.) by the 1970s—specialized audio versions of comparable analog computers used by the aerospace industry and the early investigators of computational modeling of non-linear dynamics—and 2) custom designed circuits or ad hoc recursive circuit constructions made from interlacing independent function generators, filters, mixers, tape-loop feedback, etc. In some ways this is a wonderful example of how similar human knowledge can be constructed from very different sets of cultural assumptions and approaches to phenomenal explanation. In one instance, dynamicists were creating these kinds of systems in order to study their formal properties, while musicians were using similar tools to explore the perceptual attributes (sound) of the same physical phenomena. Just a few of the musicians who have explored conceptual terrain that borders upon these issues have been: Warren Burt, Sal Martirano, Richard Maxfield, Pauline Oliveros, and David Tudor.

One of the striking characteristics of experimental music traditions of the last few decades has been a concerted effort to invent and implement generative strategies for musical composition in the sense that the composer is primarily responsible for the generative system's global design rather than the primary decision-maker for the specification of constituent details. The familiar "top-down" model of the composer as the organizer of notes on a page that signify meaningful sonic events is replaced by a "bottom-up" model of the composer as the system's designer who is responsible for the organization of a generative mechanism from which the details for sonic events can emerge. While the majority of such explorations have been concentrated within the dominant use of digital algorithmic tools and materials, my use of the term "generative mechanism" must also be inclusive of other means and resources that are not constrained by technological innovation per se. This is also true for what has now become a minority area of music technology investigation: the unique qualities and attributes of analog circuit design. Most such research has by now become merely prosaic in the sense that it largely concentrates upon improvement in the design of basic tools—amplification, mixing, and signal routing tasks—where optimization of their utility is of value. Innovation into questions of structural form, sound synthesis and generation, signal processing, spatialization, and complex organization are almost exclusively the domain of digital music research (see Lansky (https://paul.mycpanel.princeton.edu/lansky_beingdigital.pdf), *The Importance of Being Digital,* for an explication of this idea). Assuming the validity of the observation that digital systems seem vastly superior for the serious investigation of these musical frontiers, why would I pursue my investigation using what appear to be—by comparison with "state-of-the-art" resources—largely archaic tools?

The distinction between analog and digital music systems harkens back to one of the most enduring debates of Western philosophy. The nature of digital code requires a level of specificity where mathematics is, in some sense, a purer model of what we understand about the world but in its most abstract terms. It is a Platonic world where we create an experiential manifestation from the reductionist archetype that must be exact in its numeric representation. Analog circuits can only be manifest as physical entities that are prone to a range of variations dependent upon the distinct properties of their constituent parts. We can describe them mathematically and infer their abstract state as systems but their unique ontological status is as imperfect physical examples of their otherwise ideal potential. They exist as messier Aristotelian things from which we can extrapolate abstract principles.

Obviously, I have posed this distinction between digital and analog systems metaphorically for overt effect. The truth is that either type of system exists in a kind of "chicken or egg" condition that is dependent upon our temporal relationship to how and when we create or explain them. My point here is that we do experience them differently and those differences inform how we use them: with analog circuits it is easier to propagate and sustain meaningful "mistakes" that might lead to novel insights. In the particular instance of the project herein described, my choice has been constrained by the nature of the questions that I am asking. Is it possible to create machine-generated sound behaviors resembling primitive conditions of biological autonomy, and can such machines be physically manifest in their imperfections akin to those of living things?

Modeling Biological Autonomy through Sound

One of the most enduring conundrums in science has been the question of how to define life. There has never been a truly satisfying description that can account for the transition of non-life to life. Its origins remain a mystery and the tentative status of phenomena such as viruses only further complicates the issue. The reduction of life to certain outstanding properties such as the ability to reproduce seems arbitrary since many individual living systems cannot. While we certainly know a great deal about many of such properties (biochemistry and genetics), and have assembled a vast knowledge base about the specifics of many living organisms (taxonomy and behavior), we still cannot easily answer the question: what is life?

An enduring contribution to this issue was the elegant book by physicist Erwin Schrödinger whose title consists of this very same question (*What Is Life?*) (Schrödinger, 1992). Towards seeking an answer, Schrödinger poses two ideas that have subsequently become essential to science. One was the concept of an "aperiodic crystal" as a carrier of genetic information. The other was the principle of "order-from-disorder." The first idea influenced the discovery of the double-helix molecule of DNA by Crick and Watson and the subsequent revolution of molecular biology, while the second is a fundamental notion in the formulation of such concepts as "dissipative systems" and "negentropy" that are at the heart of complexity science. While Schrödinger poses no definitive answers, he does contextualize the question in an essential 20th-century manner by asserting that classical physics is simply incapable of resolving the contradictions inherent in the stability of molecular structures that are necessary to sustain a living organism.

More recently the biologist Lynn Margulis has posed the need for a more global vision when seeking answers to this question in another book of the same title (Margulis and Sagan, 2000). Without eschewing the need to understand essential mechanisms in nature, she argues for a non-reductionist approach to the question by asserting that life is matter that sometimes is capable of making decisions that not only impacts its environment but also its own evolution.

Another viewpoint is found through a nexus of biology and neuroscience and it is this perspective that largely informs the project herein represented. The theory of *autopoiesis* was originally framed by Francisco Varela, Humberto Maturana, and Ricardo Uribe in 1974 (Varela et al., 1974) and systematically summarized by Gail Fleischaker (Fleischaker, 1988). The criteria for defining autopoiesis are specifically meant to apply to the organization of living systems as they are constrained by the laws of thermodynamics. They also support the claim by Maturana, Varela, and Uribe that *autopoiesis* is an explicit mechanism of identity that characterizes the organization of a living thing and also characterizes the transition of non-life to life. The molecular and organellar components specified by autopoiesis determine all the necessary and sufficient interactions to account for the cell membrane as a system-logical boundary that is determined by its own internal mechanisms.

It is precisely this property of a self-organizing autonomy that gives special status to living systems and characterizes the theory of autopoiesis. Living systems are defined by their organization rather than by their specific material constituents. Over time, living systems establish their organization through creating and replacing their own components. While the components are subject to change, it is the organization that remains as a stable identity and invariant property of the system as a whole.

Biology has traditionally framed the study of living systems in representationist terms that stress the interactions and behaviors of a system with its environment through description of their mutual correspondences. The formal language of cybernetics has also couched the description of various dynamical systems in similar terms built upon Information Theory and homeostatic feedback mechanisms that—while largely self-regulating—are understood as *allopoietic* systems, meaning that they have as their product something different from themselves. Such input/output-based descriptions are highly appropriate for many forms of phenomena and exist in a complementary relationship to the autopoietic organization of living systems defined through their operational closure. While highly useful in understanding certain network relationships that constrain the perturbations between living systems, allopoietic descriptions tend to ignore certain features of life such as its essential condition of autonomy.

One of the behavioral domains in which this autonomous status is most critical is that of perception in living organisms. In an allopoietic explanation of perception, the nervous system receives input from its environment and acts upon it as information in order to represent the outer world. In an autopoietic explanation, the nervous system is a closed network where perception and action are inseparable. "Information" can be understood to be something imposed upon the organism's environment and cognition, an attribute of the nervous system's operational closure.

Another way of describing this more recent view of neuroscience is that perception is predominantly a construction of the brain (and total organism) rather than a direct experience that is decoded. Much of our perception of the world is a "best guess" informed more from memory and past experience than from minimal sensory data. In fact, the signals that come from our bodily sensors are amazingly impoverished perturbations to our living coherence. Our experience and perception of an "external" world is largely an inference. This understanding has profound implications for what we mean by the term cognition. It is less about having knowledge of the world beyond us, and more about how our knowledge of self arises and remains coherent through a structural coupling with that world. From this standpoint, Maturana has stated that the biological emergence of cognition is not necessarily dependent upon the existence of a nervous system per se: "Living systems are cognitive systems and living as a process is a process of cognition." A cognitive system occurs when an "organization defines a domain of interactions in which it can act with relevance to the maintenance of itself." (Maturana, 1970)

As previously stated, the organization of a living system is maintained by its constantly changing structure in a manner that both creates and conserves its relationship to its environment. It is therefore composed of two domains of operation at two different levels, 1) the component level of its physiology, and 2) the organism level of its behavior. The result is a closed and autonomous self-regulating unity (operational closure) that is simultaneously open to its environment (structurally coupled). Living systems have dual ontological status through being neither fully separate from, nor fully belonging to, their environment.

From the very beginning of the idea's articulation, Maturana, Varela, and Uribe proposed the positive value of modeling autopoiesis within a computational framework. While the realization of virtual non-living autopoietic unities went on to become one important standard in the field of Artificial Life—demonstrating the potential for exploring the abstract nature of living systems in computational media (McMullin and Varela, 1997)—the creation of quantifiable models of the concept has remained both problematic and controversial.

One important understanding about the difference between living and non-living autopoietic unities can be derived from their differing standards for the use of the term homeostasis. While we may speak of the search for—and explanation of—exclusive mechanistic processes in both circumstances, in cybernetic parlance, homeostasis refers to the ability of a system to sustain a stable state of behavior. In a biological context the term is now more often reserved to describe an autonomous and recursive network that sustains organizational stability through self-maintaining processes. While this distinction also points to an essential difference between underlying properties of open networks (machines) and those that are closed (living organisms), it also shows how soft the distinction really is regarding the capacity for machines and organisms to imitate—hence describe—each other.

Despite the contributions that autopoietic explanation has made to the biological and cognitive sciences, the dominant scientific model remains representationalist with "information-processing"—the assumption that an understanding of the phenomenal world can be essentially reduced to an exchange of pieces of information—the preferred explanatory framework. The power of this approach is undeniable and largely responsible for the current successful state of computer science and related research regimes. While this representationist paradigm of allopoietic description has traditionally been associated with the first generation of cybernetic explanations (most notably Shannon and Weaver) and an engineering bias, autopoietic descriptions have been referred to as a "second-order" cybernetics that emphasizes autonomy, cognition, self-organization, and the role of the observer in the modeling of a system. However, while it still forms the dominant paradigm for cognitive science, representationalist description of biological unities seems less complete as we understand more about the nature of how living organisms create and sustain their internal worlds. Following Varela's lead, I assert that autopoietic and information-processing explanations of the world are actually complementary views rather than antagonistic ones. Nor are they each other's logical opposites. Each can be used to describe both the open and closed mechanisms and networks of machines and living organisms but lead to important differences in perspective and methods. Varela also makes a similar point:

> Clearly these two views (input and closure) are not contradictory, but the key is to see that they lead to radically different consequences, and to

radically different experimental approaches as well … But what is clear is that in order to study life and cognition, we need to explore the almost entirely unexplored land of autonomous-closure machines, clearly distinct from the classical Cartesian-input machines. (Varela, 1984)

Within this range of explanatory appropriateness, it is unusual for autopoietic descriptions to be applied to an understanding of the behavior of machines. The instances of this approach seem rare and therefore destined to be regarded more as metaphoric or philosophical expressions than as scientific explanations. We are so used to the information-processing model as the status quo for thinking about both machines and living nervous systems that—while many observers can understand the merit of applying operational closure to the latter—thinking about machine behavior in this way seems relatively strange. It is within this context of a much less familiar research regime that this current project can be situated and understood as an attempt to create a machine-based expression that straddles artistic and scientific categories of explanation.

Technical Details of the Autonomous Circuits

While the underlying conceptual approach to the design of these circuits has been that of autopoiesis, it has been my intention to assert their creation—as an autonomous artifact—to be an illustration of both metamorphic and mechanistic description. As scientific instruments they may be seen to fulfill these criteria when applied within an ongoing research regime as part of a mechanism to contribute to the potential disorganization of communication through mechanical vibration by certain arthropods. At a second-order level of intention—where they are not applied as a direct research instrument—they may also be regarded as an illustration of pure research into the domain of autonomous analog computation.

Ultimately the project explores the global behavior of hyper-chaotic analog audio circuits. The emergent complexity of these systems results from the dynamical attributes of coupled chaotic attractors interacting in a high-dimensional phase space. The control of circuit parameters determines a range of instabilities and structural couplings between nested chaotic circuits, allowing self-organizing behaviors to emerge.

The numerical expression of the non-linear chaotic oscillators is as follows:

$$\ddot{x} = -0.5\ddot{x} - \dot{x} \pm [x - sgn(x)] \tag{14.1}$$

This is implemented with the negative sign as the following circuit [NCO]. All resistors, except R, are 1k. R is variable.

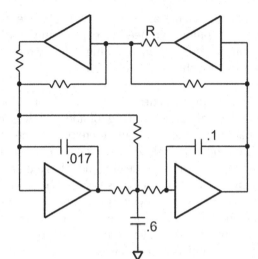

FIGURE 14.2 Circuit implementation of the chaotic oscillator.

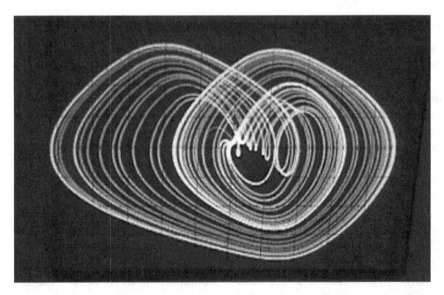

FIGURE 14.3 The double-scroll pattern generated by the oscillator.

This circuit yields the following double-scroll attractor:
Sub-circuits [with abbreviations] that make up the complete system:

4 nonlinear chaotic oscillators [NCO]
4 low frequency oscillators [LFO]
6 low pass filters [LPF]

(a)   (b)

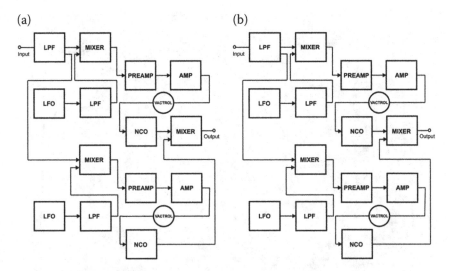

FIGURE 14.4 Block diagram of the two halves of the chaotic instrument.

- 4 line preamps
- 4 line amplifiers
- 6 multi-channel mixers
- 1 summing matrix network
- 4 opto-isolators [VACTROL]
- 2 voltage regulators

A flowchart of the two primary autonomous units that are linked through a resistance network as shown in Figures 14.4:

Layout and patching for the complete system: Figures 14.5 and 14.6.

Performance Instructions

While these circuits will continue to behave autonomously for indefinite time periods if left unperturbed, my intention has been to optimize the continuous novelty of such behavior through a coupling to their physical environment (either sensor driven or by performer influence). The following algorithm describes the approach I have taken as a performer:

1. Determine the initial conditions for the oscillators through settings of the various control potentiometers.
2. Change these settings over time so as to exhibit a form of *auditory neotaxis* (orienting behavior towards the seeking of new behavioral conditions as evidenced through sound).

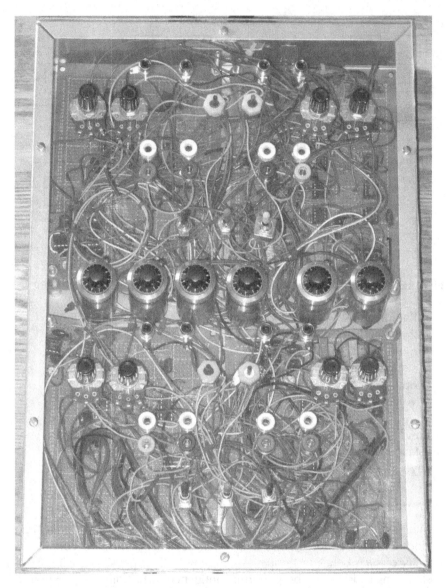

FIGURE 14.5 View of the complete built system.

3 Continue to seek out changes that optimize novelty in as many parameters as possible (frequency, amplitude, timbre, rhythm, shape, ratio of sound to silence, transition, global patterning, spatial positioning, etc.).
4 Additional signal processing may be added between the signal output of the oscillators and their final amplification. This processing should be minimal and predominantly spectral in nature rather than temporal. Any

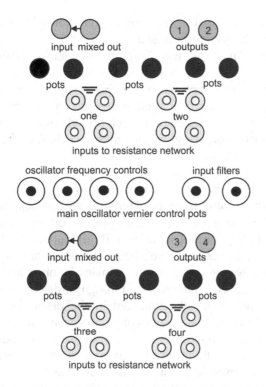

FIGURE 14.6 Schematic representation of the interface's controls.

such processing should not interfere with the generative oscillator behaviors.

5 Additional control systems can be added to the original interconnecting resistance network between oscillators (potentiometers). These systems should be limited to devices that only change resistance values such as photo-resistors or similar sensors.

6 Additional commentary on the performance practice

Questioning two recurrent tropes of electronic music practice:

1 There is a recurrent presupposition often assumed by the naïve electronic music listening public, and some not so naïve practitioners, that someone initially goes to a synthesizer to realize some sound(s) that they have imagined a priori. I personally doubt that this has ever been the case, or if so, whether it has ever succeeded. The same can probably be said for traditional music composition tools but even more so with complex electronic technology. The distance between what we can imagine sonically and what the technology allows for is a fairly large chasm. In

reality we must construct the final sounds in negotiation with our machines and the performance/listening context.

2 I question whether the term "musical" when used to describe a certain electronic music genre—as opposed to others—actually refers to something specific. It has been my recurrent experience that it is generally an invocation of someone's personal aesthetic preferences and an attempt to argue for their superiority. While the assertion of such prejudices does not always manifest in the perpetuation of well-worn auditory cliches, it often can and does.

While attempting to avoid such dubious assumptions, my approach to performance has sought to disrupt the emergence of redundancy and patterns by the hyperchaotic circuit network. It is just the opposite of the more familiar performance role of trying to instill some desired order upon unruly technology. I'm usually trying to optimize the chaotic behavior and not subdue it. While there is certainly a precedent for this thinking in John Cage's transcendence of personal aesthetic preferences through indeterminate processes, my interest is more specifically grounded in the desire for a deeper understanding of the dynamical behavior of physical systems. It's less about the expression of a communicative intent (or even ideas, per se) but, rather, the merest exploration of something yet to be discovered. My interest has been to set the stage for the experience of a domain of "scientific" and mathematical thought as directly as possible. It is certainly not an original insight to point out the difficulty that humans have in comprehending exponential change yet alone that which is chaotic. To merely assert a counter-cultural utility to the current embrace of noise as a sonic resource seems insufficient unless it can be both internalized and shared at a more sophisticated and experiential level.

I don't really see this as a musical composition per se, or even a system to compose with (eerily the circuits can be a much more interesting composer than I am). It's more a research project to go deeper into my interest in exploring machine implementations of the concept of biological autonomy. I really don't know how else to describe it. For many years I've also been interested in designing generative sound systems to placate my addiction to novel complex sound phenomena and noise and that may have much to do with the similarity that such phenomena have to the communicative acoustic web of the so-called natural world. My interest goes beyond the embrace of "generative art" or "algorithmic composition" to a desire to make auditory systems that are organized like—and therefore behaviorally akin to—living organisms.

The resultant sounds largely result from the autonomous and perpetually novel generative behavior of these circuits acting together. My performance role is not to compose a familiar musical structure but to participate in the emergence of novel behavior by perturbing the system in an intuitive and heuristic manner that encourages and retains its autonomous status. I'm often listening for the inevitable manifestation of fixated and redundant states and disrupting those by guiding the behavior towards a less stable domain. This often requires a reactive mode of listening but sometimes necessitates an anticipatory engagement. Either way the result is seldom predictable and almost always unexpected. The cue to react can be a bit akin to fishing. I wait for a subtle perceptual tug like the proverbial bite on a baited line and react by turning a pot or two a tiny increment. The change might be aurally productive or it might lead to silence. However, there are so many infinite possibilities outside of my influence—like the underwater ecology of a lake—that it is only a matter of time before new opportunities to react (or merely listen) will manifest themselves. Worst case, I can always go searching for another drastically different behavioral domain by altering the global conditions through changing multiple knobs and switches. It's a little like rowing the boat to a new location where I can drop my line into the water.

John Cage's goal of transcending personal aesthetic preferences, as previously referred to, can become a dicey construct during real-time improvisational performance. This is often an issue for some of the most interesting free improvisation practitioners regardless of what musical resources they engage with. The search for constant novelty can become its own aesthetic trap and pathway to rapid creative exhaustion. Can we ever be so open to any ontological possibility that a need for preferential action becomes irrelevant? Perhaps we are each so uniquely wired that our aesthetic preferences are revealed even when enacting complex tactics that attempt to subvert them. In the case of *Thresholds and Fragile States*, the other metaphor that I prefer to investigate is that of argument. I am engaging in an argument with a valued (electronic) colleague whose autonomy can challenge me. It requires that I sometimes assert a point of view by allowing certain conditions to persist but also a kind of tolerance to that which feels uncomfortable or confusing. Sometimes I am surprised to find the circuits sonically manifesting something so intriguingly beautiful that I try to act in a way that might optimize its persistence. Suddenly things can shift to an uncomfortable state and I am confused by my desire to object. In either case, shall I attempt to redirect or allow it to flourish? Any reaction carries with it a potential for loss or gain but, just as with any useful argument, if I only assert my will or point of view without a concerted effort to listen, nothing important can arise. I will lose the opportunity for a dialectical transformation to take place. It is both an occasion to delight in that which I find agreeable and to practice tolerance of that which I might otherwise reject without consideration.

It should be stated that my use of the term "hyperchaos" is both colloquial and mathematically strict. Most aspects of these circuits conform to the formal definition of hyperchaos as a form of chaotic behavior with at least two Lyapunov exponents but I am also using the term informally to describe the interaction of several chaotic circuits interacting in a complex fashion that may or may not always rise to the formal definition as stated.

References

Fleischaker, G. R. 1988. *Autopoiesis: System Logic and the Origin of Life*. Ph.D. dissertation, Boston, MA: Boston University.

Lansky, Paul. The Importance of Being Digital. https://paul.mycpanel.princeton.edu/lansky_beingdigital.pdf

Margulis, L., & Sagan, D. 2000. *What Is Life?* Berkeley and Los Angeles, CA: University of California Press.

Maturana, H. 1970. Neurophysiology of cognition. In *Cognition: A Multiple View*, ed. P. Garvin, 3–23. Washington, DC: Spartan Books.

McMullin, B., & Varela, F.J. 1997. Rediscovering computational autopoiesis. In *Proceedings of the Fourth European Conference on Artificial Life*, eds. P. Husbands & I. Harvey, 38–47. Cambridge, MA: MIT Press.

Schrödinger, E. 1992. *What Is Life?* Cambridge, UK: Cambridge University Press.

Sprott, J. C. 2000. Simple chaotic systems and circuits. *American Journal of Physics*, 68, 758–763.

Varela, F. J., Maturana, H. R., & Uribe, R. 1974. Autopoiesis: The organization of living systems, its characterization and a model. *BioSystems*, 5, 187–196.

Varela, F. J. 1984. Two principles for self-organization. In *Management and Self-organization in Social Systems*, ed. G. Probst, 25–33. New York: Springer-Verlag.

Part 2 Measuring Autonomy

Central to *Thresholds and Fragile* States is the concept of autonomy, a question which, for Dunn, has as much to do with electronic music as it does with the behavior and organization of living ecosystems. While Dunn explains that the piece is inspired by his experience listening to the sounds of night insects in the Atchafalaya River basin of Louisiana, it is no doubt the product of an entire career spent listening to, recording, documenting, studying, and interacting with environments, soundscapes, and living ecosystems through, and with, electronic circuitry. Dunn has explored the phenomena of autonomy for his entire career, although often under different names or metaphors, from his early large-scale, outdoor site-specific works of the 1970s, which sought to connect with the spirit of a location, or sense of place, to his later explorations of chaotic and nonlinear dynamical systems, began in the late 1980s, in which Dunn networked hybrid systems of digital computers and analog circuitry. Running throughout these inquiries is an infatuation with life and mind, concepts that are closely related, perhaps even equivalent or isomorphic, for Dunn, with autonomy.[1]

In the following section, we present an analysis of Dunn's circuitry as seen from the perspective of instrument design, documenting new work by David Kant to model Dunn's system in a digital computation framework and study how autonomy arises from it. How is autonomy instantiated in the circuitry?

How does autonomy manifest in the sound output? Which aspects of the instrument design are responsible for the "tremendous global diversity" produced by the circuitry "over extended time periods"? And what is the scope or range of this autonomous behavior? We present novel tools, both conceptual and mathematical, that may be useful, in general, for studying the emergent behavior of generative systems, as well as for describing the creative decision-making involved in working with emergent systems.

Dunn, who has worked extensively with both analog and digital systems, explains his preference for analog circuitry when constructing *Thresholds and Fragile States*, inviting the questions of whether or not it would even be possible to get such results from a digital implementation. Central to the project of digital modeling was the question, is the emergent behavior produced by the oscillator endemic to networked chaos or is it specific to this physical instantiation, an artifact of working with analog circuitry?

The Chaotic Oscillator

By far the most critical aspect of the analysis and digital model is the chaotic oscillator, not just because the oscillator is the building block of the entire system, but because constructing the analysis and digital model was meant to ask the very question, is the emergent behavior exhibited by the network endemic to networked chaos in the abstract, or is it a product of this particular physical instantiation. In Dunn's system, multiple chaotic oscillators (sometimes four or eight) are networked together in a complex and hierarchical modulation system, including various nonlinearities, filters, and gainstages as well as additional low-frequency oscillators. Individually, each chaotic oscillator is capable of producing an array of sounds, but, when networked together, new sounds emerge, sounds that are not exhibited by the oscillators in isolation but are rather a product of the complex, audio-rate modulation paths.

Dunn's oscillator has a single control parameter, a variable resistor which, when adjusted, brings the oscillator through a complex and varied sequence of sounds, from periodic sinusoidal waveforms to quasi-periodic oscillations and band-limited noise. Changes in pitch, amplitude, and spectra are all modulated together, coupled, or mapped, along this single modulation parameter, or control dimension. Contrary to conventional oscillators, which typically produce simple, repeating waveforms and are generally engineered to give predictable, or really *proportional*, changes in output relative to adjustments in input and/or parameter values, Dunn's oscillator, by comparison, can produce (but not always will) disproportionately large changes in output for even the smallest adjustment in parameter values, making it unpredictable and difficult to control. Conventionally understood, this is the paradox of chaos: "deterministic chaos"; even though such systems can be

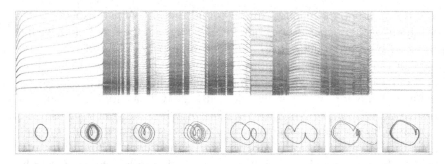

FIGURE 14.7 Spectrogram of Dunn's chaotic oscillator as the control parameter R is swept from low to high, and the sequence of fixed, single, and double-scroll attractors produced.

fully (and deterministically) described by a system of equations, such systems are actually unpredictable in practice, being far too sensitive to minute changes, changes smaller than the resolution of measuring instrumentation or the precision of digital computers, to actually predict.

When visualizing or listening to the chaotic oscillator output as a function of its single control parameter, the variable resistor labeled R in Dunn's diagram (Figure 14.2), is swept over its operating range, from low to high. Transitions between distinct sound patterns (spectra) can be seen or heard clearly in a plot or recording of that output (Figure 14.7). These are sometimes gradual, the oscillator slowly morphs between spectra, and other times abrupt, when a harmonic series spectra bursts abruptly into noise or shifts down/up an octave in a discontinuous or step-scale manner, almost like tipping off a ledge. Very tiny adjustments to the control parameter can cause large changes in oscillator output. Importantly, some regions are dense across the horizontal dimension—the oscillator quickly swings through a seemingly infinite collection of sounds, as if slowing down would unravel an infinitely deep fractaline structure. Dunn uses high-precision 20-turn poten-tiometers (normal oscillators turn only once) salvaged from Los Alamos National Lab and NASA surplus in Dunn's hometown of Sante Fe, New Mexico for very fine control of these dense regions and thresholds.

The field of dynamical systems theory has a rich set of tools for analyzing and describing the complex sequence behaviors exhibited by the oscillator. Plots called phase portraits (more commonly referred to as Lissajous plots in electronic music), graph two signals against another rather than through time (as in a waveform plot) and are common in dynamical systems analysis because they give distinct visual form to signals which can otherwise appear noisy and unstructured when viewed over time. The forms, called attractors (i.e., Figure 14.3 in Dunn's essay), represent a set of values towards which the oscillators tend to evolve for a wide variety of starting (initial) values, regardless of where the signal started. Some of the attractors are stable—they lock into regular, periodic waveforms—while others are unstable, vacillating unpredictably on the edge between two or more patterns, or noise. This occurs when the orbits, or trajectories, of two or more attractors are so close that small deviations in signal cause the oscillator to hop from one attractor to the other.

Bifurcation diagrams are another common tool in chaos theory to represent how the behavior of a dynamical system evolves as one or more system parameters are changed. Traditional dynamical systems analysis would find three distinct attractors. For values of R less than 10 kΩ, the oscillator produces a constant DC offset signal, or fixed point attractor. As R increases, the offset decreases until a periodic orbit emerges, resulting in a Hopf bifurcation at $R = 14.3$ kΩ, where the stability of the system switches abruptly to a limit cycle periodic attractor, or sinusoidal waveform in the x dimension. At about $R = 18.6$ kΩ the oscillator begins to exhibit period doubling as a double-scroll attractor begins to emerge, named for its something. From $R = 18.6$ kΩ to $R = 33.2$ kΩ, the oscillator interpolates through various forms of the double-scroll attractor, including stable and unstable periodic limit cycles as well as chaotic and strange attractors. Transitions between attractors can happen abruptly. Above $R = 33.3$ kΩ, the oscillator exhibits a single-scroll Rössler-like attractor, which eventually converges to a single period limit cycle at $R = 40$ kΩ. For low values of R, below 10 kΩ, the oscillator is unstable and bounds off to infinity.

Within this family of single and double scroll attractors, there exists a variety of sounds that, while they may not be considered mathematically distinct by quantitative measures of dynamical systems analysis, nevertheless can be *heard* to be quite different, and are quite different by perceptual measures. While tools of dynamical systems theory are usefully and illuminating, it is important, when working in a musical context, to use musically and perceptually motivated features as well. This observation motivates the use of hand-engineered perceptual features in the following musical analysis.

Modeling the Oscillator

Dunn's oscillator is based on a design by J.C. Sprott, a physicist who has written extensively on chaotic systems, having published a family of simple chaotic analog circuits (to which Dunn's belongs) as well as significant computational algebraic explorations. Sprott's designs are common in DIY audio electronics, most likely because they are simple to build, involving only relatively few components that are cheap and readily available (resistors, capacitors, op-amps), and are robust to deviations and inconsistencies in cheap component values—the Nonlinear Systems Sloth module[2], for example, belongs to the same family of jerk circuits as Dunn's. Dunn's oscillator belongs to a family of third order systems, or jerk circuits. The circuit has three integration stages, marked x, y, and z, the outputs of which are fed back and summed to form the input to the third order integrator input, w, as well as passed through a feedback nonlinearity, labelled g(x), which, in this case, implements the function sign(x). The circuit can be described by a third-order system of differential equations, as given by Sprott (2000) and reproduced by Dunn. A compact form of the circuit equation is given by:

$$w = -Az - y - Bx + C \operatorname{sgn}(x)$$

When modeling the circuitry in a variable, or voltage-controlled, context, however, the oscillator behavior deviates from the equations given, and, in order to produce the complex behaviors described in the previous section, particularly the sequence of attractors and transitions, it is necessary to modify the equations in a number of ways. The deviations are due to non-ideal properties of op-amps, which introduce additional nonlinearities such as saturation and slew limiting, as well as the choice of circuit components and topologies used to realize key mathematical operation—for instance, the use of a passive RC filter to realize the middle integration stage (perhaps in order to fit the circuit on a quad op-amp chip) and lack of buffering which yields significant self-modulation. Sprott's design was likely not intended to operate in a continuously variable, or voltage-controlled, manner, and Dunn's circuit is built with a LM741-like quad op-amp chip, a relatively cheap and low-quality circuit component, which introduces further irregularities and inconsistencies.

Among the more significant modifications necessary is the introduction of op-amp saturation, which can be modeled as a simple clipping function applied to the active integration stages. Without saturation, the variable resistor R affects only the size, or amplitude, of the attractor (coefficient C only). With op-amp saturation, however, the variable resistor changes the attractor shape as well, affecting *all* coefficients in relation to one another, allowing one parameter, R, to produce a family of different attractors as its value is swept. Additionally, the op-amps have a limited slew, or rate of change, the effects of which are most pronounced in the feedback nonlinearity sign(x) and z integration stage. These op-amps flip from positive to negative and vice versa near instantaneously. Dunn's chip, having a slew limit of approximately 0.5V/us, is a slowdown on orders of magnitude, which is responsible for producing the curly edges of the attractors, as well as many of the more intricate and fine details of their structures.

The analog circuit also exhibits a significant amount of self-modulation, due to a lack of isolation from the power supply. Interestingly, the self-modulation brightens the sound, adding even harmonics to what would otherwise be a symmetric waveform (only odd harmonics). This occurs because the modulation produces a DC offset in the oscillator signal before the integration states, which, when integrated, produces a slightly asymmetric waveform, introducing even harmonics into the spectrum and giving the oscillator a touch of warmth. Lastly, when solving the numerical approximation, it is important to realize the middle, or y, integration stage not as an active op-amp based integrator but rather as a passive RC integrator, or lowpass filter. This is critical because the frequency response of the RC filter, and thus strength of the y signal and value of the corresponding coefficient in the system of equations, is frequency dependent.

The oscillator is modeled by simulating the equations and introducing the additional distortions into the numerical approximation process. Among the challenges of modeling chaotic systems is their extreme sensitivity to error. Such systems cannot be solved exactly (analytically), rather their solutions are approximated using numerical methods, such as Euler or Runge-Kutta, which approximate the solution to a fine degree of error, although there is often a tradeoff between accuracy of the approximation and the amount of computation required to compute it.[3] Given the extreme sensitivity of the oscillator to small deviations in system state coupled with the error inherent in numerical approximation, the digital model is remarkably similar to the analog oscillator, exhibiting the same key features, including: the sequence of fixed, single, and double scroll attractors; extreme sensitivity at thresholds between attractors; and a large DC offset for low values of R. Furthermore, because the approximation error is not constant, but rather depends on the curvature of the signal and the sensitivity of the oscillator at that point, digital modeling introduces a new source of uncertainty into the system. When considering Dunn's design ethos—"the meaningful propagation of errors"—this is a new source, and one that is endemic to digital realization.

Coupling between Oscillators

In Dunn's system, multiple identical chaotic oscillators—or near-identical chaotic oscillators, given the inconsistencies and error tolerances of cheap electronic components—are networked together, organized hierarchically in pairs. As Dunn explains, the design is motivated by the principle of *autopoiesis*, a theory proposed by Chilean biologists and neuroscientists Humbero Maturana and Francisco Varela in the 1970s to explain the self-organizing phenomena of cellular organisms, and transferred to electronic circuitry in this piece by Dunn. Central to Varela and Maturana's theory is the idea of "structural coupling," which describes how a system can be both operationally closed, or distinct from its environment, internally maintaining the processes necessary to sustain itself and constitute a whole, while, at the same time, coupled to its environment, part of a dynamic system and sensitive to interaction (Maturana 2002).

Dunn's circuitry reflects this in the design of the coupling, or feedback pathways, between chaotic oscillators. The oscillator circuits do not share voltage signals directly, as a mathematician might be tempted to couple them, but rather interact through a predefined interface—the control parameter R. The oscillator are isolated from one another, seen as separate, internally consistent entities that sustain their own internal organization, but nevertheless interact with one another and with the greater environment through a limited point of contact that modulates, or perturbs, its state, but does not affect its internal structure or organization. This separation, however, begins to break down, at least in the system's analog form, due to the challenges of isolating electrical circuits from one another—the oscillator self-modulation in particular.

While the idea of control parameter modulation is widespread within the modular synthesis community, especially at present day, the approach is a bit different from how synchronized dynamical systems tend to be implemented and studied by mathematicians. Mathematically speaking, the circuit coupling allows the oscillators to control one another's equation *coefficients* (A, B and C in the equation above); this is as opposed to adding signals directly into the oscillator equations, as is common in studies of coupled chaotic systems. Furthermore, the modulation signals are not mapped directly to resistance values, in a linear fashion, but are passed through a complex signal path, including filtering, amplification, and other nonlinear operations necessary to implement the connection in analog circuitry, producing a modulation map that is complex, nonlinear, and time-variant. This, I would argue, is crucial to producing the emergent autonomous behavior, and is also something of an artifact of audio circuitry and modular synthesis. There are a few important features of this pathway that I found to be necessary to incorporate into the digital model.

The oscillators are electronically isolated from one another by means of a vactrol, or opto-isolator. Consisting of a light-emitting diode (LED) taped end to end with a photosensitive resistor (photoresistor), the vactrol is a common hack in DIY audio electronics and a cheap and easy solution to allowing voltage control of resistive circuit components. In Dunn's system, the vactrol is necessary to the modulation path, allowing the chaotic oscillator outputs, which produce voltage signals, to modulate the state, or variable resistance R, of other oscillators in the network. Both the LED and photoresistor, however, introduce additional non-linearities into the system, which are crucial to model. The photoresistor attack—that is, its response to increases in light intensity—is significantly quicker than its decay, or decreases in light intensity, a property which turns out to produce desirable audio effects, such as envelopes for optical compressors and lowpass gates. Furthermore, the attack and decay tend to be on exponential scales, and, most importantly, exhibit hysteresis. The sensitivity of the photoresistor to light is dependent not just on the current light intensity hitting the surface of the photoresistive cell but on its past light exposure as well. The photoresistor is more sensitive to brightness when previously having been in a dark environment, and vice versa. This property turns out to be crucial to the emergent musical behavior produced by Dunn's system, and properly modelling these vactrols required measurement of a variety of real components close to what Dunn used. The digital implementation of the vactrol relies on an asymetric exponential filter with control over attack and decay rates where the response curve of the filter is fitted to those experimental measurements. Hysteresis is modelled with an additional exponential filter, operating at a much lower frequency range to replicate control over the sensitivity of the vactrol.[4]

Another particularly interesting decision of the feedback path design is the inclusion of low-frequency oscillators, which are mixed in with the modulation signals, one for each chaotic oscillator, immediately before the amplification stages. The LFO, implemented using a 555-based square wave circuit, saturates the mixer, essentially nullifying the modulation signal temporarily because it does not leave any headroom for that modulator. At first glance, the use of LFOs seems to run almost counter to Dunn's desire for an emergent complexity of coupled chaos. Are the emergent long-term patterns and changes in sonic behavior a product of the coupled oscillators or are they simply the result of multiple LFOs cycling in and out of phase with one another? Dunn's decision to include LFOs is challenging; I struggled to understand it at first. However, considered in light of Varela and Maturana's theory, the LFOs serve an important role: perturbation. In Varela and Maturana's theory of autopoiesis, the concept of perturbation is central to the emergence of autonomous behavior. Autonomy arises not in isolation but from the dynamic system of exchange between an entity and its environment—how an entity sustains its internal organization in the face of external change. In Dunn's electronic microcosm, the LFOs serve this role, perturbing the oscillators every so often. Depending on the oscillator states, these perturbations can cause the oscillators to change radically and reorganize, or have little-to-no audible effect.

When coupled together, the oscillators produce new behaviors, behaviors that are not exhibited by any single oscillator in isolation but are rather a product of the coupling between them. The oscillators move through *sequences* of attractors, exhibiting patterning on a larger temporal scale than produced by a single, individual oscillator. The length and complexity of these sequences vary greatly, from fractions of a second to many minutes, and are affected by adjustment to feedback parameters, such as preamp gain or lfo cutoff frequency. The oscillators exhibit a variety of synchronization behaviors, including phase locking—this occurs when the oscillators produce different waveforms but are synchronized in phase—and near-identical synchronization, when the difference between two or more oscillators collapse, the oscillators locking into near-identical, time-aligned waveforms. The oscillators also produce synchronization behaviors typical of weak coupled systems, such as anticipation and lag, the changes of one oscillator anticipating changes in another, as well as envelope synchronization—the oscillators maintain independent patterns but exhibit periodic envelopes that loop at the same frequency. These weakly coupled behaviors, in which the oscillators maintain separate patterns of pitch, amplitude, and timbre, yet can still be heard to respond to, coordinate, and interact with one another, are perhaps those that contribute to the perceived sense of the system's autonomy.

Computational Experiments

At present, Dunn's complete system contains two boxes, each box containing four oscillators. The boxes can be used independently or networked together by means of the resistance network, a banana jack interface that allows Dunn to dynamically patch new network topologies. The number of control parameters grows quickly with the number of oscillators, as the signal path for each oscillator involves at least five or six additional controls—low pass cutoff, preamplifier gain, power amplifier gain, LFO frequency, and chaotic oscillator offset—all of which can be adjusted by Dunn in performance via corresponding potentiometers. Any knob has the *potential,* when adjusted, to affect the observed system behavior in significant ways, but it is also very likely that nothing at all will happen. This depends on the values of the entire set of controls relative to one another, the current system state, as well as its past history.

Dunn's approach to instrument design runs counter to conventional engineering epistemologies, in which electronic instrument controls are carefully engineered to correspond with semantically meaningful musical parameters; effect change evenly and predictably across the entire parameter range; and produce a possibility space that is diverse, interesting, and distinct for different parameter settings. Dunn's system is quite the opposite. The control parameters are not explicitly engineered to correspond with any particular musical features, and the systems' response to changes in parameters can be patchy and irregular, at times extremely sensitive to touch, almost beyond human precision, as well as unyielding, insensitive, and stubborn. The possibility space circumscribes a rich and diverse range of emergent sonic behaviors, but navigating it via the exposed parameters can be a challenge.

Motivated by Dunn's interest in the chaotic and unpredictable, in particular considered in light of the rich history of artists, from David Tudor and Louis and Bebe Barron up to the present day, who likewise embrace complexity, led me to wonder: *How can we reason meaningfully about the compositional decisions involved in building such systems—musical instruments that are chaotic, unpredictable, and complex?* Specifically, two main questions came to mind:

1 How unpredictable is Dunn's system? And in what ways? What *causes* the infinite, ever-changing variety of electronic soundscapes?
2 What is the scope and range of sound produced? How do we measure and describe the *character* of a generative system?

In the remaining sections, we present two computation studies designed to address these questions. The computational experiments rely on large datasets of audio produced by the digital model of Dunn's oscillator network. The digital model implements the analysis as described above, simulating the chaotic oscillator and network signal flow. One of the challenges of modeling chaotic audio systems is balancing model accuracy and compute power. The digital model is implemented in two forms. One, built in SuperCollider, is intended for real-time performance and contains a collection of GUIs to visualize and control system state. This is available for others to use and explore.[5] The second, implemented in rust in collaboration with Andrew Smith, runs offline and is used to generate the large datasets necessary for machine learning analysis.

Experiment 1: Auditory Neotaxis

One of the more notable features of Dunn's instrument is its capacity for continued change and variation over a very long-time scale, on the orders of many minutes to many hours. The instrument's behavior seems to be in a state of constant flux, slowly meandering through a diverse possibility space, without human interaction. The system locks into short-term patterns, or loops, but also morphs and evolves gradually over long time periods. What is the source of this long-term change and variation?

Generative music systems, circuitry in particular, require a way of specifying time that is different from that of written music notation, one in which form is defined implicitly as an emergent result of the systems unfolding through time, rather than written explicitly with pen on manuscript paper. Fascinated by the idea that temporal structure can be encoded implicitly in a generative procedure, the first experiment designed to study how form and long-term structure is composed in *Thresholds and Fragile States*. In this experiment, we sample a large dataset of random network settings, generate their outputs using the digital model, and use multivariate regression analysis to find correlations between systems parameters and the emergence of long-term temporal structure.

Measuring Long-Term Change

Measuring change, especially at very long timescales, is a complicated task, involving both physical aspects of an audio signal as well as perceptual notions of attention and musical value judgements about what is (and what is not) salient to experience. There are countless ways that an audio signal can evolve over time, countless features that serve as indicators of form. In this experiment, long-term change is measured using trend analysis, a feature we refer to as the *trend spectrogram*, which captures the raw change in audio magnitude spectra over time. Trend analysis is a common statistical technique for finding long-term, or seasonal, changes in time series data—such as distinguishing annual cycles in financial markets from short-term, day-to-day fluctuations. In this experiment, we apply it to audio signals.

A trend spectrogram is computed by first transforming a time-domain audio signal to the frequency-domain using any of a number of spectral transforms (FFT, CQT, Melspectrum), and then reducing the spectrogram to a one (or more) dimension(s) through time, by Principal Component Analysis (PCA), giving a downsampled signal that tracks spectral variation in the audio over time. The first few PCA dimensions are those that explain the most variation in data, and it is assumed that any significant long-term change in the audio spectrogram signal will also be reflected in the PCA dimensions. Finally, performing Fourier analysis on the PCA signals gives the trend spectrogram.[6]

Trend spectrum analysis finds significant energy in low frequency bins, with a peak at 120 seconds, indicating that there is substantial formal structure at the two-minute timescale. Listening to the recording, this low-frequency motion can be heard, as the system morphs through different sound behaviors, both in temporal patterning (when sounds are heard) as well as timbre (which sounds are heard), at roughly the speed of every two minutes or so.

Experiment Design

The dataset consists of 100,000 60-minute example outputs, generated by the digital model Dunn's oscillator network, with parameter values chosen uniformly at random from the entire range of the model. Each example consists of a vector of 44 system parameters, including both *playable* parameters, which Dunn has exposed as potentiometers and switches on the top plate of the instrument for control in performance, as well as *design* parameters, which determine aspects of the circuit design not made variable in performance, such as the vactrol attack and decay times, these being a physical component value. The experiment also includes a number of *lesion* parameters. Intended for analysis, these switch on/off aspects of the circuitry, such as gating on/off the LFOs, to help identify their impact. Each of the 100,00 examples, is rendered to produce 60 minutes of audio—data at a scale only possible using the digital model. And each example is rendered starting from the same initial conditions such that the state of the system in each differs only in the values of system parameters randomized. Finally, the trend spectrogram is computed for each example using a 60-minute window, and the FFT bins are aggregated into perceptually spaced time intervals, giving energy on timescales from 1 second to 60 minutes.

EXPERIMENT 1

Dataset = 100,000 random network parameter settings, each rendered for 60 minutes. Features = FFT trend analysis, measures energy at timescales from 1 sec to 60 minutes.

Results and Analysis

The data is analyzed using multivariate linear regression analysis. A common statistical technique, regression analysis estimates relationships—in this case *linear* relationships—between input (independent) variables and output (dependent) variables. Importantly, regression analysis can be used to reason about cause and effect between independent and dependent variables, which is how it is used here, for its explanatory power. The 44 circuit parameters comprise the independent variables and the energy of the 24 trend spectrogram bins the dependent variables. The goal is to identify which of these 44 variables correlates with the emergence of structure across the various timescales.

Several variables were found to have significant bearing on long-term change, as indicated by a low p-scores and high correlation coefficients, relative to the other parameters (Figure 14.8). These include the *vactrol hysteresis duration* as well as the *chaotic oscillator potentiometers* and the *lfo gates*. The influence of the oscillator potentiometer is easily explained as a simple on/off switch—the oscillators are silent unless the potentiometers are set above 4kHz—which produces a strong correlation. The relationship between the LFO and vactrol hysteresis is perhaps more interesting. At timescales of up to a few minutes, the LFOs are the primary determining factor of trend spectrum energy, with high coefficients and low p-scores. However, as timescales increase above a few minutes, extending all the way up to 60 minutes, the vactrol hysteresis eclipses the LFOs as the primary determining factor of energy in the trend spectrum, showing increasingly larger coefficients and smaller p-scores as the length of analysis timescale grows.

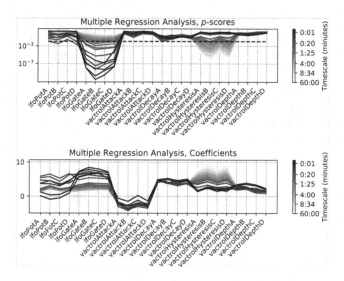

FIGURE 14.8 Multivariate regression analysis for LFO and vactrol parameters that influence long-term change in Dunn's network.

This can be rationalized considering the design of the feedback circuitry. The effects of the vactrol hysteresis and the LFOs can be rather pronounced, thus showing strong correlation, but the LFOs have a maximum period on the order of only a few minutes, not hours. The vactrol sensitivity, on the other hand, slowly adjusts over a very long timescale, up to 24 hours in some cases, providing a very low frequency modulation to the entire system, which can slowly drive the oscillator network through novel regions of behavior on timescale on the order of hours.

Interestingly, LFOs are not the only source of small timescale change. This was verified by performing a lesion study—turning off the LFOs, generating a new dataset, and performing multiple regression analysis. Small timescale changes, on the order of seconds to minutes, are still present, driven by vactrol hysteresis but also due to emergent properties of the chaotic oscillators alone. While the inclusion of LFOs is conceptual for Dunn, motivated by the theory of autopoiesis, it is likely aesthetic as well. The inclusion of LFOs increases the likelihood of rapid and pronounced changes in sound, often every five to seven seconds or so. Without them, the network tends to produce very slowly moving chaotic drones.

Experiment 2: Mapping the Possibility Space

In addition to long-term change, Dunn's instrument demonstrates a tremendous *scope* of variation in sound behavior. However, even within this limitless variation, there is character and identity—sounds that tend to happen more often than others and sounds that tend to happen not at all. How do we come to know or sense the character of an infinite, or near infinite, variation?

Such questions are reminiscent of mid-century musical thinking, exemplified by composers such as John Cage or Christian Wolf, who, by allowing chance procedures into their music, provided for infinite possibility, yet the music still retains an identifiable character. These questions are even more pressing when considering generative electronic music systems, in which the score is written not with pen on paper but in circuitry, and the performers who animate the chance procedures are electronic signal processors, not people. While dynamical systems theory, widely popular in the 1980s, provides *an* answer for finding structure (chaotic attractors) obscured amongst seeming noise and variation, the question is particularly salient to an area of computational sciences new to the 21st century: machine learning and big data—how does one come to know, identify, measure, and even *reproduce* the character and structure of a large dataset?

Experiment 2 is designed to explore the possibility space of Dunn's emergent system, to both quantity and quality it in shape and scope. The experiment consists of two contrasting datasets. For each dataset, we measure the structure of the dataset, using music similarity as a basis to compare individual examples to one another, and consider the structures of the two datasets relative to one another.

Experiment Design

The experiment consists of two datasets, the first, similar to as in Experiment 1, contains 100,000 examples, sampled uniformly at random from the entire parameter space of the model, although in this case rendered for only three minutes of audio. This dataset represents the full possibility space of the system. The second dataset, sampled from constrained parameter ranges, reflects how Dunn actually uses the system in performance. When performing on the system, there are some parameters that Dunn adjusts more frequently than others and some parameters that are left fixed, or constrained to very narrow and specific values, rather than left to range over the full scope of potentiometer values. The second dataset is sampled from constrained parameter ranges to reflect Dunn's performance settings. These ranges were arrived at through conversation with Dunn.

Measuring musical similarity is a likewise complex and multidimensional task, with challenges similar to those of measuring long-term change. Judging the musical similarity between two network outputs can be extremely subjective; there are many musical and perceptual features that can be indicators of similarity—pitch contour, rhythm, pattern periodicity, timbres—many ways in which sounds can be both simultaneously similar and different. In this experiment, our approach is to cast a wide net. We measure a large collection of features and use machine learning to find those that are most structurally significant throughout the dataset. Each example is rendered for three minutes, and a collection of audio features are extracted using the Essentia Freesound Extractor (Bogdanov et al., 2013). Features include a variety of time-varying features and their statistical summaries (e.g., min pitch, max pitch, and pitch mean), together with first and second order differences (min change in pitch, max change in pitch, as well as min change in the change in pitch, etc). Ultimately, each example is described by a list of 550 descriptor values.

EXPERIMENT 2

Datasets = two datasets, containing 100,000 random network parameter settings each, one set randomized within a constrained range as to reflect Dunn's performance practice, the other set randomized over the full potentiometer ranges. Features = large collection (about 550) of audio descriptors extracted and then embedded into 2-d for visualization (using UMAP).

Results and Analysis

The datasets are visualized by Uniform Manifold Approximation and Projection (UMAP), a machine-learning technique commonly used to reduce dimensionality of and visualize structure in large datasets (McInnes et al., 2020). UMAP finds a low-dimensional manifold, often referred to as an embedding (in this case having two dimensions) that has similar topological structure to the original data, which, in this case has 550 dimensions—far too many to visualize directly. Embeddings such as those produced by UMAP help visualize relationships between data points, showing which points are similar to one another, and can reveal topological structure, such as whether or not the points are grouped into one big connected component or scattered into small disconnected clumps. Embedding plots, however, must be interpreted carefully, as not all visual relationships are meaningful. Local relationships (connectedness) are best preserved—sounds that are nearby in embedding space tend to sound similar—but sounds that are far away are not necessarily proportionally different in sound. It is important to perform multiple embeddings across a range of UMAP parameter values and initial conditions; the structures that persist across are most reliably present in the original dataset.

Figure 14.9 shows UMAP embeddings of both the constrained and unconstrained datasets, plotted side by side for comparison. To aid the visualization, the examples are grouped into 20 sound categories, using k-means clustering, and color-coded on the plots. Comparing the two plots, it

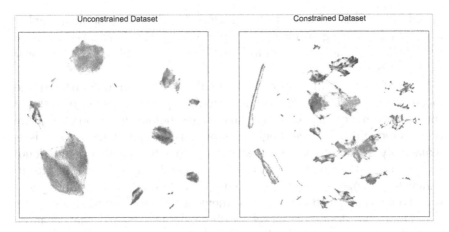

FIGURE 14.9   UMAP embeddings of the full parameters space (unconstrained dataset) and constrained network regions that exhibit greater variety and unpredictability (unconstrained dataset).

can be seen that the constrained dataset exhibits a greater degree of fine structure (many smaller clumps of points) as well as a greater diversity of sound categories (more colors). The major components of the unconstrained dataset can even be seen to be contained in the constrained dataset. These observations, aided by the visualized, lead us to believe that the constrained dataset exhibits a greater degree of sonic diversity than the unconstrained, and perhaps even a greater degree of unpredictability. Dunn, who navigates the possibility space in performance by slowly and occasionally tuning knobs in small amounts, would be more likely to stumbled upon novel behaviors when searching the constrained parameter settings, not the unconstrained settings.

In order to support these observations, we run a perturbation analysis. Perturbation analysis, common in dynamical systems theory, measures the sensitivity of a system to small changes, or perturbations. Here we apply perturbation analysis to the possibility space, essentially asking the question, given parameter settings that are very similar, how dissimilar are they in sound? Note that the analysis is performed on the data in its original dimensionality, not on the embedding features, which can distort relationships.

For each example in the dataset, we find its n nearest neighbors, as measured by the Euclidean distance between the two parameter vectors, giving a group, or neighborhood, of nearby points in parameter space. For each neighborhood, we then compute distances in possibility space, taking the distances between all points in the neighborhood, as measured by

Euclidean distance between feature vectors. Finally, we compute a number of statistics, including the mean and variance, which are intended to measure how dissimilar points are from the source point, as well as how dissimilar they are from one another.

We find that, within local neighborhoods of the constrained dataset, the mean and variance are significantly larger, suggesting a greater diversity of sounds and greater sensitivity to small changes, helping to support the visual observations above. Interestingly, the process of perturbation analysis is remarkably similar to Dunn's performance practice, which is no doubt inspired by it. Within Dunn's constrained control ranges, small changes to parameter settings are significantly more likely to produce large changes in sound outputs and lead to novelty in emergent sound behavior. This suggests that Dunn has found a subset of the parameter space that is more unpredictable and sonically diverse, optimizing his performance practice for these aesthetic preferences. Importantly, the embeddings provide a way of studying the structure of the possibility space, of identifying not just the diversity of sounds present and in what proportion, but, more importantly, of grasping its topology—which sounds are nearby to others in parameter space—providing a way of making judgements about control adjustments in performance as well as about decisions affecting system parameterization and design, or composition.

Notes

1 For an overview of Dunn's work and the metaphors of mind that run throughout it, see: Heying, Madison and David Kant. "The Emergent Magician: Metaphors of Mind in the Work of David Dunn." Sound American 18: The David Dunn Issue (2018), Web.
2 A contribution from the designer and maker of Nonlinear Systems, Andrew Fitch, is also included in this volume, see Chapter 23.
3 Numerical solvers approximate continuous curves by moving in steps along the derivative.
4 Editor's note: the digital implementation of this vactrol model includes these measurements made by Kant based on a trip to Radioshack as well as the linked asymetrical exponential filters, and is available here: https://github.com/davidkant/thresholds/blob/master/thresholds/classes/vactrol.sc.
5 Available at: https://github.com/davidkant/thresholds.
6 The code for this particular computation is detailed here: https://github.com/davidkant/temporal-features/blob/master/temporal_features/trending.py.

References

Bogdanov, D., Wack, N., Gomez, E., Gulati, S., Herrera, P., Mayor, O., Serra, X. (2013). Essentia: An open-source library for sound and music analysis. In *Proceedings of the 21st ACM International Conference on Multimedia* (pp. 855–858). New York, NY, USA: Association for Computing Machinery. doi: 10.1145/2502081.2502229

Kant, David, and Madison Heying. 2018. "The emergent magician." *Sound American.* http://soundamerican.org/theemergentmagician.html.

Maturana, H. 2002. "Autopoiesis, structural coupling and cognition: A history of these and other notions in the biology of cognition." *Cybernetics & Human Knowing* 9 (3–4): 5–34.

McInnes, Leland, John Healy, and James Melville. 2020. "UMAP: Uniform manifold approximation and projection for dimension reduction." arXiv. http://arxiv.org/abs/1802.03426.

Sprott, Julien C. 2000. "Simple chaotic systems and circuits." *American Journal of Physics* 68 (8): 758–763.

15

VIRTUAL MATERIALITY

Simulated Mediation in the Eurorack Synthesizer Format

Ryan Page

There is a tension between nostalgia and novelty in Eurorack-format modules that is related but not limited to that of digital computing. This is especially true of modules that simulate analog media. Eurorack modules are often hybrid objects that process signals digitally but interface with analog signals. The transmission of audio and control information as analog signals in modular synthesizers complicates distinctions between analog and digital media. Working from interviews with Eurorack programmers and designers, as well as my experiences as a scholar and practitioner, I detail in the following chapter how the fuzzy boundaries and material implications of modularity enable and constrain our understanding of simulation.

A module that simulates analog media is never a neutral translation. Salient properties must be abstracted and then modeled. How designers implement these abstractions will alter the aesthetic outcomes. To better understand mediation and the diverse cultures of simulation in modular synthesis, I spoke to several programmers, designers and manufacturers of prominent Eurorack modules to document their practical and theoretical approaches to transferring technical and artistic concepts across devices. I examine how the aesthetics of simulation influenced Eurorack modules' designs and how these abstractions provide altered aesthetic outcomes. By examining the abstractions used in simulations, it is possible to determine what the designer determined were the salient properties of the object they were aiming to simulate.

Initially, the only available Eurorack modules were approximations of extant analog synthesizer components, such as the transistor-ladder filter designs found in Moog designs or the state-variable designs popularized by Tom Oberheim's eponymous brand of synthesizers. The format has since

DOI: 10.4324/9781003219484-17

expanded to include thousands of digital modules, in addition to the proliferation of a variety of analog designs.[1] This association with early analog synthesizers often codes the cultural reception of Eurorack as nostalgic. Despite this, contemporary Eurorack modules often feature the same embedded microcontrollers found in consumer electronics and generate sounds using modern digital signal processing techniques rather than analog methods.

It is challenging but necessary for synthesists to come to terms with the aural and functional limitations of these simulations. For example, a user may be asked—via associated paratexts such as marketing materials, manuals, design choices, or naming conventions—to think of a module as a tape loop, despite the fact that manipulating the audio in ways similar to tape will result in sounds specific to digital audio, such as aliasing. A "tape" module may digitally simulate the mild compression, saturation, low-pass filtering, and wow and flutter of analog tape,[2] it could include analog circuitry to recreate this effect,[3] or may simply refer to the manner in which the media is stored and accessed.[4] In other cases, tape is simply an interface metaphor for a digital audio buffer that makes no attempt to reproduce any of the characteristics of analog tape machines or tape. In addition, these modules may also provide controls unavailable on analog tape, such as the ability to adjust the pitch of recorded material at fixed intervals, or start and stop playback at arbitrary points, and synchronize to additional musical material.

By focusing on the unit of the module and its status as an individual object, I aim to highlight its precariousness: modules are dependent on external input to operate and are therefore difficult to define outside of their relationship to other modules. Each module is part of a larger system for producing sound. In this context, the process of defining how simulation functions presents challenges. Many modules are engineered to achieve broad functionality. Their use is dependent on how the user connects the various inputs and outputs. For example, Maths, from North Carolina-based manufacturer Make Noise, can function as a filter, slew rate limiter, low-frequency oscillator, mixer, logic gate, gate to trigger converter, and a variety of other applications, depending on how it is patched.[5] The information received into the devices' analog inputs also changes the functionality, often in drastic ways.[6] In an analysis of simulation in the context of modular synthesis, it is critical to examine the methods designers used to simulate media in both internal mechanisms and external means of interaction.

In addition to examining the ambiguity created by the patching interface, it is necessary to consider which properties were recreated using discrete components, integrated circuits, or microcontrollers. For example, Mutable Instruments' Streams combines analog VCA and VCF circuits with digital control signals provided by an ARM Cortex-M3 microprocessor.[7] In one mode, a simulation of a vactrol—an electronic component that combines a light-emitting diode (LED) with a photoresistor—is used to convert the module into a dual low-pass gate. When analyzing the device, one must combine an analysis of the code with an analysis of the schematic to properly examine the aesthetic choices made in the process of simulation.

If a manufacturer uses harvested integrated circuits from extant digital synthesizers, such as many of the modules developed by manufacturer ALM/ Busy Circuits,[8] the potential added cost and difficulty these parts bring to the manufacturing process offer some indication of their importance to the design. There are a variety of reasons a designer might seek to use extant circuits. Manufacturers that use components from consumer electronics such as the Commodore 64 and the Sega Genesis often market their modules based on the nostalgic connection potential users feel for these deprecated technologies. In other cases, the manufacturers I spoke to were ambivalent on the relevance of these circuits. For example, Tom Whitwell of Music Thing Modular described his choice to use discrete logic integrated circuits rather than microcontrollers in his Turing Machine modules as "pure 'mojo' ridiculousness."[9]

In some cases, digital modules provide simulations of analog synthesizer components, but in many instances, the modules are simulations of communications media such as ¼" tape, compact disks, radios, and components of video game consoles. Many of the most widely used modules, such as Make Noise's Morphagene and Music Thing Modular's Radio Music, are also designed to simulate aspects of the functionality of the media they emulate. While the simulation of deprecated media technologies is not unique to Eurorack, software plugins such as Goodhertz's Lossy and guitar pedals such as Chase Bliss' Generation Loss simulate MP3 artifacts and VHS generation loss, respectively[10]—the interleaving of analog and digital information in Eurorack is distinctive, as is the ability to alter parameters of the simulation at audio rates.

While the poetics of mediation have historically had little place in musicology, there is abundant evidence it has been an important aesthetic consideration in both experimental and popular music. Compositions such as Alvin Lucier's *I am Sitting in a Room* (1969) and *Quasimodo the Great Lover* (1970), Max Neuhaus' *Radionet* (1977), Maryanne Amacher's *City Links* (1967), and John Cage's *Radio Music* (1956) are notable examples of experimental works where mediation informs structural elements of the music. Popular musicians and soundtrack composers have also adopted these techniques. Consider the soundtrack to *Doom* (Bethesda Softworks, 2016). In a video posted to *Doom* composer Mick Gordon's YouTube channel, he claims:

> I came up with this idea of breaking sound into its kind of bare-bones
> components, which are sine waves and noise. I then pulsed rhythms of

sine waves and noise through vast arrays of analog equipment. So, the idea here was not to use the analog equipment as an effects processor, even though that's sort of what it's made for. I am using the equipment to corrupt the pure sine waves and noise.[11]

The use of the word "corrupt" in this context explicitly refers to the alteration of information as it passes through a medium. In the video, Gordon can be seen using a 12U[12] modular system. While he refers to the equipment that he uses to process the signal as analog, the patch he is using also contains a number of digital modules, including digital emulations of analog equipment such as the Dave Smith Instruments Character module.

The design of a simulation blurs the boundaries between scientific and aesthetic practices: to what extent is a simulation a scientific model or an aesthetic abstraction with intentional divergences from reality? What is the relationship between these perspectives on simulation and the perception of verisimilitude? These questions should be considered an essential part of understanding the function of simulation in a given module.

The Make Noise/Soundhack Erbe-Verb: A Meta-Reverb

The Erbe-Verb is a module coded by Tom Erbe and designed in collaboration with Make Noise, a prominent Eurorack manufacturer.[13] It is ostensibly a digital reverberator, but is capable of synthesizing sound without external input. In operation as a reverberation device, it receives a monaural audio-rate signal which is then sent through a delay feedback network, as well as a series of all-pass filters, before entering an analog mixer where it is combined with the original signal and is sent to the stereo output. Rather than the traditional reverb function of creating an artificial "space" for incoming sounds, Erbe-Verb is a malleable effects system designed to morph between many reverb types. As Erbe notes:

> Most reverb processors are used as the last effects device, the room that one places an instrument in. Although the Erbe-Verb supports this use, I wanted it to be more fully integrated into a synthesizer voice. For this reason, all parameters are fully modulatable, designed to be controlled by common control-rate signal generators such as envelope generators, low-frequency oscillators and envelope followers. Each note, moment, or gesture can have its own resonant character.[14]

The module features a front panel with nine parameters that the user may adjust using potentiometers and control voltage inputs. While it is possible to produce an approximation of physical spaces by setting the parameters on the front panel or by modulating it with external control voltages, it is not the primary purpose for which the module was designed. As Erbe stated in an interview I conducted with him:

> I had a bit of a reaction to previous reverbs, since reverb is really almost totally a digital development, where almost all of the good devices are digital devices. Especially because of the economy of how things got developed at first, everything was sort of a "patch-like" system. You have a patch, then you go to another patch, and then maybe you dive down and tweak a few variables, and then maybe save another patch. Then you're just refining all these little models. So, one of the main motivations was to make something that was not like that at all.[15]

A traditional digital reverb either approximates the reverberation of sound in a physical space of a given size and shape by simulating the reflections given by walls positioned in a particular relationship to a source or simulates mechanical reverberations created by plates or springs by emulating the reflections in those devices. In either case, the approximations may diverge significantly from the physical space or mechanical reverberation technology. Many of these divergences are considered aesthetically appealing by musicians and engineers. As Jonathan Sterne notes in his essay *Space Within Space: Artificial Reverb and the Detachable Echo*, "Reverb devices achieve canonical status not because of their realism or their particular operational characteristics but because of sonic signatures that they impart to notable passages in notable recordings."[16]

Many of the commercially available artificial reverbs articulate the size of a reverb algorithm in terms of the physical length of a space and are meant to model the dimensions of a room in which that quality of sound could be achieved mechanically, from simple energy dissipation as sound waves reflect off surfaces. While there is a size parameter on the Erbe-Verb, there are no markings to indicate how each knob position correlates to a specific distance. Erbe describes the range of the size parameter in the Erbe-Verb as extending from four inches to over four hundred feet. The extreme range of sizes, the relatively coarse control offered by the single turn potentiometer of the size parameter,[17] as well as the lack of numerical feedback, make specifying an exact room size by hand unfeasible in this particular system. Instead of simply mediating the incoming sound by a specific set of parameters to create the impression of the sound reverbing in a real space, as a traditional reverb might, the Erbe-Verb's design encourages the musician to perform extensive modifications of the sound in real time.

The module provides voltage control of the size, speed, pre-delay, absorption, modulation depth and type, filtering, decay, and mix of the reverb.[18] The control voltage input for the size parameter allows for audio-rate modulation—providing automated, real-time alteration of the simulated space. This parameterization allows for modulations that would be physically impossible in a real space and creates significant alterations to the input signal. In contrast, contemporary digital Eurorack reverbs such as the Æverb from Audio Damage or the Halls of Valhalla card for the Tiptop Audio Z-DSP[19] offered fewer control parameters, and sonically relevant parameters such as size were either unavailable or fixed to specific algorithms.[20] Internal feedback featured in the Erbe-Verb's digital signal processing allows the module to function as a synthesis voice without external input; the medium becomes the message. Erbe specifically recalls an instance of this, when composer Alessandro Cortini created a patch that modulated the size parameter at audio rate to produce cymbal-like timbres.[21]

The Erbe-Verb is an extreme example of what Jonathan Sterne describes when he notes that "an artificial reverberator is more aesthetically akin to a musical instrument than a building."[22] While the continuous control of certain reverb parameters is not dissimilar to reverb pedals or controllers such as the LARC for the Lexicon 224, the ability to smoothly interpolate between reverb types is unique, as is the incorporation of voltage control for every parameter.

The implementation of voltage control adds to the Erbe-Verb's instrumentality. Voltage control allows the user to map reverb parameters onto various performance-oriented controllers or sequencers. In traditional rackmount and pedal reverbs, internal low-frequency oscillators (LFOs) are mapped to set parameters. Parameters such as "size" must be adjusted manually. An engineer creating a reverb intended to approximate the acoustic properties of a space, or a particular class of reverb hardware, such as plates or springs, would not be incentivized to provide the user with the ability to continually modulate the reverb size simply because natural and mechanical reverbs do not continually alter their size. The Erbe-Verb draws upon the aesthetics of continuous parameter modulation common in modular synthesis and connects this with an abstraction of established reverb types. Even reverb systems that incorporate MIDI control do not allow the user to route audio-rate modulation, following the limitations of the MIDI specification. Erbe claims that his goal was to "develop knobs that had a large amount of space in them"[23] and that "my criteria often was often to go a little bit too far in each direction."[24] The decision to provide a wide range of values for each parameter meant that the scaling of the incoming voltage from the potentiometers and control voltage inputs became an important aesthetic consideration. He notes, "The knobs do not make sense in any sort of way, except that I wanted to make them feel good when you perform on them. I think that's what led me to that idea of avoiding any presets; to make something that was quite performable."[25]

Classifying the Erbe-Verb is difficult. It is not an attempt to simulate a room or a specific class of reverbs. Instead, Erbe has extracted aesthetic properties from previous reverb models and abstracted these properties into a malleable reverb capable of approximating many reverb types and interpolating between them. While it is possible to describe the Erbe-Verb as a tool for simulating various reverb types, this fails to account for the emphasis on interpolation and performance. The Erbe-Verb is more accurately described as a meta-reverb instrument that abstracts various parameters of historical reverb algorithms and incorporates them into a modular synthesis environment.

ModBap Modular: "Everything Is Vinyl"

ModBap Modular was founded in 2020 by Corry Banks, a hip-hop producer and journalist who previously created the blog BBoyTechReport. Banks designs the modules in collaboration with Ess Mattisson, who is responsible for programming and engineering the modules. According to Banks, he created ModBap "to create things with the intention of Hip-hop, with the intention of that kind of beat-driven music, but in technology."[26] Throughout my interview with Banks, he emphasized the importance of providing tools designed for the purpose of making hip-hop, since, in his view, "a lot of the stuff that we use as ours, in this genre, in our culture to make this music, none of it was designed for us to do what we do."[27] The word "modbap" is a portmanteau of modular synthesis and boom-bap. Boom bap is a subgenre of hip-hop characterized by minimalist production with an emphasis on the kick and snare within a sequence or loop. The word itself is onomatopoeic, with "boom" and "bap" originating from beatboxers' vocalizations of kick and snare sounds. According to ModBap Modular's website:

> The term was created by Banks as a denotation of his experiments with modular synthesis and boom-bap music production. From that point forward, a movement was born where like-minded creatives built a community around the idea of Modbap.[28]

Banks has released two modules, with another currently in development.[29] ModBap Modular's first release was the Per4mer, a multi-effects unit featuring four simultaneous effects that the user can trigger in real-time using four arcade-style buttons mounted to the front panel. The aforementioned effects are delay, reverb, "glitch" (a short sample of the audio input repeated indefinitely), and tape stop. In addition to the performable effects, Per4mer features eight "color' processors and a compressor. The color processors were designed to impose the aesthetics of sampling and mediation on real-time synthesized sounds. While the "classic" and "lofi" modes emulate early 12-Bit and 8-Bit samplers, respectively, Banks designed Wax and Wax2 to replicate different characteristics of vinyl records.

I wanted performance effects, but I also wanted this aesthetic that I could have the vinyl sound and technique, or the sampling sound and technique with original melodies and original music, and still have that feeling. And it kind of becomes its own thing, but if you go deep enough into it, it can be indistinguishable from "yo, where'd you get that sample from?"[30]

ModBap's next release was Osiris, a wavetable oscillator featuring a unique fidelity parameter inspired by the earliest personal computers capable of sample playback, such as the Commodore Amiga and Atari ST series. According to Banks, it was important that Osiris was capable of producing high-fidelity audio so that the user could shape the sound to their liking.

The oscillator runs at 96kHz, it's just a very clean, pristine sound, but then, being able to dial in more of the character, more of this musical degradation, that was really important.[31]

The fidelity parameter introduces phase jitter, as well as bit and sample-rate reduction. These effects are based on algorithms Mattison developed to emulate the Commodore Amiga, as well as experiences Banks had with Atari computers in recording studios. According to Banks "I remember the sound was very specific, and it was great for hip hop."[32]

Banks told me he wanted his modules to be accessible to newcomers, especially those that might otherwise feel alienated by the notorious difficulty and lack of documentation prevalent in Eurorack. For example, the manual for Per4mer is 47 pages long[33] and features numerous illustrations detailing the installation of the module, block diagrams describing the signal flow and charts pertaining to the LED color associated with each effect. According to Banks:

That was a big thing for me because I had a goal of introducing Eurorack to more hip-hop producers and first-time Eurorack users. It feels odd to say that. I don't think I voice that too often, but that was a goal of mine. If I am saying, "I'm all about hip-hop, and I want to be the voice of hip-hop in everything I do and design these things from that perspective. I already knew that it would be a lot of first-time users coming in. So, I thought that the documentation had to be easy to get into and accessible.[34]

Many of Banks' customers are new to Eurorack or uninterested in it as a platform aside from his creations. Banks intentionally designed Per4mer to send and receive line-level signals and sells a Modbap Modular-branded 20HP case[35] designed by 4MS Company[36] so that producers could use Per4mer as a stand-alone effects processor.

The simulations of media in Per4mer and Osiris should be understood within the context of tailoring instruments to produce hip-hop. As a genre distinguished by its emphasis on repurposing and altering recorded media, the history of hip-hop is marked by explorations of the limitations and possibilities of media technologies, from turn-tables to samplers. The simulations in Per4mer are influenced by the gestures of disk jockeys and producers developed for these technologies, such as scratching records or performing tape stops. Like many of the other designers I interviewed, Banks' simulation practice was focused on reproducing an aesthetic experience rather than emulating components because "at the base level of everything, you're going for a sound."[37]

Radio Music: A Digital Emulation of Radio Aesthetics

Music Thing Modular is a small London-based Eurorack company founded by Tom Whitwell, a programmer, designer, hardware engineer, and former assistant editor of *The Times*. Unlike Make Noise, Whitwell does not manufacture the modules he designs.[38] Instead, he furnishes kits including printed circuit boards (PCB), front panels, and the electronic and mechanical parts necessary to build his modules. The circuit designs, PCB layouts, and code are released under Creative Commons licenses and are available for use or modification.

Radio Music is a sample player in the Eurorack format that emulates some features of radio, such as multiple channels of synchronized audio that are accessible by "tuning" to a particular "station."[39] It plays audio-rate samples that the performer alters via two knobs and a single momentary button on the front panel, as well as respective control voltage and gate inputs. Whitwell did not design the hardware interface to resemble a radio, nor did he attempt to reproduce the characteristic static, distortion, interference, or low-level noise of analog radio. However, he did preserve some of the basic interactive features of radio. The user selects "stations" by adjusting a potentiometer on the front panel in a manner that roughly approximates tuning into various radio stations on an analog radio. Whitwell was initially inspired by Donald Buchla's experiments with voltage-controlled analog radio, as well as John Cage's *Radio Music* (1956),[40] which gave the module its name. Whitwell explains that he decided to transition from a voltage-controlled analog radio to a digital emulation after attempting a performance with this analog prototype in an electrically insulated room.

> There's a modular meet-up in Brighton every year, and I went down to that. It is held in the university there, and this year it was held in a room that was essentially a Faraday cage, and the room has got lighting gantries all over it. So, I went in and plugged in this thing I was going to show everyone, and there was just nothing there.

Obviously, that was a kind of an amazing thing to happen; to go, and my radio module doesn't produce any audio, but it was a little bit annoying, and so at that point, I thought, 'okay, there is they should be a way of doing this.' So, it's the same joke, the same sort of gag, but it works in a more reliable way.[41]

The module uses a Teensy 3.1, which is an ARM Cortex-M4 development board[42] that can be programmed with a bespoke audio programming language called Teensyduino.[43] The Teensy provides micro-USB connections as well as low-resolution analog-to-digital converters (ADCs) and one high-quality digital-to-analog converter (DAC). It is considerably easier to reprogram and write new firmware onto the Radio Music than many other digital modules due to its available USB connectivity. Whitwell mentions that the fact that Radio Music was "hackable" was something he only became aware of after releasing the module, but that this became "an important and interesting part of it."[44] This, along with the considerable resources and encouragement provided to the user base by Whitwell meant that a number of alternate firmwares were created for the Radio Music by its users. Whitwell hosted many of these on the official Music Thing Modular GitHub repository, further increasing their circulation. While the initial firmware was programmed by Whitwell, the current official firmware was created by Martin Wood-Mitrovski, a customer with no affiliation with the company. As Whitwell noted in my interview with him, "My only real brief [for the outside programmer] with that firmware was that you could use it exactly as it was beforehand, without using any other features, and as far as I know, it does that."[45]

The degree to which Whitwell incorporated external modifications into his existing product strongly suggests that he was less interested in the specifics of simulating radio than he was in abstracting specific attributes as part of a sampler design. He confirmed this when I spoke to him, stating:

With the Radio Music, there are ways that you could much more explicitly reproduce a radio. You could have it so that you could tune between stations. You could have it so that you could have it untuned, and it would make the sound of an untuned radio. And I suppose I partly thought that would be difficult, so I didn't do it. And also, I think it comes back to skeuomorphism as well somehow. Having a sampler that behaves in the way it switches between stations or audio files in this slightly odd way, to me, felt fine. Whereas trying to simulate—you could imagine having FM/AM/Shortwave switch and simulate the going between them—that to me feels very skeuomorphic and kind of like rendering a wooden surface in your VST plugin. It felt like that would be kind of weird. And so it feels a bit like you are giving somebody too much of the answer.[46]

Whitwell did, however, provide a library of suggested sound files, including many recordings of government radio propaganda, lectures by philosophers and artists he admired, field recordings, and other long-form sounds. These suggested files were distinct from the preset patches and samples typically burned into the ROM of various samplers in that they were optional. However, this library of sounds became an important feature of the module because, in the initial firmware produced by Whitwell, users had to convert their files into the RAW format through a somewhat tedious process before transferring these files to the module. This made it difficult for users to transfer large numbers of small files onto the device, as the bit and sample rate conversions that enabled the translation to this format were error-prone. Rather than converting their own files, many users relied upon the available libraries that Whitwell created. When comparing the Radio Music to One, a nearly identical module manufactured by Tiptop Audio, Whitwell was quick to point out, "I don't imagine it comes with hour-long recordings of North Korean radio stations."[47]

Given the highly abstract relationship between the Radio Music and radio, it might be tempting to conclude that this connection is superficial. Other media such as television or tape loops could be used as a metaphor for the interface. In this regard, the context in which the Radio Music is presented is important. While there are now many pre-built or used Radio Music modules available for purchase online from independent builders, the module was originally only available as a kit. This meant that in order to build and test the device, users would necessarily come into contact with Whitwell's commentary on the module, suggested readings from Alvin Lucier, John Cage, and Robin Rimbaud, as well as the curated audio recordings. At the very least, a builder must spend time examining the PCB in order to determine the placement of electrical components. When they do, they will find explicit references to John Cage, [Karlheinz] Stockhausen, Akin Fernandez, Don[ald] Buchla, Nicholas Collins, John Peel, Max Neuhaus, UbuWeb, and Resonance FM amongst the labels for resistor values and diode orientations. The curation of suggested content does not determine the module's use, but it does offer strong cues to the user. These samples are also audible in the most popular video tutorials on the module,[48] furthering their association with the device.

Another distinguishing feature of the Radio Music is its emphasis on the use of long-form audio recordings. According to Whitwell, "One 32 GB micro SD card can store about four-and-a-half days of audio in the normal Radio Music format."[49] Samplers have, historically, tended to allow for polyphonic articulation of relatively short recordings. While the expense of random-access memory was the primary reason that early samplers did not allow for longer sampling times,[50] there were other complications preventing the use of longer samples, such as the challenge of synchronizing longer files. Sample manipulation becomes difficult with longer sample times, and it becomes difficult to control the time and pitch relationships between sustained non-periodic material beyond a few seconds. Even periodic material will often fall out of sync unless it is arrhythmic or manually synchronized. Most commercial samplers require that a key or pad remain depressed in order for the sound to continue playing. Audio on the Radio Music begins playing as soon the module receives power, without requiring a gate or trigger from external modules. While it is possible to reset to the beginning of the file by pressing the reset button or sending an external gate into the reset input, there is no way to stop the file from playing. Users can control where the file starts, but the resolution of the ADC prevents precise control of the position in longer files.

Radio Music is a module that cannot be defined by its functionality and interface alone. Judging from these on their own, the connection to radio is quite tenuous. However, the way that the device is situated and the performance practices Whitwell encourages by framing the building and sampling process in terms of radio art weigh heavily on its reception and use. While the musician using the device is free to ignore this information, Whitwell himself considers it to be an important aspect of the instruments he builds. During my interview with him, he claimed that,

> I always want to have some sort of story, or, I suppose, not always, but generally, I want to have some sort of narrative. For me, the narrative around the thing is important.[51]

By providing longer samples and making it difficult for the user to operate it as a traditional sampler, Whitwell encourages the user to consider simultaneity, something that many of the composers he cites were invested in as a compositional strategy. The user of Radio Music is expected to use the module as a source of indeterminate material for processing. By synchronizing all audio "stations", Radio Music allows the user to consider the dynamic relationship of various unrelated recordings over time, much like the performer of Cage's *Radio Music* (1956). It is this connection to the performance practices of radio in experimental music that makes the Radio Music legible as a simulation.

Aesthetics of Digital and Analog Systems

There was a clear consensus among those I interviewed that while there are valid reasons to be attracted to anachronistic forms of media, particularly analog media, they were less interested in exact simulations of older media than appropriating the aspects they found aesthetically appealing and musically meaningful. These conversations occurred against a backdrop of analog fetishization in the professional audio industry and the synthesizer hobbyist community. The importance of the particular characteristics of analog media varies depending on the use case and, more generally, context. A company focused on marketing its products to professional musicians will likely have less interest in emulating a particular medium's limitations than they will in extending its functionality. As Kris Kaiser from Noise Engineering notes, "most professionals we talk to don't care whether things are digital or analog, they care what the product does."[52] In addition, because these modules are products that are meant to provide value to the musicians that use them, tension may arise between the designer's aesthetic instincts and the desire to appease customers. Tom Whitwell addressed this tension in relation to his decision to use discrete CMOS logic chips to design the Turing Machine, Music Thing's shift register-based sequencer,

> You sort of think, 'obviously, you could do it quite simply with a straightforward microprocessor.' I imagine you could do it at audio rates because the CMOS chips don't go that far. If you try and do it at like 10,000 hertz, it won't work. It goes up to—I don't know—a couple of thousand hertz. For me, it just doesn't seem interesting, doing it like that - but that is pure 'mojo' ridiculousness. There's no customer end value for that.[53]

Questions such as this point to the ambiguous status of modular synthesizers now that it is possible to emulate all of the components of a modular system on a laptop computer. The software suite VCV Rack simulates the look and functionality of Eurorack, albeit with a suggested design language that replicates modern computer interfaces' flat designs. Many manufacturers offer free versions of their modules in the VCV Rack language, including an adaptation of Whitwell's Turing Machine. One of the VCV brands, Audible Instruments, shares the interface and code of the Mutable Instruments Eurorack modules. There was a sense among those I interviewed that the differences between analog and digital media were not necessarily inherent properties of either medium but instead reflected the design principles of most commercially available analog emulations. A programmer may determine that oscillator pitch fluctuations, filter saturation, noise, and non-linearities in various components' frequency response are not worth simulating.[54] Idealized circuits are often simpler to program and consume fewer resources than those that simulate circuit components at greater degrees of granularity. In my interview with Stephen McCaul, he addressed these design choices in regard to a stigma toward digital modules:

> You do still hear complaints but they are often valid and interesting. Someone we talked to recently was lamenting how accurate his clocking was in the box which is one reason he likes modular because it was

actually much messier which made it sound more human to him. Once you unpack this from 'digital sucks analog is better' (which is somewhat the way it was originally presented to us in that discussion) you see that he has a very specific and very valid reason for his preference that really has nothing to do with it being a computer or circuits or digital or analog but it is how we as a society have taught people to oversimplify this.[55]

The examination of Eurorack modules offers unique challenges compared to either the study of mass media or analysis of complete musical instruments. Analyzing the relationship between Eurorack modules and the media they simulate offers a constructive means of studying the design principles and aesthetic choices made by their designers and programmers. I have outlined these designers' specific processes for abstracting various forms of media to establish a methodology for contextualizing the modules' effect on the music created with them. While the choices made in designing a module do not determine specific aesthetic outcomes, designers use these decisions to direct the musicians' use of the devices. In some modules, such as the Radio Music, the designer's choice to emulate a particular medium forces the user to avoid particular interactions while simplifying others, such as the playback of files over an hour in length. In others, such as the Ataraxic Translatron and the Erbe-Verb, the module's sonic character is difficult to avoid.

The simulated reverbs, radios, and video-game consoles in Eurorack-format modules form only a small subset of simulated media. They do, however, offer some of the most complex examples of simulation, given their hybridity and interconnectivity. While Eurorack modules themselves are not widely used by the general public[56] these designs have an outsized influence on popular culture due to their use by soundtrack composers such as Hans Zimmer, Trent Reznor, Mick Gordon, Michael Stein, and Kyle Dixon. Additionally, Eurorack offers unique challenges to the study of simulation that complicate the study of less complex simulated media such as those found in software plugins or guitar pedals. Like the internet ecosystem in which they are promoted, discussed, and sold, Eurorack modules represent and reflect a complex relationship with culture, one that is at once highly referential and personal. Eurorack, more so than any previous modular synthesis format, is a global phenomenon with a startlingly diverse range of products and personalities. Given that the range of professionalism, musical and philosophical interests, cultural backgrounds, and personality types are reflected in module design and functionality, Eurorack offers a rich opportunity for examining how culture becomes embedded within technology.

Notes

1 Chris Carter. "Doepfer A100." Accessed 14 June 2021. http://chriscarter.co.uk/content/sos/doepfer_a100.html.
2 Strymon. "Magneto - Four Head DTape Echo & Looper Eurorack Module." Accessed 29 June 2021. https://www.strymon.net/product/magneto/.
3 Instruō. "Instruō - Lúbadh." Accessed 16 August 2021. https://www.instruomodular.com/product/lubadh/.
4 "The Phonogene is a digital re-visioning and elaboration of the tape recorder as a musical instrument. It takes its name from a little known, one of a kind instrument used by composer Pierre Schaeffer." "Make Noise - Phonogene," 12 March 2013. https://web.archive.org/web/20130312123152/http://www.makenoisemusic.com/phonogene.shtml.
5 "Make Noise Co. | MATHS."
6 A four-quadrant multiplier can act as a VCA and modulate volume when a unipolar envelope is applied to an audio signal, whereas a bipolar audio-rate signal will result in amplitude modulation when modulating another audio-rate signal.
7 "Modules – Streams – Specifications | Mutable Instruments," 7 February 2015. https://web.archive.org/web/20150207030252/http://mutable-instruments.net/modules/streams/specifications.
8 "ALM - ALM011 Akemie's Castle." Accessed 16 August 2021. https://busycircuits.com/alm011/.
9 Tom Whitwell, *Music Thing Modular Interview*, 8 January 2021.
10 The nearly ubiquitous tendency towards simulating analog media remains relevant to the discussion of the particular implementation of simulation in the Eurorack format.
11 Mick Gordon, *DOOM: Behind The Music Part 2*, 2016, https://www.youtube.com/watch?v=1g-7-dFXOUU.
12 12U describes the dimensions of a modular synthesizer's housing: for a 12U modular synthesizer case, there are four rows of modules. Each row is three standard rack units tall.
13 "Make Noise Co. | Erbe-Verb." Accessed 16 August 2021. http://makenoisemusic.com/modules/erbe-verb.
14 Tom Erbe, "Building the Erbe-Verb: Extending the Feedback Delay Network Reverb for Modular Synthesizer Use," 2015, 4.
15 Erbe, Tom. *Tom Erbe Interview*. 2021.
16 Jonathan Sterne, "Space within Space: Artificial Reverb and the Detachable Echo," Grey Room 60 (July 2015): 110–31, https://doi.org/10.1162/GREY_a_00177.
17 It should be noted that the position of a knob does not correlate to a linear increase in size. Erbe intentionally emphasized the ranges he found to be interesting.
18 "Make Noise Co. | Erbe-Verb." Accessed 16 August 2021. http://makenoisemusic.com/modules/erbe-verb.
19 Tiptop Audio. "Z-DSP." Accessed 21 April 2021. https://tiptopaudio.com/zdsp-ns/.
20 The Tiptop Audio Z-DSP is a platform for digital signal processing based on the Spin FV-1 signal processor. The user purchases cartridges with various effects loaded onto the cartridge's ROM.
21 Erbe, Tom. "Building the Erbe-Verb: Extending the Feedback Delay Network Reverb for Modular Synthesizer Use," 2015, 4.
22 Jonathan Sterne, "Space within Space: Artificial Reverb and the Detachable Echo," Grey Room 60 (July 2015): 110–31, https://doi.org/10.1162/GREY_a_00177.
23 Tom Erbe. *Tom Erbe Interview*. 2021.

24 Ibid.
25 Ibid.
26 Banks, Corry. *ModBap Modular Interview*, 2 January 2022.
27 Ibid.
28 Modbap Modular. "Our Story." Accessed 12 February 2022. https://www.modbap. com/pages/about.
29 Banks, Corry. *ModBap Modular Interview*, 2 January 2022.
30 Ibid.
31 Ibid.
32 Ibid.
33 Manuals for modules are often short or non-existent.
34 Banks, Corry. *ModBap Modular Interview*, 2 January 2022.
35 Modbap Modular. "Modbap20 Powered Eurorack Case." Accessed 19 February 2022. https://www.modbap.com/products/modbap20-powered-Eurorack-case.
36 4MS is a Chicago, Illinois-based Eurorack Manufacturer.
37 Banks, Corry. *ModBap Modular Interview*, 2 January 2022.
38 Mylar Melodies, *WHY WE BLEEP PODCAST 001*.
39 Tom Whitwell, *Music Thing Modular Interview*, 8 January 2021.
40 Ibid.
41 Ibid.
42 "17 Things to Know about the Music Thing Radio Music Module." Accessed 12 January 2021. https://musicthing.co.uk/pages/radio.html.
43 This language borrows extensively from the Arduino programming language, as indicated by the name.
44 Tom Whitwell, *Music Thing Modular Interview*, 8 January 2021.
45 Ibid.
46 Tom Whitwell, *Music Thing Modular Interview*, 8 January 2021.
47 Tom Whitwell, *Music Thing Modular Interview*, 8 January 2021.
48 Future Music Magazine. *Modular Monthly: An Intro to DIY Eurorack & the Radio Music*, 2016. https://www.youtube.com/watch?v=9g2Q0esgBuk.
49 "17 Things to Know about the Music Thing Radio Music Module." Accessed 12 January 2021. https://musicthing.co.uk/pages/radio.html.
50 The E-MU Emulator and the Ensoniq Mirage—two early professional samplers—each had 128 kilobytes of memory, approximately two and six seconds of sampling time, respectively.
51 Tom Whitwell, *Music Thing Modular Interview*, 8 January 2021.
52 Kris Kaiser, *Kris Kaiser Interview*, 2019.
53 Tom Whitwell, *Music Thing Modular Interview*, 8 January 2021.
54 Audible pitch drift is not necessarily a feature of analog, voltage-controlled oscillators. The Prophet 6 and OB-6 VCO-based synthesizers from Sequential respectively include "slop" and "detune" parameters that introduce random pitch fluctuations.
55 Stephen McCaul, *Stephen McCaul Interview*, 2019.
56 Most manufacturers I spoke to were not comfortable providing sales figures, but the general estimates I was provided with were between four to five figures.

16

INTERVIEW

Designing Instruments as Designing Problems

Meng Qi

EE: Einar Engström

MQ: Meng Qi

Translated from Chinese by Ruochen Bo

EE: It seems that interference and instability are very interesting for you.

MQ: One thing that interests me a lot is controllable chaos. Machines can make some decisions by themselves, but as a player you can control the direction of these decisions.

EE: Are you implementing chaotic equations into your circuit designs?

MQ: No. I am doing mostly a network of feedback. A lot of modulations are interwoven in my circuit design. Especially in the Wing Pinger.

EE: Can you describe the Wing Pinger? I sense a really elegant implementation of circuit interference.

MQ: The internal structure of Wing Pinger has several feedback loops. In the logic circuit there's an interweaving system—two oscillators. Every step of the output will affect how the other oscillator works, so the final result is quite unpredictable, yet controllable. So just like what I was saying, you can control the general direction of the sound synthesized, but the precise details of the sound are determined by the circuit itself. I find this kind of situation to be interesting, first of all, and second, it sounds nice. Regarding the interface design—actually I've put a lot of work into the interface design. The internal structure has a lot of data parameters, what

DOI: 10.4324/9781003219484-18

I tried to do is for the knobs (and the range they reach) to arrive at a healthy balance between musicality and practicality. For instance, I tried to make the range of the knob largely remain musical, but also in some way conducive to experiments. When you turn the knob, you can feel something natural to it. For instance, with a 10% turn of the potentiometer, one should feel that there is a proportional change in the sound: 10% as well. But that's just a feeling, as the actual change in sound is not a mere change in number—as in when going up an octave will directly result in an octave change only. But in a complex and chaotic system, a 10% turn of the knob might completely change the sound—its timbre, its volume, the dynamics. So I wanted to establish a correspondence between the knob and the sound, as well as a healthy, satisfying range of change. Also, there are ten knobs on the panel, with two for volume, which makes for an intuitive interface. One can use intuition to play Wing Pinger, without particular regard for technical details. This appeals to me. It depends not on technical understanding or mastery, but on listening to the sound while operating, and on a kind of muscle memory. This is what I wanted to make. We've discussed Peter Blasser's circuits, but I failed to mention the concept of gestural patching. Gestural patching refers to the process of quickly forming muscle memories. This makes a circuit akin to traditional musical instruments. This is what I'm interested in. Since patching is a primarily technical skill, making it gestural would be fun. I've always been interested in this direction and have been experimenting with it (Figure 16.1).

FIGURE 16.1 Wing pinger.

EE: This is really interesting. Patching, as we are familiar with it, means to draw out sound by spanning voltages across connection points. You're talking about archiving this movement of the body, freezing it in space-time for future access.

MQ: Right. It means that what you are making and performing is not only rooted in the quality of sound. For instance, with traditional musical instruments what matters is the sound itself. You use strength and power to control volume, and use fingerings or different hand gestures to control timbre and tone. But electronic music has the concept of patching. There are some knobs and when you are turning them you are making connections and controls, but not directly on the essence of the sound itself. So I wanted to make all these connections visible and manifest in the play interface, not so much as a technical interface.

EE: Within this kind of structure—where interface concerns the playful first, and the technical second—how is synthesis conceptualized? How do you define the relationship between gesture and the resulting sound?

MQ: I see what you mean. You can simply use the idea of an operation-changing-the- sound-quality to define the sound Wing Pinger makes. Because it could be a combination of so many factors. A lot of changes in various sound parameters. So in this case, in the process of designing, one would notice and pay attention to the choice of all patch-points. Then when the users are performing, they would notice a certain level of responsiveness—a responsiveness that is sensitive but not too radical or extreme. This still requires a lot of exploration and experimentation. One would need a lot of performance experiences to have a better basis for this design. Especially since this kind of instrument is not ubiquitous nowadays. So we are still on the path for further innovation and experimentation.

EE: I see. When people are playing electronic musical instruments, there's a specific type of pleasure that they are after. You turn on the machine and then immediately you have a techno track. It seems simple. From a certain perspective it is easier than playing traditional acoustic instruments. But with the interfaces and instruments that you've designed, it seems that you envision a certain relationship between the player and the music that is played. It almost seems like you are giving some autonomy and freedom to the machine. I sense that you have a certain respect for the instrument—that it may have its own will, standpoint, thoughts and life. Paradoxically, this sovereignty means that it takes a significant amount of time for the instrument's personality to be recognized and celebrated—for people and their instruments to develop a meaningful relationship. I'm interested in this as well.

MQ: There are a lot of musical instruments that have been popular. The drum machine is one such classic design. The kind of fast techno music that you were talking about has become a habit. This kind of habit in music-making inadvertently shapes how people design instruments as well. There are instruments that are made and produced to solve certain problems. For instance, the previous drum machine wasn't good enough or efficient enough, then people design newer and better ones to tackle specific weaknesses—to meet a certain demand. Of course this is a valid principle for designing. But as you mentioned, with my instruments there might be a learning curve. This is because the originary motivation for the design was not to fulfill a certain need or demand. It was, rather, to raise a new question—not answering, but *posing* a question. You can see this point being externalized and manifested in the actual existence of the learning curve.

EE: I really like this—posing a question instead of answering one.

MQ: Predictable electronic instruments have become really common-place. I want to do more experimentation with the interface. I want to see whether new interfaces, and the new correspondences between interface and sound, could bring out new dynamics and performativity. Because I really believe in the musicians. Historically we have so many examples—any kind of instrument could be used to fully exhibit and actualize musicians' creativity and originality. It is a common phenomenon. I believe in the potential of the performers and musicians.

EE: Have you been in touch with musicians and users of your designs? Do you listen to their work?

MQ: Yes—a lot of them are my friends. I read their letters and we communicate often. I truly enjoy listening to their music.

EE: Are there any long-term customers/friends who make requests for new designs? If so, have you accommodated their suggestions or needs?

MQ: Yes that does happen. For instance, one of them thought that Wing Pinger should have MIDI Channel, so I made that into my design. Practical and technical requests like this can be accomplished. But as for more general ideas and concepts … well I've always been inspired by users. As I just mentioned, I do really believe in the musicians, and the dynamic range in their own performances. To me this isn't mere faith; looking at others playing these and other instruments, it is undeniable that their creativity has brought me much inspiration.

EE: So what are some other sources for your ideas and inspirations?

MQ: There are many! Books, literature, etc. I really think all creative thoughts have similar natures. They reach each other because we are all humans. There are communal, collective, unchanging parts to the process of creating. This process includes musical instruments (for musicians), paint brushes (for painters), computers, pens and paper (for writers), because they all share the structure of having input and output. Put otherwise, perhaps it all comes down to a need to tackle anxiety, a need to express. The sources for inspiration are omnipresent—they are every day and everywhere, from life.

EE: You mentioned that you'd like your instruments to bring to people a new life, "I'm expecting instruments could enable us to lead a life that is inviting, with characters of instruments corresponding to certain parallels in our lives." I understand that as saying, "musical instruments can change one's values, attitudes, one's very sensibility." Which, in turn, means that instruments could change the process of creation itself.

MQ: Right—and that's not news. For example, if you are playing a drum machine you won't be playing a piano sonata, right? Given one inspiration, but played through various instruments, I think there will be various expressions/outputs. The tools of creation really can be crucial to creation itself. You write differently with a pen than you do with a computer, I feel.

EE: So which comes first? Creativity, or the tools for creation?

MQ: It is not a matter of order. First, the process of creation … well, different works are made/produced differently. Let's say for a sculptor, given different materials and tools, the created works are very different. Sometimes it's the materials that give you a certain push for a certain idea, but then when you are working with the materials, molding them, their own essential characteristics would always shape your next steps. So it's really hard to say which one comes first, and usually as the process starts, things naturally fall into their respective order. It's not fully determined. In some instances the tools and things would be primary, but in others your ideas and thoughts would call the shots. But in the end you can't really be certain which one dominates because, quite frankly, it takes both. So any purely rational categorization (let's say a chart that lists how much influence each poses on the project) would be incredibly difficult. In addition, for electronic music, as you know, in many cases would have the tools/instruments to be the predominant factor in creation. I don't think it would be controversial to say that the 808, 909, or 303 entirely shaped a whole generation's music. Of course, the creator's own innovation is a big part, but without these instruments, it's unthinkable. Consider, as technology develops, how many new kinds of music have been created—uncountable. They come about because these technological artifices made them possible. So strictly speaking, we had iron-making first, and only then was there sculpture.

EE: You've created so many kinds of instruments. Desktop synths, pedals, controllers, Eurorack modules, stand-alone instruments. Can you speak a bit about your experience with Eurorack? What challenges, difficulties, or other interesting phenomena have you faced when designing Eurorack modules? And why did you stop?

MQ: So it's been one of my core beliefs that stand-alone instruments can permeate into more facets of the designers' ideas. Stand-alone instruments have to take into consideration the hand gestures and the playing techniques of performers. Modules serve to open up the inner core of an instrument, and then let users become the other half of the instrument. Modular, for me, is a half-instrument. And we become the other half. Through patching, a new musical system is formed. One has to design one's own method of playing, design how to operate the instrument—considering which parameters are more important or less important, and also through using an attenuator one seeks to optimize the range of the knobs. The freedom that modular gives people comes down to half of the instrument. It can't limit function because it is built-in with the modular. The challenges for the designers are that, first of all, it provides a standard to be complied with—the designed product has to coexist nicely and compatibility with other modules. Second, to avoid hard (physical) errors for the technicians, such as damages that might be incurred. Finally, to design fun and fresh functions—this remains the biggest challenge for the designers. The advantage of the modular is to leave the other half of the instrument to the users. But for stand-alone instruments, given that they are complete and whole instruments, all we are leaving the musicians is the relationship between gestures/move-ments and sound. But modules leave it to the users to determine that relationship.

EE: Would you conceive this as a kind of power dynamic between you and the users? In terms of how much power I (as the designer) can create/leave for you (as the user) while you are playing? It's like carving out a path for them.

MQ: That's really interesting. In a sense, yes, it can be put that way. Stand-alone instruments, in some way, do restrict one's freedom. A framework is established, a range is determined. But of course, this range has to be versatile enough and adequately diverse. So that the performers can do what they need to do with this instrument. Some interactive instruments are almost akin to art products. Some designs by Peter, let's say. The relationship between input and output is incredibly complex and chaotic. In a lot of videos on YouTube you can find that the output music all sound pretty much alike. There are no obvious personal characters to them. It's like I can choose certain sounds, but it's obvious that these sounds are made from the instruments. Well, Wing Pinger is kind of like that too, but I really tried to make the dynamic range more diverse, more choices. This actually leads to another question that might be worth discussion—to what extent could people have control over the instrument? In my view, if one can't trace and track the status of the instrument *in its entirety*, he/she would be actually controlled by the instrument. So instruments would play a disproportionately big role in the process, while people are left with the minor roles of listening and choosing. So back to the modules—they do give people a sense and degree of freedom. But this freedom—should we really understand it simply in the sense that the more freedom the better? I would say no. You can tell by the designs of a lot of the modules nowadays— there's a tendency of simplifying it. All is pre-made and pre-packaged—single voices, single chords. And when you plug it in you can create a four-voice harmony; when you turn the knob you change the movement of the chords. It's like completing the harmonic series and pre-packaging them for you. So simplified. These kinds of products, of course, have their place and meaning, but are they in line with the original vision for modular synthesis? Probably not. The early modules really depended on the users' technical proficiency. Especially with Serge. The level of freedom that it left to its users was incredible—one could decide on the very function of the module itself. This is Serge's concept of patch programming. You, the user, use your own patching or the functions of a knob to determine the functions of a module—it can be an oscillator or a filter or a VCA. The complex modules like this, how many people do you think could master them nowadays? Very few, in my opinion. A high level of freedom and control, indeed, but is it a boon or bane for the creators or musicians? I think that really depends. Some of them find this high degree of freedom really liberating, and might not jibe with the presets and restrictions of more commercial instruments. Different artistic creators have different needs and desires; when we discuss the question of freedom and power that the machines or instruments offer them, it is not a simple binary in the sense that more equals merrier. When too much freedom is offered, they might actually find themselves a bit lost. It is not a matter of talent, I must say, but just of difference. Some are used to giving the freedom to the instruments, but some like to grasp that freedom within their own hands.

EE: I often feel that the people who fall in love with the Serge format, with patch programming are, to some degree, falling in love with interacting with the others who use these instruments.[1] So the sense of community plays an important role. A lot of them don't use the instruments to make music, in the sense that they don't record and release music, even if they are playing their Serge panel every day. In some sense this goes deeper than mere hobby. So I want to ask you about a related phenomenon that I've noticed. When it comes to hardware, a lot of people start with relatively affordable entry-level drum machines and keyboard synths (designed with usability and conventional music-making in mind) but eventually they move on to more expensive and complex devices (which inevitably raise the bar for ease of music-making). After some years some may move into Eurorack, or other modular formats, and perhaps some years later … if said tools don't offer enough satisfaction, they get in audio programming—Faust, Supercollider, and so on. What do you think about this pattern?

MQ: What you've summarized is spot-on, really nice. The curve from accessible to inaccessible. I think this is meeting a psychological need. There is a real importance in the community for music-making. When you buy Eurorack, and Serge, you are in some way buying into a community—by which your social status is improved. From easy to complex, affordable to expensive, it is rooted in a traditional psychological need—an externalization of one's own social status. If I play on a Serge system then I'm more badass than the person who uses a Roland keyboard, right? So coming back to the modules—for a lot of them who are playing with those, it is difficult for them to find their own path, because the degree of freedom is so high. With a guitar, I play several chords and sing a song, it's pretty fast to make a piece of music that gives satisfaction. But with Super Collider, live coding, the learning curve for those is so long and steep, so it's impossible to quickly learn to make something of one's own satisfaction, but on the other hand, one does have a "higher" social status by simply using those. So I guess the sense of recognition that comes with this is fairly important. It's not just music, right, but with cars, motorcycles, etc. the more advanced the thing you are playing, the higher social status that gives you. But of course the main presupposition for this to happen is that it has to already have become a cultural symbol in the first place. If there's only one person playing Serge, then it is impossible. But if it is something that is widely and socially accepted and recognized, then it can give people that sense of satisfaction for social recognition. All hobbies come from this kind of social phenomenon, right? That it creates something communal for people so that they can share a common object together, but it also creates the necessary distinctions between people—it distinguishes certain groups from others, and this kind of formation and grouping of social circles seems natural. Really interesting that you brought it up. It seems to point to and open up the black hole in instrument designing or making—the dark side of it. Because when you are designing, it is really difficult to not be affected by factors like these. It's like I want to create something that is really difficult to use, but if it meets a certain social demand, then this becomes a principle of designing that works. A kind of strange demand, perhaps. For instance I used to buy different camera lenses, recommended to me by my friend—I couldn't use those at all but you know, it was cool.

EE: What kinds of demands or needs, then, does Meng Qi meet?

MQ: Frankly, all of my designing has come back to this one fundamental need for me. How can the instruments and the tools become some kind of legacy or preservation of part of me? We all die, but things can last. It's akin to the sense that they continue to live as me. I feel a lot of (artistic) creators share this desire. I want to leave something to the future generations, and I want that to be something good, representing part of me. A form of self-respect I guess, the most basic need for me. Furthermore, it fulfills a need of mine to communicate and interact with others. If this thing is part of me, and others like using it, then I guess they like part of me as well. Part of me is validated—another fundamental and basic need, I guess. As for the other needs, that's easy—one has to make money. That's the secondary kind of need. So the primary need is communication and the secondary is to be able to sell the products.

EE: Setting the commercial aspect aside, let's talk about the personal expressions in a given design. All created products inevitably say something about the creator. But in the end, not all products can really be long-lasting. When they do persist through time, however, something interesting happens. To me, for instance, "Serge" is not a person, but a concept, a thought embodied as both the technical and social dimensions of Serge synthesis. In this case, when one says that a musical instrument is a success, the individual creator has already merged with the instrument as one; they are synonymous with technique. Is this a desirable outcome for you, to transcend into circuit and interface?

MQ: I think so. It's always difficult to think about eternity. Well, one simply doesn't know for how long the world will keep going on and all that. But I feel pursuing something like that is in a way a practical thing, since it satisfies that psychological need. Especially for Serge and other early figures, who have made such a difference in the field of synthesis. To me this is the highest degree of success—one can't get more successful than this.

Note

1   Editor's note: on patch programming, see, for example, the 1976 Serge Manual, https:// web.archive.org/web/20190423042216/http://www.serge.synth.net/documents/1976 sergemanual.pdf.

17

INTERVIEWS WITH FOUR TORONTO-BASED MODULAR DESIGNERS

Heidi Chan

Throughout the past six years, as the owner and user of a constantly-changing Eurorack modular synthesizer, the modular community—both locally and virtually—has been an indispensable resource for knowledge, social, technical, and musical support, and performance opportunities. In particular, I found an unprecedented level of access to modular designers, often through social media and online discussion forums, which has made me think increasingly about the relationship between design, workflow, and musical output within the modular synthesizer community that was previously unattainable with electronic musical instruments and tools from previous generations. Speaking to a number of other modular designers over the years, I also realized that—and as this interview reveals—the distinctions between designer and user, and even the distinction between expert and novice, are often blurred and temporal.

For this interview, I chose four local, Toronto-based modular designers, because of my own connections with each of them, and for their range in personal and technical backgrounds, their differing affinities to certain synthesizer systems and design lineages/philosophies, and the spaces they occupy as designers within the larger modular community. This interview was also partly inspired by a modular workshop I facilitated at Moog Audio in January 2020, when I invited Jeff Lee and Yoni Newman for a panel discussion on filter design. What began as a chat about the design of filters (and lowpass gates) became a much more involved and candid dialogue about the challenges of running a small Eurorack company, navigating the world of electronics manufacturing and global treasure-hunts for parts, personal philosophies and approaches to instrument design, why they have

DOI: 10.4324/9781003219484-19

chosen to make the modules they make, and other nuanced aspects of life as a modular designer. And, through my acquaintance with Dakota and Nyles, initially through the local modular buy/sell/trade community, and subsequently learning about their work as designers, it had been a longtime goal of mine to bring these four designers together for a conversation.

HC: Heidi Chan (interviewer)

JL: Jeff Lee

DM: Dakota Melin

NM: Nyles Miszczyk

YN: Yoni Newman

HC: I'd like to start by having you introduce yourself, your background, when did you get into modular and synth design, what are you currently into?

NM: My name is Nyles Miszczyk. I used to manage Moog Audio[1] a long time ago, and we had this customer Deadmau5, who loves modular synths, and we didn't carry any of them [back then]. He said, you get them in, I'll buy them. So, we started getting them in, and he and I became friends, and I started giving him guitar lessons for a bit. One time I went over to his house and he had this huge Macbeth panel, a giant Modcan, and just crazy amounts of synthesizers. This was [around] 2012, 2011. I was just looking at them, and he was like, yeah man, go nuts on it. I got super intimidated. I had no idea what I was doing, and I kind of wiggled a knob, and he was like, well, only the ones that are patched do anything. And I got embarrassed immediately. So, realistically, I got into synthesizers because I had something to prove.

YN: I'm Yoni. I got into the design game through hooking up with Phil Mease, who was doing Rabid Elephant stuff. I just finished grad school, and there was this synth meetup at a bar a couple blocks down from my street. So I just started hanging out with this guy who looked like a hick. I drive over to his place in New Jersey, and spent half the week there. And as we're making music and putting a [techno] set together, I'm starting to realize, this guy really knows his stuff. He's teaching engineering at the local university and he just keeps doing things that I would never think of with the gear that we have. He'd already finished designing Natural Gate, but we were still waiting on the first shipment to be done manufacturing, and most of what we've designed since then came from headaches of using preexisting gear, just stuff where it was like, there should be a smoother way of doing this. So there's a lot of cranky old man kind of stuff that goes into trying to streamline things to musical use. And that's become more and more central to the stuff we've designed.

DM: My name is Dakota. I guess my teacher for synthesizers and stuff has just been rabid curiosity. My parents had so many broken clocks and computers and stuff from when I was a kid. I got my first synthesizer in 2014. It was a Juno-106, ripped it apart immediately. By the next year I was building DIY stuff. I think the [Mutable Instruments] Shruthi was one of the first little monosynths I built. From there I went headlong into all the open-source stuff I could find, learning as much as I could, and ended up finding the Serge modular system, which has become a little bit of an obsession for me, because I really love the philosophy of how it works. Basically, it all boils down to anything I get my hands on. I've also been a musician. I had a guitar before I could talk. So, the combination of technology and mystery.

JL: I'm Jeff Lee.

Hearing your stories I realized I'm quite a bit older. I got really into electronic music in the mid 90s, and put together a studio in 2000. I was living in the UK, and I got to the end of my PhD. I'd published all my papers and I was already really into the music scene there. I was DJ-ing, I was putting on shows, and I wanted to get into production.

YN: Where were you living at the time?

JL: Cambridge. So, 20 years ago, Vintage Synth Explorer was a thing, right? It's funny to think of it, but a Juno-106 in 1999 was only 14 or 15 years old. Now, a piece of gear from 2005 is what that is now. But 1999, 2000 was kind of the first wave of vintage obsession. You couldn't get a Minimoog for 100 bucks anymore. I basically bought up my vintage studio in the first wave of vintage mania, so that's how I got started in making music. I made music, off and on, from about 2001 to 2012, and then I had too many children and had to pack up my studio.

I was actually a cell biologist by trade, a research scientist for 20 years. But, like what Dakota said, as a kid, I was taking apart my parents' clock radios and TVs, and I took an electronics course in high school. But I kind of got back into electronics with synths, although the funny way I got into it was I had this period where I was really obsessed with vintage bicycles, and the thing that I was most interested in was the lighting systems. Back in the 1940s, designers at the time, particularly, European designers, took lighting very seriously, and they developed these hub dynamos. I basically got obsessed with designing modern LED lighting systems that were in vintage style housings, so I had to design the LED circuits. So that's actually kind of how I wet my feet with design work, like [how] Dakota figured out how to get circuit boards made in Asia for not a lot of money.

HC: Nyles, at some point, you made your own cases as well, right?

NM: Yeah. Around 2013, I couldn't believe the exorbitant prices of getting cases across the border and there were no Canadian case manufacturers at the time. So I took Andrew Kilpatrick's open-source Dintree power supply, threw a bunch of those into some wooden cases and thought, okay, let's get some cheap cases going on in Canada. I made 500 of those and they sold in three months. And at the end of that three months, there were 10 new case manufacturers in Canada, so I stopped. And then I was like, this is *fun*, I like building and designing. I started working in sales at Kilpatrick Audio for a while, and I was very thankful for the experience with Andrew. He taught me so much, and watching him go through the R&D process and talking about design philosophy with him was what inspired me to start designing my first series of modules, which I'm now hoping to develop at Toronto Metropolitan.[2]

HC: You hadn't designed anything before?

NM: I have some failed designs. There was some research and development. I made one that was basically magnets that switched on and off really fast like an Ebow, but with an extra pole that worked like a guitar pick up, and had a piece of resonant string with a little bar that was voltage-controlled moving across the string to change the pitch and … it wasn't great.

DM: Can we get a show of hands for people with failed designs? [everyone raises their hand]

NM: I would like to just point out that so far, all of my designs have been failures.

HC: And what's your favourite thing right now to use?

NM: Okay, favourite all-time, Modcan Quad LFO, the Eurorack version. Unbelievable. I love the Modcan stuff so much. I love that as you get to know it better, it reveals the hidden synth inside the synth. It's like a Matryoshka doll of learning, it's amazing. But lately I've been really into what Eli Pechman [of Mystic Circuits] is doing. All the things he's doing right now I find very, very exciting.

YN: Talk about failed designs. I got into the whole synth thing back in, I think, 2008. A guy who used to work at Moog Audio, like most locals have at some point or another, Chris Popovski, wanted to start a band with me, and we're both guitarists. A couple of weeks in, he was like, yo, let's just not use guitars at all. Let's just fuck around on synths. I was like, all right, I'm a student so I can't really afford any. So I just started building DIY stuff and circuit-bending, and that was how I got into the whole world. It's actually how I met Jay [Lemak].[3] He was working at Steve's[4] fixing gear, and I made this [MFOS] sequencer. That was [one of] the first few things that I've built. But I knew zero about electronics, like, nothing. So I went to the wood shop, built a *big* chassis. And you opened up the panel and it was just a bird's nest of wires. [laughs]

NM: As it should be.

YN: This was probably two months after I started doing anything. And I brought it to Jay and I was like, I can't get it to make any sounds. He's like, it's a sequencer. And I was like, yeah, I can't hear what I'm doing. And—this is one of the great things about Jay—he's so kind to people who don't know anything.

DM: To a fault. And to his own injury at times.

HC: [Dakota,] when I first met you, five years ago when I got into all of this, you had also been working with some other designers in the community, in either updating their circuits or modding things?

DM: Yeah, through an over-abundance of confidence. I got into fixing synthesizers, and I started looking into how you manufacture electronics. And the next small company guy I was talking to, who was still making everything through-hole, was manufacturing it all by hand. I offered, hey, I could convert that to surface-mount technology for you, and then you could just get a factory in China to run them off, which worked out fine. I do not like looking at the layout I made for him, because from what I know now, man, that [was] terrible. Once I learned that you could get stuff prototyped in China for pretty cheap, it really expanded what I was able to try and tackle. [In] the last year, I basically copied clones of the entire Serge modular system for my own education. It can't be released publicly or anything like that—just trying to learn new ways from people who try to keep them secrets.

YN: Have you built one or you just sort of worked out the schematic for it?

DM: I built one. I think I'm at seven panels of original Serge, STS stuff. Some of it vintage, some of it modern, some of it really modern, like, built last year. I took the original boards, and all I made were some translator panel pieces that have voltage regulators and a couple of other things that Serge was just really terrible for. There are a couple of weak points in the Serge system, and one of them is the power distribution. So I'm just basically copying all the panels that I have for myself and trying to learn how it works as I do it. And I've come up with some new ways to do some of the things, and I've had the opportunity to talk to Serge about some of the changes as well which is super fun. And occasionally, he gets super mad at me.

JL: I had a Prophet-600 that, on paper, was a good synth, but was terrible in use. And I was doing some Arduino-based art projects with friends. So, I basically started to reverse-engineer the Prophet-600, and started to write a new operating system on a new microcontroller. I wrote these incredibly verbose blog posts about how the USART in the Prophet-600 for handling the MIDI data works, because the Prophet-600, along with the Jupiter-6, were the first MIDI synths, so all of the peripherals in [them] are really archaic. It required a lot of detective work to figure out how they worked. So I had a basic working version where I had replaced all the basic features of the Prophet-600 with a modern microcontroller. The envelopes updated faster, the ADC was higher resolution, so you didn't have the stepping and all that stuff. Digging into the Prophet-600 forced me to learn how analog hybrid polysynths work. [And] now that I knew how to do this, I wanted to make a programmable synth. So I started doing all this design work, planning out my fantasy dream polysynth. I needed analog VCOs, analog VCF, I needed VCAs. I need [ed] the whole system of ADCs and DACs to control everything. And so, every time I developed one of these modules for my polysynth, my friend Bart was like, I want to put together a Eurorack system, why don't you just port these designs into a module? That's how I got started. Basically, I was doing some DIY stuff for my friend Bart. He designed and hand-screen printed the panels, and we did an envelope generator, a mixer, a VCO, a VCF, and I made five of each of them.

DM: I made one white panel, based on your original things that I actually gave to one of Heidi's friends, I think. It's a Jove, but I based the panel on your original panel with the circle. I think it says "System79" in the centre.

JL: That's right. System79 was the precursor to System80.

NM: I got the System79 SVF. I was going to ask if that was your first available module.

JL: Yes, that was the first. There are only five of those, hand-screen printed by Bart, which is why the paint chips off them so easily because he just rattle-canned them. And the intervening story is, I was getting into doing DIY synth stuff. A friend of mine who lived in the neighbourhood was helping Bruce Duncan out after he had a stroke, doing Modcan stuff. So, he said, well, you know a little bit of electronics, you should help us get Modcan off the ground again.

NM: Was that Byron [Wong]?

JL: Yeah, exactly. Byron basically financed Modcan for two and a half years. We set up a new shop and got a couple batches of Quad LFOs out, and that's where I got to go through the whole manufacturing process for the Quad LFO. Bruce had all of the systems in place for getting circuit boards made for the surface-mount assembly. So, I was like, I should do this for Jove. Going through the Quad LFO production taught me this is what I need to do to run 50 [Jove] modules and get them assembled locally, get panels made, all that kind of stuff.

Around the same time—this is about five years ago—I designed this 808 in Eurorack, which everyone has gone bananas about. Again, that started out as kind of a design challenge, let's see if I can fit an entire 808 into a 60 hp Eurorack module, for the hell of it. I already have an 808, I didn't need a clone for my own purposes. And against my better judgment, I put it in production. And now, four years later, we're still working on the second round. So that's kind of how System80 got started. It's interesting hearing Nyles talk about how much he loves modular synthesis and wants to learn and know everything about it. And Yoni is like, I need to improve the interfaces, because I don't like the way these ones work. Dakota is like, I want to learn how everything works. All three of you have indicated this level of obsession. I grew out of it. [laughs] I'm too old now, and it sounds terrible to say it, but now I have a technician I need to pay. I need to get modules to customers. All of a sudden, I don't care how this works. I just need it to work. I need to get these designs out the door. I've completely lost that passion. I'm in this production grind where I need to design things and release them to the world.

YN: It's bad. We built the last run [of Natural Gates]. We did the final builds.

NM: Jeff, what was the impetus to make Jove open-source?

JL: That's a good question. When I was first learning all the ropes, I was relying on open-source designs. And the Jupiter-6 filter is basically just ripped from the Jupiter-6 service manual, right? Some brilliant Japanese engineers already figured this out 35, 40 years ago. It required a little bit of, how do I match up the CV mixing and audio mixing, how do I get the levels right, and then how do I manage the filter switching, the mode switching? In my view, it wasn't mine. I'm into open-source because I'm not an original designer. I don't imagine a design for a completely new filter topology, spend months or years dedicated to perfecting it, and then release it to the world and then want to keep it to myself. That's not the kind of designer I am. I like these old designs, vintage gear, I want more people to have access to vintage stuff. So it doesn't seem to me like I should be taking designs from service manuals and then having them closed. That said, it is a little annoying when people start to take your open-source designs and sell them in kind of like a cottage industry, kind of what happened with Mutable.

NM: Oh, you don't like that?

See, I'm super *pro* that. I think intellectual property is poisonous to society, personally. "Here's a design, world. Go ahead. If you need a job because you can't play music this year, build some synths."

JL: I definitely felt like that when I started, but ultimately, what happens is that your design becomes your product. Your product becomes the face of your company. And when someone bungles it up, makes a crappy panel, puts ugly knobs on it, and then takes all your copy from your website, and puts it on Reverb and says I've got 10 of these available...

DM: ... And then also claims that they changed it because there was too much "distortion."

YN: Then gets in touch with *you* to fix it. [laughter]

JL: Yeah, exactly. To Nyles' point, things like the Ornament & Crime, which is basically *the* post-capitalist module, there's clearly a huge amount of engineering expertise that went into that. And that's awesome they can do that. If it was just me doing this, I have the ability to make really high-quality panels, I get all the quality control sorted out.

The other issue is, I had the luxury of not needing to do this for a living, [and] my philosophy changed when I hired a technician to work for me. At times, he needed some work because he had to pay rent. It kind of made me pause and realize, there's this person that needs the capital generated by these designs. The other thing is, the Jove design has a non-commercial clause. So, technically, that some person could take my design, run off 10 of them at a local maker space, put them up without asking permission—that's not allowed.

NM: I didn't know that.

DM: I built lots of Mutable stuff when I first started out, because I was new to everything. I was young and I needed money. And slowly my approach to it changed, where within the first year I was telling people, I'm not building anything that's still in production by Mutable Instruments, because I just did not like what was happening. The other stipulation was that I will build the adaptations. If you want the original design, I was encouraging people to go to Mutable Instruments, because you see what was kind of seeping into everything through Mutable. And what it has done to my own view of open-source is I think that IP should be open to the point at which it doesn't stop the person making it open from being able to develop more IP. When it's open-source to the point that the person who actually came up with it can no longer support themselves or continue to come up with those good ideas, that's the point at which it balances out the argument that the community has an opportunity to add to it. So the line I ended up with is similar to Jeff where, when I start releasing things—documentation, schematics—people should be able to see that so they can record their stuff after I'm gone. Secondly, so they can see what I learned, because I did not learn this stuff by myself. But as to the actual production details, I'm going to keep that a little closer to my vest.

YN: When Phil and I met, we were still after that holy grail of everything you need in a performance case, your whole music-making machine, the thing you perform on, your studio. Fast-forward two years, and we're just looking for a way out of modular. We've realized the limitations of what this format [can offer]. At some point it was like, well, what are you trying to do? Why is it modular? For a long time, that wasn't even on the menu. [But] we started realizing that a lot of other bits of gear just [had] far more practical ways of doing certain things, and we actually began to get very irritated with modular. I think the last two or three modular designs we did—and I guess they don't exist yet on the market for me to be proven wrong—but I think they're much better designs. Because … I think nostalgia is really bad for design. You don't want to ignore it entirely, because it should be encouraged to make something that people love so much that they develop emotional attachments to. I think that's probably a sign of the success of the thing you did, but to design off of that feeling, I don't think it leads you anywhere terribly interesting.

HC: That's an interesting point, and it's partly why I brought the four of you together, because I know that Jeff and Dakota are both into vintage Roland stuff. And Dakota and I have chatted about old Korg circuits. Meanwhile, Nyles, you've been [working on] your own designs, and Rabid Elephant has designed a couple of things outside of standard synth [building blocks]. So I think this is a really interesting conversation to have.

YN: A big thing for us is that the design process is very, very rigorous. As soon as we have a good design idea we try to kill it. It goes through about four or five cycles of it just getting destroyed. Everything is called into question. And by the time it's something we want to release, the fact that it was inspired by some patch you had going on, that initial burst of creative energy, the genesis phase of the thing, whatever led to making that thing is barely even there anymore. But you have to be able to completely destroy your cultural sentimentality for it because you can ask more profound questions about the thing [that way]. You're making something that isn't already there. You get into this real strange realm of conceptualizing. It gets very abstract. I guess that's what I'm in it for.

NM: That feeling that you're talking about, that initial creative burst, that's the feeling I'm always chasing. And it's funny, I also find the opposite direction of what you were talking about. I find so often the life cycle of a design for me goes from, oh, I really like this idea as a patch, I would like to make a module that includes all that. And as I refine the design, I'm like, is this an integrated circuit? And when the answer is no, then I'm just kind of like, nah, I've gone too far in a weird direction.

YN: That's the thing, the border of where the idea starts and finishes. That's not an obvious thing at all.

NM: Yeah, it takes a lot of soul-searching to invent an experience that is fun and worthwhile. You got to ask yourself, what am I doing, what's the point of this? And it becomes a really philosophical question really fast.

HC: How does this play out in Dakota or Jeff's work?

JL: Well, it's funny, listening to Yoni and Nyles with the sort of new frontiers innovation, "nostalgia is bad," because, of course, I take the opposite approach where I really like this thing and I'm very attached to it. I want to make more of this so that other people can have it too. And listening to you guys talk, it makes me realize—I've said this about my own work before—I'm very conservative that way. I'm not actually a very creative person. I had a burst of creativity for a few years where I made music, and then I literally have no desire to write music anymore.

So I definitely design around nostalgia, and there's room within the Eurorack market to do that.

YN: Of course. I just want to be clear. I don't reject nostalgia. I want to place it where it belongs in the chain.

JL: And I admire the innovators. I admire the Rabid Elephants and the Mutable Instruments of the world because, to be frank, I'm just not that creative. Here's something that someone's done before that I like, I'm just going to do more of that.

HC: And the modular community does seem to be able to hold space for all of that. It holds space for the System80s, the clones, the Serge, but also the Nonlinearcircuits and the Rabid Elephants, *and* also the Intellijels. Everyone just seems to be into everything.

JL: And that's what fascinates me about Jove. Jove is just a straight-up subtractive analog multi-mode filter, but people will run crazy things through it. I see all the patch examples online, and I'm like, that sounds amazing. I would just run a VCO through it, into a VCA, modulate it with an envelope, just create a subtractive analog synth voice. That's all I do with it. The last point I wanted to add about Rabid Elephant and kind of wanting to get out of it, is [that] I did this crazy thing where I put an 808 in Eurorack, and some people thought that was an amazing idea, but most people thought that was a stupid idea. And that doesn't really bother me because I didn't do it because I'm going to sell tens of thousands of these. But what's crazy about it is that because of demand, I built a desktop enclosure for it. And now when people preorder it, they have the option to buy the desktop enclosure.

YN: [laughs] You've gone full circle.

JL: Yeah. It's kind of funny that [with my] flagship module, 90% of my customers just want to put it in a desktop case. That's the funny thing about modular, my big module that launched the company, most people don't want to put it in their system.

DM: In terms of what has pushed me to make my own designs, we have so many more parts and so much more accuracy available to us now. I always had the thought [that] I wanted to do an idea that's done before, but can I do it better? There was an article written on the 1047 state-variable filter in the [Arp] 2500, and basically, he didn't have the tools to make the filter he envisioned at the time. I've got more parts available to me now, let's see if I can nail the specs he was trying to go for.

    The other thing that drives me is I'll have musical ideas that the tools don't exist for me to explore. One of the things I've been really obsessed with is making a complicated, harmonically dense, sound-design kind of piece, [using] very specifically tuned filters' resonant banks and different EQ-ing to pull the polyphony out of a more complicated single tone, so that I could articulate parts of that. You know, a modular synth, you can make something so complex that you can actually pull polyphony out of that complicated [tone]. That's what I was really obsessed with. So, [I'm] basically inspired by older ideas, older stuff, nostalgia, and seeing if I can make it better, or seeing if I can make it do the thing I want without leaving behind the thing that makes it special.

NM: I've been toying around with this idea of a vactrol-type circuit that is based on Bell's Theorem, using polarizing light filters. The working title for it is "Four Rooms." If anybody's familiar with Bell's Theorem, you can completely cancel out light with blackness at a 90-degree angle of two filters. But then when you put filters beyond the blackness, you can actually increase the amount of light from there, which is really interesting, to be able to go not all at the same time, being able to render multiple voltages off the same LED, and using a vactrol-type circuit to render these multiple voltages. And ideally, I'm thinking about maybe reinforcing whatever that level is for each of these four rooms in order to make kind of a quad-peak single filter that you move using light phase as opposed to frequency selection. That's my filter idea.

JL: That sounds very heavy.

NM: It's turned into this idea of now actually creating three polarizing light filter arrays. I want to try to design them as little hats for light-dependent resistors, so that you can start drawing extra voltages out of light-dependent resistors. Ideally I'd want these hats to be addressable in some way. Like, here's 90 degrees with a 22 after, so you get 100%, 0%, and 85% is what that actually tends to be. But having that variable filter at the end of it, it's just basically a way of rendering lots of voltages off a light-dependent resistor by feeding it several different light amplitudes.

YN: Cool. Have you prototyped anything?

NM: Actually I'm going to be working on the first prototype, which I'm going to make huge. I want to do it with a one-foot tube lined with Black 3.0 ink to really limit that light, get actual photo filters from a camera, build it into this large thing. Ultimately I think nobody will have any idea that these things are changing based on light phase, but … I'll know. [laughter]

JL: And from a track-writing perspective, unless you're making ambient, where the timbre of the patch becomes really important, if you just want something to fit into the rest of a track, then a lot of filter typologies will do the trick.

YN: That was my big takeaway. I did this experiment [with] 14 filters. I kind of put something out on the [discussion] boards and everybody pitched in a few filters, and I wanted to have a big massive filter shoot-out, just to see what I could find. I rigged up a few cases, took each one of these 14 [filters] through the ringer, and trying to see how close I can get them to sound to each other. And I found that *while* I was using them, I didn't think I was getting very close. That's probably because I would put that in the "nostalgia" camp, something about the physical experience of that panel. How it looked, how the knobs felt, probably made me experience the filters all sounding incredibly different. And [then] I did a blind listening test. I don't know that I've the best ears in the world, but they're reliable for hunting around for different things happening in different frequencies. And there was only once or twice out of the 14 where I could tell which filter I was listening to. Once it was just recorded and part of a track itself, it was *way* more negligible than I thought. Some filters for one reason or another were very, very hard to dial in. That sweet spot became tiny and you're kind of just touching the knob, and you go a fraction of a millimetre too far and you've lost it. So there's a lot of that. But sound-wise, I could get them all sounding pretty similar, and a lot of the things that made them characteristic just had to do with gain staging, essentially.

JL: The last filter I really listened to a lot, and it drove me crazy, was when I was helping Modcan out. It was the Modcan Discrete OTA VCF or something like that, a Eurorack module. One of my first jobs when I started helping him out was he had a box of 10 of them that had not passed QC, so I had to troubleshoot them, figure out what was wrong with them. I think I got about 9 out of 10 of them working. But what drove me crazy about them was because it was this discrete OTA design, there were 4 OTA cells, and they were all set up in a way so you can have your bandpass, highpass, lowpass, and a notch. These filters, no matter how you dial them in, all sounded slightly different, like when you're doing that critical listen. Especially the resonance response was *very* different for all of them. I never really studied the design well enough to understand why. But from a troubleshooting and calibrating thing, I got to the point where it's like, I like this one, this one is good, this one could go and be sold. But I'd get another one and I'm like, it's working like a filter, it's doing everything it's supposed to do, but I *hate* the way it sounds. And I went through the testing procedures, took all the measurements. Technically, these have passed QC, but some of them sound way better than others. And it must be components, because they were discrete, so, component tolerances, maybe they're not all the same temperature, I don't know what it was. But it was weird.

NM: That's actually one of my favourite things about manufacturing. It's sounding like almost all of us have a background in guitar-playing, but you know how sometimes you pick up that cheap Epiphone, that $200 guitar that rolled off a line in the factory and it plays *stunningly*? I live for that gear.

DM: The reason why I really pushed to clone the Serge system and learn as much about the Serge system for myself, was very much because that was the first time I sat down and had so much fun patching, and it just didn't seem like there were any limits. With Eurorack, you always run into "it doesn't do this thing," "doesn't do that." With Serge, if you could figure out how to patch it, you can have that thing. When I started using some Buchla systems, very similar, it's all there. And it taught me that the design of the interface is a lot more important than the circuits that are inside. And that jives with a quote—it might be apocryphal, you don't know if Don Buchla actually said this—but apparently Don Buchla was far more protective of his panel layouts and his actual physical design of how he wanted people to use his instruments than the circuits.

NM: Well, that was his whole thing, was design the usage first, and then design the circuit after that. That's the way I design too. I think about what is the experience that I want to have, and then how do I get from where I am to that point.

JL: Yeah. I call those fantasy panels. I have probably like a dozen fantasy panels of designs that I'll never make, or by the time I get the panel done, I'd be like, I don't want to write all the code that's required. But I love designing fantasy panels, because you're using Illustrator and just moving knobs and switches around and thinking about how you're going to use it. That's my favourite part. And then, I'm just like, I don't think I could actually build that, or if I did, nobody would buy it.

DM: Well, you're following a grand tradition, because Buchla had fantasy panels in some of his promotional materials.

YN: That's how we do it too. We have the same designer that Mutable and Hex[inverter] uses, Hannes. We do just a basic Illustrator mock-up and then send it off his way, and he's usually working on that while the circuit changes and develops.

JL: And would he move knobs and jacks around?

YN: Yeah. We kind of give them free rein, but also we have pretty long conversations with him about the usability of the module. So he usually doesn't veer too far from what we do. But he's also a user and he's obviously very familiar with the community, so he also gets to chime in about how stuff turns out.

DM: My wife is a UX/UI developer for websites and stuff. And I've been totally copying her homework, because I'm looking at how she figures out layouts. So, I'm thinking about filters in a new way. The way we approach a filter, is that the most gestural way to control harmonics in a signal? Is that the most natural way to do certain things?

YN: Two modules that we have, a VCO and a sequencer. I don't know if this is going to happen in the end, but right now there's a little LED squirrel [on the sequencer]. And if the squirrel is on, it means all of the standard quantizations and things have been shut off. You're allowing for all the crazy shit that unregulated analog circuits do. And if you need it to do something that's a lot more precise, you just turn the squirrel off.

HC: Did you say *squirrel*, as in the animal? [laughter]

YN: Yeah. So if you don't like the pitchiness, the weird dips and things that happen, turn it off. Now you have something that's perfectly sterile, which I tend not to like, but you need it for certain things. It's like taking the safety rails off the thing.

HC: Sequencers are something that everyone probably has a lot of insight about. For me, it's like this icon of electronic music, of automation, the centrepiece of a system for many people. When you're designing sequencers, or when you're using a sequencer, or using something that's inspiring you to design a sequencer, what kind of stuff do you look for? For me, it's all about the fluidity. I like everything that makes it *not* 16 steps or 8 steps or very square patterns.

NM: I personally am of the school of thought that most of it's been done with sequencers. The thing I try to get away from is control and any sort of feedback, in the sense that, I like it when sequences are just being pulled from the ether, more than me telling it what to do. I've been working on this eight-channel sample-and-hold quantizer, and I'm working on it with a few friends, just as far as the conception of it. This one is more specifically about eight ins, eight outs, but it has a quantizer in-between, for the sample-and-hold outputs. But I want its natural state to be loaded with really obscure scales, like, the Wikipedia page of bizarre scales, but I do want to make it updatable with .tun files or Scalar. So, out of the box, I'm not going to label any of the selections of what you're in; you just kind of have to feel it out. I find that "step one is going to be this value" kind of boring. It doesn't really play into my style. I have to admit I'm one of the only people probably on the face of the planet that's not that fussy about a Cirklon. I like randomness and lack of control.

JL: I'm on the wait list for a Cirklon, but I have no idea how to use it. When my number comes up, that's when I would do some research and figure out if I want it. The people that like it *really* like it, but I'm kind of with Nyles. If I really want to sequence something very deliberately, then I would use a DAW.

YN: Exactly. A Cirklon is just a slower way of doing what a DAW does really easily.

HC: I'm going to mention Elektron here, which has kind of set a standard in recent years with what people want in a sequencer, the per-step probability, per-step automation. When I've used it at Moog Audio, I started to appreciate it more, but it's still pretty rigid in its way.

YN: Yeah. That's Elektron's thing. I do respect the balls it took for them to go, "Look, it took us five years to design this thing. We've tested it out the wazoo, we know it's good, we know it's unique." To this day it'll do things that you'll have a hard time finding out of any other hardware thing. I got an Octatrack, it sat on a shelf for a year. I didn't touch it and then finally, one day, I started reading the manual and fucking around with it. In a couple of weeks it had replaced what I was doing on Ableton. But I also agree with you. I'm kind of of two minds with the Elektron stuff.

JL: One sequencer that I really like, which is super basic, is just the Korg SQ-1.

HC: I was going to bring that up too. I love that thing. It's so good.

JL: It's frustrating, but it doesn't have any memory. When you turn it off, all of the gates, the note on, off, all the four parameters you can set, they all just reset to default. I've literally got sequences where I'm like, "I love this sequence, I love the way it sounds." I literally am taking a photograph of each of the four modes that shows me what's going on. I'm just making sure I don't touch the knobs.

DM: You're taking Polaroids of something with a USB jack.

YN: The problem is though, you still get some sort of perspective and foreshortening. When I first started doing Eurorack I'd take pictures of my set-up, just so that I'd know where to dial things in. You get within a few degrees, but the photograph doesn't give you enough information to go on.

DM: That's what really annoys me about Serge. Because Serge was obsessed with attenuverters, you immediately have half the control on everything, because every pot is cut in half. I was actually talking this week with some of the things I'm working on with Charlie. It's like, get rid of your attenuverters, man. Put a little button that switches the phase of what the knob does and then you have the full sweep. People are not going to like that, because Serge always has attenuverters, but play it, it's better. Just believe me.

NM: That is a capital idea. My favourite sequencer that I've encountered in a long time is the Arturia MicroFreak. I absolutely feel like this is a $2000 synthesizer for $300. I'm in love with it. Specifically it's got this Spice and Dice thing where it starts to randomly animate other parameters of your patch besides pitch. It'll go to filter cutoff and your envelope decay and a few other places. It's kind of random, [but] you start to get an idea, [with] the Spice factor, my rhythms stay pretty basic in the first few degrees and then it starts to separate from them and then they're really spaced out at the top. Then the random variable, the Dice factor, I just let it write songs for me at this point.

HC: I've just picked up the Westlicht Performer. I was at the lab playing with it for a couple of hours and I was pretty much able to get through most of the functions without a manual. It's kind of Elektron-like in that you can have per-step probability for everything from just gate on or off to note range, so it'll randomize between a range of notes, but the very cool thing is that it has an internal routing matrix, so there's eight tracks and four CV ins, and you can assign any of the outputs of the eight tracks to modulate a parameter on another track. Also, each of the tracks can be a note track or a "curve" track, so you can dial in shapes for each step instead of a note. I've been mainly playing with the routing, so I'll dial in some sort of LFO envelope thing to modulate the clock division of another track. That second track is just constantly speeding up and slowing down and doing really interesting things. Somehow I just kind of connected with it because of that routing feature. You can modulate start and end step, sequence direction, clock division of each track, etc. That's kind of the fluidity that I've been looking for.

NM: I love being able to re-window a sequencer start and end steps. When I was working at Kilpatrick, the Carbon was the latest product, and that was the first sequencer I ever played with that had this start and end select. The implications of your reset button, or the reset input, becomes so much more exciting than just back to one, because now it's kind of like you can almost set up a sequence and the reset button together to start changing through which step the sequence starts on and suddenly, it's almost like I've got weird fractal patterns of one larger sequence.

DM: The original Serge N-Step style sequencer has a reset that is separate from preset. Your sequence always restarts at the last key you actually hit and that's your preset key. Separately, on the PCB board, there's a pad that you can reset to one. Reset always go to one, whereas preset goes to whichever key you hit last. When I made my version, I was really looking at some old boards and old schematics and I was like, "I need to have this. This is ridiculous it's not on a jack." On the sequencer I designed for my Serge set-up, I added reset different from preset and I added LED indication to show where your preset was at all times. There's select input for all eight steps and there's gate output of all eight steps, so you can window any grouping. You can have your preset coming in at different times and your reset will always jump you out of your loop. When I finally was playing with it I was like, "There we go. All right. I'm happy with this."

JL: One of my favourite compositional methods, which I only discovered within the last three years or so, is the SH-101 sequencer. You hit Load, you play some notes, you add rests, you can add some legato, you can add some bends. As a non-musician, a non-player, everything that I write is kind of non-intentional. With the SH-101 sequencer you play a bassline, for instance, but then because you can use something like an 808 to trigger or advance the steps, you can play eight notes, but you can add all the syncopation because you can use an 808 trigger output so that you're actually programming the number of steps and how they're being controlled. Basically, when you start jamming with an 808 and a 101, you start to make all this techno and electro that you heard from the late 80s and all throughout the 90s and you realize this was …

YN: How they were doing it.

JL: What I love is that, take some sequence of notes, and obviously that sequence will just repeat itself over and over again, but what you can change up is the timing of it. You literally can add, create the rests and what notes are grouped together. It's super basic, but as a compositional tool I still love that way of sequencing.

YN: For us, there are two sequencers that I'm working on right now that I'm pretty excited about it. We've been through so many different designs. But a bunch of stuff became very apparent, one of which being, at a certain level of powerful feature sets, you're not going to do better than a piano roll and a DAW. We've tested it. It's actually what killed the Elektron stuff for me, mostly. It reduced my enthusiasm for the Octatrack by a good 50%, because I was trying to do something and it took me 45 minutes to get the whole compositional idea flushed out. It was like, okay, now go do it in Ableton. Of course it took five or ten minutes. Then you start to really think, if I'm not making techno, how usable is this thing with all of its powerful feature sets? I know you hate the screen and the mouse, but how long do you want to spend on this thing? How long do you think your enthusiasm will last for whatever you're doing right now?

JL: I'm on my second sequencer design iteration. And I get to the point where the concept is done and realized, and then I look at it, I'm like, I don't think I'd actually want to use this.

YN: The thing that became consistently apparent was, when you're using any version of a step sequencer, it's pretty easy to come up with something that sounds pretty decent to your ears. You just turn the knobs for a little while and eventually your intuition and your "adventuring" will land you in some loop that you like. Then the question is, how do you create meaningful variation? The probability thing, we played with that for a while, but I found when I was actually working on tracks, it was a little bit too unruly, because it's random, it's not *musically* influenced. Sometimes it's bang on, and then sometimes it just sounds wrong. The big thing with hardware for me is, I'm in the ballpark now, can I maneuver in a musically meaningful way? We just started focusing more on what the most performance-friendly way to design an interface around these things is.

A couple of things came up. One, we had this three position "chain sequencer." There's something like a Turing Machine, just spits stuff out, [and] when you hear something you like, you catch it. I guess you could think of it as layers in Photoshop, masking layers, right? It feeds a second sequencer that takes whatever it's feeding it and changes it. It could be a shift register or a scoop or a number of different ways of transforming the step that's feeding it. The other thing that fell in my lap was Scott Campbell's sequencer. Do any of you guys know Scott Campbell? He's presented at Frequency Freaks.[5] He's doing this little touch sequencer.

JL: He's a local guy? I just look at his beautiful videos and sounds on Instagram, and I'm like, that guy's too amazing to be from Toronto. He can't be from here. [laughter]

YN: He's a local guy. I started helping him with his sequencer called the OK200. The original usage for what ended up being this four-channel quantizer sequencer thing was when he first started using modular, he had a bunch of oscillators, and he'd just tune them by ear into a sort of drone machine, a chord machine. Wouldn't it be nice to have something that could do this for me? So he starts doing this thing, and he's got a pretty slick-looking prototype, really, really cool, touch buttons, contact buttons that turn into sequencers and quantizers. So he gives it to me and I'm playing around with it, and I do a couple of videos and send it back to him. And there's this thing that Phil and I had been talking about for a long time, and I've been trying to put it on basically everything we've designed, [and] it keeps not fitting the designs that we have, which is, I wanted to put a whammy bar on one of our modules. I wanted the way the Bigsby ones feel, just a big spring action thing, because it's a really nice way of modulating certain things. It's a much better metaphor for certain things that you want to do with musical pitch. [Scott] had this sort of slide knob thing, and I was like, this isn't very functional, this feature doesn't really work great. So I've convinced him to look into it and I'm thrilled. It's already a really wonderful super intuitive usable musical thing, but with that extra whammy action—you know the old Nord Leads that had that ...

JL: Oh yeah, that little wooden block.

NM: I *loved* that little wooden block.

YN: [Nord] hired a guy who's done this PhD work on tensile steel, to work out the tensile properties of that piece of metal and then the circuit integration. That's no mean task. That's a really involved thing. I'm pretty sure I've set back [Scott's] manufacturing deadline by half a year or something. But I leaned on him so heavily for it, because I was just like, listen, this is going to take this thing that you have right now—which is already very nostalgia-worthy—it's just going to give it this extra little tactile [experience].

NM: That sounds amazing. Definitely my kind of thing.

HC: That's a really fresh take on a sequencer.

YN: It's really tactile, so for performing, you can set the bend amount per time, per whammy bar. Just choose your intervals on the four different outputs and then you can set how many semitones up or down, so you can transform the chord. You can have maybe three different chords just by taking the set and throwing a whammy up or down from a centre detent. To me, that pays off the hardware investment just because of its tactility. It's almost like you're playing a guitar now. Performance-wise, that takes you really far. I'm more curious these days in how far you can meaningfully go, and whether this interface lets you do that in an immediate way.

JL: It kind of factors into this whole idea of authenticity, right? I don't know if it's still the case, but 15 years ago, there was a huge amount of discussion on forums about what constituted an authentic, live, electronic perform-ance. There were a lot of people making terrible music, but their big shtick was that it was all hardware, all live. Nothing is pre-recorded. It became a performance aesthetic. Even if it sounds terrible, this is the way I do it and you, as my audience, will appreciate the fact that I do it this way. The people that had wanted to die on the hardware hill were basically saying, "If you use a DAW, you're not doing it right. I've spent all these years and time and money figuring out how to get my hardware sequencer set up to work and this is the way I do things and it's better than the way that you do things."

HC: I tried the live performance route one time. I was trying to be all Imogen Heap and trigger, play drums and loops on my Roland pad.

Then after a while I'm like, I'm still playing pre-programmed tracks [underneath]. The question was, what am I really trying to do? I think at the time, I really wanted to retain my identity as an instrumentalist, so I was exploring how I can do that with my electronic music.

YN: Yeah. I think that's the source of that. The self-identifying with the gear, to me that's just purposely rigid. It's just the wrong thing to identify with. None of those distinctions are creative or artistic distinctions.

NM: A guy I went to highschool with is working at the Magenta labs for Google, in their AI neural net and machine learning music department. Apparently they're working on algorithms that will actually take the rules of counterpoint and things like [that] into account for a sequence, and this algorithm is going to spit out three other sequences that are not rigidly attached to your sequence, but in fact thought of by a machine to compliment your initial input, which I think is the exciting new feature of sequencers. Again, I think the most exciting thing about a synthesizer is that I'm going to sit down and have a conversation with a machine. When the conversation is completely unidirectional where I just tell the machine what to do, it's boring and isolating and I don't do anything fun. If I'm not having fun, why am I here? I love the idea of all this neural net and the fact that Western music's a pretty simple and rigid system that is, in comparison to other neural networks, probably relatively quite small. Hopefully it'll mean that everybody will realize Western music's boring now, because you can just get a machine to spit it all out.

DM: I looked through the [Doudoroff sequencer comparison] list[6] that you sent, and it made me think that I would do different categories. I kind of consider sequencers to be track-based sequencers or kind of DAW-replacement sequencers. Then there's instrument sequencers, where the focus is a full set of tools. I much prefer an instrument style, single track, multi track, whatever, but not attempting at all to replace a DAW. Like Yoni said, a DAW is better at DAW things. I don't like the tendency of Eurorack to almost become hardware plugins, where you just keep adding things and try to replace everything with Eurorack.

HC: That's a good segue into another question I had. I'm interested in your lenses on how modular design has evolved in the last five to ten years. For me, what I see is an increase in size and complexity, which I find is in direct inverse proportion to a module being a tool for patching. I find the more menus there are, the less you learn to patch with it. That's my take on it.

NM: I think I totally agree, because like you were saying [in another conversation] about feature requests, I remember we made the joke about "how does your synth module fit into my limited understanding of synthesis"? I feel like that's where it all goes, where people just want a plugin.

JL: Does anyone know who Alex Ball is? He's a YouTuber. He's one of these guys that can make a track out of anything. The videos I'm most compelled by from him are related to modules and functionality and what's changed. I have a pair of System 100 M cabinets, which only have basic, what we would now call "vanilla" modules—VCOs, VCFs, VCAs, sample-and-hold, envelopes, LFOs, and a sequencer. That's it. In today's Eurorack market, people would be like, "That's boring." When I tend to patch my 100 M, I tend to just recreate subtractive analog synth voices with it. But what Alex does is he makes these amazing patch examples. I'm like, I've had this "basic" modular system from the late 70s. It doesn't have any esoteric modules, it doesn't have anything weird, and yet he does all this amazing stuff with it. It makes me think of this trend towards more complexity, modules with higher functionality and here's this guy who's making amazing sounds with this system that I have. I look at that, and I'm like, I don't need to buy any more gear, I don't need to get a new module, I just need to better learn the thing that I have.

DM: It's interesting that the functions that you make easiest to access in a device affects what music is going to be made with that device, because people are feeding off of your imagination that you put into it. It's like you like to do this little thing, so you make a thing that does that thing easier, and then other people find that thing and it also comes into their music. I like Serge [because] everything is right there and all of the tricks are in the patching. All of the mistakes, everything is in the patching.

HC: I think that's why one of my favourite Eurorack designers right now is Joranalogue. Dakota and I have talked about this. His modules are still modules you have to learn to patch with. They're just VCAs and filters and oscillators, basic building blocks, but with a really fresh twist.

NM: His function generator is gorgeous. I'm kind of obsessed with function generators. I have an abnormal amount of them, I like to make entire synth voices just in functions, so I would say, it's between him and the Falistri for my favorite ones to use as VCOs, as far as ones that track really well. The Falistri I can get eight octaves of beautiful tracking out of it, which is quite nice. I can't quite get that out of the Joranalogue, but the Joranalogue, it just has so many nice and subtle features. One of the only ones where you can modulate the slope value, which is basically a waveshaper, instead of just these static values in something like Maths. Yeah, his approach to design, he definitely sees the opening in the field and takes it.

HC: I've been in modular for five years, and even five years ago the options are so different, the popularity of certain modules, the most talked about modules are so different in five years. I find there's a lot of heavy duty all-in-one multifunction modules today. I guess there are different factors, right? People kind of oscillate between having large systems and smaller systems, so if you have a small system then you probably want something a little bit more multifunctional.

JL: Certainly as a designer, the pressure to make things smaller is real. My first module was 14HP and it was just a filter. It made total sense to me. It's been a popular, best-seller, but the number one criticism is that it's too wide. I'm like, but if it's any smaller I can't get my big fingers in. Certainly what I've noticed is, especially in the last couple of years, a trend towards ever more compact modules, and then cheaper modules. People are on budgets and, because there's this little blast of dopamine you get when you buy a module, you can do that for something that's small and fits in your system and it's cheap and you can do it more often.

NM: Rather than thinking about that from the perspective of the designer and manufacturer, thinking about that from the perspective of the collector, it's like, I want one or two super function-dense modules that are kind of the core of my system. A lot of people want a really function-dense sequencer. The super dense thing that is kind of the centrepiece of my modular would be the [Modcan] Quad LFO, where it's like, I've got wavetables, I have a lot of other weird uses for it other than the LFO, probably 15 different central patches that I use often with it. Other things, I like simplicity. VCAs, for instance, I have the Linux because there's a lot of VCAs in a low footprint. Simple, easy to use, small footprint makes a lot of sense for a lot of utilities, and then having that one thing that's just where you put the little piece of paper on your tongue and go and see through space and time.

HC: A few years ago, Qu-bit made a bunch of really big modules with a lot of space in between [knobs]. I remember there were a few customers [at Moog Audio] that really liked them. They're like, "That's what I want in my system. I love how spacious those modules are. That's what I need."

YN: Yeah. I understand that. The conversations are ridiculous when we're doing UI stuff. Phil and Hannes and I, we'll agonize over the slightest things, and it's mostly to make stuff feel inviting to look at, and it's not too crammed when you put your hand on it. You don't want to dime the wrong trimmer because your hand is on one knob.

JL: Probably 80, 90% of the people that order an 880 now want the desktop enclosure for it. They don't want to actually put it in their system. I kind of have to agree with them, because it's an XOX-style step sequencer. You need to be able to access all of the steps unimpeded and then you also want to be able to adjust all of the tone controls on the drum voices. I'm a sloppy patcher, I will never pick an appropriate length cable, I will never plan my patches. I'm just like, "Oh, I want to modulate this." So all of a sudden my 6U rig is just this spaghetti mess and I can't even access the sequencer on my drum machine, because it's all covered in cables, which I then have to push out of the way. Yeah, this idea of interface, usability. As Yoni was saying, some designers, manufacturers, they put so much attention to it. Some users appreciate it. I mean there's a whole cottage industry around taking wider mutable modules and scrunching them down, shaving two to six HP off of them, right?

DM: Very Toronto-centric cottage industry, funny enough.

NM: Boy, I would like to do this every week. It's been great talking to you guys. This is inspiring. I kind of wish that I didn't have all sorts of crap to do.

JL: I really like hearing from passionate creative people because I don't possess either of those qualities anymore. [laughter]

YN: The succubus children have leeched it all out of you.

NM: Well hopefully we can have more exchanges like this.

YN: For sure. I can't wait to hear some of the results of your filter experiments. That sounds like really interesting stuff.

DM: I'm having fun.

Notes

1 Music retail store founded in Montreal, with a store in Toronto, that specializes in electronic music equipment—synthesizers, DJ gear, digital audio/studio equipment. The Toronto store closed in May 2020.
2 Nyles is currently Research Associate with the Responsive Ecologies Lab at Toronto Metropolitan University. Heidi is a Graduate Researcher and Lab Technician there.
3 Jay Lemak owns and runs Synths When, a vintage synth repair shop in downtown Toronto.
4 Steve's Music Store is a musical instrument retailer with stores in Montreal and Toronto.
5 A monthly Toronto modular/synthesizer meet-up. www.freqfreaks.com.
6 https://doudoroff.com/sequencers/.

18

INTERVIEW

Dave Rossum

AK: Andreas Kitzmann

DR: Dave Rossum

AK: How would you describe your approach to sound and synthesis design? Where do you draw your inspiration from when it comes to creating new modules or synths?

DR: I'd say my approach to sound is a combination of scientific and intuitive approaches, having studied two areas that embody those approaches. My degree in biology from Caltech included a lot of neurophysiology and related subjects. So, I understand the neurophysiology of the human hearing system, at least as it was known 50 years ago. And I try to keep up somewhat with that field. But Caltech also required six terms of calculus, physics and quantum mechanics; there aren't many biologists that end up with that kind of education. All of that informs a fairly scientific understanding of the nature of sound and its perception.

Additionally, I largely think in terms of the evolution of human hearing, which is to say the principles that created our sense of hearing and have shaped it as we have evolved as creatures. All of these aspects inform my intuitive understanding of what we can hear, and what we find interesting to hear. Also, I think every designer has their own unique pair of ears, although mine are getting old. They were awfully good when I was young. An aside: One of my first childhood electronic projects was building an audio oscillator. I then went around and tested everybody's hearing and found out that mine was really, really good. And that explained why I could tell, when walking up to the front door of

DOI: 10.4324/9781003219484-20

somebody's house, whether their television was on, because I heard the horizontal sweep at 17 kHz. I could also hear ultrasonic burglar alarms because they weren't ultrasonic to me. When I was a youngster, I remember sometimes I'd put my fingers in my ears when visiting stores. My dad, also a scientist and Caltech grad, would patiently explain to the shop owners that I could hear their alarm, and ask them to turn it completely off.

So, the combination of my education and personal experiences informs my approach to sound design. What I try to do, what really delights me, is to come up with analogue and digital algorithms that give a musician a palette of control over the nature of the sound they're making. What I'm seeking are algorithms, modules, and configurations that allow musicians to engage with an instrument in a somewhat predictable way. That's the other thing that I strive for in my modules and my approach - I like the controls to behave coherently, and be labelled accurately, so that when a musician creates a patch or turns a knob, he has at least some idea of what's going to happen. There certainly are cases where chaotic behaviour is appropriate, so sometimes what the musician expects to happen is "who knows what", because that's the purpose of that particular control. But to me, that's a very explicit thing that just doesn't happen randomly, but rather as a specific result of the intended algorithm. My favourite algorithms tend to be those that have varying modes, like the "zing modulation" in our Triton oscillator. When you use it in one way, it gives you a wide variation in timbre that's relatively unpredictable on the gross scale, but you can also carefully tune it to give a specific result. Or you can tune it among a handful of related timbres, and it becomes quite predictable within that range. When I stumbled across that particular algorithm, I was very excited.

AK: And where do you start then in the design process? Do you start with an abstract idea like human physiology, or do you start with a circuit idea, or do you just stumble across an algorithm and see where that takes you?

DR: "Stumbling across" is probably the best description. For instance, when I figured out the idea of "zing modulation", I used synchronous ring modulation as a starting point. That idea was very logically arrived at. I was thinking specifically about ring modulation and how it creates non-harmonic frequencies, and that makes it problematic for a lot of tonal music. And then I asked myself, what happens if I synchronize the inputs? I realised that if the modulating oscillator was forced to be periodic with the fundamental of the carrier oscillator, then all the intermodulation products would be forced into the carrier's harmonic spectrum; they'd be strictly periodic. And once I thought of that, then I'd built it up, listened to it, and it was more exciting than I ever expected.

This brings to mind another project - the Linnaeus module - another example of my process. The principle behind Linnaeus is through-zero modulation of a filter's frequency. John Chowning, back in the 70s, introduced us to through-zero frequency modulation of oscillators, which became the "FM" of Yamaha's DX-7. Back in the 70s, I had designed a unique state variable filter for the E-mu modular system, and many folks since then had encouraged me to re-create it for Eurorack. But I kept dragging my feet - I wanted to do something new, rather than just a rerun of that filter. It's much more fun to invent something new. One day I was talking to Gary Hull, my lab mate at Universal Audio, about this, and all of a sudden I said out loud, "Could through-zero modulation work for filters as well as oscillators?"

My first thought was simply about linear, rather than exponential, frequency modulation of filters, which also is relatively rare. I'm not sure even that has ever been marketed before. Most filters are controlled by a one-volt per-octave input. If you exponentially modulate a filter's cutoff frequency when it is highly resonant, the exponential curve is going to bias the perceived resonant pitch away from the unmodulated pitch; it will actually shift the perceived resonance upward. But if you modulate linearly, that perceived pitch stays in the same place, which is much more musically useful. Now we're talking a matter of seconds here from when I went from that initial thought of how through-zero linear modulation works for oscillators, to the question of whether it could work for filters? And then after a couple more seconds of thought, I was led to the revelation that, yes, it should work for filters! But then how the heck would you do it? That began the inventive process where I had to figure out how you actually implement such a thing in a circuit or digital algorithm.

Next I wrote a little MATLAB simulation and listened to the result, and it sounded really great. Then suddenly I was taking off and building prototypes and listening to them and learning more about what else you can do and putting different knobs on it, for the musician to control all the various aspects of such a filter. That was the evolution of the Linnaeus module. Another thing: I have had instances where I've come up with design ideas in dreams. All throughout my life I've remembered my dreams and I honestly do wake up in the middle of the night thinking, "Oh my God, what a great idea!" To make sure that I remember it the next morning, I will write down some notes, to take it from dream memory into real memory.

AK: And I suppose some of these dreams or inspirations don't lead anywhere.

DR: Yeah. And you know, sometimes I come up with inventions and think, "That ought to sound good." But then it doesn't.

AK: Some inventions or ideas could be more interesting conceptually. Or maybe it's a technical challenge that is worth pursuing but not necessarily a musical one.

DR: I've got a couple that I'm working on right now. Both are technically challenging, and I haven't quite determined whether they're really interesting or not. I have the feeling they are, but I haven't had the chance to play with them enough to determine that for sure. But another past example I can think of was using a sampler with phase modulation, modulating the actual playback of one sample with the waveform of another sample. Again, a sudden idea - "I wonder if that would work?" I did a quick digital mockup of the thing, and gee, it sounded great! Next, I implemented it in real time so that we really could explore it. That's generally the progression: here's a good idea, let me do some mockups, play around with it, see if what I hear excites me. If it does, then I dive into the real design process where I have to make a manufacturable and reliable product and also expand it as much as I can to make it the best tool for the musician.

AK: Do you test out your prototypes with musicians or friends?

DR: There's a small group of us within Rossum Electro, and some former Muons,[1] who I'll drag into the lab when I have a circuit on the breadboard and have them listen to it. Once in a while, I'll make a recording and email it to somebody. And then as we get into things, depending on the module, we have a cadre of beta testers that we recruit. And often, depending on the module, the testers may be mostly looking for bugs and things like that. Or they may get in early enough that they can influence the controls and the panel and so on. Very seldom do we have beta testers involved in the actual algorithm design.

A typical example was the Assimil8or's voltage-controlled aliasing parameter, because people wanted to be able to get really grungy "aliasy" pitch shifting and also really, really super clean pitch shifting. But when we got the first version out in the field, the response was, "Yeah, but it doesn't get bad enough." Now I had to come up with a new algorithm. I knew what we wanted was something that, in a continuous way, took us from the clean sampling of high order interpolation through the drop sample nature of the SP-1200 and Ensonic Mirage, but then to just keep going further in that sonic direction. Having clearly stated the goal of the algorithm, it was next just a matter of it having to be dirty enough, and it didn't have to be very sophisticatedly dirty. It didn't take long to get there. So in this case it was the beta testers who were the ones who drove it down that direction and the ones who helped decide the final spec.

AK: Are you still inspired by past synthesis techniques? You mentioned, in an earlier conversation with me, Don Buchla and Bob Moog, and how you're going to be appearing with their descendants at the MIDI Lifetime Achievement award ceremony.

DR: That's becoming more interesting to me. This comes in part from being a member of the Universal Audio Algorithm Group. I'm coming to an understanding of what is pleasing about many of these vintage synthesis techniques. And I've gotten to play around with a few of these in my work with Universal Audio, as well as to talk to the algorithm guys there who have become so expert at understanding vintage studio equipment: what makes a piece of vintage studio equipment sound good? Yes, that's definitely an area of interest to me. I don't think in my own career I'm going to be replicating other people's vintage instruments. But I absolutely want to understand why vintage instruments are popular, so I can build on those principles. It'd be fun to take one of Bob Moog's or Don Buchla's designs or Serge Tcherepnin's designs, and further improve the state of the art. I've actually got one project like that in the back of my mind now, but of course I can't talk about the details yet.

AK: Would you describe yourself as someone who's attracted to more conventional or unconventional approaches to musical synthesis? Where do you get your inspiration, or what kind of sound palettes are you drawn to?

DR: I think a lot of people would describe me as conventional in the sense that I very much appreciate the talent and skill of people with traditional music training. You might even say I'm a bit of a snob, in that I am totally awed by outstanding musical talent. Now understand, I absolutely support anybody who enjoys playing with sound and music, and I've said many times, "Music is whatever the person involved wants to call music." But I also will caution more traditionally oriented friends asking about new music concerts, "I don't think YOU wouldn't call that music, but it's music." Nonetheless, as I said, I like making instruments that are understandable, predictable, and learnable. I really personally don't subscribe, as much as many other people in our industry, to the proposal of making the user interface deliberately confusing to encourage people to randomly explore. That's why I think other people would call me conventional.

I love to explore, I love to do the unexpected, to connect things in ways that you wouldn't expect. In fact, in the original E-mu modular, we very deliberately departed from the Moog synthesizer's strong distinction between audio signals, triggers, control voltages. In the Moog, the envelope generator triggers even had a unique kind of plug. ARP led the way to what we would do, by using just 5 Volt digital signals as the gates, so at least there was only one kind of plug, and they also standardised the signal levels so that control voltage and audio were compatible. We took it further by doing things like giving most modules a constant 1k Ohm output impedance, thus making it so that when you plug two outputs together, the signals would mix or average. We also made it so that if you plug gates or triggers or other digital signals together, they logically "AND" with each other. One of the big reasons for this was to make it so that when you did the unconventional, what happened was understandable.

That reminds me of another story. I love to go backpacking. But for the first two years when we were starting E-mu Systems, each summer, a fellow who was kind of my second father, and who taught a course in Mountain Ecology at UC, Santa Cruz, would invite me to be the mountaineering instructor on that month-long course. And I responded, "I can't go, I've got this business we're building up." The third year I started to say the same thing, and suddenly realised the whole purpose of having my own business was to be able to do what I wanted. And here I was not doing what I wanted, being a slave to the business. So I told Scott Wedge, my business partner, that I wanted to go backpacking. Scott was fully supportive of this, so off I went.

About halfway through the month-long backpack trip, my mind started to wander back into synthesizers. I had not thought about work for two weeks. Before I left, we were considering designs for a sequencer as part of our modular system. On the backpack trip, with my mind cleared out, I came up with this idea: "We've got a modular audio system, why not build a modular sequencer, which is to say, build a set of modules which somebody can patch together into their own custom sequencer." That idea was brand new, totally unconventional at the time. We created digital building blocks - modules like AND gates and OR gates and flip flops and so on, as well as sophisticated counters and even memories so you could record keyboard control voltages. Then you could mix and match things. As far as I know, nobody ever copied that idea. It was very popular with the people who got into it; they just loved it. It's a pretty good example of how I'm very unconventional in one sense. But these things were digital modules, they were really quite conventional. I wasn't shrouding them in mystery or naming them funny things. An OR gate was just called an OR gate and so on.

AK: So you are really driven by innovation then?

DR: Yeah, that's what I love. I love exploring new things. I love doing something that nobody else has done before. That's the unconventional aspect of my approach. But I also love doing things thoroughly and really understanding them. Also, I absolutely don't want to be controlling of the musicians. I don't want to tell them what to do. In fact, the other thing that absolutely delights me is when musicians take an instrument and do something I never would've expected them to do with it. I mean, the greatest thing in my entire career is the inspiration that the SP-1200 had for the hip-hop artists of the early days, being able to take both the beats off of other records and recombine them and also take urban sounds and things like that and sample and make them into the expressive percussion sounds for their music and their story. I think that is just so cool.

AK: Which you couldn't have anticipated.

DR: No, I never would have. And the other thing that absolutely delights me is when people turn the tables on me and become innovative with my instruments. So I certainly try to design instruments that allow folks to do that. But I very much avoid telling people what to do. And I'm somewhat at odds with people who imply that by being confusing, you can foster that type of creativity. I just don't think that's the way human creativity works. That's just a personal opinion. Remember at PatchUp!,[2] when we did that exercise where we collectively designed a synthesizer front panel, and I kind of got put down for suggesting we ought to begin by defining our module paradigm - things like our convention in our modules for the visual cues for signal and control flow. I feel making a module having a pleasing visual symmetry is secondary to having its function clear from its appearance. And I think that speaks to how I am different from many other designers in that respect.

AK: Perhaps that's the scientist in you.

DR: Oh, absolutely, it's the scientist, engineer and teacher in me. Did you know I almost went into education? That's another funny story. There was a very innovative education programme at the University of Massachusetts, Amherst. I flew out there when I was looking for grad schools coming out of Caltech. One of the things they had me do while I was out there was take a personality test to see if mine was appropriate for a teacher. In that test there were several dozen questions that used the phrase "cutting up." And I'd never heard that phrase before; it was not used as slang where I grew up. So I guessed it meant something like self-harm. After the test, I asked what it meant, and when I learned it meant goofing off, I thought "Oh my God, will they think I'm warped", because I am absolutely a goof off! I think I ended up flunking the personality test.

AK: Well, good thing because you gave the world all these wonderful instruments.

DR: But in my role at E-mu and at even Universal Audio now, I do a fair amount of mentoring and educating other people.

AK: Does the general modular synth community play a role in your design decisions? And by that I mean either the forums or some of the meets that you might go to or just some more informal connections, whether it's so-called "synthflunencers", other musicians or other designers,

DR: Musicians, certainly. And also when I go to trade shows in particular and talk with people. I always try to pick up ideas that'll inspire me without treading on anybody's domain.

I get in trouble sometimes with my staff here because I have my own tastes. For example, our modules have aluminum knobs. I just like the feel of aluminum knobs, which makes the modules much more expensive than they otherwise would be. They can add a couple of hundred bucks to the cost of a module if it has a lot of knobs on it. But I love how they feel, and for me, it creates a sense of being the precision instrument that I've worked so hard to create. That's part of my gift to the musician - that feeling. So, in that sense, I want to inspire musicians. I truly believe that the emotional relationship between the musician and their instrument participates in the artistry, though probably in unpredictable ways.

But I get conflicted when my colleagues say, "why can't we make our modules cheaper?" We could use plastic knobs and cheaper controls on them, but that's just not who I am. Similarly, our silver and blue front panels have been carried on from the original E-mu Modular aesthetic. We also have musicians that write to us and say, "I won't buy any of your modules unless you make 'em black." While we talk about this fairly often, it's a difficult thing to accomplish because I'm not going to stop making silver modules, so now I would need two SKU's and all the planning and inventory issues around that.

Our latest module, Locutus, is a MIDI interface module; it plugs into the Assimil8or and enables extensive MIDI control. We have great internal arguments about it. We hear that if we make the LEDs any other colour than red, we're going to get people complaining because "All LEDs should be red." But my thinking as an engineer is that the MIDI activity LEDs should tell at a glance if Assimil8or was sending or receiving MIDI, and different colour LEDs make that easier.

The prototype has an amber and a blue LED. I do know that there are people that don't like the brightness of blue LEDs, but this one only very occasionally blinks. I went out on the forums and asked what colours people like and so on. And I was unable to find anybody saying, oh, it's got to be red LEDs. But just to test, we're going to show the prototype at NAMM and get some feedback. So that's the sort of thing that sometimes influences the design process, but it tends to be fairly minor. In conclusion, I'd say I'm mostly self-driven but definitely listen to the forums. One of the places we do use customer feedback is when we release a module including software, and we hear something in the forums that is both practical and sensible to change.

AK: Such as firmware updates?

DR: We're very careful with firmware upgrades. This is just company policy. We won't announce anything until we've done it and we know we can do it. There are some companies who will even do it as a form of test marketing: "we're going to do a software update with this, that, and the other thing!" And then they don't do it because nobody was that interested in it, or were more interested in something else. We won't do that. I don't spend that much time on the forums. I've got our marketing guys, and that's their job. So, I rely on those people.

AK: So, talking about the market then, how would you position your company aside the bigger players in the synthesizer industry, such as Korg, Behringer, etc. in terms of innovation and interesting products overall?

DR: I think each of those companies has their own trajectory, quite different from ours, particularly the big companies. I'm not sure it's even appropriate to try and characterise them. I wouldn't mind Rossum Electro becoming a bigger company. It's challenging to do it as we are, more or less, bootstrapping.

When we built the Emulator One, we hired the same company that did the enclosure for the Apple Two and the mouse for the Mac, a company called Hovey Kelley Designs. Working with them was a lot of fun. The E1 enclosure they designed was built like a tank. Once, David Kelley was visiting and told me, "The reason for having your own company is so you don't have to work with jerks." He used a rather more colourful word than jerks, but you get the idea.

Rossum Electro-music began as a result of a few ex-Muons who got together and said, "Let's start another company and have some fun." We are not a company that was founded to make a lot of money; it was created to allow us to do what we love and be able to work with our friends and hopefully bring in more friends, as well as some new young folks. It's not an entrepreneurial vision. I think this is often true in the Eurorack world of small companies, whereas I think the larger musical instrument companies, by necessity, end up being much more entrepreneurial visions. I'd love for Rossum Electro to grow, but I don't want to grow in a way that causes me to lose control of the company ethos.

I don't mind losing control of the company ownership. The truth is, back at E-mu, Scott Wedge was the president and he actually owned a hair more of the company than I did. But we did that specifically because he was going to be president, and that seemed appropriate. But the two of us were very closely aligned about what kind of company we wanted E-mu to be, so there was never any issue.

And we went on to hire outside CEOs and get investors and board members. Technically, we had "lost control" of the company - I think I had about 11% ownership when we were acquired by Creative Labs. But the truth is that the natural leaders in the company set the ethos, and I guess I'm a natural leader. So I'm not very worried about losing control.

AK: Control is perhaps the wrong word, but I guess it sounds like you have a community actually within the company which can start to unravel if you grow too large.

DR: Yeah. Well, you have to be careful. E-mu, all the way through to the very end, kept that sense of community. And Universal Audio has become quite large, yet they still are doing a great job of keeping that sense of community, particularly given their size. It really is a challenge. But I feel part of my job at Universal Audio is helping maintain the community. Bill Putnam's a good friend, and the engineering management frequently relies on me, even though I'm a "technical fellow" with nobody reporting to me. I think they understand that's how you keep a company dynamic and innovative, and how you hold onto the people that will really contribute to that.

We always kept that sense of community at E-mu, even after Creative acquired us. With the Creative acquisition, we also put together what was originally called the Joint E-mu Creative Technology Centre., eventually renamed as the Creative Advanced Technology Centre. There we managed to hire the best brain trust that had ever been put together in audio. But Creative Singapore management directed our efforts, and they really didn't know what to do with all this talent. They didn't realise what they had and what a powerhouse it was. During that period my efforts were directed primarily to Creative's chip design, and with the frustration with Creative's lack of vision, the sense of community at this Tech Centre dwindled.

AK: When you go to these trade shows such as NAAM, what excites you most? I mean, if you're just walking the booths, what are you drawn to?

DR: I recall one instance where I walked throughout NAMM's Eurorack ghetto. It was a lot of fun because I had my name badge on it and I was wandering through looking at the various modules and gadgets, and, for example, enjoying seeing the Doepfer booth - all the things he'd done and the great variety of modules. And also seeing lesser-known companies with quite interesting and novel approaches. Then every so often I'd walk into a booth, and someone would say, "Oh my God, you're Dave Rossum." That was really, really fun, seeing how few people recognised me. I enjoyed not being recognised as well as occasionally being recognised and, and once in a while being recognised by folks with eyes popping out of their heads.

That was a very special NAMM for me, quite different from my more recent experiences. Now, as a small company, I end up staying in the booth most of the time. A lot of time other folks from other companies come to our booth and chat with me because they know I'm happy to share ideas and consult with them. And also, about the time we started Rossum Electro, we were establishing SSI, the chip company which I also represent at some of the shows. So I'm specifically there both to help other engineers design their circuits, as well as creating general interest in our chips. I enjoy that a lot. I really enjoy meeting enthusiastic customers. I'm pretty shy in some ways, so I don't tend to really get out and hang out at other people's booths. That's just who I am.

AK: You touched on this before, but one thing that we're interested in is exploring how technological innovation shapes new forms of community learning and creative practice in music and art. And you've touched upon that when you mentioned your influence on the hip-hop community. I wonder if you have any more thoughts on that and how maybe in particular, modular synthesis is special in some way?

DR: When we first started E-mu, modular synthesizers were rare as hen's teeth. We would go to parties, and we'd bring a modular synthesizer with us. It was by far the most fun when there were kids at the party, kids of an age where they were interested in the synthesizer, but completely unafraid of it. Adults tended to be afraid of the modular synthesizer.

Adults were self-conscious as well. They didn't want to break it. Of course, you can't break it; it's designed to be impossible to break. There's nothing you can do on its entire front panel, besides taking a hammer to it, that would break it. And the E-mu modular was even designed to resist that fairly well. So just go for it! Even telling the adults this, you still couldn't get them to go for it. But if there were kids around, the adults would be drawn to it. So - be childlike, be silly, unafraid, like a kid is. And that promotes learning when they are around - people who are unafraid and unassuming.

So, I think that's special about modular systems. We loved bringing the modular system to parties and just watching people play with it. And sometimes it would just sit in the corner and we'd be doing other stuff, and sometimes Scott or I or Paula would drag people over and try and get them started with it. But that was fun.

I've always loved the idea of people playing with sound. I get to play with sound all day. But for others, coming home from a day at work, kicking back, pulling out your synthesizer and doing fun stuff, making sound for the purpose of making sounds that you love - I think that's a great thing. As I mentioned earlier, some people might not think it's music, but if anybody thinks it's music, it's music.

AK: And there's something freeing about that. Most people wouldn't pick up a guitar unless they know how to play it or a piano, because there's this pressure, around what piano playing sounds like. People might get frustrated, but with synthesis you're exploring sound in a more pure, non-genre specific form.

DR: Karen and I are big supporters of the Moog Foundation and the Dr. Bob's SoundSchool, which is an elementary school experiential curriculum teaching the science of sound through exploration. I think that's a very wonderful thing. I was recently interviewed for a book on music synthesis, and we were talking about the early 70s. I'd forgotten that, back then, the inspiration for a lot of the companies, including E-mu Systems, were schools that wanted synthesizers in order to get kids involved in music. It never went anywhere, but in spring of 1970, right after the name "E-mu Systems" was born, and before we'd ever built any kind of a credible synthesizer, the San Diego School District was looking for a company to build custom synthesizers for them.

I was in my first year of grad school at that point. The two other guys I was working with on the synthesizer were still in their undergraduate career down at Caltech. One of our original targets was to build some synthesizers for the San Diego school district. The school board was smart enough to not contract with completely untested people like us, because it would've taken several years for us to build a credible synthesizer (which it did). It was 1972 before we actually got the E-mu Modular into production. Nonetheless, that was very popular thinking back then, and I think it was right on. But it was also a different world. At that time, California schools in particular had a lot of budget, before a political movement called Proposition 13 changed the property tax structure, which funded the schools in California and as a result California schools went from being the best in the nation to being one of the worst. Now we're somewhere below the middle.

But still, I think the modular system is one of the best ways to get people to be creative about sound and music. And I think that spreads out into other parts of their lives. That's one of the reasons that I am drawn towards making modules that have paradigms and nomenclature that you can understand. I like the idea that you're not only teaching people how to be creative, but also how to organise their thoughts about creativity, because the world is understandable. I guess I'm a scientist, right?

AK: I think there is something about modular and DIY synthesizers that help to make people comfortable with complex technologies that are not contained in a box where you don't know what's going on inside. I mean, you still don't know exactly, but it forces you to deal with some of the principles around waves at least, and relationships and maybe some of the math behind it, even if just in a very indirect manner. So perhaps that offers one way into science and engineering for people who might think, oh, that's not for me. Especially for kids, which I think is important.

DR: As an aside, back when we started with electronic music, women in the field were very rare. We had Wendy Carlos and Suzanne Ciani, but very few others. I really like both of them a lot, the women and their music. At Synthplex this last fall, Suzanne was honoured, and gave a concert. It was an absolutely wonderful concert, and wonderful to spend the evening with her and the other honouree, Tom Oberheim. And the talent and the joy in the room was just palpable. So that was pretty cool. But nowadays it seems there are a lot more women involved, and I really, really like that the modular system may be helping to break down the perceived barriers to women in technology.

AK: Yes, definitely. Now, this goes back to an earlier thought. How do you strike a balance between what you think is a really great idea that you're really excited about for whatever reason, and making something that at the very least can cover the development and production costs?

DR: It can be challenging, though I don't think it's anywhere as difficult for me as it is for some people.

Some of that is because I've become quite good at what I do. And so the things that really excite me generally have a fair amount of commercial appeal. The challenge for me comes when I have an idea for a synthesis technology or an actual synthesizer. When does it come out of the idea stage and to where we can say, "This is what we're making", how do you cross that line? Scott Wedge, my business partner at E-mu, and I always had this tension going on. He was the business guy, even though he was very technical, and he'd often be on my case, saying things like, "Okay, Dave, it works, get it out of the lab and into a product." And I'd reply, "Nah, it's not quite ready." For me, there's a feeling about a product, I think Scott calls it "balance." But it's basically when it goes from being a feature or a technology or an algorithm into a true product, typically an instrument, although in modular systems, it's a module. It's when all the pieces come together in a harmonious way. Sometimes a module does one thing and that's the centrepiece, but even though it does one thing, there are other aspects about it. All those aspects have to come into balance. That's very intuitive for me: at a certain point I decide "Yeah, it's ready." That doesn't mean things can't change a little bit after that, but it's finally crossed that line.

A good example is our SP-1200 reissue. From the day we announced the existence of Rossum Electro, we were getting one or two emails a week about reissuing the SP-1200. So that was always on our list, but we didn't know what we were going to do, whether we were going to improve it, and if we're going to improve it, what we were going to add, what we're going to take away, or maybe we should just keep it the same.

AK: That is a pressure for these kinds of legacy instruments.

DR: Exactly. And it wasn't until Isla Instruments announced the S2400 that the right direction for us became clear. Brad at Isla chose a direction for the S2400, and I looked at it and said, "That's not the direction I want to go." That crystallised our thinking - "Let's do a true vintage reissue." Now I could enumerate the few things I'm going to change and all the things I'm going to keep the same. We chose to make a premium product that people are going to love. It was targeted to be so well built that everybody's going to feel wonderful about it and not be upset with the fact that it's expensive. So that was when it all came together. We'd been waiting three or four years for that crystallisation to happen, and it happened in a matter of a couple of weeks.

When we founded Rossum Electro, we knew of four modules we were going to do: the Evolution analogue filter, the Morpheus Digital Filter, the Control Forge, which was the function generator out of the Morpheus module, and a sampler we hadn't named yet. Actually, Morpheus was the only one we knew the name of when we started. Each of those modules had to go through this same process - we know what it does, how do you make it into a final product? Going back to your original question, we knew from the beginning that they were saleable products. Even right now I've got three or four product concepts that are nowhere near completion. I know what they are going to be as products, but there isn't that tension you were talking about between it being a great idea but will it sell? The question for me is more "Where do I sell it and how do I price it so that it creates its own niche in the marketplace?"

AK: Do you have any modules that you personally love but don't have the kind of uptake that you feel they deserve?

DR: I've been a little disappointed with the Linnaeus through-zero filter because I think it sounds really cool. So I would say that's the only one of our modules that hasn't been taken up as much as I would've guessed it would've been. But I certainly get feedback from people about how much they love it. One of the things about it is that because it's a digital filter, there are analogue addicts who don't want it. They don't want it because it's digital, even though it really is a digital simulation of an analogue circuit. So it sounds analogue, it behaves very much like an analogue implementation, even more so than the Morpheus filter.

But the Morpheus filter has been extremely popular. So being digital isn't the whole problem, though I think that is some of it. We announced Linnaeus and Panharmonium at the same time, and Panharmonium has been more successful than we expected. We did that kind of as an experiment - can we successfully announce two new modules at once? What happens if we do? And the answer pretty well came back "No!" The marketplace is such that the dealers have a budget for Rossum Electro, and if you give 'em three new things, they'll buy a third of as much of each as they would have of one new thing, because they can't change that budget very much. The money doesn't come out of nowhere. And similarly, I think throughout the Eurorack world, there's a fixed rate of growth and Rossum Electro gets a fixed percentage of that. Changing your market share is always really hard without doing a revolutionary type of thing.

AK: One of the reasons that people often give for their interest in Eurorack and hardware synthesizers in general is to get away from the screen, to get back to reality and to touch those wonderful knobs that you have. What's your relationship with DAW and screen-based synths?

DR: I had fairly strong opinions about this in the early days of E-mu. First of all, we were involved in the launch of Digidesign, which was called Digidrums when we first met founders Evan Brooks and Peter Gotcher. Their first product was custom drum sound ROMs for the Drumulator. There was a time when I suggested, "Gee, maybe we should merge with Digidesign." This was just when they were changing their name to Digidesign and doing computer stuff; making products that talked to the Emulator II and performed real time DSP. But Scott was not interested at that time in merging our company, so we didn't go down that path. So we chose not to pursue the professional computer sound world, leaving it to Digidesign and other companies like that. Digidesign, or course, went on to become a huge success story in that arena.

I think that helped develop my love of standalone instruments. As the general purpose computer took off, it became pretty obvious that if you were willing to put up with the downsides, you could get more sound for fewer dollars out of a piece of software on a personal computer than you could from a standalone instrument. And not just a little more, no, 10, 15, 50, maybe 100 times as much sound making capability for your dollar. Eventually, computer music software became close to freeware, and that drove the standalone instrument almost out of existence for a while. And my strong opinion at that time was, "It's going to swing back at some point." I genuinely believe, and I've said this so many times in public forums, that the process of making music is a matter of dealing with emotional content. You're expressing emotion, and part of that emotion is the relationship with the instrument that you're playing. And I find it hard to have an emotional relationship with a mouse and a keyboard and a screen compared to instruments that have panels and knobs and buttons and other tactile stuff, instruments designed specifically for making sound. So I very strongly believe, as an instrument designer, that the tactile and visual nature of instruments is very important, almost ultimately important.

Yet we now live in a world where so many of the electronic products we interact with don't have switches and knobs. So, I've always been fascinated with, but never really been able to dive into the aspect of controlling instruments. I've always wanted to bring new technologies into a musical instrument keyboard to make it more playable. I've had lots of ideas, but I've never been in the right corporate environment. I've never been able to work out the details that would allow it to become an actual product. I'm not a competent mechanical designer; I do mechanical design and it's okay, but it's not my forte. But I can work very well with mechanical designers, and often inspire them.

My feeling is that making music on a screen and a mouse and a QWERTY keyboard is very different in terms of the emotional experience than making it on almost any other kind of standalone or modular type of instrument. Computers are very much a part of our life today. So, I don't have any problem with, say, our Assimil8tor module. You can edit samples on your computer, you can download them from the internet, then you can transfer the sounds from your computer onto a memory card, stick it in the Assimil8or, and load the ones you want. But Rossum Electro is not currently providing any software apps that would ease this process. I certainly would support anybody creating librarians or editors and things like that for our products if it's appropriate, because such things are now a part of our world. But the actual musical instrument itself, I feel, shouldn't be a computer. In contrast, I've been in recording studios using DAWs, and really liked how that worked. And unlike my feelings about musical instruments, I think the DAW is a valid use of the computer. You're not killing the emotional aspects of the recording studio by putting everything on a computer screen.

AK: You just make it more efficient in some ways.

DR: Right! Personally, I think a combination of a computer screen and a physical console might be optimal. But again, like a mechanical engineer I'm not a competent recording engineer; I can only fake it.

We've got the OLED screens on some of our modules, and I think we're striking an appropriate balance in the modular world in keeping the analog feel while taking advantage of digital technology.

I haven't put a touchscreen on any of my modules. I'm not crazy about touchscreens, but I certainly can see that they do have utility. So I'm constantly debating, am I going to jump into that or not. I haven't gotten there yet, and I really can't tell you how close I am or not at this point.

AK: What I find satisfying about hardware-based synthesis is that you are forced to deal with limits. Whereas in computer-based synthesis, there's a million different options and you're constantly scrolling through and you can spend more time looking at all the various options instead of using them. At least I do. Maybe I'm not disciplined enough.

DR: Both the early Emulators, the Emulator I and II, and in particular the SP-1200 many, many artists have said that what was special about these instruments were their limitations. There are a lot of the early hip-hop artists that talk about the crazy things they did to get everything into the 10 seconds of sound in the SP-1200 and how that created its own sonic footprint in the music that then became part of the genre and gave it its identity. I love that kind of stuff. And I agree with you that as an instrument designer, I often try to minimize limitations, but I love it when the musicians embrace the limitations and do all these original things around them, in terms of figuring out workarounds that would never occur to me. It's magic to me.

AK: What do you think will be the greatest opportunities and challenges to independent modular and synthesizer designers in the years to come?

And how do you imagine the future for your company or in the modular synthesizer industry in general?

DR: One challenge I would've thought would happen, and so far it hasn't, and I hope it doesn't, is the commodification of modular synthesis. When we got into Eurorack, I was afraid that we'd end up with a few large manufacturers controlling the market because of cost efficiencies. And that hasn't happened. In fact, increasingly it seems that we're getting more and more manufacturers joining the genre. And while a lot of products are sort of "me too stuff", there is also some great innovation. There is also a spectrum of cost, trade-offs, appearances, etc. And I'm delighted that's happening. In terms of challenges, I would ask, is this sustainable? Because I think Eurorack has become a wonderful community and works really well. Another concern in terms of potential challenges relates to that famous statement by the US patent office in the 1890s where they proclaimed that everything that could be invented has already been invented. I do have some concern that we're going to run out of innovation in the modular world and we will wind up spending more energy rehashing as opposed to innovating. I think the real challenge is how can we be making modules that are affordable, understandable and appropriate and that aren't just rehashes of the same thing.

When Rossum Electro got into the market in 2016, there were questions of whether Eurorack had reached its peak. Then the pandemic hit, and folks spent even more time in their home studios. Right now, with the pandemic becoming less of a threat, are people going out into the world more? Will they suddenly not want to buy Eurorack modules? Will we have a drought of sales? We don't know, but I very much do suspect that we're going to see an ebb and flow in the long term for this industry. I was there when modular systems made a good business in the late 70s, to the mid-80s when you couldn't sell a modular system to save your soul. I recall how my friend Gary Hull spent his life savings of $700 to buy a Moog 55. That's now worth $40,000 to $50,000. His father thought he was crazy to spend his money. Gary purchased it from a university somewhere where they had it mothballed. He's restored it to its former beauty. And you wouldn't believe what some of the original E-mu modular systems have sold for.

The pendulum goes back and forth. I'm one of a very few people who stayed in business through that whole experience. E-mu was, I believe, the only American synthesizer company that made the transition from analogue to digital synthesis. We probably were more experienced in digital technology with the E-mu modular than other analogue synth companies, and we were one of the first in sampling. But Sequential didn't make it. Moog didn't make it. Buchla didn't make it. Arp didn't make it. And many smaller companies as well.

I think however the pendulum moves next, the real challenge for the Eurorack community to overcome is somehow keeping together its community nature, the joy of innovation and of playing with sound through whatever technological disruption comes next in this marketplace. Let's be sure to keep the diversity of manufacturers and the nature of our community, where we all realise we're not competing with each other but instead we're building something together.

AK: Yes, it's collaborative as opposed to competitive.

DR: Right! So, looking at Rossum Electro, we've branched out already into the SP-1200, and I expect we are going to do more branching out. We're no longer an exclusively Eurorack company, and I have ideas for several more innovations given that we have made a transition in our contract manufacturing situation.

One of my agenda items is how our new relationship with contract manufacturing can allow Rossum Electro to continue to prosper in the Eurorack market, while we also continue to go look at other things we can do akin to the SP-1200 reissue. I've got plenty of really fun ideas in that, both with and beyond the Eurorack space.

AK: So that's something to look forward to.

DR: And I think long term with Rossum Electro, I'm an old fart. I'm going to be a really old fart in not too long. So, I'm definitely thinking about the long term trajectory of Rossum Electro and my name. Obviously because we're associated with Universal Audio, that's the likely place for us to land. Though I'm definitely open to other possibilities and I've had a few other conversations going on in that area as well. So I'm hoping the Rossum name will stay out there. But I'm nowhere near ready to retire. I'm hoping for another decade, or maybe two, of designing synths, because I just have so much fun doing it.

My wife Karen has asked me, when are you going to retire? I said, "If I retired, I'd just be doing the same thing but not getting paid for it, so I might as well keep working." That's my planned direction, in addition to all my mentoring activities, and keeping alive the concepts and lessons I've learned in how to make weird music.

AK: Or to help other people make weird music.

DR: Exactly. How to enable people to make weird music.

Notes

1 Muons refers to employees of E-mu Systems, Rossum's previous company.
2 Rossum is referring to the workshop "PatchUp! A Workshop on Synthetic Sound and Modular Thought," that was held at the Responsive Ecologies Lab at Toronto Metropolitan University in collaboration with York University. It ran from April 1st to 3rd, 2022.

19

INTERVIEW

Paulo Santos

ET: Ezra J. Teboul

PS: Paulo Santos

ET: In your previous interviews you mention your dad was an electronics technician, and that you got into synthesizer music from listening to Kraftwerk's *Autobahn*, then that you studied electronics in school... What was your community like at each of those stages? Did you and your dad build DIY radios and speak with amateur radio operators? Were there other people your age to talk about *Autobahn* with as a teenager? Were there other people building musical electronics in engineering school?

PS: Unfortunately, in the beginning (70s and early 80s) I was alone because my friends did not share my passion for this field. I remember my dad and I building some phasers and spring-reverbs from electronic magazine projects, so my dad, with his passion for music and electronics, was my biggest encouragement.

It was only after I graduated from engineering that I encountered some friends who showed me the world of synthesizers (keyboards), which was a technology that I hadn't had much contact with until then.

The year that I enrolled in engineering school, which was 1986, was a turning point for me. I managed to buy my first synthesizer, a Roland JX-3P and my first drum machine, a Roland TR-707.

My friend, Tony Moreira, who is a music professor at Belmont University in Nashville - Tennessee, showed me the, at that point, new

DOI: 10.4324/9781003219484-21

MIDI technology and some main synthesis techniques that really opened my eyes to this new universe.

Yes, my dad lived in a time when radio amateur operators built their own equipment. I remember, as a child, at the age of 6–7, seeing him building large radio amplifiers and antennas, and this was a huge stimulus for me. From that point I realized that those real efforts could result in practical changes in how you would be able to enjoy your life. My dad spent hours on weekends contacting people from all over the world by radio and that left me amazed. By the way, my father's station's prefix was PY2BEB.

ET: Generally, do you see the process of becoming a synth builder, musician and collector a solitary process, or more community based?

PS: An absolutely solitary process.

ET: Today, how much interaction do you have with other synth designers and builders in Brazil and South America? How active is the electronic music community around you? Do you ever collaborate with others, hire apprentices, share designs online and at meets?

PS: I think that over the years I got used to working alone, so I never had any contact with other hobbyists/builders. Sometimes a customer asks me for some feature in my modules or comes up with a brilliant idea that I immediately use to make changes or design a new module. What consistently drives me is the desire to build electronic instruments that allow me to give form to my musical or sound ideas. Sometimes I listen to an electronic piece, and I think, I can build a module that will allow me to implement this sound, and that's it.

ET: You also mentioned you use all through-hole components, is that still the case? Do you model circuits before building them, or do you breadboard everything? This is quite different from engineering school; how did you start being closer to classic synth DIY philosophy than electronics commodity philosophy?

PS: YES! The main idea behind EMW is to build modules that bring back the sound character and form of the old synthesizers and I absolutely believe that the components used to build them have a noticeable influence on that.

Yes, I work with protoboards and sometimes spend weeks changing the circuits until I reach a sound that matches what I had in mind. When a sound from a VCF, as an example, gives me the same sensation of being played on any of the finest synthesizers I had over the years, then I know I'm getting close to finalizing it.

ET: What is your musical background? Did you learn signal processing before learning to play instruments?

PS: I had three years of piano lessons as a child, besides that, I didn't have any other serious musical education. The little I can "play" is what I managed to reach with some practical training by myself. Not too much, believe me.

ET: In your previous interviews you mention being inspired by classic American synths like the ARP or Moogs. What are some other influences you might be willing to share?

PS: My designs were mainly influenced by EML, ARP and Moog equipment. I found through the EML line the kind of "harsh / raw" sound that always caught my attention in the early electronic music pieces. In a certain way this is what I always liked and tried to reproduce. It may seem easy but it is not. Designing a circuit that is capable of creating or letting the highest harmonics to pass through with an embracing brightness is not a simple thing to achieve.

   This is one of the things that distinguishes the older machines from the great majority of the brands on the market today.

   Just listen to an ARP 2500 VCO and anyone with perfect hearing will note that there's something more in its sound compared to a standard VCO.

ET: You offer a significant number of modules, with a selection of analog circuitry that rivals Doepfer, but also a number of digital modules. This represents a lot of work, over a long period … Do you follow your instincts for what to do next and how? Was there / is there a trajectory you are following for your brand and its products overall?

PS: In most cases I create what I need to implement the sound and resources that allow me to realize my sonic ideas. I also often listen to great ideas from customers and make modifications to our modules. There's always someone that feels the need for some extra pulse or sync input, and these suggestions are always welcome.

ET: It is common these days to combine multiple functions in a single module, but your approach seems more granular. Is that a fair thing to say about your designs? If so, can you tell us more about why?

PS: A very relevant topic!

   This is closely related to people's idea of what a modular synthesizer system is. What it is made for, how it will be used, and mainly, how we expect our interaction with the machine to be. It's almost a philosophical matter for me.

Let me put it this way, for me a synthesizer is an instrument that, like any other, must be played by hand. When you play an instrument, your mind must be tuned to emotions, not to logistical issues.

The right side of our brains, where creativity and emotions reign, must stay focused during the process of creating sounds and playing on a synthesizer.

If you eventually need to stop your creative process to remember what the fifth press on a button while holding others will do, this activates your brain's left side and you will risk losing all your inspiration.

There's no way around this, modular synthesizers are big machines, they must be. They must be simple, there's real reasons for this. This is why EMW makes simple modules. Yes, we have the technology to put a thousand commands and/or functions on a single button, we can also choose to make all digital circuits with small components (SMDs) and double layer printed circuit boards, but we are making analog synthesizers, doing our best to reproduce the sound of the old machines, not from cellular phones.

ET: What is the advantage of learning electronic music on a computer with VCV Rack, Pure Data or Max/MSP, vs. modular? Do you use both these days?

PS: I think that for learning purposes, the virtual modular synths are excellent tools! I bought a complete system from Cherry Audio a few years ago and I enjoyed it very much. The sounds are very good and very useful on a mix, but for a purist like me still very far from hardware analog machines.

The interaction for sound creation is obviously limited by the interface and as expected this makes a huge difference.

ET: Do you do all your assembly in house, or do you outsource any part of the manufacturing process?

PS: Our company opted from the beginning to keep all the processes involved in the equipment's manufacturing in house, with no use of outsourcing.

We recently made a huge investment in two high power laser machines for cutting the aluminum panels and steel. We are also migrating to the use of laser etched front panels instead of using special printers, like we had done so far, which will considerably increase our modules quality and lifetime expectancy.

We also possess our own facility to do the panels electrostatic painting, printed circuit boards manufacturing and parts machining.

ET: How has COVID affected your work these last couple of years? Is the current chip shortage affecting you more than shipping delays?

PS: The only issue that interfered with our workflow was the fact that some countries have imposed restrictions on receiving packages from Brazil, mainly for the standard shipping methods, so we were forced to pause small quantity orders.

We use Microchip brand chips on some modules and have had no problem in buying them.

ET: What does your musical setup look like these days? How about your manufacturing setup? Does where you play music overlap with where you build your modules and instruments, and do you see these two tasks as separate projects or two moments in the same process?

PS: I have a studio in my residence where I keep a big EMW system for playing, testing modules and eventually recording some demo videos. In my company there's also a studio room where I keep a smaller EMW system and most of my keyboard synthesizers.

In the home studio I also have some synthesizers, which I usually change from time to time. I really like to analyze the work of the major brands, so I often purchase the synths that I find most interesting for a more detailed in-person review. I recently acquired a Roland Jupiter-X, an ASM Hydrasynth and the Modal Cobalt-8.

In relation to old machines in the home studio, I am always changing as well. Recently I've been playing a lot with old sampler/synthesizers, more specifically the Korg DSS-1 and the Casio FZ-1, that I really like! My approach nowadays is more related to sound analysis and an effort to understand what exactly determines the sound character of these synthesizers.

Although I haven't been involved in any musical projects recently, the ideas and the desire to implement something never leaves me.

Yes, I believe that I see the development of the EMW modules in some way overlap with my ideas for creating electronic music.

ET: You share some block diagrams and frequency responses / waveform measurements on your website but are any of your designs open source? If not, why not?

PS: We don't have any open-source projects so far, but we haven't ruled out that possibility for the near future. Offering some assembly kits are also in my plans.

ET: What are some interesting patches or signal flow configurations that you might be willing to share? What are their inspirations? Are there any patches that rely on specific EMW designs that would be hard to implement otherwise?

PS: I believe the biggest difference between what you can achieve with an EMW system compared to others is a certain distinctive sound quality and also a way of working and creating your patches that directly results from the ideas behind the modules implementation.

Yes, there are patches, somewhat 'unique', that I come across during the process of researching and using the modular, but unfortunately I end up not keeping them.

For example, the use of our ANALOG DRUM SYNTH in conjunction with the NOISE STATION, ADSRs and VCAs always surprises me and results in very breath-taking sounds.

On the digital side, our MIX SEQUENCER together with our ALIAS DVCO, ARCADE NOISE and a like the VCF SP-12 or VCF OB-12 can also create very interesting sounds.

ET: You've been an active circuit designer through a couple of decades of "modular resurgence" - what lessons did you learn in that time, in terms of what sounds, and functionalities customers are looking for?

PS: For me, the main lesson learned was that analog audio circuitry is not as simple as it seems. Anyone can build a VCF, but not everyone reaches the point that matches an original ARP or Moog VCF, that's why to this day we worship these old synths and we still keep making comparison videos between old and new machines. If you think about it, this challenges technological progress. Nobody makes videos comparing the image of televisions from 30–40 years ago with those of today. Audio synthesis technology should, in principle, be way ahead, but this is not the case.

The beauty of this area is exactly that, technology can evolve quickly, but your ears do not evolve at the same rate. In this area, what was very good 40 years ago is still very good today.

That has been my goal, to do my best to reproduce the sound of these old machines and at the same time try to maintain an attractive price range for the consumer.

The other modules not directly involved in the audio path and the digital ones were thought to have an interface as simple as possible, allowing for a synthesis process without the need of manuals and particular procedures in its functions.

ET: What was the biggest surprise in terms of module design?

PS: No huge surprises until now, just a substantial amount of dedication and research, always observing, comparing, studying and learning.

ET: Are there any themes or central concepts in your circuit development work? Your very wide range of modules feels like it has a very pragmatic and clear approach to synthesis, as evidenced by a mixed use of analog and digital components and subsystems where most appropriate/interesting, but we are curious to know more about your underlying threads

PS: I am a practical guy, I study the circuit's operation theory until a certain point and try to implement within the design stage any new ideas I may have had. Once in that practical stage things almost never go as I have imagined, so if this is the case, I start again from the beginning.

Sometimes it's incredible, as some small detail on the circuit can bring an entire new dimension to the sound or control range/dynamics.

If the circuit involves programming, as is common with most digital modules, the process is more mental. I spend days thinking and outlining the steps that will have to be completed for me to have a good operation. This translates into a development time that demands more attention and patience.

ET: Other designers refer to current academic papers on signal processing concepts, or refer to old engineering textbooks, as the basis of their circuits, do you do this as well?

PS: I do not follow any particular rule, I read and learn all I can find, and sometimes go look back in my old ideas and circuits to check if there's something good that I can use in new designs.

What I always study on a regular basis is everything related to microcontrollers and programming. I love this area, but I don't always use its full potential, so it doesn't take me away from my design principles, but it's definitely something that's a hobby for me.

20

INTERVIEW

Corry Banks

ET: Ezra J. Teboul

CB: Corry Banks

ET: Can you tell us a little bit more about your module design process where your musical interests and technical expertise came from?

CB: While performing around town and coming up in the chicago hiphop scene and culture, I decided to go and get an electronics degree because I knew I wanted to be around music and music equipment no matter what was to come in my lfe. We were fortunate young men, myself and my group mates, we were able to be in professional recording studios and not just basement attic studios and bedroom studios. We were really working towards something big, opening up for the who's who of hiphop when theyd come to town, freestyling and appearing weekly on college radio shows. We were living that life, hiphop and more hiphop. And although we enjoyed it, I started to realize like, yo, we broke, we broke ... with no end in sight. So, I started to game plan on my future in more ways than one.

I realized that I needed to expand my skills and pursue multiple opportunities by having more than one iron on the fire. While in the studio, I noticed they were using either an Atari or an Amiga as a sequencer, along with the typical huge professional studio mixboard, reel to reel tape machines, DAT machines and on and on - all electronic. I was just as fascinated with that stuff as I was with the creative side. I foresaw a need for someone to maintain and repair this equipment and if nothing else, I was just intrigued enough to want to learn something deeper about

DOI: 10.4324/9781003219484-22

the inner workings of all of it. It wasn't enough to know signal flow for recording, I guess. I decided to learn more about electronics. At the time I guess I thought of it as a fall back plan but the reality was that it was something that I had a deep curiosity and eventually a passion for it. My initial thought process was … Hey, this way, even if my music career doesn't pan out, I could still be involved in studio life one way or another. Because who's gonna be the tech guy when needed.

So the long and short of it is that I went on to get a degree in electronics and I felt like I had found a new world of super cool stuff that aligned with my childhood tinkering and such. Because I'd been around so much music equipment in professional studios, I somehow came across Dr. Bob Moog on the internet. I learned about the Theremin and how he invented the synthesizer and sort of revolutionized electronic music through technology. That was mind-blowing to me. Eventually, I convinced my friend to build a Moog Theremin from scratch with me for our Senior Project. He was down to do it but we never got a chance to do that in the end, but it always stuck in my mind. In our final semester, just before graduating with an electronics degrees, somebody said, "Hey, for your senior project, you can build something, or there are some schools that are taking volunteers to fix up their computers and that will count as your credit for senior projects." So I never built the theremin for school. I eventually did build a theremin, many, many years later, but I did not do it for the senior project. And I think had I done that then I probably would've been building music equipment and circuits and stuff since then.

Funny thing about that Moog Theremin? When I started blogging a decade later, having lived this dual tech by day hiphop artist by night sort of lifestyle, I built BBoyTechReport.com as a place where hiphop and technology converged. I'd review music equipment and customize drum machines, etc. I'd also write about the latest hiphop projects and news. I'd interview makers, designers, sound designers, etc. One day I told the story within a couple of tweets about my failed goal to build a Moog theremin. To my surprise, Moog tweeted back at me saying, we love that story and we are going to send you a theremin kit to build. It was dope. It felt like a full circle moment.

I think I met Upright and Voltage Controller at NAMM, and those guys were actually into modular before I was, but we were all hip hop and we were all beat makers and we were all kind of into music tech. The more I did reviews, the more I got into these relationships of knowing different people. I started doing beta testing for different things because I forged relationships with folks and whether they're at this company that company, my opinion became such that people wanted that opinion as part of testing teams. Hey, tell me what you think of this. And so that's how that kind of all came about.

Eventually, I'd caught the bug and began keeping note books while learning modules. All the while beta testing and reviewing music equipment. Then I'd started sketching things up and writing out specs for my own ideas.

So yeah, having been educated in electronics and understanding circuit design, etc. was one thing but not being in a situation where i could build on those skills on a daily basis was another thing. My path just went somewhere else. IT was my path. I was no longer working on a component level. Nonetheless, I recall talking with my friend at one point and we were like, "yo, we should design a drum machine." And we sketched up drum machines and made plans, but we were in Chicago at the time and it was like, "yo, how the hell do we get a drum machine built?" You know what I'm saying? We were so far removed from anyone who were even remotely doing anything of the sort that we just had no clue.

Until I got into this music tech industry as press with the formation of bboytechreport that I'd start to make these relationships with folks in the industry of electronic music manufacturing that I started thinking about that again. When you are beta testing and reviewing music equipment, you naturally deconstruct processes and features and find bugs. If you're beta tester, you are actually responsible to find bugs and report bugs. And if you're really good, you should be making suggestions on how to make things better. That becomes a regular part of doing beta testing and even reviewing. I think at some point I started to realize, "yo, it would be dope if I could make my own stuff!" And the more I thought that, the more it became real to me. "Well, why don't you?" So I began to take things more seriously eventually. The confidence grows when you see products on the market that has a feature in it that you suggested or something's working better because of some bug that you found.

My good friend Glen Darcey brought me in while he was working at Arturia to do some drum patterns for their latest product at the time. It was a drum machine, the DrumBrute. He got me to do all of the hip hop drum patterns for the drum machine. It may have been half of the patterns on the drum machine. And so this drum machine ships even to this day with drum patterns that I programmed on it. Now I know that most folks just erase them all immediately to make room for their patterns but the idea that this is a thing that I can do was huge for me. So when you start to see these sort of things, I started to think, "Hey, actually I'm adding a little value here." It felt like a badge of honor. Then I start to think more and more like "I could probably do this." It would be cool if I could do this for myself and make my own things.

ET: Just to place some spaces and some years I was hoping you could name whatever venues you remember from Chicago, LA …

CB: Going back to when I was performing a lot in the nineties in Chicago, we performed at all of the venues, even the bigger events. So we performed at we called the Riv in Chicago, which was the Riviera, opening for a lot of acts. There was the Oak Theatre … there was Subterranean … We even performed at some like Hebrew Israelite Temples on the south side back in the day. I performed with my group at festivals for the likes of NAACP and Rainbow Push. We were performing a lot from maybe 93 to maybe 98. And distinctively 98 was the year when the group kind of fell apart a bit. A few years later, I'd began producing myself and gathered a good deal of solo material. Eventually I went out and formed a band. We started performing together in the early 2000s at venues around Chicago including loft parties and random clubs. I think the highlight was performing at the Wild Hare, a very popular reggae club on Chicago's north side near Wrigley Field.

　　When I moved to LA in 2005, I performed at all sorts of places many of which i dont recall the names but I performed a good amount at the Airliner, a few places in Santa Monica and Long Beach.

ET: Beta testing started before or after you got your master's degree?

CB: Well, I got my master's degree in project management in 2015. I was already blogging on bboytechreport and working my day job as an IT Manager. Somehow, I was able to do all of that simultaneously and at some point thereafter I was asked to beta test for a couple of companies as a result of blogging and equipment reviews.

ET: The equipment reviews on the bboytechreport website are really extensive, and now I understand why you've been doing that in a really dedicated way for a few years now. I thought talking about the MPC One that you reviewed recently was a nice way to talk about the history of drum machines, the way that they interface between hip hop and techno and other genres. And just thinking about the MPC and how the history of it as a beat making device could eventually see it get re-designed with the connectivity required to interact with modulars. Have manufacturers been generally going in that direction of combining these different influences in their new hardware? Is it something you talk about with them when you reach out for review products and better taste thing and all that?

CB: I dont think drum machine makers think in terms of culture or the driving forces of culture (i.e hiphop, djing, beatmaking). I think technological trends drive changes to the standard features. There was a point when vintage keyboards and synths were very much sought after but there were no new options available. Then there was a resurgence in analog synths that came with things like CV connectivity. Eurorack modular synthesis experienced growth simultaneously and these sorts of things began to drive change on grooveboxes and even some controllers. Soon there was an emphasis on integration and connectivity and boom … CV shows up on MPCs and the like.

One larger manufacturer invited me to beta-test a comparable device because I had become known for my beatmaking infused with modular synthesis. That's how Modbap came about through those kinds of explorations and experiments. So, when the time came for me to provide feedback on a sampler instrument with CV I was already deep into the patch cables with it, so to speak. That's a value add to the team. I had that instrument probably a year or more before it ever hit the market.

That opportunity came about largely because I had very much made it known all over the web that I was about that sampler life, and I was about that Eurorack life. So I was a part of that sort of world that was starting to converge. And I don't know if there was anybody else on the team at the time that knew that world. There are always a lot of beta testers all the way around, but I definitely think that I was the only Eurorack hip hop guy on that team at the time.

Anyway, and from me digging into modular, I came up with this term Mod Bap, because obviously when I was doing just samplers and synthesizers, I called it Synthbap. I released a album called Synthbap, and then from there I got into modular. So naturally I called it Modbap, and the thing caught on. The name stuck. Going back to Voltage CNTRLR, Ken Pierce and Upright, these guys had incorporated modular into their beat making already, but we all have our own sort of style of what we do. So I called what I did Modbap, and eventually it just kind of became the thing that we all do, the umbrella that we sort of collectively decided to reside under. And it kind of flourished from there.

ET: To clarify, you didn't suggest the voltage control capacity on the device you beta-tested?

CB: No. Not at all. It was there, and I don't know why they decided to put it there, but I definitely was kind of the guy on the beta test team that was considered knowledgeable enough to give feedback about what we are doing here with this. When it's all said and done, I really appreciate being invited to give feedback. It also gave me a great sense of pride because I felt that this was the eve of something I'd been championing … making music with samplers and Eurorack with seamless and intentional integration from the design perspective on up through practical use.

ET: The Modbap Osiris module fits kind of nicely in this discussion of different conceptions of sound. This time it's more hi-fi, and I was just curious if complimenting the sampler discussion you wanted to talk about something more recent.

CB: This world of hi fi / lo fi bridging with Osiris, started being developed in 2021. My first module, Per4mer, was released in 2020, development started in 2019. But Osiris was a response to my desire to do my primary oscillator in the system that I'd designed. I needed this oscillator to be able to do a wide swath of sounds in a relatively compact space without compromising usability with the user interface. I also want it to be something a little more special than what anyone else would make in relation to what was already available. My thought was "How do I pull this into this hip hop frame of mind?" I'd ask myself How's this fit in as a puzzle piece to that system considering the style of music that I make?" Then I linked up with Ess Mattisson of Fors, formerly of Elektron. He's the guy who designed the synth engines for the Elektron Digitone and the Elektron Model:Cycles. When I heard that he'd left, I was like, well, I know Ess and Per4mer's doing well enough for me to move on to the next product.[1] It would be dope if I linked up with Ess and hired him as a developer to help develop my oscillator's synth engine. Subsequently, when we started talking, everything just kind of fell into place. I'd known him before. He'd been on my podcast and everything when other products had come were released. So we just started talking about the possibilities and he was very much open to working together.

I had this mock up and I shared my thoughts on interface and what I'd want to achieve in the end. For this particular product, I want something that's going to give me this wide variety of sounds, and we discussed the idea at length, realizing this would work out well. In that conversation I mentioned that I want to have a lofi control on the oscillator. Ess was basically like "oh, okay! I'd been messing around with this different way of adding sample jitter and some bit reduction or resolution reduction, etc. to oscillators." It's something he had been experimenting with for a while and it aligned with my needs and my design. Also, I liked that Ess said "it's not just bit crushing."

And so the thing about Bi-Fidelity, the Hi-Fi and Lo-Fi, is that I wanted Osiris to have a lo-fi control. He had already been messing with this sample jitter and sort of this idea of making his synth voices sound similar to old Amiga music computers, where the screen scans a certain way and sample jitter gets added to the audio depending on what you're doing. And that was just kind of the way it was back in the day. When Ess said "yeah, I was experimenting with some of that and I never did anything with it." I said "well, yo, that's the type of atypical lofi control that I want." So when he gave me an example of it, I was floored and we were off to the races. The first iteration of this thing, during development and testing, I fed it some MIDI notes and modulation to see how it sounded and responded, and it was incredible.

I realized as we continued to talk and build on these ideas, there were so many similar concepts. It was clear that we had common interests. So, ultimately, Osiris became a wave table module that smoothly morphes between wavetables, includes wave folding and FMing for extensive timbre expression. Includes a sub oscillator with continuously variable waveshapes, separate sub out, added sub v/oct for sending a secondary v/oct for bass lines if needed. The idea of smoothly sweeping through timbre modes, sub osc waveforms, wavetables, the added fidelity parameter and the inbuilt VCA all of this came to be what is now known as Osiris, the 12 hp powerhouse bi-fidelity wavetable oscillator. I just really liked the collaborative environment that was created by working on Osiris together. I would speak of a feature or requirement and Ess would most times be like "yo, I've experimented with something similar before." It made for a very easy working relationship.

We started with the wave folder and the phase modulation, and we just kept building on it. The development process was like a volley back and forth. Wave folding done. "And what about the Lo Fi?" Ess was always on point like, "all right, I got that done. Here is something to test while I work on the next piece." It began as software that I could load and test. I was planted on a computer for probably months or more, and it just sounded so pristine, I mean, it was such great quality. We created the firmware to go onto the Daisy Patch, which is a testing environment made by Electrosmith. We had it on a Daisy Patch so that you can now take this software, drop it down to firmware and throw it on a Eurorack module and test with it. But it didn't sound the same. And I thought "but wait, the Daisy is capable of very high quality, you know what I mean?" So we iterated a bit more realizing that the Daisy was capable of 96 kHz / 24-bit audio hardware. So we took advantage of that by running it at 96khz which added so much emphasis to the fidelity feature doing what it does to take this high quality sound and degrading it in this pleasing musical manner. The juxtaposition of the two just worked out very well to have the module running at 96khz and being able to add sample jitter and bit reduction to the waveform to create this sound that I was shooting for the whole time.

When it came together, I knew immediately that was the sound that I liked hearing when it was on my computer during testing. It was very pristine, and I wanted it to be the highest quality. Thats what I'd wanted to hear but I also knew that I liked the lower fidelity. So we combined the best of both worlds there. That was super important for me. In testing it worked well for my style of music and I knew that t would be useful for other styles like ambient or techno, etc. I think it's a great primary oscillator with many tricks up its sleeve.

And so that's where this whole thing came from. It became something that was very much a part of what I wanted Osiris to sound like or be capable of. And then just before getting towards the end of development, I really wanted to drive the point home that, yo, I feel like this fidelity thing is something that should be highlighted. I can't just put this out and say it's the Osiris, and it's a wavetable oscillator, and that's cool. There are wavetable oscillators out there, but the "character" of this wavetable oscillator is what makes it special. That has to do with the timbre modes, the wave folding and FM modes. But another big part of it is the fact that it's very pristine sounding, but you can dial in this lo fi, so you get all this detail and nuance. You know what I mean? So, I came up with this term, Bi-Fidelity, the same way I'd come up with Modbap. To me, that's what really stood out with what we developed is this idea that high fi and lo fi could exist in one place and Osiris' character and sound had as much to do with the bi-fidelity as it did the timbre modes. I knew from that point I'd certainly welcome the opportunity to work with Ess again.

ET: You said briefly you wanted to maybe talk at some point about funding and things like that. Feel free to mention that here, but I'm also just curious about the process with your collaborator. Are you willing to go into the details of the module itself?

CB: Yeah. I mean, I am not a programmer. I design my modules and contract a developer. We then work together to bring the thing to life. The designs that I create come from user interface and musician perspective first. I ask myself "how would I like to interact with this thing?" Then I pull in a developer that can help me bring that to fruition based on the design and spec that I've created. I dont imagine thats all too uncommon in product development. The designer and developer are often not the same person but it helps to have a great working relationship between the two because things tend to evolve during development.

So my modules have come to live in a couple of different ways. The Per4Mer, Meridian, and HUE were developed using C++. All of my modules are based on the Daisy Seed. That's kind of the backbone of my modules. There will be a couple of analog modules as is Transit, the analog 2 channel stereo mixer with ducking inputs, but for all intents and purposes, most of my designs are digital modules. And so that gives me a good deal of flexibility with the developers that I work with. In terms of how it gets developed and what environment it gets developed in C++ is the most common approach so far.

However, Osiris was completely different in terms of programming. It was developed in the Max/MSP environment and that was sort of a big deal for us. Developing an instrument for the Daisy platform was done in the Max/MSP environment because we had to have a way to translate and convert that over to the Daisy platform. Fortunately, Graham Wakefield had been working on the Oopsy which does just that. Osiris was the first commercially available product to use the Oopsy in development. It came with its challenges, being the first down that path for a commercial product, but it ultimately worked out perfectly in the end.

I still have a version that will run in Ableton. It's funny because I sort of previewed the audio of this on my Instagram account and nobody knew that we were developing the Osiris. I was literally playing it in my beats and showing the controller that I used for tweaking it, the DJ Tech Tools MIDI Fighter Twister. I would map that controller to the parameters.

It gave me these kind of different environments to hear the thing in. So it's one thing when it's coming out of Ableton in the software version of it as is being developed. And then it was another thing when we exported it to put on a daisy patch while we developed the hardware, so we could test it in the actual Eurorack environment. So we bounced back and forth during development.

A Daisy will run at 96 kHz / 24-bit. The specs make it a really powerful and capable platform and electronic music equipment specific processor. It took a lot of optimizing to run it at 96khz for that pristine sound considering all of the features that Osiris has. It's able to smoothly morph between these wave folding modes and this phase modulation. Also, you essentially have two oscillators in one thanks to the sub that can be used independently, and then you have the lo-fi control. So to have all of that processing happening all at once and running at 96 kilohertz. Ess did a wonderful job at optimizing the synth engine without compromising the sound quality and feature set. We were essentially beating the crap out of the daisy at this point. The whole idea was to not lose the core functionality, like the bi-fidelity and the wave folding and the smooth morphing between the wavetables, etc. It all became very important and all of those features were essential to the character of what we'd built.

I've been fortunate, over the years, I've made a lot of different friends and acquaintances and gotten to know folks, forged these relationships with different folks in the industry. And one of the folks is Andrew Ikenberry from Qubit, and the guys over at Electro Smith and Electro Distro, they distribute Modbap Modular. So Electro Smith, they were developing the Daisy when I came to them about developing and launching the Per4Mer. I approached them regarding development initially. As we talked about all the details, they pulled out the daisy and showed it to me. We talked about the specs. It was one of those moments where I was all in like "yeah, this is going to be the key!" I was initially thinking I'd use Arduino or something like that but this seemed to make more sense for my situation. I mean, they developed the Daisy and I was hiring them as developers for my first product, the Per4Mer. This is how I launched the whole thing. We eventually forged such a relationship that Electro Distro became my distribution partner for my Modbap line. So Per4Mer was out there and it did really well straight out of the gate. I did not want to nor did I intend to stop there. I wanted to move on to the next thing because the intention was always to establish a company and a a brand and not just a one time product. I didn't start it to be the Bboytech guy that made a one time cool module. That's when I kind of reached out and said, "yo Ess, I want to hire you to develop to collab with me to develop this module, and here's my mock up and here's what the requirements are, and here's what I need."

I think it was just great timing that the Daisy seed and Daisy patch was launching and coming along at the same time. It sort of steered me in a direction that was perfect for what I wanted and needed to happen. Ess developing in Max/DSP and Graham working on Oopsy. There was some discussion and development towards a process for converting Max patches for use in a Eurorack world with the Daisy. And so Osiris, due to the timing, Osiris was the first commercial product that was developed with that process. It's funny that it's basically called the Oopsy, as in Oopsy Daisy.

ET: Do you do your panel design?

CB: Yes. I'm a very visual user interface designer. I'm very specific about how I want to interact with the modules that I design. I think of what the interface may look like according to how I want to interact with the module. I mocked up the Per4mer with that interactivity and perform-ability in mind. Initially it was a little different in terms of button placement but then as I sat with it and tweaked it a bit, I landed on the current design. As I mentioned before though, as I work with my developers I tend to iterate on the design for various reasons. Sometimes its a matter of placement for functionality and workflow. Other times we iterate on the design to accommodate some sort of technical spec or pcb layout, etc. It always begins in a note book, or now a days its mostly in a sketch app on my ipad. Then I take to illustrator to mock up designs, and that's where things start to come together visually for me. I'll write out specs intially and iterate on them as I go but typically the core functionality is there from jump. I mock up everything that's in my portfolio. I mock them all up in Illustrator before then discussing what is and isnt possible with my developer. We, myself and the developers, really have to work closely together at this stage. They let me know what can and can't be done and what's feasible from a technical perspective. From time to time I'll tell them well this has to be feasible for whatever the reasons is. I mean, you have to know what the intention of the product will be and if we can't do XYZ then it could turn into an entirely different thing that veers away from the original concept. Sometimes that happens and it just becomes something better than we initially thought it would be but most times we seem to align with the original concept. But the key is that we are flexible in the development stage as long as we dont loose the overall objective, which is usually performability, intuitive user interface and fun to use.

It's very much a collaborative thing once we enter the development phase. I'm fortunate enough to have such incredible subject matter experts to help steer my vision into the realm of real world application.

ET: Has the pandemic, the pandemic affected your supply chain?

CB: Per4Mer definitely felt some delays in prototyping because of early pandemic stuff. It would've been out earlier than October 2020 but still, the delays were very apparent initially. Fortunately, for those of us that use the Daisy platform, Electro used their super spidey 6th sense to insure that we would have what we need to continue development and bringing products to market with as little hardship and delays as possible. We felt it a bit but I think I've probably been sheltered from it a good deal.

ET: Not many manufacturers have statements like the one on your website, and in our early emails I had tried to see if you would simply write a longer-form version of it?

CB: Sure. For me, it's important to keep hip hop present in everything that I do. I am from Chicago and I grew up in the city in a time where hiphop culture raised me. I am of hiphop culture and so I need that to be present in all that I do. It's as much a part of the fabric of who I am as is the fact that I am from Chicago having grown up as a black youth in the 90s fresh off of all that happened in the decades directly preceding my youth. Subcultures are the tapestry of who we are as people and how we are built and how we react, speak, walk and dress. How we view things around us. I very much operate from that space and I am conscious of it. And it's just something that I always want to keep present. It's something that has always been at the center of everything that I've done with every endevor and passion of mine. I explained with BBoyTechReport, Beatppl, Modbap and everything that I've touched sort of stems and or leads back to hip hop. And it's important for me to keep that culture present and to assure that the hiphop perspective and voice is present and represented wherever I am. The tech space is no different.

Modbap Modular as a company was founded on a few principles that all sort of have to do with hiphop, DJing and beatmaking. Also, exploration and experimentation in terms of creativity and synthesis. Performability in terms of how my products are intended to be used is very important. And finally community is really important. I see it as more of a movement than anything. I want to be able to make it ok to be experimental in our beatmaking while never loosing sight of the foundational aspects that relate to traditional hiphop, techno, etc.

Modbap Modular was really born out of my passion for modular synthesis blended with beatmaking but also my willingness to invest in myself. When my day job asked that I move to another state for work, I immediately replied with a big NO THANKS. I had no worries about what that meant for my technology career. I had just the weekend before organized and spoke at a Modbap panel discussion at Synthplex in Burbank. The room was packed and there were loads of opportunities coming my way. Developing my own modules were always on my mind and plans were always in my notebook. So I figured now was the time and I wanted to not be concerned with conventional approaches.

So, ultimately I got a severance package and that became the seed money for Modbap modular. I stayed there another 120 days. And by the time I left there, I had already began development on the Per4mer and began investing into developing my business officially. Incidentally, I kind of overlooked the idea that it would apparently be the first black owned Eurorack company until I had a discussion with someone who impressed upon me how important of a detail this is.

I wanted to spell out my principles on my site. I should never lose focus of why I'm developing what I'm developing. When I do synthesis, I want to be able to blend those two worlds. And the brothers who I roll with also have that sort of mentality. So all of the things that I put up there were all really about having some foundation to stand on.

Our values include integrity, Community, Culture, Experimentation in beatmaking including Sampling, Synthesis. On the site it's noted as "We value community and enjoy bringing folks together. Together we learn and grow. The "We" perspective is infinitely more powerful than the "I" perspective. HipHop culture is where modbap was born. Our aim is to add to the culture and bring like minds together via music-tech. In HipHop we have always used music tech tools that weren't designed by us or for our genre. Modbap is the first black owned and designed Eurorack synth company. Our tools are made for hiphop. Modbap was born out of experimentation and pushing the boundaries. We encourage experimentation in the spirit of creativity, art, technology, teaching, and learning. We love sampling and synthesis equally. As a core component of hiphop and boombap music, sampling is a core production technique.

We embrace it as such. We love synthesis and sampling equally. As a transformative, sometimes technical, and highly creative form of expressiveness, synthesis is anything and everything imaginative. We proudly enjoy syncopated and quantized beatmaking as much as we love off-grid sloppy drum programming.

So many things are not diverse. And I wanted to really spell out the diversity of what I had began building. And I was just telling somebody the other day, it's like when I was growing up, KRS-One would say "yo, you are not doing hip hop! You are hip hop!" It was like a light bulb experience when he used to say that. And then he would go further to say, "wherever I am, hip hop is there because I'm not doing hip hop. I am hip hop … If I'm there, hip hop is there." You represent the culture wherever you go. And so that was my whole sort of train of thought, if I'm going to do this synthesis thing, I'm automatically going to fold it into what I'm already doing and what I'm already about.

ET: Is there anything you want to add?

CB: No doubt, I think about this stuff all the time. It's important to me that I create from this perspective of city kid, hiphop culture, beatmaking and DJing. I always say that the turntable wasn't made to do what hip hop does with it, you know what I mean? It was meant to set it and forget it. Put record on, put the needle down, walk away, time to flip it, come back, flip it, put the needle down, walk away. Somehow DJing took hold and hiphop was born. This idea of playing the best parts of the record and being able to spin it back and forth was revolutionary. And it wasn't made for hip hop to do what we did with it, but we did it. And now a days folks would probably think that it was designed for that purpose because it's the most prevalent form of use of the turntable. Same thing with drum machines and samplers. Roger Lynn designed the MPC but when he created that drum machine, it was not for the purpose of doing what all my heroes did with it, the Premieres, the Dillas, the Mally Malls of the world. The fact is that he made the MPC because he is a guitar player who wanted a backing beat, a drummer, and he didn't have a drummer, so he was able to create this drum machine to give him what he needed. The use case become something far more extensive that a backing beat in lieu of a drummer.

And I'm assume that's a super, super simplified, watered damn version of it. But the fact is, that wasn't made for us to do what we eventually did with it. Hiphop has always sort of taken these situations, turned them into cultural moments in urban areas. Hiphop was born out of that. Make the best of what you have or whats available and repurpose things to express ourselves as best we can.

Art was being systematically dismantled and removed from the classrooms, including music. And so this is where you have this situation where if there ain't no instruments, we ain't about to be doing jazz like my uncle. My brother, he played in a marching band but these are the generations before me. These things start disappearing in the school system. Then cats started using what they had at their disposal, turntables and cheap drum machines that nobody wanted. Everybody hated 808s and 909s when them things were released. But because they were cheap, people who did the hood music, like hiphop, house music, and electronic music in Detroit, Chicago, New York, and LA and all of those places, folks used what was accessible and made magic. Culture shifting magic in the form of music unlike anything anyone had ever heard before.

So I think we kind of have always in hip hop taken things that were not made for us and created music with it, developed techniques and shifted things with creativity. Thats were I want to always have as my foundation when I create things. For me, I've never seen the difference. When I look at my Eurorack system, I look at that and think all that shit is vinyl. You know what I'm saying? And I've said that so many times before. But the reality I looked at, I felt this connection when James Brown said in his biopic, and he was talking to his band, he was trying to get them to understand the concept of starting on the one and the funk, the downbeat. And he was like, yo, what is that that you're playing? And the dude was like, this is a trombone. He was like, what you playing bass? What you playing, lead guitar?

All of y'all wrong. You're not playing those things. You're playing a drum. And they were all kind of looking at 'em weird. And he was like, all of these things are drums. You do what he do, pointing to the drummer and everybody on the one. No, this whole fluid sort of thing you all are doing, it's going to be rhythmic and on the one. And when I saw that scene, I instantly thought about synthesis and modular and stuff because that's how I feel about modular. Yeah, it's all electronic music, but to me, it's all vinyl and its all rhythmic. What I do with that is the same thing that we are doing in hip hop with vinyl sampling, flipping it, chopping it up, creating longer bits. I use those same principles. When I say that on the principles of that with sampling, that's where I'm coming from.

I just want to create from that foundational place but also begin to flip the status quo. We used things that were not designed for us to create our music but now I want to design things from the perspective that this is what I'd want from my cultural vantage point.

ET: You said you were an MC. Was your first instrument the turntable, or was it the voice?

CB: Well I've never been a DJ. It was definitely the voice. I began as an MC and Lyricist. I grew up around musicians. Like I said, my brother played the trumpet, and he's also a drummer. My uncle is a trumpet player, and he's played with a lot of well known people. So I've always had music around me and I've always had a distinctive voice. So one day my uncle said, why aren't you rapping? I was 12, and I've been emceeing ever since and that subsequently got me into beatmaking and producing which led to synthesis and eventually Modbap. It's all in the journey.

Note

1 Editor's note: the Modbap Perf4Mer was Corry's module released before the Osiris.

21

MODULAR SYNTHESIS IN THE ERA OF CONTROL SOCIETIES

Sparkles Stanford

Sitting in my studio, asked to write about modular synthesis, I wondered to myself "what modular synthesizers do I actually have?" Sure, there are the modular cases sitting on the table, but are they each a distinct system, or do they count as just one? What about hardware modules that contain software modules inside of them? It's a single unit, but it contains within itself its own set of modular components. Is it then its own system unto itself, or is it just part of the greater whole? One of the hardware cases consists almost exclusively of connection points directed outside of the synth, integrating it into my effects rack, my desktop, hardware sequencers, even the external audio input of a Yamaha CS-10. Where does it stop being modular synthesis? Why is it so hard to talk about it as *a particular modular synthesizer* and not *modular synthesis* in general?

Explaining categories of sound synthesis like additive or granular synthesis can be rather straightforward: a specific process is outlined, we describe a starting state, the constants, the variables and their general effects, and generic examples of past deployments of the process. Additive synthesis might begin with a bank of oscillators, they tend to have a constant set of waveforms to select from, we vary how many oscillators are working, we vary the ratio of frequencies and amplitudes between the operative oscillators, and these build up complex timbres out of simple components. Historically, we might note that the Hammond organ is a particular technical deployment of a lineage of additive synthesizers. But modular synthesis isn't identifiable in this kind of way as a particular category of sound synthesis.

Is it as simple as saying "is it connected to a patch bay or something equivalent?" What if the heart of your studio is a patch bay? I could route the oscillator of a CS-10 through a digital clone of the Korg MS-20 filter or a

DOI: 10.4324/9781003219484-23

hardware clone of the same filter in the modular rig. At what point does it stop being modular synthesis and start being composition or production? Does the very presence of a brand-name or standardized format within a system as a potential set of connections make the whole thing modular synthesis, even if it's not actually implemented? Or would routing various synthesizers through one another via a rack-mounted patch bay to mix and match filters, effects, and sequencing constitute modular synthesis? The answer is less than clear.

This connects into a long history within philosophy of the problem of individuation: how does an individual become a singular thing distinguished from the environment around it? The French philosopher Gilles Deleuze serves as a valuable patch point to connect an analysis of modular synthesis to the problem of individuation for several interconnected reasons. First, Deleuze's own approach to individuation is influenced by philosopher of technology Gilbert Simondon, who is concerned with the relationship of individuation to technical objects and their capacity for continuous variation and modulation. Second, Deleuze, in conjunction with philosopher and experimental composer Richard Pinhas, take cues from Simondon to conceive of a certain technological and musical lineage from the blacksmith to the electronic musician related to this capacity for modulation. Finally, Deleuze sees the latter as definitive of our society's current mode of political and social organization by means of control. In this chapter, I show how modular synthesis as a compositional practice is inextricable from the politics of what Deleuze calls our "society of control."

In a joint exchange between Deleuze and Pinhas during one of Deleuze's seminars in 1979 titled "Metal, metallurgy, music, Husserl, Simondon," they intertwine the introduction of brass instruments into Western Classical music with the influence metallurgy has had on conceptions of the nature of materiality to flesh out a link between the musician and the blacksmith which exists "not simply because the forge makes noise, [but] it's because music and metallurgy find themselves obsessed by the same problem: namely that metallurgy puts matter into a state of continuous variation just as music is obsessed with putting sound into a state of continuous variation" (Deleuze 1979). For Deleuze and Pinhas what begins in this rise of brass music, particularly in Berlioz and Wagner, is an incorporation of a new approach for putting the materiality of music itself into variation which reaches an inflection point with Edgard Varèse and his early electronic music. Christoph Cox in "How Do You Make Music a Body without Organs? Gilles Deleuze and Experimental Electronica" sees this trajectory of metallization embodied in the history of 20th-century sonic experimentation, particularly in experimental and electronic music (Cox 2006). Here I want to limit this scope solely to modular synthesis as a paradigmatic case of the intersection of this continuous variation of the materiality of music and the continuous variation found in social and political control.

Simondon gives a history of individuation which explores several ways in which classical models of philosophy conceal key aspects of the process of individuation, and where metallurgy occupies a privileged space in the history of materialism because it brings to the surface features constitutive of matter that were otherwise ignored or concealed by philosophy. Deleuze sees metallurgy as making "what is ordinarily hidden in other matters rise to sensory intuition" (Deleuze 1979). One example Deleuze takes from Simondon to illustrate this is Aristotle's theory of hylomorphism, which proposes that substances or individuals are created by imposing form on a substratum of matter. A brick is formed by taking clay, putting it into a mold, and imbuing it with a particular geometrical structure.

On Simondon's account, hylomorphic molding is a quite sensible way to think about how technological operations work. An oscillator produces a sine wave (sound or electricity as matter) which possesses certain effects and properties. Running this through a filter unit (a mold with a particular geometrical structure describable in terms of a circuit diagram), out comes a new sound (substance) with different qualities and properties than the original matter from which it was made. While this might be a sensible way of thinking about regular objects, it is a static view of an underlying model: modulation. Deleuze describes this underlying model as "molding in a continuous and variable manner. A modulator is a mold which constantly changes the measuring grid that it imposes ... or molding is modulating in a constant and finite manner, determined in time" (Deleuze 1979). Molding as a model conceals the underlying chains of technical operations where changes (shifts in temperature, relative changes in pressure, forces of reaction, material elasticity, metastable states of energy, and so on) are constantly occurring and which have to take place. Raw matter is never simply subjected to form, but rather matter is always in some capacity pre-formed and form is in some capacity materialized, and the process of individuation is a chain of operations which modify already formed sections.

In metallurgy this constantly changing motion and modulation underlying bodies comes to the fore. In *A Thousand Plateaus*, co-written by Deleuze and psychoanalyst Felix Guattari, they provide us with two contrasting examples: the crucible steel sword and the iron sword. For crucible steel, iron is melted at a high temperature, through successive processes the carbon is removed, and finally the highly uniformed crystal structure of the melted steel solidified on cooling in a cast or mold resulting in its hardness, sharpness, and finish. In contrast, the iron sword is forged not molded, it is quenched rather than air cooled, and while it appears similar to steel, the designs are produced by inlay rather than the internal process of crystallization. The taking form in metallurgy thus doesn't happen in a single instant, but rather involves successive operations. While the brick in the mold actually requires a whole series of operations to happen within the mold as it sets, these operations don't necessarily rise to the level of visibility. Nothing compels us to understand how a brick is formed in the mold from clay at the molecular level, but for functional results metal *requires* that we intuit it as continuous variation and development, forging vs. quenching, carbonization and decarbonization, and so on. The process forces us to necessarily think of the materiality of metal and its successive operations in order to do metallurgy at all.

Deleuze and Guattari use modulation to understand how the material bodies of objects affect and are affected by other bodies. Thus, for Deleuze and Guattari a body is constituted by the specific capacities and limits for affecting and being affected by other bodies. As Brent Adkins notes, for Deleuze and Guattari "To understand a body is to know what affects it is capable of. For a human body this understanding can only be achieved through experimentation. This same way of understanding also applies to politics and economics. One can always ask, what affects is this political organization of human bodies capable of?" (Adkins 2007). The particular organization of the parts of a body with respect to motion and rest or change and constancy determine both the affects it is capable of and the limits of that body.

Take as an example a dog. This dog has some affects which are possible for it (running, leaping, sitting, and so on) and some which are limited (flying, climbing trees, and so on). The same dog is also capable of being affected in certain ways (it can hear in the 20–45 kHz range) which humans are not able to be affected by. But the organization of a body is always somewhere on this spectrum between constantly changing and being a finite constant: operations change the affects bodies are capable of, sometimes in fixed ways and sometimes in variable ways. For each affect Deleuze and Guattari designate a threshold at which that affect is possible. The minimum threshold of intensity, dubbed degree zero, is when an affect that is possible to be affected by occurs in such a way as to functionally be non-existent. For example, in the absence of a dog whistle or other animals which make those high-frequency sounds, the dog's capacity for being affected by 20–45 kHz is at degree zero. Similarly, when an affect is fully intense, at the limit of which the body is capable of handling it, it reaches the maximum threshold of intensity. Similarly, we can describe a synthesizer at any given moment in terms of modular intensity, from a degree zero stand-alone oscillator to a maximum threshold taking advantage of every possible patch point and connection.

Seen from the perspective of metallurgy, Deleuze wants us to understand that matter is not fixed, static particles which come together to form objects, but rather matter is a flow carrying "turning points and points of inflection"

(Deleuze 1990, 63) he calls singularities. Singularities are not qualities of matter, but rather the fact that at a certain state of energy the iron melts and crystallizes or when the rate of growth of a tree's circumference outpaces the growth of a branch then a knot is formed. Singularities are not directly visible but rather give to matter its capacity to be operated on (decarbonated, cleanly cut) and express affects (more or less porous, more or less hard). The singularities operated on to produce a crucible steel saber rather than an iron sword result in different affects: one works for piercing rather than hewing, you attack from the front rather than the side, one can be manufactured in number and the other only piece-by-piece. This model is particularly important for electronic music because Deleuze, following Simondon, notes that once electronics develop, it is the constantly changing state of matter which becomes the normal state of matter itself. It is this attention to singularities which is absent from models like hylomorphism and molding but is essential to understanding the modulation which underpins them. From the point of view of this model modular synthesis is a constantly changing fixed synthesis, while fixed synthesis is modular synthesis in a constant and finite manner determined in time. In short, all synthesis is semi-modular: somewhere on the spectrum from constantly changing to a finite constant.

Corresponding to these models are two modes (amongst many) of intuition by which consciousness grasps matter. Distinct from an intellectual or conceptual relationship to an object that can be acted upon, an intuition is a pre-conceptual grasp of that which is prior to, or constitutive of, both the object grasped and the person capable of reflecting on the object. For example, Immanuel Kant describes space as a form of intuition because it is not itself an object or quality but rather "it issues from the nature of the mind in accordance with a stable law as a scheme, so to speak, for co-ordinating everything which is sensed externally" (Kant 1992 [1770], 397) For Kant, we can imagine a space that is empty, but we can't imagine an absence of space, precisely because it is part of the very mode by which we organize external objects. But intuition also determines the subject doing the intuiting: we also cannot imagine what it would be like to intuit the external world in any other way than spatially.

For Deleuze, the two modes of intuition for grasping matter or materiality are "matter as movement" and "matter in motion" (Deleuze 1979). Deleuze defines intuiting in the form "matter in motion" as when we grasp it as a set of individual objects or parts (*this* table, with *these* legs) bearing specific qualities (it is a more or less resistant table, it is more or less porous wood). This approach, corresponding in general to molding, presents matter and materiality to us as static objects *in* motion. Applying Deleuze specifically to synthesis, we might point to a TB-303 clone. The synth has buttons which can be pressed that are more resistant than an Akai MPC2000. It has specific interfaces that can be affected: knobs can be twiddled, buttons pressed, it can be set in edit mode or performance mode, it has a fixed kind of pattern sequencing, and so on. Certain aspects of the device are so stable, so fixed, that we don't even grasp them as changeable (or, if they were to change, it would no longer be identifiable as the kind of object it purports to be). Rather, we grasp what can be changed *relative to this stability*. This particular stability and possible affects give it its '*TB-303-ness*' or TB-303-form even if that's not the name on the machine.

On the other hand, we intuit in the form "matter as movement" or movement-matter when what we are grasping is the continuous variation of matter itself, not as it presents itself in a sensory mode of more or less present qualities, but as a stream of operations which we may or may not intervene in, which may or may not be enacted. Rather than stable qualities, this mode of intuition grasps the singularities: a cutoff frequency, the passage from a low-frequency oscillator to an audio oscillator, or the point at which a reverb takes long enough that it becomes a delay. Here matter presents itself as the operations themselves which express affects, inextricable from our desire to present one singularity over another, to bring certain singularities into convergence to reveal other affects. Here we are not concerned or engaging with particular or static objects, distinct individuals, but rather the operations which are constitutive of matter in general and (when modulating in a constant, fixed manner) specific kinds of objects.

The transference of the intuition of matter as movement from the perception of metal in metallurgy to the perception of sound in musical composition produces what Deleuze and Pinhas call "metallic sonorities" after a phrase used by Varèse (Deleuze 1979). For Deleuze, Varèse's piece "Ionization" is "at the root of electronic music" (Deleuze 1979). because, as Varèse notes in his commentary (relayed by Deleuze) on the work that at "a certain moment, there is a sudden crack ... the musical scene changes completely. There are now only metallic sonorities." (Deleuze 1979). The sound of metal brings with it into the compositional process metallurgical considerations of timbre trans-forming what was otherwise a constant finite conception of percussive instruments into matter as movement. The work organizes instruments by their materiality (reverberating membranes, wood resonance, metallic sonori-ties, sirens), grouping them by their affects, but for Deleuze at this moment in the work when only metallic sonorities (gong, tam-tam, triangle, anvil, and the siren) remain, it heralds what will become electronic music's focus: no longer the affects, but the singularities, points of inflection, and the operations at work that precede affects and expression. As musical technology passes from brass, to metal, to electronics, these successive operations and variability become the normal state of sonic materiality.

Taking as an example the piano, Deleuze argues that metallic sonorities introduced "a totally new problem of orchestration, orchestration as a creative dimension, as forming part of the musical composition itself, where the musician, the creator in music, becomes an orchestrator. The piano from a certain moment on, is metallized." (Deleuze 1979). A "new" intuition of the piano emerges as a stream of operations to be intervened in. This is new only in the sense that it was always the case for the piano-maker, and to one extent or another always the case for the pianist who notices the wear and tear on the felt and the strings as they de-tune. The intensity of these elements was never at degree zero, but with these changes it is now risen to the Nth degree and on the path to the maximum threshold of intensity.

What gets incorporated into the musical composition itself is the act of, not simply creating the timbre from the instrument, or choosing a particular instrument for its timbre amongst other instruments, but of creating timbre directly or making an instrument for the timbre and incorporating this process directly into the composition. No longer a matter of selecting a piano for its timbre, but of transforming and manipulating the timbre mid-composition or mid-note, such that composition is "no longer a question of imposing a form upon a matter but of elaborating an increasingly rich and consistent material, the better to tap increasingly intense forces" (Deleuze and Guattari 1987, 363). Here John Cage's prepared piano marks a point along the trajectory from Varèse to the synthesizer: his piano is metallized, operating on material components of the piano previously intuited as matter in motion so as to incorporate into the piano-form (those elements that constitute the stable "piano-ness" by which we recognize a piano) a continuous variation of the literal metal aspects of the piano (steel wire, screws, and so on) akin to Varèse's own explorations into ionization. What were originally stable elements of the piano-form whose intensity were at zero degree are opened up by Cage, incorporating variation into the piano-form itself.

This incorporation of orchestration goes further when the piano-form develops into the electronic piano for which the field of expressible timbres of a piano, or even the expanded field of the prepared piano which modifies a relatively stable material structure of wire, wood, felt, and so on, are themselves the object of transformation. The set of possible fields of expressive timbres is itself up for transformation by the composer within the act of playing itself, but this act requires intuiting the piano form as movement-matter. A shift from one preset sound to another isn't a change in timbre qua affect, but a change in the system producing the possible field of timbres: it concerns the singularities, points of inflection, operations, and the methods by which the affects are produced and expressed. This is a different understanding and intuition of sound composition than when orchestration changes occur external to the compositional process, even if on the level of affect they are largely similar.

Further along this trajectory we move from the electronic piano to modular synthesis. The piano-form disappears entirely, it is no longer recognizable as stable elements identified by qualities or the affects its capable of. Rather only the underlying singularities and dynamic modulations which at one point were stable in the form of the piano but whose "body" is now capable of so much more. This shift to focusing on singularities also can result in the complete destabilization of these otherwise stable types, epitomized in the aesthetic styles of people like Peter Blasser and instrument-maker JMT. As Blasser notes, a key feature of JMT's instruments is a lack of text or instruction, which requires "a good deal of intuition and a period of be-friending to learn their tricks" (Blasser 2015, 9). He aligns his own design strategy with this approach, noting that he will "often renounce text as superfluous to the synthesizer; if it is necessary to name a control" (Blasser 2015, 9) then he will "find a way to eliminate that control" (Blasser 2015, 9). This design aesthetic is a turn directly towards the singularities and a renunciation of the recognizable, stable types raised to the Nth degree, refusing to articulate the logic of the instrument in affective or enunciative terms.

This should give us some sense then as to why outlining what constitutes additive or granular synthesis is so much simpler than modular synthesis. Modular synthesis is not one kind of relative stability amongst others, as we have seen with the piano-form or the TB-303-form. It is rather related to this mode of intuition which conditions a whole field compositional perception and its functioning, by rendering possible the constitution and application of both relative stabilities and continuous variations. Seen from the perspective of matter in motion, an additive synthesizer is a particular kind of timbral field, effective at invoking certain affects (early speech synthesis, organ sounds, etc.) Seen from the perspective of movement-matter an additive synthesizer is just a temporary state or form that a set of oscillators, amplifiers, and filters can be made to coalesce into to produce a possible timbral field, as the operations pass into new states.

For Deleuze and Guattari this framework makes artistic and musical production no longer centered around matter and form but rather become centered around capturing and making visible the forces at work in the continuous variations of matter. Here Deleuze and Guattari specifically name the synthesizer as the exemplary form of this process, noting that "the synthesizer makes audible the sound process itself, the production of that process, and puts us in contact with still other elements beyond sound matter" (Deleuze and Guattari 1987, 378). It achieves this precisely by taking heterogenous singularities or points of inflection, putting them in contact with one another, and harnessing them to transpose movements of force from one element to another. This "musical machine of consistency, a *sound machine*, (not a machine for reproducing sounds) … ionizes sound matter" (Deleuze and Guattari 1987, 378). It is with modular synthesis that this process is raised to the Nth degree, marking the passage from a machine for reproducing sounds (playing a TB-303-form because it sounds like a TB-303, for its affect and timbre) to a sound machine which makes visible the constantly changing state of matter from which temporary systems like the TB-303-form emerge.

For Deleuze the key to achieving the process of ionizing sound matter or making audible sound processes is the kind of relationship produced between two elements, intervals, or forces, such that the two elements taken together do not collapse into a single element nor that they remain so unconsolidated as to be unable to communicate. He calls the degree to which this synthesis of disparate elements occurs *consistency*, or the holding together of heterogeneous elements, arranging them, intensifying them, but without homogenizing them. If the connections between the elements are too tenuous or contingent then the machine "ends up reproducing nothing but a scribble effacing all lines, a scramble effacing all sounds" (Deleuze and Guattari 1987, 379). The result is ambiguous, and nothing is able to emerge. Running noise into every element of a patch bay isn't exactly going to produce particularly aesthetic results or allow for specific forces to emerge and express themselves in a coherent fashion.

On the other hand, if the connections are too fixed, too integrated, the result becomes standardized or sterile. In Guattari's work *Machinic Unconscious*, published the year before *A Thousand Plateaus*, he notes that one of the byproducts of capitalism in the rhythms of life itself in the 20th century was the homogenization of elements, an impoverishment of timbres, basic melodic cells, and the reduction of complex rhythms of life in favor of the same fixed meters and cadences for everyone. Between the zero-degree threshold of homogenization and the maximum threshold of ambiguity are syntheses of disparate elements which achieve consistency, with enough internal necessity to pass force from one element to another and enough inequality and contingent elements as to allow for disparate but related rhythms and variations to emerge. If all of the changes in forces are too similar to a pre-existing model, nothing new will emerge. Patching in changes to a TB-303 to modulate its filter and accent isn't enough to reveal the forces that underlie the form, something more complex must be changed to reveal what was previously conceived of as a fixed element. As Guattari's reference to capitalism qua homogenization suggests, the logic of consistency, the shift to singularities, and the model of modulation are not limited in Deleuze and Guattari to their analysis of music. It also takes on a central role in Deleuze's later political works, particularly his brief essay "Postscript on Control Societies" and his book *Foucault*.

In these works, he draws on French philosopher Michel Foucault's analyses of kinds of power, forms of society, and the principal technologies and institutions that correspond to these models. Foucault is associated with his analysis of disciplinary societies in the 18th and 19th centuries, societies whose institutions are organized by sites of confinement. While prisons are the model example of confinement, Foucault showed how the family, the barracks, the asylum, the hospital, and the factory all follow the same basic ideal institutional structure, while the individual passes from one of these closed institutions to another. In the factory, everyone is brought together in the same place, time is clearly organized and regulated, it is clearly delimited from other institutions, and the customs, habits, and practices of production are all heavily regulated. Yet, the disciplinary society, in succeeding the sovereign societies of earlier eras, was itself already being replaced by, as early as the 1960s (Foucault 1980) by what Deleuze comes to call control societies.

Most notably for Deleuze, the difference between the principal technology used by disciplinary societies and those used by control societies is that "confinements are *molds*, different moldings, while controls are a *modulation*, like a self-transmutating molding continually changing from one moment to the next, or like a sieve whose mesh varies from one point to another" (Deleuze 1995, 179). Deleuze's key example here is the factory in the disciplinary society, which clearly delineates itself from the home, delineates work from home, relies on the enclosure of the environment, and maintains the maximum flow of capital by a fixing "a body of men whose internal forces reached an equilibrium between the highest possible production and the lowest wages" (Deleuze 1995, 179). In control societies, the factory is replaced by the business, which, like modular synthesis, takes previously stable or fixed constants and puts them into continuous variation. The clear delineation between work and home becomes muddled into spaces like "the home office" and "playbor" and the relationships become variable: working at home is work when it comes to needing to respond to emails at any time of day, but it's home when it comes to who bears the responsibility for paying for the office supplies in a home office.

Rather than fixed equilibriums meant to be sustained over long periods of time, these forms of continuous variation operate according to metastability. Consider as an example the concept of a price tag in a store. While it historically has served as a fixed point (the right equilibrium between what the customer is willing to pay and the total profits earned on the commodity) between the store and its customers, in the advent of online shopping and mass data collection, price points can now fluctuate based on individual consumer metadata, purchase history, and time of day, ensuring that a cost has become undulatory, in continuous variation, and ultimately rendering consumers unable to negotiate as a mass or to mobilize resistance against price gouging. For Deleuze this is the result of a mutation in capitalism corresponding roughly to the 1960s and the rise of speculative finance capital and high-frequency trading where markets "are won by taking control rather than establishing a discipline, by fixing rates rather than reducing costs" (Deleuze 1995, 181). This control is produced precisely by infinite modulation's capacity to make a connection contingent or ambiguous in one moment and necessary the next. It is no longer the site of confinement and its specific barriers which control access to one point or another, rather the site of confinement becomes a form which can be implemented anywhere, as when the prison is replaced with home monitoring.

Just as the synthesizer was the archetypal machine for Deleuze with respect to aesthetic processes of making visible the forces at work in matter, "control societies function with a third generation of machines, with information technology and computers, where the passive danger is noise and the active, piracy and viral contamination" (Deleuze 1995, 180). By noise and contamination Deleuze is here referring precisely to the threats present to consistency: ambiguity and homogenization. Political struggle becomes one to produce a political consistency, to articulate a clear set of forces that constitute the political by means of synthesizing disparate elements without falling into ambiguity or homogenization while forces like capital make use of these very same modulations to contaminate these struggles or reduce them to incoherent noise. We see this clearly at work in "identity politics," where white supremacist groups will often equate some aspect of whiteness to an element of another racial identity in order prevent that identity from being able to explicate its own constitutive elements as clear and distinct, or if a particular event is used to typify a certain kind of violence aimed at a minority identity, these groups will multiple the issues of identity and disparate elements to the point where the event appears to be an incoherent picture. In short, the logic of composition in modular synthesis bares a clear structural similarity to the field of political struggle.

There are two connections to explore here between societies of control and modular synthesis. On the one hand, all technology is always inextricable from the social, political, material, and economic regimes it exists within. As Patricia Pisters emphasizes, "Deleuze's universe is full of ideas that challenge our received conceptions of traditional oppositions between materiality and immateriality ... They allow us, for instance, to see how the materiality of the earth is connected to our digital screen worlds and thus to our collective consciousness" (Pisters 2015). For example, all our media, electronic music included, is closely connected to the history, ethnography, and geopolitics of mining, or the to "elaborate long-distance commercial circuits" (Deleuze and Guattari 1987, 455) into which the raw metals, the electronic components, and even the fully completed artisanal or boutique synthesizers are connected in to. The social and cultural practices of modular synthesis operative in technical expositions, YouTube channels, boutique stores, bedroom studios, dilapidated former auto-repair shops hosting regular noise and electronic music shows, audio production classrooms, and so on are directly integrated into control societies, these larger material circuits and all of these things have impacts on nature of the materiality of modular synthesis. Thus, the compositional process that constitutes modular synthesis finds itself connected into extended chains that lead from mines to the machines themselves, and again to the very speculative finance which constitute them.

On the other hand, the modes of intuition operative in modular synthesis and composition make it such that we see how this same transformation from disciplinary to control societies happened within music and musical consciousness itself. As Cox carefully and systematically lays out in his analysis of 20th-century electronica, experimental music of all kinds has been systematically subjecting the molds, confinements, regulation and organization, institutions, and social practices of music and composition to continuous variation. One example here would be the audio production studio: it's a clear disciplinary institution, organized and regulated, materially and immaterially separating out the sound in that space from the sound outside of it, for the purpose of the production of albums, tracks, songs, and so forth. Cox details how, for example, "electronica tracks are not closed 'compositions' or 'songs' but open-ended, provisional assemblages that, at any level, can be connected with, plugged into, other assemblages" (Cox 2006, 19). By their very nature which often incorporates aleatoric elements, generative algorithms, and gradual musical processes, having a definitive, delimited, fixed end product doesn't make sense. In other words: an album is a particular form of individuation which captures a set of affects and expressions, but it fails to capture the singularities, operations, and variability which are the real "object" of modular synthesis or experimental music generally.

What Cox leaves out of his analysis here is the connection this has to capital and societies of control. We can contrast the disciplinary audio production studio with the Eurorack technical exposition. In Blasser's ethnographic exploration of the Tokyo Festival of Modular Synthesizers and Control Voltage Fair in New York, he describes the layout of such expos as having a variety of booths competing for attention, where venues provide speakers "to each table, where the proprietor would patch and tweak a sonic presentation" (Blasser 2015, 3). These expos usually include a set of performances by artists like Richard Devine or Keith Fullerton Whitman who often have prominent online presences, YouTube channels, or appear as guests on modular-themed podcasts where they demonstrate their systems, post videos of testing specific equipment, and otherwise serve to attract audiences. These performances are often, explicitly and intentionally or not, showcases of particular modular systems and their open-ended, provisional, connective structure. As Blasser rightly notes, this is "a three-way relationship between presenters, performers, and visitors at the event" (Blasser 2015, 11) which does not generate production but rather vies for attention, sales, and products, with performers serving as an intermediary, a blend of "creative personality with informative salesperson" (Blasser 2015, 11) which, like working from home, exists within these kinds of dynamic spaces where the performers are artists when it benefits gathering attention and brand representatives when a sale is possible.

Yet the logic at work here is more abstract than simply replacing one site (production studio) for another (expo center). The relationship itself, between attention, sales, and products can be made to be manifest anywhere, akin to home monitoring. YouTube channels bring together a similar relationship between the central videos, the surrounding video advertisements marketed to the channels' audience, and the comments section. It is extremely common for home-studio demonstration videos to be littered with comments about what equipment is where, what it's good for, and comments about having gone and purchased a particular product because of the video. While there is often a certain social stigma against doing so, even at noise venues modular performers can often be swept up in this logic when audience members come up to ask about the modules they use.

In disciplinary societies selecting a place to play music is quite straightforward. You don't do it at work, you do it at home or in a concert hall or the recording studio, or other areas where music is instituted, organized, and regulated. When you break these conventions, it is often because that site is in the process of experimentation that is constitutive of the transformation from discipline to control. Forms like the recording studio make less and less sense in control societies and for modular synthesis, this is partially why it's so hard for me to answer the question "how many modular synthesizers do I actually have?" In a certain sense, the very form of the studio collapses into a temporary state of modular composition, and yet my home, like many, has a home studio, but even this, as we have seen, is more of an abstract form than a clearly delineated space (epitomized by the blurred line between play and labor for the bedroom producer). What this shows us is that, seen from the point of view of modular synthesis, that is, from the mode of intuition of matter-movement, the distinction between political sites and composition collapses into singularities and dynamic operations. The placement and exhibition of modular compositions is directly incorporated into the larger social power dynamics that constitute every space and the compositional decisions become political ones.

Modular compositions, if we are to think of them with respect to consistency, to making sound processes audible, then also need to be considered as works which make political identity audible, which make audible the implementable forms of control that can be instituted anywhere, without falling into the traps of ambiguity and homogenization, or without reverting back to the infinite reproduction of those forms which already exist and serve to perpetuate those systems of domination which constitute control societies. Deleuze notes at the end of "Postscript on Control Societies" that organizations like trade unions, designed to deal with the clear delineations and confinements of capital in the factory, no longer serve a clear role in control societies as they are not equipped to handle the variation and gaseous nature of capital in the business. In the end he asks whether one can "glimpse the outlines of these future forms of resistance, capable of standing up to marketing's blandishments?" Perhaps this is a role modular synthesis can begin to contribute to, precisely because it is a mode of intuition capable of making audible the singularities that give rise to forms of control. It is, rather, those forms which reproduce the systems of domination that modular composition should be working actively to force into ambiguity.

What kind of new forms can we produce, rather than, for instance, infinitely recreating the same forms again and again? In what way can we take a modular system and use it to identify the singularities that constitute the "expo" form and disrupt it? How can we make audible the singularities which condition comment sections to cultivate markets and sales and make them into sites for social support? How can we plug these structures into modular synthesizers, to introduce ambiguity into the gathering of attention and produce new social sites? These are not questions we can give direct answers to because they aren't answers that have clear laws to follow. The subject at hand is variable, an endless stream of dynamic operations. What we can do with these questions is find a degree zero intensity, patch a knob into it, and try to make audible its points of inflection, its states of energy, and its rates of growth.

Bibliography

Adkins, Brent. 2007. *Death and Desire in Hegel, Heidegger and Deleuze*. Edinburgh: Edinburgh University Press.

Blasser, Peter. 2015. "Stores at the Mall." MA Diss., *Wesleyan University*.

Cox, Christoph. 2006. "How Do You Make Music a Body without Organs? Gilles Deleuze and Experimental Electronica." unpublished manuscript, published in German translation as "Wie wird Musik zu einem organlosen Körper? Gilles Deleuze und experimentale Elektronika." In *Soundcultures: Über digitale und elektronischeMusik*, ed.Marcus S. Kleiner and Achim Szepanski [Frankfurt: Suhrkamp Verlag], 2003, 162–193.

Deleuze, Gilles. 1979. Cours du 27/02/1979, 'Metal, Metallurgy, Music, Husserl, Simondon.' Transcribed by Timothy S Murphy. https://www.nettime.org/Lists-Archives/nettime-l-9809/msg00122.html

Deleuze, Gilles. 1995. "Postscripts on Control Societies." In *Negotiations 1972–1990*. Translated by Martin Joughin. New York: Columbia University Press.

Deleuze, Gilles. 1990. *Logic of Sense*. Translated by Mark Lester with Charles Stivale. New York: Columbia University Press.

Deleuze, Gilles and Felix Guattari. 1987. *A Thousand Plateaus, Capitalism and Schizophrenia Vol. 2*. Translated by Brian Massumi. 177–182. New York: Continuum Press.

Foucault, Michel. 1980. "Body/Power." In *Power/Knowledge: Selected Interviews and Other Writings 1972–1977*. Edited by C. Gordon. 55–62. New York: Pantheon Books.

Kant, Immanuel. 1992. "On the Form and Principles of the Sensible and the Intelligible World [Inaugural Dissertation] (1770)." In *Theoretical Philosophy: 1755–1770*. Translated by David Walford and Ralf Meerbote. 373–416. Cambridge: Cambridge University Press, 1992.

Pisters, Patricia. 2015. "Deleuze's Metallurgic Machines." Los Angeles Review of Books. November 8, 2015. https://lareviewofbooks.org/article/deleuzes-metallurgic-machines/.

22

RANDOMNESS, CHAOS, AND COMMUNICATION

Naomi Mitchell

Introduction

What is the primary way that users interface with a modular synthesizer? There's the tactile nature of turning knobs and pressing buttons. But there's also the interface between individual modules using patch cables. These patch cables enable modules to talk to each other, an essential part of modular synthesis—the flow of electricity and user intention flowing from module to module. You take ownership of your patches; you are the designer of this unique instrument interface. Depending on this interface's complexity and the patch cables that comprise it, this may or may not be reproduced. It exists in the moment. It's ephemeral music. It exists within its own time-space, never to come together in the exact same way again.

History

Randomness as a tool for musicians in the modern era has links that extend over a century ago. Uncertainty gripped Europe following WWI. So many young men in the prime of their life marched heroically into battle, only to be killed or bear the lifelong scars, both mental and physical. It was out of this angst following WWI that the Dadaist art movement flourished. As a response to the absurdity of life in the early 20th century, many Dada artists employed what seemed like nonsensical elements. The artists rejected traditional bourgeois politics that inform much of the academic, artistic world; cutting up and pasting together seemingly unrelated visual elements, displaying everyday objects as pieces of art in a gallery, and using chance to inform the direction of artistic work.

DOI: 10.4324/9781003219484-24

One of the best-known artists who emerged out of the Dada art scene was Marcel Duchamp. Originally a painter, he eventually began to display everyday objects in a museum context, calling them "readymades." In addition to his paintings, collages, photographs, and readymade sculptures, he composed two pieces of music using chance operations and a conceptual piece that outlined a musical happening (Raymond 1994; Kotik 2008). Duchamp conceived the first Erratum Musical as a piece for three voices (Philadelphia Museum of Art, n.d.). Duchamp made three sets of 25 cards, a set for each voice. Written on these cards was a single musical note. These were mixed in a hat, drawn out randomly, and Duchamp documented the notes to inform the composition of the piece. There is no indication of how to perform the three parts.

The second compositional piece by Duchamp was included in the notes[1] that led to the piece entitled *La Mariée mise à nu par ses célibataires même. Erratum Musical."* (Philadelphia Museum of Art, n.d.), better known now as The Large Glass. It is an unfinished work never realized in Duchamp's lifetime, but there is enough documentation to make some sense of the piece. Duchamp described two separate parts in the written manuscript, a score for a mechanical instrument and a compositional system, titled "An apparatus automatically recording fragmented musical periods." The compositional system consists of a funnel, an unspecified number of convertible cars, and a set of numbered balls. This set of 85 balls each represents a piano key, as pianos traditionally had 85 notes before the standard 88-key piano we see today. The cars pass under the funnel at different speeds, and the balls fall into them. The composition runs until the funnel is empty of balls, which completes the piece. In some realizations of the work, a toy train, wagon, or similar device plays the cars' role, making it a bit easier to perform.

Duchamp never finished the piece for mechanical instruments and used numbers instead of standard musical notation. However, he left notes detailing the significance of this system, allowing musicians to realize the work. Duchamp wrote that the instruments intended to perform the piece are "player piano, mechanical organs, or other new instruments for which the virtuoso intermediary is suppressed." Although I have never seen it realized on a modular synthesizer, this piece seems ripe for reinterpretation using modular synthesizers. They are adjacent to the category of mechanical instruments, and the systems of notation used in those instruments translate to a sequencer in a modular synthesizer. The rolls of player piano notation used to automate the playing of the instrument may translate to the sequencing lanes of a complex CV or MIDI sequencer.

John Cage sought to create music detached from rules, artistic taste, or memory that references other work using chance operations and precomposed charts of musical notation. The groundbreaking 1951 piece, titled *Music of Changes*, is a direct reference to the translation of the *I Ching*, the Book of

Changes (Wilhelm 2011).[2] He used this traditional tool of divination rooted in Chinese culture to perform the chance operations. It uses a type of divination called cleromancy. This ancient method of future telling uses the outcome of a random event that indicates divine will or forces. Cage used a coin-tossing method where he flipped a set of three coins six times. These coin tosses create one of 64 possible primary hexagrams that make up the *I Ching*. Each hexagram has an associated meaning that informs the divination. Cage's methods for calculating random events using the *I Ching* differed from traditional methods and were made to suit his needs, balancing chaos and order, rational and irrational. The Dadaists, on the other hand, sat squarely in the realm of the irrational. Cage's work stresses the relationship between random and control, something that informs my artistic practice as well (Cage 1961).

Duchamp's interest in artistic work waned over the years, and his true passion moved to chess. He devoted himself to the game, writing once in a letter, 'Nothing in the world interests me more than finding the right move (...) I like painting less and less.' In 1968, Marcel Duchamp and John Cage performed the piece "Reunion," which took the form of a chess game played on a specialized chessboard. Duchamp had taught Cage to play chess, with Cage using that opportunity to get to know the aging artist. Cage eventually suggested they play a public game of musical chess.

Cage approached Lowell Cross to create a specialized chess board for the game. This board contained 64 photoresistors housed under each of the spaces on the board. Covering or uncovering the spaces triggered four sound sources that played different compositions from David Behrman, David Tudor, Gordon Mumma, and Cross. In addition, contact microphones placed on the board amplified the motions of the pieces as they moved across the board, more for the audience's perception of the game happening than for their acoustic properties. Duchamp, a chess master, beat Cage within half an hour, even with a handicap of only one knight. A second game played between Cage, and Duchamp's wife Teeny lasted until 1 AM, with both contestants equally matched, eventually coming to an inconclusive ending (Cross, 1999). While not random in the strictest sense, the possible lower bounds of possible chess moves according to the Shannon number is approximately 10^{120}, or a one followed by 120 zeros (Chessdom 2007). The Shannon number comes from a paper written by American mathematician, Claude Shannon, who came up with this number based on the assumption that there are approximately 10^3 possibilities for a pair of moves of white followed by black in a game of approximately 40 moves (Shannon 1950a,b). This number is sometimes contested depending on who you ask, but it gives a rough approximation for the order of complexity of chess. The estimated number of atoms in a universe is on the order of 10^{80}, so there are, according to Shannon, more possible moves in chess than the number of atoms in the universe.

If we're looking at randomness in modular synthesizers, one of the first places to look would be at a pioneer of the esoteric, Donald Buchla. The first iteration of the Buchla modules, the 100 series included the model 165 Dual Random Voltage Source (Buchla 1966). It is very simple compared to his later works, made up of two channels, each with a trigger input and two random voltage outputs. Interestingly, Buchla did not use LEDs to indicate when the module received a trigger pulse, but relays that would make an audible click every time it received a clock pulse.

The move from 100 to 200 series of Buchla modules increased many of the modules' complexity, bringing together years of experience and artist feedback to design the next generation of modules.[3] This line included two modules; both called The Source of Uncertainty, models 265 and 266. Model 265 consists of three pairs of noise outputs, stored random voltage, and random voltage output, later called fluctuating random voltage. This module uses sample and hold circuits to generate the random voltages. A sample and hold is a circuit containing a clock input and a sampling input. When the clock input receives a pulse, the sample and hold looks at the voltage level present at the sample input. It then captures the voltage level and holds it in its memory. Sort of like taking a snapshot of the voltage with a camera that can hold a single picture.

This voltage level is passed to the output and held at that level until the circuit receives another clock pulse, at which point it samples the new voltage, replacing the voltage currently in the memory buffer with a new voltage. While the module has several noise outputs, the source for the sample and hold circuits on the module are a 100 Hz triangle-wave synced to white noise. The rationale behind this was that the distribution of white noise is more concentrated in the middle of the possible states. In contrast, a triangle wave synced to noise offers an equal distribution across the frequency spectrum.

The fluctuating random voltage uses an internal VCO to drive a sample and hold whose output is low pass filtered. This low pass filter slews the random voltages and results in a continuously fluctuating smooth random source. The period of the VCO ranges from .05 Hz to 50 Hz,

The stored random voltage section samples the same noisy triangle waveform but outputs the kind of stepped random voltage you normally expect to find from a sample and hold. However, Buchla implemented a correlation control as another way to shape the random outputs.

This passive mixed control is a crossfader between the noisy triangle and the output of the sample and hold. All the way up, the sample and hold only samples the noisy triangle; all the way down, it only samples itself, completely eliminating the output. In the middle between those two extremes, the sample and hold's output is influenced by its previous states, but it still gets new information for the sample and hold from the noisy triangle wave. This combination of randomness and control is what makes the Buchla school of thought on random so influential.[4]

From there, the model 266 Source of Uncertainty developed into an altogether new entity while borrowing some elements from the original 265. The random voltage output, finally labeled as fluctuating random voltage, offers a pair of varying random voltages with independent control of time/low pass filtering.

The noise sources on top of the module are white, labeled +3 dB/oct, pink, labeled musically flat, and reciprocal white noise, labeled -3dB/oct. The reciprocal white noise seems at first glance to be filtered pink noise, but the energy is redistributed, instead of filtered.

The quantized random voltage section generates two complementary random voltage outputs, whose timing is determined by a trigger input. The quantization control sets the possible number of states for the two outputs, labeled 1–6. N stands for the number of possible states. The two outputs are $2^{\wedge}N$ and N+1. N is the numerical value set between 1–6 by the quantization control knob. If N=1, then the N+1 output has two possible outputs, 1+1=2. The $2^{\wedge}N$ also has two possible outputs, $2^{\wedge}1=2$. As the value for N increases, the total number of outputs proportionally increases. With N=2, N+1=3, but $2^{\wedge}N=4$. At N=3, N+1=4 and $2^{\wedge}N=8$. N+1 scales linearly, while $2^{\wedge}N$ scales exponentially. Think of the increasing scale that a pyramid scheme needs to operate, and how every member needs to recruit several more who in turn also need to recruit additional members.[5] The distribution of random value for the N+1 favors the middle of the spread, while $2^{\wedge}N$ is spread equally across the possible values.

The stored random voltage section, although it is different from the 265 submodule of the same name, nevertheless also outputs random voltages when it receives a pulse. It has a pair of outputs, linked by a common trigger input. The top output generates random voltages, whose distribution is spread evenly through its possible range. The bottom output, on the other hand, has control over the distribution spread. With the probability control in the lower ranges, the output will mainly favor lower voltages. At the upper ends the output congregates in the upper reaches of the voltage range. In the middle between these two extremes, the voltages are spread across the voltage spectrum.

Gracing the bottom of the module are an integrator, pulse divider, and a trio of sample and hold outputs. The integrator is a slew generator with control over the slew time both manually and with external CV control. Slew is another word for glide or portamento, a sliding between two notes instead of a sudden jump.

This changes any of the many stepped random outputs into a slewed or smoothed random output. The pulse divider has a clock input and two outputs. The outputs swap back and forth between high and low states. At a clock pulse, one goes high, while the other goes low. And vice versa. The trio of samples and holds are clocked from the clock input and the swapping pulse outputs.

Buchla's practical implementations of complex, yet controlled randomness informs the work of synthesizer designs to this day. A pioneer in many ways, Buchla paved the way for randomness that exists out of the simple paradigm of noise into a sample and hold, the most common random implementation. This method is simple and in many ways, an inelegant solution to random generation, with very little in the way of control besides modulation depth and rate. Let's now look at a few other methods for generating randomness with modular synthesizers.

Random Methods: Shift Registers

Randomness was always the most exciting part of modular synthesizers for me. I spent a lot of time on the Casper Electronics site, run by Peter Edwards, when I started exploring synthesis and electrical design. However, this treasure trove of information has since been taken down, accessible only through the Internet Archive (Edwards n.d.). I found schematics for Rob Hordijk's Benjolin on the site, and I thought it was the coolest thing that I had ever seen (Hordijk 2010a,b).[6] It was an unpredictable, chaotic system in which small adjustments to the controls gave way to massive changes. It consists of two oscillators, a filter, and a circuit called a rungler. The rungler is a shift register that makes pseudo-random patterns.

A little on how a shift register works- there are two inputs, clock and data. When the shift register receives a clock pulse, it looks at the data input, and if it's high, a high bit moves into the first stage of the shift register. It only expects binary information, i.e., off or on. Whatever is in the first stage of the shift register, a bit that is either on or off, moves sequentially to the second stage, whatever is at the second stage moves to the third, etc.

In certain configurations, like the rungler, when a bit reaches the last stage of the shift register, it is fed back into the data input and is XOR'd or exclusive ORd with the incoming data. XOR is a type of binary logic operation in which the output is high if one input or the other is high but low if both or neither are. Active if the inputs do not match, and inactive if they do. The three stages of the shift register run into a lo-fi digital to analog converter (DAC). The DAC used is an r2r DAC that uses two resistor values, R and 2 R, or a value and another value twice as large. The placement of the input signals determines their overall weighting, with each input changing the output by half as much as the one above it. This turns them into stepped random voltage. The output of that stepped random voltage is fed back into the two oscillators and the filter.

The two oscillators also modulate each other's frequency in addition to the rungler CV output. The two oscillators are fed into a comparator, and that output is fed into the filter. The FM between the oscillators and the Rungler to each CV destination has its attenuator, which is crucial to this patch. The attenuators provide a way to fine tune the chaos and reign it in. Making even small changes to the levels of the attenuators may cause rather sweeping changes as the Benjolin finds itself locked into a new feedback configuration.

The rungler CV output only has eight states, i.e., 2 to the power of 3, the number of stages that make up the DAC. Because of this, it gets stuck into repeating patterns, as there are a limited number of possible states that it outputs. Little changes in the attenuation of these CV signals ultimately have a massive impact on the output. A slight turn of a knob can send the rungler off into a new ever-changing state until it falls again into a pattern. It is a network of discrete memory cells with feedback. It has a network of feedback, arranged for making music and experimenting with relationships. The relationships between the two oscillators frequency modulating each other; as well as the relationship between the frequency of each oscillator and how that clocked and seeded data evolves over time … .

The Benjolin and, more specifically, the rungler circuit informed the design of my first module, the Dual Digital Shift Register. Breaking out the shift register from the prewired Benjolin configuration allowed for greater flexibility and a more modular way to approach. I included some of Pete Edwards' Benjolin modifications, including external clock and data inputs and individual gate outputs for the shift register stages. These modifications to the basic shift register design allowed the DDSR to be configured in a wide range of configurations, including but not limited to a rungler-style setup.

The DDSR was my first module, released in 2018, and like most first modules, there was room for improvement. Flash forward to 2021, and I decided that I wanted another crack at a shift register-based module. Three years of thinking about my original design and working on new designs gave me new insight into how to make a module. I added a number of features and changes to the basic shift register configuration, and what grew out of that came to be known as the Cascading Register. This new module took inspiration not just from runglers and digital shift registers such as the DDSR; but also from linear feedback shift registers (LFSRs) and analog shift registers.

The Cascading Register is a single-channel eight-bit digital shift register with an internal voltage-controlled clock oscillator, eight gate outputs, and three CV outputs. It retains the external data input of the DDSR, but also includes a button to manually seed data into the data stream. In addition to those data sources, three of the stages of the shift register are XOR'd with each other, the seed button, and the external data input. This addition of the internal data sources comes from LFSRs.

While some shift register configurations rely on external sources of data, the most common implementation of a shift register-based source of pseudo-random behavior is an LFSR. An LFSR, or linear feedback shift register, is often used in both hardware and software implementations of pseudo-random number generation because you cannot generate true noise in software. Instead of using an external oscillator for a data source; multiple stages, called taps, are XOR'd with each other and fed back into the data input. The configuration of which taps are used determines the length of time it takes before the pseudo-random pattern repeats. The tap configuration that generates the longest period is called the maximal length. A maximal length shift register will have a length of 2^n-1. So two to the n, or length of the shift register, minus one. It is minus one because an LFSR cannot have all stages at zero, or it will never recycle data, terminating the output. An eight-bit LFSR has a length in its maximal configuration of 255, while a 24-bit shift register can run for 16,777,215 clock periods before repeating. Sometimes what humans perceive as randomness just has to do with numbers beyond the basic bounds of comprehension. However, the Cascading Register is not a maximal length configuration on its own but can be made into one by patching the fourth bit into the external data input.

The three CV outputs are derived from stages 1–4 of the shift register, then stages 3–6, and finally 5–8. This means the CV outputs use the same information within the shift register, but each is shifted by two clock pulses. The particular configuration took inspiration from analog shift registers. An analog shift register is essentially a series of sample and hold circuits configured in a way that mirrors a digital shift register. However, instead of passing binary information through the stages of the register, the sample and holds contain variable analog voltages. While the Cascading Register does not include sample and holds, the core idea of moving information to a series of outputs remains. The CV1 and CV2 outputs include attenuverters, while CV3 includes an attenuator. An attenuator acts much like a standard volume control, off when all the way counterclockwise and fully on when all the way clockwise. An attenuverter (attenuator+inverter) sets both the amplitude and polarity of a signal. In the center, the output is off. Turning clockwise increases the output positively while turning counterclockwise increases the output negatively. The module also includes normalization between CV3 and the CV input on the internal clock oscillator.

Recipe: To Make a Pseudo-Rungler With a Shift Register: I use the all-powerful Make Noise Maths module as the source of the two oscillators, but you can use anything you'd like. Take one square wave oscillator, EOC or EOR from Maths, and use that to clock the shift register. Use the gate output from the other channel of Maths to provide data to the shift register. Route the CV outputs to the CV inputs on the oscillators, and at the same time, take the variable output of the oscillators and route them to modulate the other.

The fact that Maths has built-in attenuverters and multiple CV inputs on each of the function generators makes fine-tuning the CV and stacking modulation sources possible.

Now, this is just the template for a Benjolin-style rungler setup, but there's nothing stopping you from using any configuration of modules to create this sort of random-sounding setup. Use the gate outputs or the internal clock oscillator to trigger envelopes or to advance a sequencer. Add slew generators to smooth out the stepped voltages. Add filters, waveshapers, delays, or any other kind of timbre shaping modules you have into the signal path. This is simply a stepping stone to help generate new ways to use your modular, but these are not rules, just suggestions.

Shift registers have a particular sort of attraction to many people in synthesis. Maybe it's the idea of a 1-bit delay line; everyone likes delay, but why not have it at a smaller interval? An eight-stage 1-bit delay, what a novel concept! This allows for a complex generation of signals in a predictable or aleatoric way. Using two related signals, such as clock divisions of 2 and 4 or a non-maximal LFSR configuration, can result in simple repeating sequences. Using more unrelated signals, such as two unsynced oscillators or 1-bit noise from a comparator and a noise source, can result in non-repeating patterns that grow and evolve over time. Although, they do sometimes repeat, as it is a limited set of possible values. But they don't repeat the same way every time, they grow and evolve, getting caught in loops, and moving on to new patterns over time. Shift registers have been used in many synthesizers that have captured the imagination of modern synthesizer users. Rob Hordijk's Benjolin, the Ciat Lonbarde Plumbutter (Blasser n.d.), Lorre Mill's Double Knot (Schorre n.d.), the list of these synthesizers have been held up as shining examples of aleatoric synthesis.

Shift register-based random and pseudo-random generation is only as random as the constituent parts and their relationships. The placement of the taps and length of an LFSR, the frequency of two oscillators that provide clock and data, the way the individual stages of the shift register are combined to generate the pseudo-random signals. Next, we delve into other sources of unpredictability and my thoughts on my design process.

My Design Process and Between Everything and Nothing

The first consideration that goes into a module is, "Is this something I need in my own system?" Does it accomplish goals or execute functions that I have in mind for my own work? Is this something that fulfills some niche in my own modular setup that does something useful? Selling the module is secondary to me enjoying the module.

Knobs are crucial to the interface of a modular synthesizer. I personally use the color of the knobs as an indication of its function. Black knobs are for panel settings, white knobs are for CV levels, both incoming and outgoing. I don't use mini pots. I don't like mini pots; they aren't, generally, bolted to the panel, which leaves them floppy and cannot be turned quickly. Try turning a mini pot with one finger; you can't do it. Designating function by knob or jack function is a must-do when designing a modular synthesizer. In omiindustriies modules, inputs have white jacks, while outputs are left black. Unfortunately, there is no eurorack standard for highlighting or not highlighting inputs and outputs, and it may even change within a single manufacturer.

The main inspiration for another module, which I called "Between Everything and Nothing," for my usual standard of over-complicated names. We're just going to refer to it as the BEAN. It is a combination sample and hold, VCO/VCLFO, chaotic voltage source, and a comparator that takes inspiration from Buchla, Ian Fritz, and LZX Industries designs.

The correlation control on the Buchla 265 captured my attention when I first researched it. The control between the internal noise source and the output of the sample and hold is such a novel idea. It's a simple and effective way to make the randomness generated by the module slightly less random. This control causes clustering in the output, particularly in the lower ranges of the control. This kind of feedback works to make the module's overall output more predictable, as more patterns tend to emerge when feeding much of the signal back into itself.

As I got more into video synthesis, I started building the cadet line of LZX modules. I came to find that a comparator in video synthesis is a hard key generator, which controls the amount of luminance of an incoming video signal and can define the hard outlines of an incoming signal with voltage control over the key amount. This is perfect because I've searched for a module that would allow me to generate patterns over a piece of video. For a time, I got really into animation, and the process of video synthesis has brought back that love and interest in visual art. So I used video rate op-amps and configured the BEAN to be used with audio or video levels.

I originally conceived of the r2rawr just as an audio module. However, I started getting into video synthesis between modules and understood when I need to upgrade the op-amps I use in module designs. I don't think every single module will use high-speed op-amps.

As I moved towards audio and video synthesis, some of my more recent modules have filled both niches. The Between Everything and Nothing and r2rawr were my first two modules to use this type of high speed op-amp configuration. Video synthesis interests me deeply as my background is more as a visual artist than a musician. While I do consider myself a musician, I have more training in visual composition than music composition.

Initially, the Between Everything and Nothing used a noise source to generate the randomness. I used a transistor to create the noise and then amplified it and used op-amp integrators to change the noise's character. Noise is generated in a modular synthesizer by abusing components. Put enough electricity into the "wrong" end of a polarized part like a diode or transistor and after a certain threshold is passed, stray electrons are discharged from this junction, creating white noise (Hordijk, n.d.). This has to be amplified to get up to a level usable for modular synthesizers. In modular synthesizers, signals take a range of characteristics that determine their perceived randomness. This ranges from a pure sine wave with its clean lack of harmonics to noise containing all audible frequencies at equal distribution, with every conceivable waveshape between the two.

I enjoyed this noise-based design at first, but as I continued with the development, I had another module that I wanted to make: a source of chaotic voltages. A chaotic system consists of an oscillator that drives a chaotic feedback loop with a non-linear element. I referenced designs from Non-Linear Circuits as my initial inspiration from the module but have since moved to a variation of a circuit designed by Ian Fritz, a circuit implementation of a Double Well equation. Chaotic circuits do not use noise and are mathematically deterministic, although their behavior appears to be random on the surface. Chaos varies from random with the underlying concepts that go into making it.

Early on in my journey as a synthesist, I came across the double well circuit as a PCB project. It wasn't exactly a Eurorack format module, but I designed a panel for this project and put it in with my small selection of Eurorack modules. I think I still have it around somewhere in my studio. The makeshift panel broke and I kind of retired it. Wait, hold on, let me dig it out and give it a look over. The PCB is not strictly speaking a Eurorack PCB. The power connector is not the familiar 10 or 16 pin connector, but four solder points, +12 V, −12V, and two ground points, that I soldered on a makeshift power connector. The thing about this circuit is that it's not really random. It's mathematically deterministic and often falls into repeating patterns. Depending on the frequency and amplitude of the driving oscillator as well as the feedback and damping controls, it may delve into repeating patterns, more like an LFO. I took this as a basic reference point that helped inform my early work with synthesis and transformed it to fit my needs. I added some additional integrators to get the signals to a place that I wanted.

I looked at a module that I had once bought a DIY kit for but never completed, Ian Fritz's TGTSH re released by Elby. It's a combination sample and hold and comparator, intended to be used with Ian Fritz's chaotic circuits, providing a stepped random output voltage from these continuous sources of chaotic voltages. A comparator takes in incoming voltage and compares it against a reference voltage. If the signal passes above the reference point or threshold, the output of the comparator goes high. If the signal passes below the threshold, the comparator output goes low. This reference voltage may be set by the designer as a static voltage level. Or, in some cases, changed with manual control or with external CV signals. In the TGTSH and the BEAN, when the comparator goes high, it sends a trigger pulse to the sample and hold to sample the incoming voltage.

Patch programming adds another level of functionality to the module. Patch programming is a concept pioneered by Serge Tcherepnin, the inventor of the Serge modular system. Using this methodology, a module's behavior can be changed by patching within a single module, from outputs to inputs. One classic example of this is to create a self-cycling envelope or slew generator. Many envelopes and slew generators in the Serge system contain a trigger output that becomes active at the end of the envelope's cycle. Patching from this trigger output into the input of the envelope of a slew generator creates a looping envelope. Patching from one of the chaotic signal outputs into the threshold CV input of the comparator periodically interrupts the clock signal going into the comparator, bringing the threshold above the incoming clock signal level.

However, there are some pre-patched connections you can override by inserting a cable into the corresponding jack. One of the outputs from the chaotic circuit runs into the sample input of the sample and hold. The driving VCO's square wave output clocks the sample and hold via the comparator. The output of the sample and hold modulates the VCO's frequency.

The BEAN is not a set and forget random module; it requires tuning and a give and take relationship with the module. It emphasizes the concept that is important to me with modular synthesis, which is a conversation between the user and the module itself. The chaotic circuit may be unwieldy at first. It often pauses the voltage present at the sample and hold, and does not change until it is excited by an outside force.

Feedback is an essential part of a synthesis, from the operational amplifiers to resonance on a filter, to complex patches where the constituent elements all modulate each other. The overpowering sounds of a microphone placed next to a speaker is the first way most people encounter feedback. But feedback is such an in-depth underlying process of the universe, systems of feedback make up nature, self-contained systems, and connections spread out over masses of space and time. Tapping into this powerful and divine source of complexity rewards exploration, and you need an eye and an ear for the unusual.

Randomness is all around us; it informs the unforeseen ways the universe plays dice with our lives. Many artists have incorporated elements of chance and chaos within their process, to add a contradictory, humanizing, and imperfect element to their compositions. (Figure 22.1)

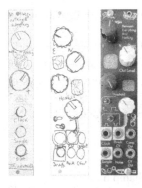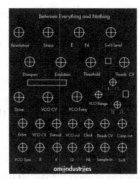

FIGURE 22.1   Evolution of the B.E.A.N. module, from initial sketch to production. Photo courtesy of Naomi Mitchell / Omiindustriies.

Teaching

When I'm teaching a class, I try to connect with participants on a fundamental level. I stop for questions regularly and try to explain succinctly how modular synthesizers work. Education is crucial to me, as someone who caught the synthesizer bug long before she could afford them and then poured through endless hours of synthesizer documentation. I want to be one of those sources of knowledge that I poured over with enthusiasm. I want to give women and non-binary people knowledge; I want to be the person I could've used ten or even five years ago. Not to say that my work as an educator has only featured non-men; they are part of the demographic. But if I could've seen a confident transgender woman spreading knowledge of synthesizers when I was first starting, it would have been inspiring.

Education in synthesis does not start and end with learning about the parts of a synthesizer. It's history; it's imagining ideas for the future, it's developing a deeper understanding of not only yourself but the world around you. This incredibly freeform and customizable instrument exists in thousands of configurations. The way a person uses a modular synthesizer or any instrument for that matter, reflects some underlying aspect of the musician's personality and their current emotional state. However, the modular synthesizer uses a different skill set than a traditional instrument. It's less about muscle control and more about understanding connections and being a sonic curator.

Is the instrument made of collective parts and individual modules? I would define the instrument of the individual configuration of modules available to the user at a given moment and the connections between them. A patch is akin to a preset on a vst. It's a single-use configuration of modules and their connections that make the instrument's current configuration. I love to watch people who know just a little bit of synthesis approach a modular synthesizer for the first time. They come up with the most interesting ways of thinking about synthesis because they haven't been taught the "correct" way to do modular synthesis.

There's an idea that informs the way I live my life. To leave the world better than you found it and to be the kind of person an earlier version of yourself could have needed. I strive to impart knowledge and new ways of seeing synthesis and the world around us. Maybe this is a lofty goal for a class on synthesizers, but I do hope my impact is more than surface level.

Notes

1 http://www.artesonoro.net/artesonoroglobal/MarcelDuchamp.html
2 Cage was initially given a copy of the *I Ching* by fellow composer Christian Wolff (Cage 1961, 17). For more, see Jensen 2009.
3 For more on Buchla's history, see Chapter 2, "Buchla's Box" in Pinch and Trocco 2004 as well as Gordon 2018, specifically Chapters 1 and 2, as well as the Interlude.
4 "Uncertainty is the basis for a lot of my work. One of the important dichotomies of music to me is predictability versus uncertainty. One always operates somewhere between the totally predictable and the totally unpredictable and to me the source of uncertainty, as we called it, was a way of aiding the composer. It's not fair to say that it's totally random because we allowed constraints (…) An interesting story I could tell now is that Ussachevsky bought three identical systems from us very early in the game, in '65 or so, to outfit the Columbia/Princeton electronic music studios. He was very disturbed by the random module and taped them over. He didn't actually disassemble them but in the two graduate studios he taped them over. In his own studio he allowed the randomness to be used but he did not want to assess compositions made with the random voltage generator." (Vasulka and Buchla 1991, p3)
5 pyramid scheme https://www.britannica.com/topic/pyramid-scheme
6 An open source software model was later made in Pure Data by Derek Holzer (2016).

Bibliography

Blasser, Peter. n.d. "Plumbutter." Ciat-Lonbarde. Accessed May 11, 2021. https://web.archive.org/web/20210216233905/http://ciat-lonbarde.net/plumbutter/.

Buchla, Donald. 1966. "The Modular Electronic Music System." Buchla, Incorporated. Berkeley, California.

Cage, John. 1961. *Music of Changes (Score Volumes 1–4)*. New York: Editions Peters.

Chessdom. 2007. "The Number of Shannon." April 15, 2007. https://web.archive.org/web/20201111233257/http://mathematics.chessdom.com/shannon-number.

Cross, Lowell. 1999. "Reunion: John Cage, Marcel Duchamp, Electronic Music and Chess." *Leonardo Music Journal* 9 (December): 35–42. 10.1162/096112199750316785.

Edwards, Peter. n.d. "Casper Electronics." Accessed May 11, 2021. https://web.archive.org/web/20120731211228/http://casperelectronics.com/.

Gordon, Theodore Barker. 2018. "Bay Area Experimentalism: Music and Technology in the Long 1960s." University of Chicago.

Holzer, Derek. 2016. "Pure Data Benjolin." Macumbista.Net. November 24, 2016. https://web.archive.org/web/20161124141118/http://macumbista.net/?page_id=4690.

Hordijk, Rob. 2010a. "Electro-Music.Com:: View Topic - Benjolin Schematics." Forum. Electro-Music.Com. May 3, 2010. https://web.archive.org/web/20100503025806/ http://electro-music.com/forum/topic-38081.html.

Hordijk, Rob. 2010b. "Electro-Music.Com:: View Topic - Benjolin PCB Layout for the Real DIY." Forum. Electro-Music.Com. May 4, 2010. https://web.archive.org/web/ 20100504205704/http://electro-music.com/forum/topic-40834.html.

Hordijk, Rob. n.d. "Noise, Randomness and Chaos." Accessed May 12, 2021. https:// web.archive.org/web/20210226073805/https://rhordijk.home.xs4all.nl/G2Pages/ Noise.htm.

Jensen, Marc G. 2009. "John Cage, Chance Operations, and the Chaos Game: Cage and the 'I Ching.'" *The Musical Times* 150 (1907): 97–102. https://www.jstor.org/ stable/25597623.

Kotik, Petr. 2008. "Marcel Duchamp." January 29, 2008. https://web.archive.org/web/ 20080129025818/http://www.artesonoro.net/artesonoroglobal/MarcelDuchamp.html.

Philadelphia Museum of Art Archives. n.d. "Marcel Duchamp, Musical Erratum, 1913." Accessed May 12, 2021a. https://web.archive.org/web/20210510101243/ https://pmalibrary.libraryhost.com/repositories/3/archival_objects/151083.

Philadelphia Museum of Art Archives. n.d. "Marcel Duchamp, The Bride Stripped Bare by Her Bachelors, Even: Musical Erratum, 1913." Accessed May 12, 2021b. https://web.archive.org/web/20210512152414/https://pmalibrary.libraryhost.com/ repositories/3/archival_objects/150423.

Pinch, Trevor, and Frank Trocco. 2004. *Analog Days: The Invention and Impact of the Moog Synthesizer.* Cambridge: Harvard University Press: 978-0-674-00889-2.

Raymond, François. 1994. "Marcel Duchamp, La Musique Et Les Machines." *Horizons philosophiques* 5 (1): 1. 10.7202/800961ar.

Schorre, Will. n.d. "Double Knot (V2)." Lorre-Mill. Accessed May 11, 2021. https:// web.archive.org/web/20201114191902/http://lorre-mill.com/doubleknot/.

Shannon, Claude E. 1950a. "Programming a Computer for Playing Chess." *The London, Edinburgh, and Dublin Philosophical Magazine and Journal of Science* 41 (314): 256–275. 10.1080/14786445008521796.

Shannon, Claude Elwood. 1950b. "A Chess-Playing Machine." *Scientific American* 180 (2): 48–51.

Vasulka, Woody. 1991. "Interview With Don Buchla." Vasulka Online Archive. December 1991. https://web.archive.org/web/20130513031544/http://www.vasulka.org/archive/ RightsIntrvwInstitMediaPolicies/IntrvwInstitKaldron/61/BuchlaTranscription.pdf.

Wilhelm, Hellmut. 2011. *The I Ching or Book of Changes.* Princeton University Press: 978-1-4008-3708-3.

23

INTERVIEW

Andrew Fitch

EE: Einar Engström

AK: Andreas Kitzmann

ET: Ezra J. Teboul

AF: Andrew Fitch

EE: I imagine that your work is very inspiring to what appears to be hundreds of people at the WAMOD (West Australian Modular). Much of my training intellectually is rooted in sociology and so I'm really interested in hearing more about how these meetings come together, how frequently they occur and your motivations for holding them. Are you the sole organizer or does somebody work with you and probably most importantly how do these meet ups inform your designs or your philosophy as an engineer of musical instruments?

AF: The workshops started about six years ago and it was actually a suggestion by some friends. I had been living overseas and when I came back from Japan one of my really old friends introduced me to a few people who saw what I've been doing and they said that I should do a workshop. The workshops are really run by Nathan Thompson and we might talk about him in more detail later because he was the designer of the CellF. Nathan and I worked together on quite a few other different projects. He organizes the workshops and he gives us access to the hacker space called the Artifactory and they have been a big supporter from day one because this is exactly the kind of thing they like to do. Perth is a small isolated city where

DOI: 10.4324/9781003219484-25

everyone knows everyone, at least in the music scene and in any sort of subculture.

Word gets around pretty quickly and from the start we were getting 20 to 30 people at each workshop and it has been steady all the way through. At the start I was not interested in Eurorack, I was into 4u and Serge stuff. But Nathan insisted that if we were going to run a workshop, it had to be Eurorack and he was quite right. So I started designing Eurorack modules and very much designed with the philosophy of DIY rather than trying to create a high end and perfectly in tune module with CV or something. I wanted things to be simple and I tried to use 100k resistors as much as possible, more than any other value, and no special chips or anything like that and I still carry through with that today. Preparing circuits for the workshop had a big influence on my design. When I design stuff, I think about how I can make this module so it's easy for me to pack a kit which is a rather weird parameter to think about.

The people at the workshops ... Well there have been a few that have been coming from the start and we have all become great friends and people have started bands from meeting up at the workshops. A lot of people come for one or two years and in that time they build a decent modular and build their confidence up and then buy modules from elsewhere and then they might disappear and then just show up when we do a module that interests them. Like when we did the Big Room reverb which had 45 people making 25 kits.

EE: Yeah, that is a surprise. No one expected a reverb from Nonlinearcircuits.

AF: Yeah well I did one in the "CellF " which has two reverbs and I got a lot of feedback from that and it's a really nice sounding one.

EE: So the feedback from workshop participants somehow makes its way into later designs?

AF: Yeah, earlier stuff definitely. We need LFOs, we need VCOs, we need filters, sequencers and so on and all those early modules were really just designed for the workshops. And that was the other thing ... because we held the workshops every month, every second Tuesday of each month to be specific, meant having to come up with a new module at least once a month and maybe four or five months in advance of the workshop in terms of logistics and getting the parts ordered in time for the workshop.

I was going pretty hard back in those days and that was fun because I didn't have kids then. But now I tend to inflict on the workshop any module that I feel like I need and I try to balance it, like on the one we just had on Tuesday. We did a very simple model which was a voltage controlled gate delay which most people built in under an hour. The month before we did the Hypster, the hyper chaos module, which is a pretty complex build so we kind of go very hard and then an easy one but at the workshop we use a stencil and a reflow oven so even beginners can come along that first night and build a pretty complex module.

EE: Amazing, that definitely explains things. Apart from the general allure of NLC circuits, the ease of building would help explain the large numbers of people you bring in. I've been to a number of workshops around the world but I've never seen that kind of attendance, so it is very impressive.

AK: I think that speaks to the local scene, because you referred to it as being quite a close knit community.

AF: Yeah, Perth is a funny city. I mean these days with the internet and DHL you can get anything from anywhere but I think growing up here we didn't have any of that and so we had to make our own music. You know, developing our own styles and culture because Perth was so isolated from anywhere else and we never got international touring bands very often, maybe once a year. So Perth had its own little weird music scenes and style of punk music and I'm going back to the 80s here, but I think that mentality has carried on a lot with people.

EE: Just a follow-up question. How would you describe your commitments to the online DIY community? I realize that you have a very active presence on the NLC Facebook forum and you had this fantastic website which I miss dearly although I appreciate the smooth interaction provided by the interface of the new website. In any case, your online presence is also remarkable in terms of participation in connection with users and builders and so if you don't mind talking about that aspect of Nonlinearcircuits and how that informs your work as well.

AF: I do answer a lot of emails every week and I just get a real kick out of people getting circuits working and the joy they get from them and the satisfaction in knowing that when you successfully build something it's another step in your journey (which is a word I don't like using). But you know you're learning a bit more and always improving and I think there's a lot of pleasure and self satisfaction from DIY and those aspects and I just love seeing people being able to do something themselves instead of just buying it from a shop.

When I was starting out I got lots of help from people like Ken Stone and a lot of Japanese DIY synth people. There was a forum in Japan that I was a member of (this is going back to 1998–99). Those guys would do anything to help you out to get your design working and to get parts to you or anything like that so I think just that whole attitude of DIY as a community based venture is something that I've always subscribed to. That's just how it works for me. It's a lifestyle.

AK: Do you think the attitudes of DIY lead to different models of innovation? I live in a town where innovation is the buzzword of choice. It is a high tech town, a kind of mini silicon valley driven mainly by corporate agendas with profit margins as the guiding principles. I don't think it started out that way but that is what my town has become to a large degree. I recall in one of your emails that you have a "try this folder" often based on mathematical papers and that seems to speak to a way of creating that is truly experimental and free from the constraints of the industry. I don't know if you have thoughts about that.

AF: Sure, I don't sit down and think what kind of design could I sell thousands of. It just can't work that way for me. Whatever bubbles up is what I want to try this week and often it'll be something from a research paper I read years ago and I might go back and read it a few times and something triggers an idea and I have to try it. I've got a huge box of bad ideas behind me and it does not work every time. If one out of 10 is successful then that's good. Of course I have to balance making a living because I don't have any other job. This is what I do in addition to the community side of things but it's working pretty amazingly at the moment. So I have no complaints, but I do work very hard. Easy 50, 60 or even 70 hours a week.

So yeah, I'm not particularly profit driven although of course I need to make a living. But some ideas, some modules really gel with people and sometimes it is unexpected what becomes popular, yeah it's really unpredictable. Sometimes I think I really like this module but no one really buys it. But it may catch on in ten years or maybe not.

ET: Can you name some of the Japanese forums you referred to because I'm just really curious to know kind of what was going on in the 90s in Japan. Do you remember the names of people?

AF: I think it was something like Analog Synths Japan. The main guys that I became friends with and they're still around. They are Motohiko Takeda and Osamu Hoshuyama who has a website called Unusual

Synthesizer Circuits.[1] I should warn you, a lot of the stuff he puts on there does not work. Not because it is a bad design, but mainly because he has just thrown up the basic design and it's up to you to complete it.

ET: I'm really interested in that process of sharing schematics that are incomplete or are part of the way done.

AF: Yeah, so he's certainly got some of the most novel synth DIY ideas out there. They are really sweet guys. We still meet up. When I lived there we'd go in to Tokyo for an electronics meet up and they'd take me to all the little parts shops and then we'd go to an Izakaya, which is a kind of bar, and we'd order beers and food and they'd get out soldering irons and boards and start soldering stuff at the table.

ET: Do you have any favourites? You said earlier that there were modules that you thought were going to catch on and didn't but what are your favourites that you are still waiting on becoming more appreciated?

AF: Probably GENie which is really just a triple Neuron which I think is catching on with people into noise and also Signum Hyperchaos which was originally Primal and Hyper Chaos Deluxe and it is sort of based on them. I'm actually pretty obsessed with the Signum circuit at the moment. I like that when you bring it into a synth module you can sort of hack into all the ins and outs of it and start injecting signals into the chaos in quite different ways and go in and out of your normal signal into chaos. So yes I like Signum but I probably sold like three of them in two years. It probably needs a cool name.

ET: That is a perfect transition to my next question. I'm really interested in your progression from reading a research paper to making the multiple variations. In past work I looked at different circuits or musical circuits that were inspired by each other and compared them to try and identify the differences between each step so that I could really identify what was important to the designers. I tried to do that a little bit with the Primal and the HyperChaos in the Li paper[2], and I was wondering if maybe for a start you were willing to tell a little bit about your technical background and your interest in chaos. Did you start with Li paper or was it from before that? Let's start there and then I'll ask specific questions about the circuits.

AF: Firstly, I did not get into electronics before I was 32. Before then I was into motorbikes! And traveling a lot. I spent a lot of time in various Asian countries like China, Japan, India, and Thailand. But I always liked my synthesizers and then when I went to live in Japan I started finding old vintage synths really cheap in junk shops that needed repairing and I was not about to pay someone to do it for me, which is a Perth mentality by the way. So I started Googling it and discovered synth DIY that way. So I didn't actually hold a soldering iron until I was 32 years old. I did that for seven years in Japan and wanted to go further with it. After my job wound up in Japan, I went back to Perth and enrolled in a tech college in electro-technology and thought I was pretty good. So at the ripe age of 40, I started doing an electronic engineering degree, which was five years of hell to be honest. I reckon the math I was doing hadn't been invented when I went to school. But I got through. I didn't go out and just had to study for five years and when I didn't study, I built synths. I mean it was great, I loved it. But anyway, during the degree, I met one professor, Herbert H.C. Lu, who was looking for students interested in chaos. I mean I knew about chaos at that point from the circuits of Ian Fritz so I went and saw Herbert. I walked into his office and said I'm interested in chaos and he grabbed my arms and said sit down and he mentored me through my final year and I ended up doing my final year project with him. I designed an analog computer but one that you could sort of patch up to make chaos and Lorenz circuits and that sort of thing.[3] In my final year he sponsored me to go present at the NOLTA electronics conference in Japan. Yeah, Herbert was great. He basically got me a scholarship to do a PhD and his research at time was on memristors so we did the chaos thing in memristors and I was lucky enough to get in with a really good team with Herbert supervising us. Another fellow, Dongsheng Yu was excellent at math and software whereas I could solder so we had a good little group going. These guys were pretty keen to publish because they liked to keep their jobs safe. So we got a lot of papers out and basically all those papers became chapters of my thesis and we shuffled them into a book as well. So that was the educational and technical side of it. But in the background of it all I just kept building synths and working on designs and stockpiling papers that were interesting. I didn't really have a plan on becoming a synth manufacturer. I expected to do my degree and get a job when I came out. But I came out with a degree and no jobs came and unless you are going into academia you are a little bit unemployable around here. And I wasn't that keen on academia at that point. But I also managed to drag out my scholarship for nearly five years but we really wrote pretty much the whole thing in the first 18 months and got all of the papers published. So for around three years I actually just spent most of my time working on synths and Herbert wasn't too worried. I should also mention that my university was quite old for Western Australia and they just chucked out stuff like you wouldn't believe. So I collected stuff like you wouldn't believe.

ET: For the chaos modules specifically can you tell me a little bit about the steps between reading the lead paper and then replacing one of the parts? I think there's a signal operation with an external input, and I wonder if that was an experimental decision or did you decide that would sound good and then you tried it? I'm curious to know if it is based more on experimentation or if it is planned and what the balance may be.

AF: Pretty experimental. When I'm designing circuits it can go two ways. If I'm fairly confident about it I'll actually just go straight to it. I'll do the panel first and then straight to the circuit board. Maybe I'll just simulate some sections on the computer. It's not always a success to simulate chaos with the LTSpice or something like that.[4] It doesn't always work and what works on Spice doesn't always work on the circuit. But I will use Spice to simulate one section of the circuit just to see what it is doing. If I'm not as confident about a design well, I'll go to a breadboard which is sometimes the best because you can just have it next to your synth then just grab it and connect it to a jack and just jam in a resistor and see what happens. It's a good way of working I think. Some designs take years to get out. Squid Axion, I reckon took me about six years before I was happy. I went through it in all sorts of different ways and again, that was based on a paper. And that is another thing I should point out. A lot of these papers, when there is a circuit in them, well, they often don't work. So you are kind of hacking them in some way. They often use expensive chips and for me an expensive chip is two dollars. I like chips that cost less than twenty cents. So when you blow them up you don't care. You want to get a useful range out of it. Will it work at audio or CV rates? And again it comes back to designing with very generic components. I want everything to be 1k or 10k or a 100k. That is a perfect circuit to me. If it's 1k, 10k, 100k and maybe a 100n cap and 1u or 10u caps then I feel like it's elegant.

ET: So for the Primal, did it go straight to panel and pcb or did you breadboard?

AF: That's another thing. I always design the panel first. Then you get an idea of your functions. And then I can design the pcb. I think it worked pretty much straight up. I just had to tweak a few values. Again, that is something you need to work on. If you look at the equations for it, it is just +A or -A, and A just means it's a stepping voltage. So just need to get your resistors so that voltage is always in a range with a circuit that is going through an oscillator.

So, say if I put 100k resistor it may suddenly stop when the pot is turned half way but if you put in a 450k or 1M the circuit will work no matter what range you turn the pot. Primal and HyperChaos deluxe are good circuits because you can drop in almost any capacitor value and they will still work.

ET: I'd like to talk about the evolution between two chaos modules. There is quite a bit of difference between the Primal and the HyperChaos and I'm curious to know how you came up with the idea for the switching chip because that seems to be pretty different from the Li paper and then you added a whole section to the HyperChaos module I believe, compared to the Li paper again and so yeah these are more significant differences and so again I was curious about the motivation or the experimentation behind that.

AF: I suppose the switching of the chip is inspired by switched capacitor filters and also I've done a similar thing with the Feague which is a quad oscillator where again a switching chip is switching between four different capacitors. On HyperChaos Deluxe, one of the ideas is that you could feed it a fast clock signal and get it switching so that you would actually create new capacitor values. You'll get a different range of frequencies out of the circuit.

The principle is based on switch capacitor filters but also just so you could have two different modes and you can have it running fast or have it running slow. Directionally you can get that switch capacitor effect but you have to get the right frequency range though it's not that hard but you have to play with the bit to get it right.

So using a scope or you can just see what the LEDs are doing on the panel and that will show you what is going on. Your other way of controlling the frequency of the chaos circuit is using your classic OTAs like the VCA or voltage-controlled resistors throughout. That's a little harder for this circuit. It would just blow to some ridiculous levels. I think you would need to add in maybe another 8 chips and 40–50 resistors but the current version is a little simple, but it works.

And there are other things that are quite curious when you translate it. With all these chaos papers it really comes down to three or four equations if it's regular chaos, maybe five if its hyperchaos and you can look at those equations and it's really just building blocks of circuits and it comes down to how you refine them or slim them down to minimize the number of components because again it's not hi-fi audio, you just really want to basically get that equation implemented into the circuit. Say an antilog function, you could make it as simple as a diode on the input of an op amp and the same with the Signum switching circuit. You can sort of use that as a multiplier and in reality that is not what it is doing. But for the function of the circuit it's almost the same result.

ET: And that was the motivation behind the original paper by Li? Replace multiplier chips with this absolute value operation to have another kind of chaos? But you were saying that you were almost thinking of it as a multiplier?

AF: If you look at it closely, it's not. But in terms of the circuit and how the circuit works it is basically doing the same job. And that is the fun part of designing. How can you slim these down and trim it right back.

AK: There was a line in the document you sent which intrigued me. You said I tend to think that digital circuits are very good at doing what they are designed for whereas analog can be pushed into regions where unexpected things happen. That struck a chord with me because that's what attracts me to modular, given that I can plug things in the wrong way and interesting things can happen and perhaps that forms a more natural relationship with the technology than what I have with my sleek Apple computer. If I plug something in the wrong way, then usually nothing happens.

AF: I have not really kept up with all the new digital modules that have been coming out but they seem to be good. I've tried them but as soon as there is some sort of menu or pressing buttons something switches off in me. There is a flow with using analog that, as you say, you can just patch it in and see what happens. At least from my experience, but then I have probably not read the manual. I'll admit that it is probably a romantic notion and nostalgic as well but I like the analog side of things.

    For me analog feels natural and that is probably what it comes down to. I'm sure you can do a lot more with digital modules and different things that you can't do with analog but it's just not my realm.

ET: I was thinking about your experience with analog computers and imagine that would have informed your ideas about circuit implementation of mathematical formulas. Is that where some of your inspiration comes from or even some of the circuits?

AF: Very much so, yes. In a sense an analog computer is like having a breadboard with all the sections of the circuits present. It's a very quick way of creating a module. They were fun to use. I was just chatting with Forest Baer who designed my new website and I'm working with him to design what is basically an analog computer. The only analog computers these days are for universities with university pricing so we want to come out with a range that you can build yourself for $100. Each module would be a fairly simple circuit but it would be a proper analog computer and it'll be a different look than the standard NLC stuff. He's going to design the panels and he's got a nice aesthetic and I enjoy working with him.

AK: Would these analog computers be used for musical synthesis or general purpose or experimentation in general?

AF: All of that, I think. For the standard modular user these won't be a lot of fun. You really need to learn how to use them. With a random patch you are not gonna get much happening. But we want to do them in Eurorack and use regular power supplies and patch cables so that if you want to use them that way you can. But also because all that stuff is accessible. I would like it if it gets picked up by some universities. I wrote a paper in my final year about an analog computer for electronic engineering and education. I basically developed a whole course based around getting students to build their own analog computer and then learn to use it. I thought it was great but no one ever picked it up. I still think it's a good idea. But I would make these computers in a way that you can patch up, say chaos or filters and oscillators if you like. But it certainly wouldn't be as easy to use but there's always people that like to dive in deeper.

AK: Yes, I would imagine that a kind of mini community could develop around the device and share ideas and patches. But switching topics, we are interested in the CellF project. I know that the pandemic stopped it in its tracks but are there any follow up plans or projects?

AF: At the moment, CellF has been set up at a gallery in Bilbao. To construct CellF, we ship it in five separate crates which all together weigh over a ton and it's a four person job to put it together. And Nathan, who designed it has to supervise it and there are some quite delicate moments in terms of making sure it does not collapse in on itself. At the back of CellF, which is not that well-known, we actually have a full bio lab built into it. And that takes four people to pick up and install. We had CellF in storage in Amsterdam but this gallery in Bilbao asked if they could display it. And so, Nathan supervised the unpacking and the assembly using Zoom which he reckons was the scariest thing he has done in his life. So at least it is sitting there on display. They don't turn it on. They don't have any neurons for it or anything. Alvin Lucier is still very keen and he is working very closely now with Guy Ben-Ary on a new project which doesn't involve synths so I'm not directly involved although I might develop some electronics further down the track. But I probably should not say that much about it. Let's just say that Alvin will live on forever. And it's going to be really big and make a lot of noise. They are going to be harvesting Alan's blood and using it to create neurons to do things with. If we can travel again, the first thing we're doing is going to New York to perform. We'll hook Alvin straight up to CellF and see what happens from there.

ET: I was curious about CellF and David Tudor's Neurosynth. Tudor and Forrest Wartham had a machine called the neural synthesizer and he made a CD out of some of the recordings. I was curious if there was any influence from that machine and it's sparse but existing documentation for the CellF.

AF: That rings a bell but no, I am not aware of it. Guy Ben-Ary is who I am working with and CellF was his idea. He is more of a bio artist and the synthesizer side of things is my contribution. He was originally just going to do it with a computer but I explained to him that it wouldn't really be very exciting. You need more! The end product is definitely a matter of teamwork with ideas from Guy and Nathan Thompson and myself and also Darren Moore. Darren is a musician and he teaches at LaSalle in Singapore and he very much influenced what modules were used in CellF and the types of sounds he wanted. But Guy has done some pretty weird designs like a glass penis growing penis cells he bought off a Russian website. He's a pretty unusual guy. He is fun to work with.

AK: Are there any recordings of CellF music?

AF: Most of the shows get recorded and some of them are online. The one I like best was a very impromptu performance when we were in Slovenia at the Kapelica Gallery, which is just an amazing gallery. All the art involves blood or guts or something and there are a lot of stains on the floor. And it is also a bit of hacker space as well. There was a DIY synth band called Kikimore and they are all female and the name translates as "noisy witches." They showed up one night and then we got to talking and they asked, where's CellF? So they started unpacking their instruments and said, let's play. And that was probably one of the best recordings for me with this band just absolutely pounding those neurons for about two hours. They recorded that session. So some shows went great. Usually when you're there it's very enjoyable and intense. Some performers really get into interacting with the neurons.

The worst show, and this was no fault of the performer, (sorry, I forgot his name). This was at the Mofo festival in Tasmania and there was a trumpet player and he was really trying to coax the neurons but they would not respond. And the poor guy was really trying but it was just five minutes of silence from CellF.

AK: I guess it's a living entity and maybe it needed a rest.

AF: Another good one, and I don't know if you could find a recording, but it's with a cellist Okkyung Lee and she also played with CellF at this MoFo festival. She was just amazing. She can really play the cello.

AK: Have your experiences with CellF influenced your designs or given you some new ideas?

AF: The dish we grow the neurons in is a small plastic dish with rows of electrodes at the base and I have often thought how I could implement that using circuit boards. Eventually that is something we'd like to try. With the idea that you could use any medium. Throw in some rotten meat, let it grow moldy and see what happens, for example. Eventually I'll take them into the lab for Nathan to play with and see if it is a suitable base to grow stuff on. The other difficult side is that when you grow cells you need it to be 37 degrees and 5% CO_2. So inside CellF we've got heating pads and a big CO_2 bottle to pump gas into the cells. I don't think I could do that on a module. But I'm not sure what modules may have come from the CellF. Maybe the matrix mixers which are standard modular stuff anyways. With CellF, when we play we have 16 speakers in the room around CellF and each one gets an individual sound and we have four matrix mixers and use quad oscillators to move sounds around the room. The idea also is to have what's happening in the neural array being reflected into the PA so we try and patch it up so that the control signals coming from each section of the neural array in the dish is the one controlling the speaker corresponding to that section.

You have to be able to be there to really experience it. Hearing it afterwards is not really the same. But when you are at a show and hearing the sounds moving from over here to over there, it is quite a good thing to experience.

We developed a few other modules. A lot of stuff for interfacing with the artists who play with frequency to voltage generators or converters and other modules so that we can get signals from the other performers and convert those signals into a format that we can feed into the neurons and make them respond.

A lot of that stuff never got released as a module because I'm not too not sure if it's useful for anyone else.

ET: Do you find that there are parallels between the bio art or biohacking community and the DIY synth community at all through this project?

AF: Yeah there are plenty of people who are into both. I have not gone deeper into the bio art stuff beyond CellF whereas Nathan and Guy are working on heart cells and the most recent project is called Bricolage where they grew heart cells on silk 3-D structures. The idea is that because the structures can fold and bend the heart cells could move around in the dish. These things are big enough to see with the naked eye so they made this amazing environment for the heart cells. It's all ceramic and metal. The first show for that was down in Fremantle and I think it was open for three weeks and then COVID hit and everything was shut down. It was meant to go to New York and Florida as well.

Notes

1 http://www5b.biglobe.ne.jp/~houshu/synth/
2 Li, Chunbiao & Sprott, Julien Clinton & Thio, Wesley & Zhu, Huanqiang. (2015). A unique signum switch for chaos and hyperchaos. https://www.researchgate.net/publication/318206297_A_unique_signum_switch_for_chaos_and_hyperchaos
3 E. N. Lorenz, J. Atmos. Sci. https://doi.org/10.1175/1520-0469(1963)020<0130:DNF>2.0.CO;2 20, 130 (1963).
4 LTspice® is a high performance SPICE simulation software, schematic capture, and waveform viewer with enhancements and models for easing the simulation of analog circuits. Included in the download of LTspice are macromodels for a majority of Analog Devices switching regulators, amplifiers, as well as a library of devices for general circuit simulation. Source: https://www.analog.com/en/design-center/design-tools-and-calculators/ltspice-simulator.html

24

FROM "WHAT IF?" TO "WHAT DIFF?" AND BACK AGAIN

Michael Palumbo and Graham Wakefield

Introduction

Jorge Luis Borges might have unwittingly anticipated version control systems when he wrote the following:

> I leave to all future times, but not to all, my garden of forking paths... In all fiction, when a man is faced with alternatives he chooses one at the expense of the others. In the almost unfathomable Ts'ui Pen, he chooses—simultaneously—all of them. He thus creates various futures, various times which start others that will in their turn branch out and bifurcate in other times.
>
> (Borges and Boucher 1948, 106)

A modular synthesizer is an electronic instrument which is made up of individual components, called modules, with each serving a particular function. Each module has one or more elements of its electronic circuit exposed as part of its interface such that the performer can make arbitrary connections or "patches" between these elements to restructure the instrument, hence its *modularity*. The decision of which modules to include in one's system, and where, is as much a part of the artistry of playing the instrument as is creating the patches and controlling its interfaces.

In a hardware context, a specific and precise configuration of modules, cable connections, and knob and switch positions is very difficult to accomplish more than once. For some, including this chapter's authors, the ephemerality of the instrument is a highly appealing aspect, in particular for its tendency to encourage experimentation. In software-based modular

DOI: 10.4324/9781003219484-26

FIGURE 24.1 Still taken from player view of a Mischmasch patch with several modules connected with virtual patch cables. A player's wand is controlling the frequency knob of a sinewave oscillator.

synthesis it is possible to save the state of the instrument, which affords a different approach to composition. However, these saved files do not capture the changes made in between modifying one patch into another.

For our project, titled *Mischmasch*, we wanted to use virtual reality (VR) to achieve both features that cannot be realized in harware, and have thus far been overlooked in software (Wakefield, Palumbo, Zonta. 2020, 547). The purpose of creating *Mischmasch* in VR is not towards recreating a world, nor of skeuomorphism as in the case of rendered wood panels in many software emulations of 'classic' of "classic" synthesizers -- see Figure 24.1 (Palumbo, Zonta, and Wakefield. 2020, 18). Rather it is about taking advantage of a malleable environment, which speaks to the heart of the improvisational experience.

We introduced the use of a real-time version control system within a VR modular synthesizer to store all patch changes: parameter changes, adding and removing modules, and adding and removing patch cables. In a multiplayer session, the system synchronizes patch changes between peers so that players see and hear each others' playing. Having successfully built and demonstrated *Mischmash* for multiplayer collaboration using a real-time version control system (Ibid, 13) led us to consider other creative applications, which this chapter explores in both practical and philosophical terms. A modular synthesizer with a real-time version control system can introduce novel creative techniques for the electroacoustic improviser, bringing them closer to the "... unfathomable [ability] ... to choose all paths simultaneously" (Borges and Boucher. 1948, 63) that Borges imagined.

Transformation Deltas

Software developers make use of version control systems (VCS) in order to collaborate remotely and manage conflicts between changes made by different authors on shared files, facilitating co-creation (Kalliamvakou et al. 2014, 1). VCS use a utility known as *diff* to compare two versions of a file (Nugroho, Hata, and Matsumoto. 2020, 790). The result of a *diff* operation is a *delta*, which indicates what had been added and/or deleted between one file version and the other version that it was compared to.

In a VCS, a file's version history is a sequence of deltas originating from the earliest version. Developers can create an index to a new version of a file using an incremental record called a *commit*, which contains a list of any new *deltas since the previous commit*. Therefore, an earlier version can be recalled by locating its commit, and applying the inverse of deltas in the commit history to the current version until the desired version is arrived at. The version history remains intact throughout this process: a change made to an older version would be recorded as a new commit.

Version control systems are limited in their ability to capture a fine degree of detail in terms of very small and incremental changes, given that the decision of when to commit the changes made to a file is up to the discretion of the developer. What is captured does not typically include atomic edits, such as adding a letter or removing a word, and so the precise ordering of micro-level edits that proceed from commit to commit is lost. One system which addresses this limitation is the *Operational Transformation* (OT), which records every individual edit made as a delta, making it suitable for real-time versioning (Sun et al. 1998, 102).

A primary use-case of the OT is in online shared document editors such as *Google Docs*, where, for example, the OT ensures that when two or more users make edits to the same word at the same time, all changes are recorded and conflicts are automatically resolved through an operation called a

TABLE 24.1 Some GOT deltas and their inversions

Player Action	Delta	Inverse
add a new module to the patch	newnode	delnode
removes a module from the patch	delnode	newnode
connects a cable between two modules	connect	disconnect
removes a patch cable from two modules	disconnect	connect

rebase. Each edit made by a user generates a delta which is first applied to their local document in the browser and then sent to a server which verifies it against the ground truth of the shared document before sending the delta to other shared instances of the document to be applied.

The virtual modular synthesizer *Mischmasch* (Palumbo, Zonta, and Wakefield. 2020, 18; Wakefield and Palumbo. 2019) uses a real-time version control system called the *Graph Operational Transform* (GOT)[1] to facilitate real-time, online, and multiplayer editing of a shared virtual modular synthesizer. In *Mischmasch*, all edits made to the synthesizer—a patch connection, an instantiation of a new module, or a button press, for example—are each recorded as deltas—see Table 24.1. In a manner similar to how the OT first applies a user's edits to their local context, GOT deltas in *Mischmasch* are first applied to the local version so that the player sees and hears their changes immediately, with fellow performers receiving that player's changes over the internet shortly thereafter.

Patch Histories

A modular synthesizer does not produce sound unless a cable is patched from a voltage source to some kind of output, like a speaker. Going from such a *blank* state to a sound-producing patch is its "...*ontogenesis*, that is, how, from being a mere object, it becomes, for an individual, an instrument to be played" (Boesch. 2007, 6). When performing in *Mischmasch*, the list of GOT deltas is referred to as the patch history. The patch histories of *Mischmasch*, comprising GOT deltas, are the ontogenesis of the synthesizer performance. Interacting directly with the patch history in real-time opens up several creative techniques such as transcription, transposition, looping, and creation of *ready-mades* (Sawyer. 1996, 285) for use in sampling or mapping to gestures. It is therefore compelling to consider procuring a version history for use as source material for further experimentation in real-time improvisation.

From "What if?"

"In Ts'ui Pen's work, all the possible solutions occur, each one being the point of departure for other bifurcations. Sometimes the pathways of this labyrinth converge." (Borges and Boucher. 1948, 107).

During development, you may have an idea for something, a "what if", so you make a feature branch for something. In a VCS, a feature branch is a new trajectory of commits which allows one to iterate on a particular idea. Feature branches can be started from any commit point in the version history, thereby enabling parallel experimentation and without overwriting the earlier history. A feature branch allows one to explore a new idea with the intention of eventually either taking those changes and merging them with a commit on another branch somewhere else in the history, or otherwise abandoning.

A multiplayer virtual synthesizer which records all player actions as deltas would support such an intertextual practice of playing with and on top of previous ideas, and also speaks to a "reflexive indexicality" (Sawyer. 1996, 290).

To "What diff?"

"The book is a shapeless mass of contradictory rough drafts. I examined it once upon a time: the hero dies in the third chapter, while in the fourth he is alive." (Borges and Boucher. 1948, 106).

A patch history can comprise many feature branches containing performative and ancillary gestures from controllers, patch changes, abandoned ideas, and more.

In the following example (Table 24.2), a performer turned a knob, and this gesture was recorded on *Branch A*. The performer returned to the delta which preceded this gesture—*Delta1* which gave a value of 0.5—and started a new branch from it named *Branch B*. The performer then turned the knob differently which recorded a new gesture:

- *Initial Value* is the version of the knob which results from *Delta1*, or the delta from which both branches originate.
- *Branch A*'s position value begins at 0.5, is changed to 0.6, then to 0.7, and back to 0.6
- *Branch B*'s position value also begins at 0.5, is changed to 0.4, then to 0.3, then to 0.2.

And Back Again

"I thought of a maze of mazes, of a sinuous, ever-growing maze which would take in both past and future and would somehow involve the stars" (Borges and Boucher. 1948, 104).

TABLE 24.2 Two branches containing edits made to the same knob where:

Branch	Initial Value	$Delta_2$	$Delta_3$	$Delta_4$
Branch A	0.5	add +0.1	add +0.1	add −0.1
Branch B	0.5	add −0.1	add −0.1	add −0.1

The result of a diff comparison could be used by a *merge* strategy for determining how to combine deltas from one branch into another. One application of this technique as a music improviser using the GOT would be to analyze the differences between two gestures performed on a knob and recorded in GOT deltas in order to produce additional gestures with varying degrees of similarity. Crucially, by creating a branch to make changes to a single parameter, a subsequent merge would preserve changes to other parameters not edited in the new branch. The result is a modification of one knob gesture among potentially many, to help maintain a sense of "… musicality, [where] the performer is required to perform something that retains musical coherence with the emergent" (Sawyer. 1996, 293).

The captured transformation deltas of a virtual synthesizer performance would offer a novel piece of recording, which can be used in multiple ways: a reperformance would entail playing the delta history back in-time, akin to a MIDI score; it would also engender in audiences a means with which to create their own variations, a resolution of control which far exceeds the more common practice of remixing a song using just its tracks; it would also aid in the educational context, where students could playback and experiment with notable regions of a patch history from a performance.

Future Paths

"I leave to various future times, but not to all, my garden of forking paths" (Borges and Boucher. 1948, 107).

A synthesizer built from a playable VCS can engender an open work in which players engage with each other's ideas in an act of co-creation (Harries. 2013, 3). Capturing topological transformations made to a virtual synthesizer provides the performer with the means to continually explore the multiple futures of an idea, to return to the context(s) which led to a particular idea and branch out with yet another ideation. In this system, there is no undo in the destructive sense, so the archival copy of the performance does not change. It is a means of remixing, splicing, and riffing on ideas from past, present, and future.

As the performative actions of playing a virtual modular synthesizer include edits made to its structure and parametric state, and these edits are captured as a process history, then following an ecological approach the process history may become both a productive force acting on other agents of its environment while also being subject to modulations wrought by these agents. Hubert Gendron Blais states that "the contribution of process philosophy to the thought of the musical experience leads us to consider the musical event as an ecology: each musical event is the result of a multiplicity of human and non-human agents forming an immanent *agencement* of productive forces" Gendron-Blais (2020: 285). We imagine the potential of a patch history that can then be connected to a sequencer for looping specific regions of patch history -- an autonomous agent not only recalling particular edits, but also modulating them and creating new patches on feature branches.

Note

1 Graham Wakefield and Michael Palumbo, 2019 www.github.com/worldmaking/ gotlib 2

References

Boesch, Ernst Eduard. 2007. "The sound of the violin." *Discovering cultural psychology: a profile and selected readings of Ernest E. Boesch.*

Borges, Jorge Luis, and Anthony Boucher. 1948. "The garden of forking paths." *Ellery Queen's mystery magazine.* 12 (57), 102–110.

Gendron-Blais, Hubert. 2020. "Music thinking process: Unfolding the creation of the piece *Résonances manifestes.*" *Organised Sound* 25, no. 3 (December): 282–291. issn: 1355-7718, 1469-8153, accessed December 6, 2020. 10.1017/S1355771820000230. https:// www.cambridge.org/core/ product/identifier/S1355771820000230/type/journal_article.

Harries, Guy. 2013. "'The open work': Ecologies of participation." *Organised Sound* 18, no. 1 (April): 3–13. issn: 1355-7718, 1469-8153, accessed March 18, 2020. 10.1017/S1355771812000192. https://www.cambridge.org/core/product/identifier/ S1355771812000192/type/journal_article.

Kalliamvakou, Eirini, Daniela Damian, Leif Singer, and Daniel M. German. 2014. "The code-centric collaboration perspective: Evidence from github." *The Code-Centric Collaboration Perspective: Evidence from Github, Technical Report DCS-352-IR, University of Victoria.*

Palumbo, Michael, Alexander Zonta, and Graham Wakefield. 2020. "Modular reality: Analogues of patching in immersive space." *Journal of New Music Research* 49, no. 1 (January 1, 2020): 8–23. issn: 0929-8215, 1744-5027, accessed January 8, 2021. 10.1080/09298215.2019.1706583. https://www.tandfonline.com/doi/full/ 10.1080/09298215.2019.1706583.

Nugroho, Yusuf Sulistyo, Hata, Hideaki, & Matsumoto, Kenichi (2019). How different are different diff algorithms in Git? *Empirical Software Engineering*, 25, 790–82310.1007/ s10664-019-09772-z.

Sawyer, R. Keith. 1996. "The semiotics of improvisation: The pragmatics of musical and verbal performance." *Semiotica* 108 (3). issn: 0037-1998, 1613-3692, accessed November 28, 2020. 10.1515/semi.1996.108.3-4.269. https://www.degruyter.com/ view/j/semi.1996.108.issue-3-4/semi.1996.108.3-4.269/semi.1996.108.3-4.269.xml.

Sun, Chengzheng, Xiaohua Jia, Yanchun Zhang, Yun Yang, and David Chen. 1998. "Achieving convergence, causality preservation, and intention preservation in real-time cooperative editing systems." *ACM Transactions on Computer-Human Interaction* 5, no. 1 (March 1, 1998): 63–108. issn: 10730516, accessed January 11, 2020. 10.1145/274444.274447. http://portal.acm.org/citation.cfm?doid=274444.274447.

Wakefield, Graham, and Michael Palumbo. (2019). www.github.com/worldmaking/ gotlib 2

Wakefield, Graham, Michael Palumbo , and Alexander Zonta . Affordances and Constraints of Modular Synthesis in Virtual Reality. Proceedings of the International Conference on New Interfaces for Musical Expression. 2020. https://www.nime.org/ proc/nime20_106/

25

INTERVIEW

The Mycelia of Does-Nothing Objects

Brian Crabtree

Brian Crabtree is largely responsible for a number of left-field, programmable instruments, some of which are devices for modular synthesis. This interview was conducted over raw text files shared through Dropbox.

EE: Einar Engström

BC: Brian Crabtree

EE: As an ensemble of composers, instruments, and community, monome is notably rooted in the world of open-source code, programming, and sophisticated virtual technologies. But as a company, it is notably rooted in the material earth. I understand you and your partner live on and operate from a farm in rural New York state, where you "tend apple orchards, shiitake stacks, and forest paths ..." And from afar indeed, sonically, textually, and visually, monome is so often peppered with birdsong, leaves, wet grass, clouds and sky ... in your and community members' creative output, nature repeatedly emerges as a focus of attention/locus of intention. Could you speak about your dedication to the land, and how/whether it informs your practice? I'm particularly drawn to mushroom farming; if I were less averse to hyperbole I'd describe monome as so many hyphae, of composer-performer-programmers ...

BC: I would say nature is foundational to our way of being. Words like "ecosystem" and "community" and "emergent" appear frequently when describing the musical systems we've designed, and I do believe that language and stories shape the energy surrounding any project.

DOI: 10.4324/9781003219484-27

I grew up in a planned suburb surrounded by wilderness. The streets all looked the same yet were different in superficial ways, whereas the mountains all looked different yet somehow present as unified and whole. This might be a stretch but the dichotomy is too good: an awareness of manufactured vs. organic embedded itself early. Splitting time between computers and the woods cross-influenced each curiosity.

The open-ended grid is a reaction against fixed-function manufactured instruments, coming from (which was for me) a golden age of DIY prior to capitalism absorbing DIY as yet another consumptive marketplace. I was more interested in creating and customizing tools in service of an idea, rather than engaging in the repeated cycle of acquiring-exhausting-abandoning. DIY represented agency and repurposing and transformation: to become capable.

Books like Christopher Alexander's *Pattern Language* (1977) deeply influence Kelli and my impulse to shape architectural space—it emphasizes how our surroundings impact daily life (like an operating system). So focusing on that was a logical place for our energy to go. We left the city (of Philadelphia) for upstate New York. This promised more opportunities to bend our reality.

Farming is not like programming. It's not determinate or quantitative and there's no debugger, no pausing execution. Masanobu Fukuoka's *One Straw Revolution* (1975) proposed "do nothing" farming: that nature knows how to best operate, and that "work" should not suck. The first part is noticing, practicing patience, and learning: not forcing ambitions or a system onto sub-optimal conditions. The second part is that "work" and "life" should be less at odds: not "working for a living" but rather "living for a living." What might be seen as "work" might approach actual satisfaction, experimentation/curiosity, meditation/spirituality, etc. In that light, some kinds of farming are a sacred act.

Many of our machines aim to turn work-like computer-stuff into art-facilitating playground-ecosystems. The designs attempt to harness what technology is good at and provide a joyful way in. These objects can become what the player imagines: instrument, composition, score, tool, game, art. (Figure 25.1)

The do-what-you-want agency the machine provides mimics the independence we've been living outside the (necessarily-bureaucratic) cities. Both provide ranging opportunities. And in each there is also a fundamental quality of interdependence. For monome there is an international community of artist-programmer-musicians who continue to regenerate ideas. Rural life somewhat surprisingly facilitates connection as well: perhaps because we all experience the land, and having fewer people around can make for unusual groupings. But there's another community—the non-human one—with which we spend the most time. Trees, birds, weather.

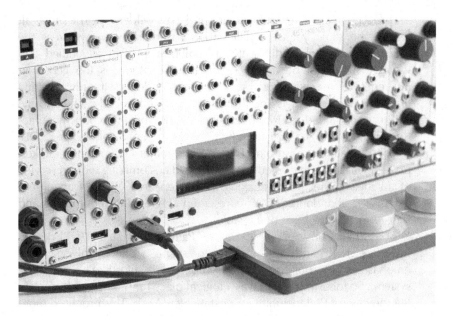

FIGURE 25.1 Monome eurorack modules and Knobs controller. Image courtesy of Brian Crabtree.

In the beginning we enthusiastically tried to combine the two in a somewhat superficial way. Recall the rooster-on-grid photo. But the internet has changed, and not to say we're collectively all more cynical, but corporations have co-opted basically all forms of expression as an advertising tactic—and that means those of us over-sensitive to advertising just sort of seize up when it comes to sharing. So rather than project an ideology by saying things like "sustainable" and "handmade" blah blah, we just continue to assert our way of living through action (alongside a peppering of inscrutable poetics).

> *the blue heron's daily visits*
> *light always changing*
> *anticipating the yearly arrival and departure of barn swallows*
> *over a decade witness a seedling become a presence*
> *see an approaching thunderstorm from miles away*
> *track the brook's changing waterway over time*
> *mycelium bursting from windfall*
> *taste wild blueberries the moment they ripen*
> *journey to the place where the hermit thrush's song might be heard*

So, returning. Nature is fundamental. After living in the forest for over a decade it is no longer an abstraction. Seasonality is visceral:

anticipation and participation. Noticing being alive. We exist amongst multitudes, part of a larger scene. A gradual and inevitable decentering occurs (the positive sort) where priorities are more clear, unpredictability teaches resilience, wonder springs endlessly, and meaning is created (however shifting).

If you want to join my cult, there's a table of literature at the exit. (lol)

EE: I detect at least two persistent threads in the above words. One: a concerted re-orientation to and within lived, physical space, that begins with a departure from planned and bureaucratic domains and demands constant, constructive interaction with the environment at multiple scales. Two: an active working with the contemporary technological landscape—as opposed to a reaction against, or worse, a passive annexation by some technological imperative—where computing is re-engineered to facilitate pleasure instead of efficiency and profit. As you imply, neither follow the path of prefabricated ways of being provided by mainstream capitalist infrastructures. They are perceived as difficult commitments, requiring sacrifice, determination, and all sorts of (cognitive, technical, political, manual) labor. Here, these stances consummate in an atypical relationship to sound and music-making. And so I'm curious as to the "sonority of things"—a term I borrow from Alphonso Lingis, which formulates the intensities that in our immersion in the world, as waves of sound, "touch us without being seen"—as they appear in the monome ecosystem. I sense the seeking of touch.

Could monome, as an abstract and embodied set of sensibilities, exist if its community did not deliberate and derive from the sense of hearing? (Inversely, to play with the easy dichotomy that Jonathan Sterne terms the "audiovisual litany": Could such a community emerge through the more dominant media of image-making technologies?) Or: how and why does this particular (open-source, digital, analog electronic, sonic) dwelling in nature translate into acts of hearing, listening, or voicing?

BC: After years immersed in the community of artists (linked by virtual means) it's become evident that while sound brought people together, the expressions, inspirations, and techniques emerge from and bleed into other forms: visual media, mythology, play, social experiment, etc. A phrase that's been occupying my mind for years is "ways of being" (which you just happened to write in your last response). I've been imagining it as a sort of variation of Berger's *Ways of Seeing* (which deconstructs visual culture (massive over-simplification!)), evoking an animist mindset to draw closer relations within our living world. So how does this relate to inspiration and capability?

People have been making songs about nature forever. Poetry and stringed instruments persevere while modular synths are a relative newcomer. Inspirations may be a scene or sensation or perceived pattern. To explore how these translate, consider some specific physical encounters:

- old tree roots which have captured a boulder
- rain/water/drone/wind
- erosion of geologic formations and revealed layers of time
- particular birdsong which approaches aleatoric yet retains a structure of permutation
- willow trees along a creek, growing tall and slowly falling towards the opposite bank, whereupon the crown takes root and the process repeats, falling the other direction, forever

After being mesmerized by observing pollinators criss-crossing, I could write a poem and some chords, but I might also imagine sonifying the visual motion or thinking of it as a generative score.

To capability: if limited to clip launching or piano roll or complex LFOs you can only get so far, which is where the code comes in. Consider an approximation of a natural pattern: actors moving point to point, some amount of "work" in between, capacity per node, ambivalence/desirability per node by species. This might be captured with a handful of lines of programming and seeded with a collection of careful (or reckless) numbers.

The numbers or patterns or gestures that are fed into a system may be made dynamic and performative by adding an interface (such as a Grid, with tactile/visual/audible response). As instrument-making emerges we enter new territory with our originating pollinators—an encounter/identification/becoming with otherness.

And yes, tools are amazing and at times inspiration can come from the tools themselves. It's good in small doses as an opportunity to improve the craft through studies, but I'm wary of an over-focus which leads to a sort of academic formalism and/or hyper-consumptive distraction. I'm interested in tools which facilitate rather than provide: the low level tools which make the higher tools/instruments/compositions. I rarely sit down and think "I want to do some programming" or "time to modular synth." These are secondary to another impulse or idea.

To draw another weird analogy: like a mycelial network negotiates energy trade between the forest lifecycle, a Teletype or Crow (both modules that deal in code) may re-animate and transform left-for-dead (by boredom?) modules while making various other modules redundant and ready for harvest (because their function is so easily recreated with code). And then there's the more literal fact that Teletype/Crow facilitate

FIGURE 25.2 A barn in upstate New York. Image courtesy of Brian Crabtree.

communication between receptive modules with an extensive digital bus, opening new possibilities for complex behaviors through programmability. (Figure 25.2)

EE: Following this thought, we also encounter a clash of techniques: programming-centric devices such as Teletype and Crow are anomalous—even blasphemous—in modular synthesis, a practice so popularly (albeit problematically) associated with the rejection of "the computer" in favor of electronic music by "screen-free" or "menu-less" manual patching. Where a user chooses the path of [digital] programmability, the negotiation with one's system may ultimately end in the revaluation (or outright obsolescence) of single-purpose modules. Yet I would like to propose another effect, of another negotiation: of the user reconceptualizing their own "energy trade" with the landscape of the technoscientific present. To be less cryptic: learning to use modules that "deal in code" is to oneself learn to deal in code; to ride the fabric of the networked world closer to its source; Teletype and Crow encourage composers to become programmers. Is this an intentional side-effect, or just something unforeseen that you welcome and celebrate? (And when does interest in low-level tools become over-focus and obsession?)

BC: There's a curiosity in the desire to escape computers, wherein the move to modular and recent market boom have brought the proliferation of hundreds of modules which are, well, computers. So now rather than using one computer you can use 30–50 at once! Granted, most of them hide their identity with an opaque standard interface of knobs and jacks, but a growing number inherit anti-patterns directly from somewhat incompatible DAW territory.

But I suspect the anti-computer sentiment is perhaps misplaced. Given all the tiny computers, the problem isn't so much computers themselves but the things that modern computing has become: distracting, fragile, inefficient, and wasteful.

Teletype is a computer but has little to do with the contemporary experience of stereotypical computing (i.e., an operating system update interrupting your anxiety shopping on the web). The process, gesture, and language of Teletype (and Crow) places musical expression in the foreground, and to me this is all compatible and complementary to patch cables and knob turning.

To elaborate (and I'm by no means claiming to be the first to say this): I believe the programmer/composer/performer to be false divisions. Each may signal a different activity but these have increasingly blurred in empowering (and even compelling) ways.

- CODE AS SCORE
- CODE AS INSTRUMENT
- CODE AS ENVIRONMENT

I'm curious if we've arrived at the point where more humans understand rudimentary JavaScript than any form of musical notation. That is to say, a score is a set of instructions. It is so obvious as to be banal that code can be a score, yet this attitude isn't held so much outside a relatively small crowd. Using code to make music means one less translation layer (from notation) and the availability of a vastly weirder palette when dealing with every dimension (time, pitch, timbre), inevitably calling into question these categories themselves.

When people say "programmer" they tend to mean some sort of engineer who deals with obscure technical specifics. But when a system provides sufficient transparency and capability, what was once a score can become an instrument: code that gets played. Creating a software instrument could be called pre-composition (or pre-improvisation?) where the "playing" is simply the manipulation of meta-variables or tables or hardware connected to such.

But the third area is perhaps the least established but growing: the environment where code itself is performative. Real-time systems that allow radical reconfiguration become an instrument on their own. The process of livecoding is itself a spectacle.

I see these as a continuum rather than distinct approaches, a vision which coalesced more in recent years. Forever ago the grid fundamentally aimed to empower the instrument designer style of programming, but more recent additions to the ecosystem have very deliberately encouraged code of all colors.

EE: The monome ecosystem is interesting precisely because it expects little of inhabitants other than they occupy *some* point, any point, along the continuum. One may be a composer, performer, or programmer and still find a role to play within the community. But to the extent that monome hardware devices are designed with the expectation of input/development from users, programmers may be perceived as disproportionately crucial to the ecosystem. I may quip that monome could only exist in a world where more humans understand rudimentary JavaScript than traditional Western musical notation. Could you speak about how your physical instruments come together, and how individuals contribute to the codebase of said instruments? Who is involved, and how do they get involved? What sort of skills are required? What programming languages do you develop in, and why these?

BC: Forever ago, creating the grid itself was a strictly engineering task: electronics and low-level obscure programming of protocols and memory management and bit packing. This is not the same as creating an instrument or composition, which in the beginning happened largely with Max/MSP. The non-apocryphal story is that the day we released the first grid is also the day I decided to open source my personal performance software, because people needed something to do with this does-nothing object.

The community (who gather virtually at the "lines" internet forum, http://llllllll.co) has sustained substantial momentum based on a willingness to share and cooperatively actualize ideas: this is where programmers and musicians met and began cross-contaminating their designs and skills and identities. While we promoted and celebrated creations of the community, the process remained decentralized and uncontrolled as we had no desire to be perceived as capitalizing on the creativity of others. Yet it's curious that the first "clip launcher" was grid-based free software that became a highly commodified, standard musical gesture.

Moving into modular presented a problem for this level of openness: the "musical" language of a grid-enabled module (like Ansible) was written in "low-level obscure" C which is certainly more programmer than musician. While the code was available to all, the knowledge requirement meant less accessibility. Furthermore, the development environment was not interactive: playing with code was not the goal. Code was in service of creating an instrument.

Teletype emerged from this. With a tiny screen, a keyboard, and code as the medium, we attempted to blur the music-programming barrier. Teletype is an extremely simple esoteric language which aims at staying small and approachable.

The secondary and perhaps more profound outcome of Teletype's introduction was the dramatic improvement of the firmware by contributors with specialized knowledge. Teletype was very basic when first released. It's been extended and optimized by volunteer effort, while maintaining the essential spirit of simplicity.

This set the precedent which would come next. Norns was created as an open-source project with collaborators participating before the public release. The goal is to use a more capable yet accessible language (Lua) to facilitate the SCORE-INSTRUMENT-ENVIRONMENT continuum and serve as a learning platform.

It's impossible to imagine these platforms existing without the code contributions of Ezra Buchla, Greg Wuller, Trent Gill, Artem Popov, and scanner darkly.

EE: Speaking of code contributions: It doesn't take much reflection to realize that the name "Teletype" points back to the history of computing. For my part, I've always sensed an affinity between monome modules and the early history of computer music. Before processing power allowed for real-time synthesis in-the-box, languages were developed as interfaces with analog oscillators and filters (perhaps most famously, as with the GROOVE system at Bell Labs). Could you speak about historical awareness in the monome ecosystem, about lineage and heritage in your work designing such modules?

BC: There's certainly an awareness (if not a lineage) given that I studied with Miller Puckette, Tom Erbe, and many others in school. The earliest grid-based phenomenon is named "mlr" to continue naming things after your predecessors (as Max is), despite the fact that a super-gestural performance sample-cutter doesn't really fit into the academic canon.

Teletype references a historical moment of less technological complexity, when it was realistic to understand a system as a whole. Laurie Spiegel's writing and music very much inspired me toward mathematical musical operations, facilitated by generalized computing. (A loaded sentence for sure, which is just to indicate some kind of antithesis of the linear-piano-roll-in-a-DAW.)

Naming an object or module or software weaves a story (however inaccessible) that has power to shape use. The module Earthsea (whose name quotes Ursula Le Guin's universe) is essentially an isomorphic keyboard, but added key combinations (called "glyphs") summon control-voltage variations (with possible slew) which make the grid a sort of spell-casting surface. This hopefully suggests something beyond data entry.

Making art requires distance from feature lists and datasheets and rationalist prescriptions.

So perhaps what I'm trying to say is that there are multiple lineages and paths being traveled herein.

EE: Your education is interesting here, not merely because you studied with such luminaries, but because your contributions to music cultures are, as you suggest, out of sync with the academic canon that houses these names. Their influence is not immediately apparent in monome technologies and techniques. Spiegel and Le Guin, however, are visibly acknowledged, and often very publicly honored in the user forums. There is much admiration for these figures, and I sense the affinity is due to a love for world-making. Put another way: for bringing the fantastical and the out-there into immediate relations with the living body, bodies of listeners and readers.

BC: Absolutely. At its core the monome project attempts to represent "other ways of being." It's been important to have some guiding and grounding minds who've revealed parts of the path, and satisfying to see their ideas resonate so widely within the community.

EE: In fact, many users develop scripts that explicitly explore "other" ways of being. The Norns platform is rife with musical instruments based on genetic algorithms, cellular automata, L-systems, wave mechanics, orbital patterns, even the stock market. The artistic appropriation of scientific thought—is it a cause or consequence of thinking code as environment?

BC: Likely both cause and consequence. Just as a guitar or piano have idiomatic playing styles (which of course can be subverted) undoubt-edly code has similar gravitational impulses and hello-world sounds, sonification of data or algorithm being one of them. The mathematical legibility of so much of science makes it ready for off-kilter creative exploration. Translating Newtonian motion or exploring different types of noise can be incredibly musical.

But just because computers are good at flipping coins doesn't mean that brilliant John Cage pieces just spill out—code (like a composed score or played instrument) reflects the artist/engineer doing the typing. So while the medium may suggest certain approaches (randomness being the power chord of livecoding) there is endless territory to explore beyond. This is perhaps what I mean by using techniques in service of an inspiration or musical thought, rather than tools themselves dictating natural paths of curiosity.

EE: Can we speak for a second about hardware components? I'm thinking of a curious feature to monome devices, in that they may be interconnected via i2c protocol, common in consumer electronics but obscure among musical instruments. On the one hand, I'd like to point out that this is a lineage, too: with consumer electronics and the telecommunications industry. It's worth considering that the presence of such ubiquitous, seemingly benign components in one's art-making machines reduces the art's distance from datasheets and rationalist prescriptions, at least on a material level.

I'd also like to point out that in the case of modular synthesis this is a fascinating design decision, due to its implications precisely for modularity in practice. Patching becomes patchless; patch cables become lines of code; patch gestures become algorithmic functions; and single-purpose modules begin to seem wasteful or redundant. Could you speak a little bit about the thought to incorporate i2c, and some interesting discoveries you've made along the way?

BC: Ah ha, I sometimes worry that analog purists feel I've poisoned the system. Essentially the thought was, if a modular synth contains a bunch of computers, why not make them be able to communicate more effectively? It's a controversial proposition for those who see modular as knobs and cables, with "visible" flow of information.

A grid-based instrument has the potential for a huge number of interactions and controllable parameters. For example, the White Whale sequencer could have its play head position dynamically controlled—say with a CV input and trigger. But if we were to expose even a fraction of these control points as analog connections the panel size would grow massively (in addition to expense, resource usage, and psychic footprint). Instead, Teletype could be configured to target any number of White Whale parameters in interesting (and flexible) ways. This was the original sin.

But the developments that followed I do feel offer tremendous value. It's astonishing to me that an extremely dynamic polyphonic synth voice like Just Friends can enliven a very small modular system, with hardly any patch cables.

Perhaps unexpectedly, the ability to interconnect everything at once has made me disinclined towards massive systems. I'd prefer to work with a small, focused set of modules assembled as an instrument. Digitally controlling other digital modules inclines me towards keeping the whole thing inside a single computer, hence the creation of Norns. And as component scarcity and climate impact increasingly factor into the manufacturing of new machines, old abandoned laptops are looking like a very interesting platform for musical-code

FIGURE 25.3 A rooster on a Monome Grid controller. Image courtesy of Brian Crabtree.

exploration. This is perhaps an uninspiring attitude for some, where a certain flavor of newness reigns supreme: so much remains to be expressed with tools already available to us now.

EE: One person's poison is another's pleasure. And doesn't pleasure emerge, eminently, as a psychic footprint? Amidst a culture that fetishizes "massive racks," the reduction to a focused set of modules may appear ascetic, a synthesist's koan. Indeed, Teletype users have been known to share their code as haiku. The ability to condense musicality into so many lines of code, is a curious test of a dual technological aptitude: at once for the musical and for computing. But I'm just as interested in the *proclivity to do so.* Do you sense a role for pleasure—for joy, or rather, the satisfaction in taking joy—in the monome ecosystem?

BC: Absolutely. And (as you mentioned) the sharing of a creation's source code is another aspect of vivifying a practice—an anarchic act of setting free, sending off, and waiting to see how an idea or instrument may evolve in the hands and minds of an international community. This is a way of being towards which many of us are compelled: that personal interest and joy has the capacity to be expanded to collective investigation and celebration.

Part of the journey is not knowing where we'll go next.

INDEX